MULTICOMPONENT PHASE DIAGRAMS: APPLICATIONS FOR COMMERCIAL ALUMINUM ALLOYS

MULTICOMPONENT PHASE DIAGRAMS: APPLICATIONS FOR COMMERCIAL ALUMINUM ALLOYS

NIKOLAY A. BELOV
Moscow State Institute of Steel and Alloys, Moscow, Russia

DMITRY G. ESKIN
Netherlands Institute for Metals Research, Delft, The Netherlands

ANDREY A. AKSENOV
Moscow State Institute of Steel and Alloys, Moscow, Russia

2005

ELSEVIER

Amsterdam – Boston – Heidelberg – London – New York – Oxford
Paris – San Diego – San Francisco – Singapore – Sydney – Tokyo

ELSEVIER B.V.
Radarweg 29
P.O. Box 211, 1000 AE
Amsterdam
The Netherlands

ELSEVIER Inc.
525 B Street, Suite 1900
San Diego
CA 92101-4495
USA

ELSEVIER Ltd
The Boulevard
Langford Lane
Kidlington
Oxford OX5 1GB, UK

ELSEVIER Ltd
84 Theobalds Road
London WC1X 8RR
UK

First edition 2005

ISBN: 0-080-44537-3

♾ The paper used in this publication meets the requirements of ANSI/NISO Z39.48-1992 (Permanence of Paper).

Printed in Great Britain.

Introduction

Aluminum alloys take obviously the first rank in nonferrous materials from the viewpoints of both production and consumption. Despite decades of extensive research and application, commercial aluminum alloys are still poorly understood in terms of the phase composition and phase transformations occurring during solidification, cooling, and heating. Numerous phase diagrams have been published over the last 100 years, yet they are still hardly available for multicomponent alloy systems. This book aims at the application of multi-component phase diagrams to commercial aluminum alloys and gives a comprehensive coverage of available and assessed phase diagrams for aluminum-based alloy systems of different dimensionality.

We very much hope that the reader will find the book useful for the following reasons. First, most of commercial aluminum alloys are multicomponent and, what is even more important, multiphase, which requires the knowledge of corresponding phase diagrams when analyzing either as-cast or heat-treated materials. Second, most of the reference books on phase diagrams and physical metallurgy of Al alloys (e.g. those by Mondolfo (1976) or Hatch (1984)) do not correlate phase diagrams to specific alloy compositions, which decreases the practical value of these books. Moreover, there are no reference sources for multi-component phase diagrams of aluminum-based systems published within the last 20 years. Third, the representation of phase diagrams is usually not convenient for a user who wants to get quick and specific information, for example: what is the solidus temperature of a 2024 alloy containing the higher limit concentration of copper and the lower limit concentration of magnesium, or what is the phase composition of a 2014 alloy in the T6 state?

The application of phase diagrams to commercial alloy compositions demands for a high qualification of a researcher. The larger the dimension of the phase diagram, the better should be the knowledge and the experience of a scientist. Even four-component phase diagrams require additional work on their interpretation. In the case of more complex phase diagrams, the problems appear both from the lack of data and inadequate graphic representation. The use of thermodynamical calculations requires vast databases (expensive and not readily available) and special skills from the user.

In order to make the book acceptable and applicable for a wide range of readers, the authors present a lot of sections and calculated data derived from multi-component phase diagrams.

Throughout the book, the data from phase diagrams and on phase composition is correlated with the chemical composition of commercial aluminum alloys and some other materials (rapidly solidified or composite).

The book includes the most recent versions of phase diagrams. In writing this book, we referred extensively to a famous book by Mondolfo (1976) that, although being published more than 25 years ago, remains the most complete source of phase diagrams of aluminum-based systems. We also frequently cite another reference book published almost at the same time by a team of Russian scientists (Drits et al., 1977) that give some additional information, especially on multicomponent phase diagrams. However, a considerable part of the data presented in this book is a result of authors' research and has been previously published only in periodic journals. In addition, some new experimental results on phase diagrams appear in this book for the first time.

The important feature of this book is the data on nonequilibrium phase diagrams. Such information can be very rarely accessed from other publications. Meanwhile, most commercial alloys are produced and used in the nonequilibrium state, which resulted either from solidification or heat treatment. The phase composition predicted by the equilibrium phase diagrams can vary essentially from the real phase composition and structure. The information given in this book is unique and will greatly enhance the knowledge of a potential reader.

We have done our best in covering all groups of commercially important alloys and materials, though in some cases the lack of experimental or calculated data on some multi-component systems prevented us from doing this in the most complete manner. We should also note that the microstructures given in the book are just illustrations to the specific phase diagrams, and by no means cover all the diversity of structures that can be observed in real commercial alloys. For more complete information, the reader is referred to special books like those by Bäckerud et al. (1986, 1990).

The commercial alloys are mainly designated using the USA standard (four digits for wrought alloys and three digits for casting alloys), which is the most well-adopted one in the world. However, some commercial alloys which have no international analogs such as a Russian grade 1420 (Al–Mg–Li) or a Russian grade 1570 (Al–Mg–Sc), as well as rapidly solidified and composite materials retain their original designation.

The authors would like to express their deep gratitude to Mr. Victor Selivanov who translated most of the book in English and to Ms. Natalia Avxentieva for her indispensable help in the preparation of figures.

CONTENTS

Chapter 1

Alloys of the Al–Fe–Mn–Si System

This chapter considers alloys that, apart from Fe, Si, and Mn, contain no other elements capable of significantly affecting the phase composition. First and foremost, these alloys are represented by commercial aluminum (1XXX series) and some alloys of 8XXX (e.g. 8111 and 8006) and 3XXX (e.g. 3003) series. In addition, the Al–Fe–Mn–Si phase diagram can be used to analyze the effects of Fe and Mn on the phase composition of casting Al–Si alloys of the 4XX.0 series. In many cases, this quaternary diagram solely makes it possible to answer the question as to which Fe-containing phases can be formed in a particular commercial alloy.

1.1. Al–Fe–Si PHASE DIAGRAM

The Al–Fe–Si system is the basic system for the structure analysis of commercial aluminum alloys of the 8111 type, and binary Al–Si alloys which, as a rule, contain an iron impurity (Table 1.1).

The aluminum corner of the Al–Fe–Si phase diagram is considered in detail by Phillips (1959), who gives the isotherms of liquidus, solidus, and solvus surfaces, as well as intermediate reactions. Numerous subsequent studies of this system have not introduced any significant changes into the constitution of the aluminum corner, and it is given according to Phillips in all major reference books on aluminum-alloy phase diagrams (Mondolfo, 1976; Drits, 1997). The generally accepted opinion is that the phases (Si), Al_3Fe, Al_8Fe_2Si, and Al_5FeSi that can be involved in the invariant reactions (Table 1.2) are in equilibrium with the aluminum solid solution.

The solubility of silicon in Al_3Fe is from less than 0.2 up to 6%, and that of iron in silicon is negligibly small (Mondolfo, 1976). The Al_8Fe_2Si phase (31.6% Fe*, 7.8% Si), which is also designated as $Al_{12}Fe_3Si_2$ (30.7% Fe, 10.2% Si), $Al_{7.4}Fe_2Si$, and α(AlFeSi), exists in a homogeneity range of 30–33% Fe and 6–12% Si. It has a hexagonal structure (space group $P6_3/mmc$) with parameters $a = 1.23–1.24$ nm and $c = 2.62–2.63$ nm; its density is 3.58 g/cm^3 (Hatch, 1984). The Al_5FeSi phase (25.6% Fe, 12.8% Si), also designated as $Al_9Fe_2Si_2$ and β(AlFeSi), exists in a homogeneity range of 25–30% Fe, 12–15% Si. This phase has a monoclinic crystal structure with parameters $a = b = 0.612$ nm, $c = 4.148–4.150$ nm, $\beta = 91°$. It has a density of 3.3–3.6 g/cm^3 and a Vickers hardness of 5.8 GPa (Belov et al., 2002a). Besides these

* Here onwards, wt% if not mentioned otherwise.

Table 1.1. Chemical composition of some commercial alloys whose phase composition can be analyzed using the Al–Fe–Si phase diagram

Grade	Fe, %	Si, %	Other	
			Mn, %	Cu, %
1199	0.006	0.006	0.002	–
1095	0.040	0.030	0.010	–
1080	0.150	0.150	0.020	–
1085	0.100	0.120	0.020	–
1070	0.250	0.200	0.030	–
1075	0.200	0.200	0.030	–
1055	0.400	0.250	0.050	–
1035	0.600	0.350	0.050	–
1045	0.450	0.300	0.050	–
8111	0.4–1.0	0.3–1.1	0.1	–
8079	0.7–1.3	0.05–0.3	–	–
413.0	2	11–13	0.35	1
443.0	0.8	4.5–6.0	0.5	0.6
B443.0	0.8	4.5–6.0	0.35	0.15
C443.0	2	4.5–6.0	0.35	0.6
444.0	0.6	6.5–7.5	0.35	0.25

phases, two more ternary compounds – Al_4FeSi_2 (25.4% Fe, 25.5% Si) and Al_3FeSi (33.9% Fe, 16.9% Si) – can occur in Si- and Fe-rich Al alloys under nonequilibrium conditions. The former phase, also designated as Al_3FeSi_3 or δ(AlFeSi), has a narrower homogeneity region than those of the phases α(AlFeSi) and β(AlFeSi). This phase has a tetragonal structure of the $PdGa_5$ type with parameters $a=$ 0.607–0.63 nm and $c=0.941$–0.953 nm. The density of the phase is 3.3–3.36 g/cm^3 (Belov et al., 2002a). The compound Al_3FeSi or γ(AlFeSi) has a monoclinic structure with parameters $a=1.78$ nm, $b=1.025$ nm, $c=0.890$ nm, $\beta=132°$ (Mondolfo, 1976).

Monovariant eutectic reactions involving (Al) and excess phases (Table 1.3) show that the (Si) phase, in contrast with the Fe-containing phases, forms at a virtually constant temperature. A general view of the Al–Fe–Si phase diagram, the projections of the liquidus and solidus surfaces in the aluminum corner in the diagram are given in Figure 1.1. These data suggest that a decrease in the liquidus and solidus temperatures (in Al-rich alloys) is primarily due to the concentration of silicon, the effect of iron being much smaller.

As distinct from the other phases, the composition of (Al) greatly depends on temperature. This concerns mainly silicon content, as the limit solubility of iron does not exceed 0.05%. The solubilities of these elements in (Al) at various temperatures for the three-phase regions are given in Table 1.4. Apparently, when the Al_3Fe is in equilibrium with (Al), much less silicon can be dissolved in (Al) as compared with the alloys in the (Al) + (Si) + β phase region.

Table 1.2. Invariant reactions in ternary alloys of Al–Fe–Si system (Mondolfo, 1976)

Point in Figure 1.1b	Reaction	Concentrations of elements in phases (from left to right in the reaction)								*T*, °C
		1 (L)		2		3		4		
		Fe, %	Si, %	Fe, %	Si, %	Fe, %	Si, %	Fe, %	Si, %	
E	$L \Rightarrow (Al) + Al_5FeSi + (Si)$	0.7	12	0.01	1.6	25	15	0	99.8	576
P_2	$L + Al_8Fe_2Si \Rightarrow (Al) + Al_5FeSi$	1.7	6.5	33	7	0.04	1.1	25	13	611–613
P_1	$L + Al_3Fe \Rightarrow (Al) + Al_8Fe_2Si$	2	4	36	0.01	0.05	0.6	33	7	629–632

Table 1.3. Monovariant reactions in ternary alloys of Al–Fe–Si system

Reaction	Line in Figure 1.1b	*T*, °C
$L \Rightarrow (Al) + Al_3Fe$	e_1–P_1	655–629
$L \Rightarrow (Al) + Al_8Fe_2Si$	P_1–P_2	629–611
$L \Rightarrow (Al) + Al_5FeSi$	P_2–E	611–576
$L \Rightarrow (Al) + (Si)$	e_2–E	577–576

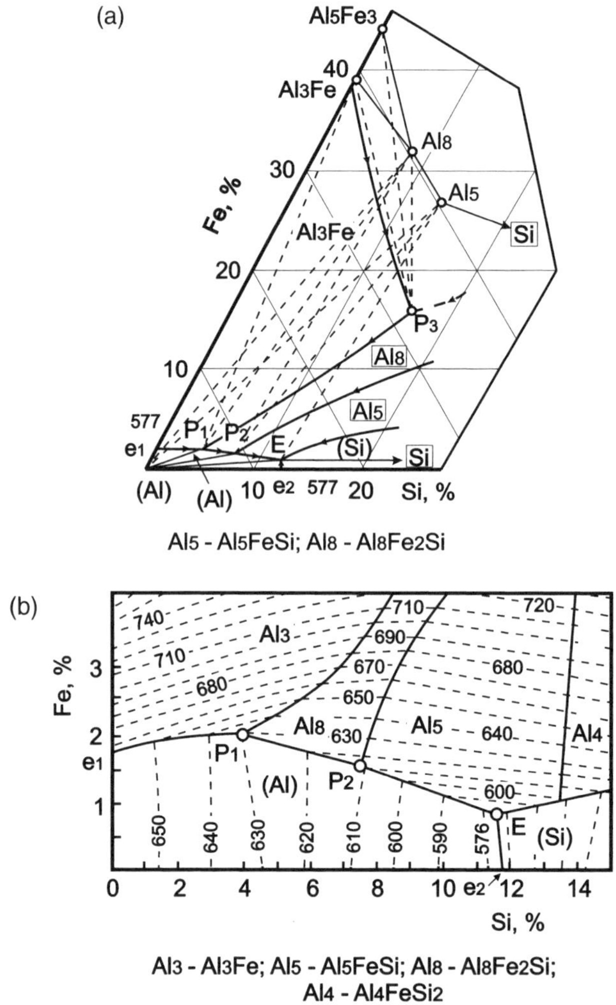

Figure 1.1. Phase diagram of Al–Fe–Si system: (a) general view; (b) liquidus; (c) solidus; and (d) solidus details (Phillips, 1959).

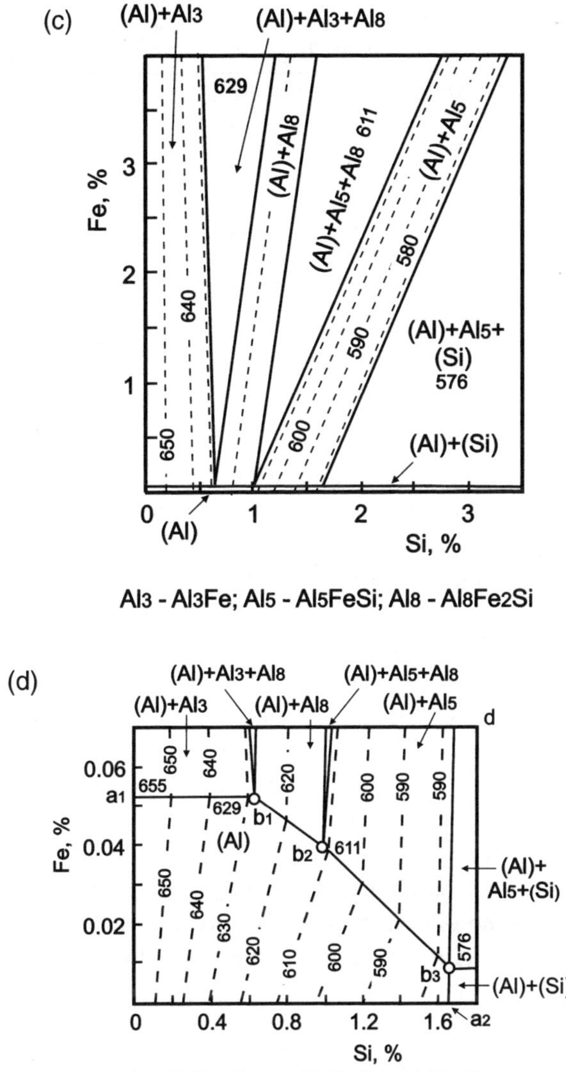

Figure 1.1 (*continued*)

The Al–Fe–Si phase diagram is very complex. The ternary phases in the solid state exist mainly outside the fields of their primary crystallization; therefore, numerous peritectic reactions should be completed for the equilibrium to be achieved. As a result, real alloys produced at commercial cooling rates can have the Al_3Fe, Al_6Fe, α(AlFeSi), β(AlFeSi), and δ(AlFeSi) phases co-existing in their structure

Table 1.4. Limit solid solubilities of iron and silicon in aluminum in three-phase fields of Al–Fe–Si phase diagram (Drits et al., 1977)

T, °C	(Al) + Al_3Fe + α(AlFeSi)		(Al) + α(AlFeSi) + β(AlFeSi)		(Al) + β(AlFeSi) + (Si)	
	Fe, %	Si, %	Fe, %	Si, %	Fe, %	Si, %
629	0.052	0.64	–	–	–	–
611	–	–	0.04	0.82	–	–
600	0.033	0.4	0.033	0.82	–	–
578	–	–	–	–	0.01	1.65
550	0.016	0.2	0.016	0.42	0.008	1.3
500	0.009	0.11	0.008	0.22	0.005	0.8
450	0.004	0.06	0.004	0.11	0.003	0.44
400	0.002	0.03	0.002	0.06	0.002	0.30

(Mondolfo, 1976). Identification of the phases based only on their morphology can often lead to a mistake because the same phase can have different morphologies depending on its origin: primary crystals or the product of peritectic and eutectic reactions. In addition, silicon and other stable and metastable binary and ternary phases can precipitate during the decomposition of supersaturated solid solutions or upon cooling of ingots or castings. Some of the phases are also known to undergo transformations during heat treatment.

As the Al–Fe–Si system is among the most important, there are quite a few studies suggesting its nonequilibrium variants. For example, the phase-field distribution in the as-cast state given by Phillips (1959) shows four- and five-phase regions (Figure 1.2a), which is the most evident feature of the nonequilibrium structure.

A shift in the primary solidification fields of the Al_3Fe, Al_8Fe_2Si, and Al_5FeSi phases depending on the cooling rate during solidification (V_c) is reported by Langsrud (1990) who shows that these fields drift towards a lower Si concentration with the increasing V_c. As a consequence, the Al_3Fe formation is less probable at high cooling rates, even in alloys containing 2–3% Fe at 2–3% Si.

At low cooling rates ($V_{c1} = 10^{-2}$–10^{-1} K/s), the onset of solidification can be analyzed with sufficient accuracy by the equilibrium phase diagram (Belov and Zolotorevskii, 1995). Binary eutectic reactions shall take place after the primary crystals are formed. However, due to the inhibition of peritectic transformations, some alloys can simultaneously form all three binary eutectics with participation of the Fe-containing phases. The nonequilibrium solidus of most alloys is equal to 576°C and corresponds to the ternary eutectics L ⇒ (Al) + (Si) + Al_5FeSi. Only those alloys whose compositions are within a narrow region close to the binary systems complete solidification by the binary eutectic reactions. The inhibition of the peritectic transformations L + Al_8Fe_2Si ⇒ (Al) + Al_5FeSi and L + Al_3Fe ⇒ (Al) + Al_8Fe_2Si leads to the following result. With the Si concentration increasing, the

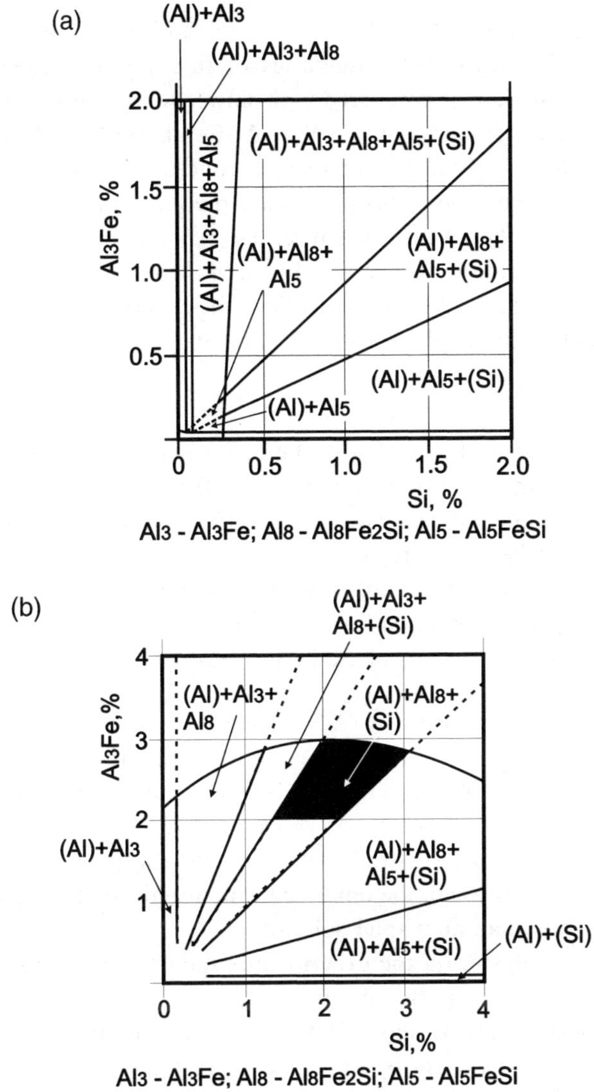

Figure 1.2. Phase fields in Al–Fe–Si system in the as-cast state: (a) $V_c \sim 10^{-1}$ K/s (Phillips, 1959) and (b) $V_c \sim 10$ K/s (Belov et al., 2002a).

sequence of phase regions (not taking (Al) into account) in slowly solidified alloys containing more than 0.5% iron will be Al_3Fe, $Al_3Fe + Al_8Fe_2Si$, $Al_3Fe + Al_8Fe_2Si +$ Al_5FeSi, $Al_3Fe + Al_8Fe_2Si + Al_5FeSi$ + (Si), $Al_8Fe_2Si + Al_5FeSi$+(Si), and Al_5FeSi + (Si). This is in very good agreement with the distributions of the phase regions proposed by Phillips and shown in Figure 1.2a.

At higher cooling rates ($V_{c2} = 10^0$–10^2 K/s), a noticeable swing of the liquidus surface (Figure 1.2b) shifts the boundaries of intermediate reactions and phase regions in the as-cast state as compared with slower solidification. Apart from these changes, the eutectic reactions L ⇒ (Al) + Al_3Fe, L ⇒ (Al) + Al_5FeSi, and L ⇒ (Al) + (Si) + $AlFeSi_5$ are hindered at certain concentrations of Fe and Si. As a consequence, the Al_8Fe_2Si and (Si) phases are present and the Al_3Fe and Al_5FeSi phases are absent in the as-cast structure. To explain this experimental fact, we assume that under conditions of fast solidification there is a significant undercooling ΔT that is not the same for different eutectics. The experimentally observed absence of the Al_3Fe and Al_5FeSi phases at 2–3% Si and 2–3% Fe can be due to the following reasons.

1. The eutectic reaction L ⇒ (Al) + Al_3Fe has a markedly larger value of ΔT as compared with the eutectic reaction L ⇒ (Al) + Al_8Fe_2Si; therefore, the latter reaction is thermodynamically more favorable than the former reaction and the formation of Al_3Fe is suppressed.
2. In the presence of a considerable amount of the earlier formed phase Al_8Fe_2Si, the undercooling ΔT required for the formation of the Al_5FeSi phase increases and the solidification of this phase is suppressed.

As a result of the suppressed L ⇒ (Al) + Al_5FeSi reaction, the solidification of the eutectics L ⇒ (Al) + Al_8Fe_2Si continues. Moreover, the ternary eutectics L ⇒ (Al) + (Si) + Al_5FeSi can be replaced with the hypothetical reaction L ⇒ (Al) + (Si) + Al_8Fe_2Si. Due to the low concentration of iron in the ternary eutectics its structure is degenerated: colonies of (Al) + (Si) or veins of the (Si) phase (alongside Al_8Fe_2Si crystals formed earlier through the binary eutectic reaction) at the boundaries of the dendritic cells of the aluminum solid solution.

Using such an analysis and the experimental data on the phase composition of cast alloys, one can plot the distribution of phase regions in the as-cast state as applied to chill casting (V_c is about 10 K/s). This distribution (Figure 1.2b) significantly differs from the variant by Phillips (Figure 1.2a).

Literature data indicate that, besides the stable phases, various metastable phases are formed in commercially pure aluminum and low-alloyed materials containing up to 0.5% Fe and 0.5% Si, at cooling rates typical of industrial casting, see e.g. Dons (1985), Skjerpe (1987), and Ghosh (1992a). In their chemical composition, these metastable phases are close to the stable phases but differ in crystal structure (Table 1.5). At low concentrations of copper and at cooling rates above 8–10 K/s, the metastable Al_6Fe phase can be formed during solidification. The complete account of metastable phases occurring in Al–Fe (see also Section 9.3.2) and Al–Fe–Si alloys can be found elsewhere (Belov et al., 2002a).

Table 1.5. Metastables phases occurring in Al–Fe–Si alloys

Phase	Structure	Lattice parameters				Reference
		a, nm	*b*, nm	*c*, nm	β	
$\alpha(Al_{12.7-12.9}Fe_3Si_{1-1.5})$ (27–35% Fe, 4–8% Si);	Primitive cubic or bcc	1.256	–	–	–	Cooper (1967), Dons (1984), and Skjerpe (1987)
$\alpha'(Al_{12-12.6}Fe_3Si_{1.6-2})$	Hexagonal	1.23	–	2.62	–	Robinson and Black (1953), and Dons (1984)
α'', q_1	Tetragonal or orthorhombic	1.26–1.3	–	3.70	–	Westengen (1982), and Skjerpe (1987)
β', β^*	Monoclinic	0.89	0.49	4.16	92°	Westengen (1982)
$\alpha_v(Al_{14.8-12.4}Fe_3Si_{1-2.1})$	Monoclinic	0.847–0.869	0.635	0.610–0.632	93.4°	Simensen and Vellasamy (1977), and Dons (1984)
α_γ(29% Fe, 5.5% Si)	Monoclinic	2.795	3.062	2.073	97.74°	Dons (1985), and Skjerpe (1987)
q_2	Monoclinic	1.250	1.230	1.970	111°	Skjerpe (1987)
19–27% Fe, 0–3% Si	Rhombohedral	0.89	–	–	111.8°	Ghosh (1992a)
		2.082	–	–	95.2°	
29.2% Fe; 11.3% Si	Hexagonal	0.836	–	1.458	–	Ghosh (1992a)
19% Fe; 1% Si	Hexagonal	1.776	–	1.088	–	Ghosh (1992a)

Metastable phases can also precipitate from the aluminum solid solution during homogenization of ingots. Among these phases, metastable modifications of the equilibrium Al_8Fe_2Si (α) phase are well documented (Table 1.5). If the homogenization temperature is sufficiently high, these metastable phases are, as a rule, transformed into respective equilibrium phases. One should bear in mind that, due to the low solubility of iron in (Al) (under typical casting conditions), the maximum amount of Fe-containing phases of secondary origin cannot exceed 0.1–0.2 vol.%.

1.2. Al–Fe–Mn PHASE DIAGRAM

Using this phase diagram, we can analyze the phase composition of 8006- and 3003-type alloys at a low concentration of Si (Table 1.6). Consideration of the Al–Fe–Mn phase diagram is topical also because iron and manganese occur in many commercial alloys where they form various phases of solidification origin. Without the knowledge of this ternary system, it is impossible to analyze more complex phase diagrams involving iron and manganese, e.g. Al–Fe–Mn–Si.

In the aluminum corner of the Al–Fe–Mn ternary system, only two phases – Al_3Fe and $Al_6(FeMn)$ – can be in equilibrium with (Al) (Mondolfo, 1976). Manganese substitutes for iron in the Al_6Mn phase, up to $Al_{12}FeMn$ (12.85% Fe, 12.64% Mn). The limit solubility of manganese in Al_3Fe corresponds to the formula $Al_3Fe_{0.88}Mn_{0.12}$ (4–5% Mn). The phase $Al_6(FeMn)$ has an orthorhombic crystal structure (which is isomorphic to the Al_6Fe and Al_6Mn phases) with parameters $a = 0.75518$ nm, $b = 0.64978$ nm, $c = 0.88703$ nm (Ran, 1992). According to Mondolfo, the lattice parameters of this phase are as follows: $a = 0.7498$ nm,

Table 1.6. Chemical composition of some commercial alloys whose phase composition can be analyzed using Al–Fe–Mn diagram

Grade	Mn, %	Fe, %	Other			
			Si, %	Mg, %	Cu, %	Cr, %
8006	0.3–1	1.2–2	0.4	0.1	0.3	–
3102	0.05–0.4	0.7	0.4	–	0.1	–
3107	0.4–0.9	0.7	0.6	–	0.05–0.15	–
3003	1.0–1.6	0.7	0.6	–	0.05–0.2	–
3207	0.4–0.8	0.45	0.3	0.1	0.1	–
3012	0.5–1.1	0.7	0.6	0.1	0.1	0.2
3014	1.0–1.5	1.0	0.1	0.1	0.5	–
3002	0.05–0.2	0.1	0.08	0.05–0.2	0.15	–
3103	0.5–0.9	0.7	0.5	0.3	0.1	0.1

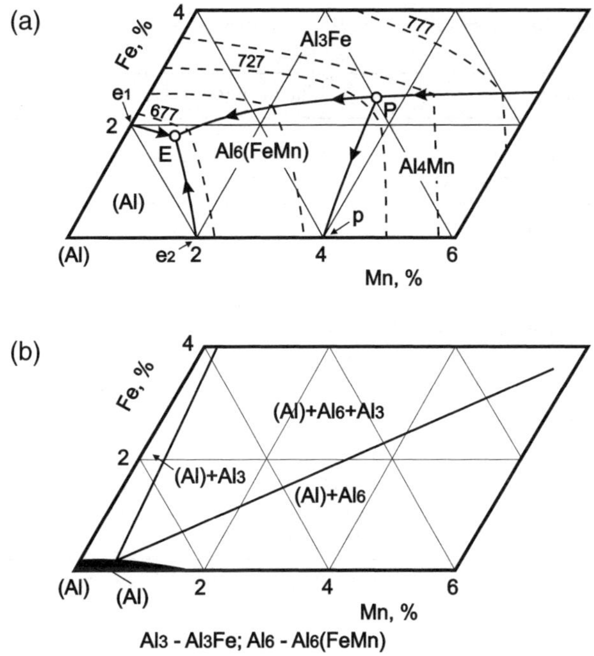

Figure 1.3. Phase diagram of Al–Fe–Mn system: (a) liquidus and (b) isothermal section at 627°C (Mondolfo, 1976).

Table 1.7. Invariant reactions in ternary alloys of Al–Fe–Mn system (Mondolfo, 1976; Ran, 1992)

Reaction	Point in Figure 1.3a	T, °C	Concentrations in liquid phase	
			Mn, %	Fe, %
$L + Al_3Fe + Al_4Mn \Rightarrow Al_6(FeMn)$	E	727	3.45	2.5
$L \Rightarrow (Al) + Al_3Fe + Al_6(FeMn)$	P	654	0.75	1.75

$b = 0.6495$ nm, $c = 0.8837$ nm (Mondolfo, 1976). The projection of liquidus and the isothermal section at 627°C are given in Figure 1.3. Tables 1.7 and 1.8 list invariant and monovariant reactions, respectively.

Two invariant transformations, one of them involving (Al), occur in Al-rich alloys (Table 1.7). Denholm et al. (1984) report that the composition of the ternary eutectic is as follows: ~0.43% Mn, ~1.7% Fe. In this case, the vertices of the eutectic triangle correspond to Al_3Fe (36.9% Fe, 4.6% Mn); $Al_6(FeMn)$ (19.6% Fe, 7.1% Mn); and (Al) (0.044% Fe, 0.23% Mn), i.e. the composition of the $Al_6(FeMn)$ phase differs from that given elsewhere (Mondolfo, 1976). In line with these data, the solubility of

Table 1.8. Monovariant reactions in ternary alloys of Al–Fe–Mn system

Reaction	Line in Figure 1.3a	*T*, °C
L ⇒ (Al) + Al_3Fe	e_1–E	655–654
L ⇒ (Al) + Al_6(FeMn)	e_2–E	658–654

Table 1.9. Limit solid solubilities of iron and manganese in aluminum in three-phase fields of Al–Fe–Mn phase diagram (below 654°C approximated values are given)

T, °C	Mn, %	Fe, %
654	0.23 (1.8)*	0.044 (0.05)*
627	0.13 (1.0)	–
527	0.05 (0.42)	–
427	0.026 (0.2)	–

* Solubilities in binary systems are given in parentheses

manganese in (Al) in the three-phase region should be significantly lower than that in the binary Al–Mn system. The estimates obtained by extrapolating the data from the corresponding binary system are given in Table 1.9.

The positions of phase fields in the as-cast state depend on the cooling rate. At higher rates (approximately $V_c > 10$ K/s), we would expect only one phase Al_6(FeMn) to be present alongside (Al). Iron considerably decreases the concentration of manganese in supersaturated (Al) in the as-cast state, as part of Mn is bound in the Al_6(FeMn) phase of solidification origin.

1.3. Al–Mn–Si PHASE DIAGRAM

Although commercial alloys with Mn and Si but without Fe are virtually nonexistent and one can only speak of alloys with minor iron concentration (Table 1.10), consideration of this ternary phase diagram is required for the analysis of more complex systems, in particular Al–Fe–Mn–Si and Al–Fe–Mg–Mn–Si.

Besides the phases from the binary systems (Al_6Mn and (Si)), the $Al_{15}Mn_3Si_2$ compound is in equilibrium with (Al) (Mondolfo, 1976).

The $Al_{15}Mn_3Si_2$ phase (26.3% Mn, 8.9% Si), also designated as $Al_{10}Mn_2Si$, $Al_{12}Mn_3Si$, $Al_9Mn_2Si_{1.8}$ or α(MnSi), exists in the homogeneity range of 25–29% Mn and 8–13% Si. This phase has a cubic structure (space group *Pm*3, 138 atoms in the

Table 1.10. Chemical composition of some commercial alloys whose phase composition can be analyzed using Al–Mn–Si phase diagram

Grade	Mn, %	Si, %	Other			
			Fe, %	Mg, %	Cu, %	Cr, %
3102	0.05–0.4	0.4	0.7	–	0.1	–
3107	0.4–0.9	0.6	0.7	–	0.05–0.15	–
3003	1.00–1.6	0.6	0.7	–	0.05–0.2	–
3008	1.2–1.8	0.4	0.7	0.01	0.1	0.05
3207	0.4–0.8	0.3	0.45	0.1	0.1	–
3012	0.5–1.1	0.6	0.7	0.1	0.1	0.2
3009	1.2–1.8	1.0–1.8	0.7	0.1	0.1	0.05
444.0	0.35	6.5–7.5	0.6	0.25	0.1	–

elementary cell) with parameter $a = 1.265$–1.268 nm (Mondolfo, 1976) or 1.260 nm (Zakharov et al., 1989b). Its density is 3.55 g/cm^3; microhardness, 8.8 GPa. Silicon only slightly dissolves in the Al_6Mn phase. The solubility of manganese in the $Al_{15}Mn_3Si_2$ phase is 0.7–0.8%.

The aluminum corner of the Al–Mn–Si phase diagram is shown in Figure 1.4, and the invariant and monovariant reactions involving (Al) are given in Tables 1.11 and 1.12, respectively.

Table 1.13 shows that in the region (Al) + Al_6Mn + $Al_{15}Mn_3Si_2$ the solubility of manganese in (Al) is approximately the same as in the binary Al–Mn system. At the same time, the solubility of silicon is very low even at high temperatures. Contrary to this, in the other three-phase region – (Al) + (Si) + $Al_{15}Mn_3Si_2$ – the solubility of silicon is at the same level as in the binary Al–Si system, while that of manganese does not exceed 0.1%.

It should be noted that the isothermal section in Figure 1.4b does not agree with the data from Table 1.13, showing much greater solid solubility of Mn at 527°C (1% instead of ~0.4% in the (Al) + Al_6Mn + $Al_{15}Mn_3Si_2$ region). The calculated isothermal section at 550°C given in Figure 1.4c corresponds to the solubility data much better (Du et al., 2004).

Under nonequilibrium conditions of solidification, in addition to the already considered phases, the Al_4Mn and $Al_{10}Mn_3Si$ phases can occur in aluminum alloys containing more than ~2.5% Mn, due to the incomplete peritectic reactions (Table 1.11).

In alloys based on the Al–Si system and containing up to 2% Mn, all alloys (irrespective of the solidification conditions) fall into the phase region (Al) + (Si) + $Al_{15}Mn_3Si_2$. According to our data, with an increase in cooling rate the concentration of manganese in (Al) in the as-cast state can reach 1% Mn in alloys

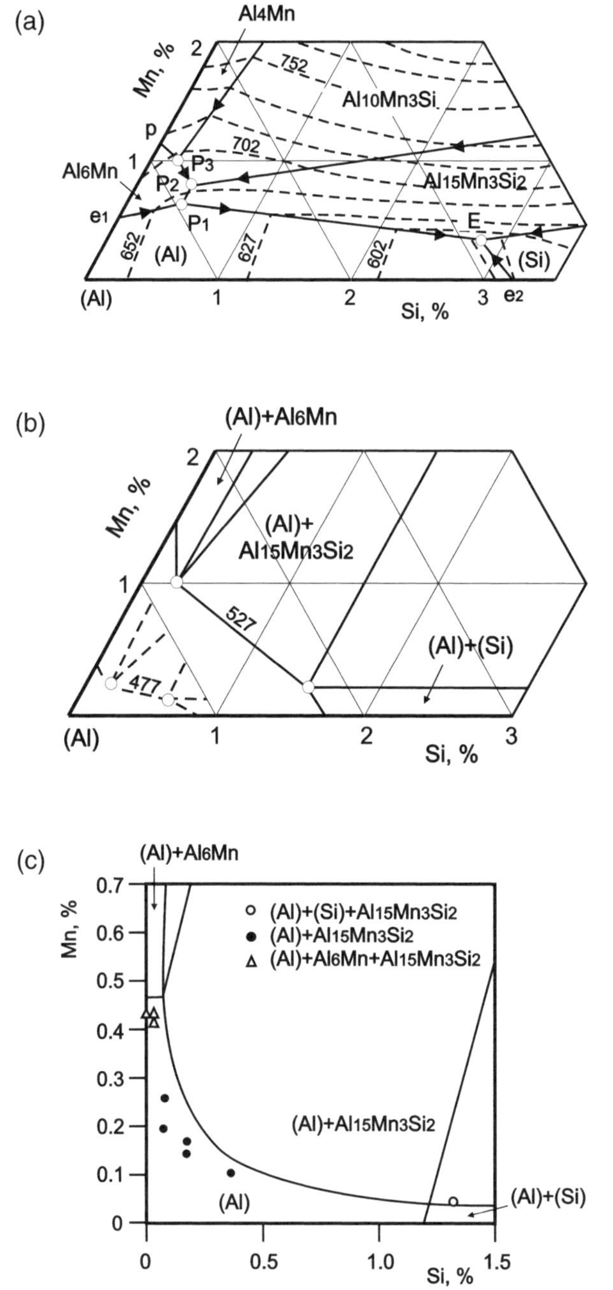

Figure 1.4. Phase diagram of Al–Mn–Si system: (a) liquidus; (b) isothermal sections at 527°C and 477°C (dashed lines) (Mondolfo, 1976); and (c) calculated isothermal section at 550°C with experimental data (points) (after Du et al., 2004).

Table 1.11. Invariant reactions in ternary alloys of Al–Mn–Si system (Mondolfo, 1976; Drits et al., 1977)

Reaction	Point in Figure 1.4a	*T*, °C	Concentrations in liquid phase	
			Mn, %	Si, %
$L + Al_4Mn \Rightarrow Al_6Mn + Al_{10}Mn_3Si$	P_3	690	3.4–3.8	0.5–0.7
$L + Al_{10}Mn_3Si \Rightarrow Al_6Mn + Al_{15}Mn_3Si_2$	P_2	655–657	2.7–2.8	1.3–1.6
$L + Al_6Mn \Rightarrow (Al) + Al_{15}Mn_3Si_2$	P_1	648–649	2.5–2.8	1.5–1.7
$L \Rightarrow (Al) + (Si) + Al_{15}Mn_3Si_2$	E	573–574	1.0–1.2	~12

Table 1.12. Monovariant reactions in ternary alloys of Al–Mn–Si system

Reaction	Line in Figure 1.4a	*T*, °C
$L \Rightarrow (Al) + Al_6Mn$	e_1–P_1	658–649
$L \Rightarrow (Al) + Al_{15}Mn_3Si_2$	P_1–E	649–574
$L \Rightarrow (Al) + (Si)$	e_2–E	577–574

Table 1.13. Limit solid solubilities of manganese and silicon in aluminum in three-phase fields of Al–Mn–Si phase diagram (Drits et al., 1977; Phillips, 1959)

T, °C	$(Al) + Al_6Mn + Al_{15}Mn_3Si_2$		$(Al) + (Si) + Al_{15}Mn_3Si_2$	
	Mn, %	Si, %	Mn, %	Si, %
649	1.3	0.1		–
600	0.73	0.09	–	–
573	–	–	0.08	1.66
550	0.44	0.08	0.07	1.36
500	0.25	0.08	0.06	0.85
450	0.15	0.08	0.05	0.45
400	0.06	0.08	0.04	0.25

containing up to 4% Si. In Al–Mn alloys, the nonequilibrium eutectic involving (Si) is observed at Si concentrations as low as 0.2–0.3%.

1.4. Al–Fe–Mn–Si PHASE DIAGRAM

The phase diagram of this system is the basic diagram for analyzing the phase composition of wrought alloys of 8006 and 3003 type (Tables 1.6 and 1.14). With respect

Table 1.14. Chemical composition of some commercial alloys whose phase composition can be analyzed using Al–Fe–Mn–Si diagram

Grade	Mn, %	Fe, %	Si, %	Other		
				Mg, %	Cu, %	Cr, %
8006	0.3–1.0	1.2–2.0	0.4	0.1	0.3	–
3102	0.05–0.4	0.7	0.4	–	0.1	–
3107	0.4–0.9	0.7	0.6	–	0.05	–
3003	1.0–1.6	0.7	0.6	–	0.05	–
3014	1.0–1.5	1.00	0.1	0.1	0.5	–
3009	1.2–1.8	0.7	1.0–1.8	0.1	0.1	0.05
413.0	0.35	2.0	11–13	0.1	1.0	–
A443.0	0.5	0.8	4.5–6.0	0.05	0.3	–
B443.0	0.35	0.8	4.5–6.0	0.05	0.15	–
C443.0	0.35	2.0	4.5–6.0	0.1	0.6	–
444.0	0.35	0.6	6.5–7.5	0.1	0.25	–

to Al–Si alloys (4XX series), this diagram is required, first and foremost, in order to understand and analyze the modifying effect of manganese on the morphology of needle-shaped Fe-containing particles, i.e. Al_5FeSi.

The phase diagram of the Al–Fe–Mn–Si system was a subject of debate regarding the presence or absence of a quaternary phase. Initially, it was believed that a continuous sequence of solid solutions existed between Al_8Fe_2Si and $Al_{15}Mn_3Si_2$. Later, this assumption was rejected based on the fact that these compounds have different crystal structures, hexagonal and cubic, respectively.

In the accepted version of the phase diagram, a broad range of solid solutions exists based on the compound $Al_{15}Mn_3Si_2$, extending towards the Al–Fe–Si face (Figure 1.5) (Mondolfo, 1976). In this variant, iron is substituted for manganese in the ternary compound to the composition 31% Fe, 1.5% Mn, 8% Si, and the broad region of homogeneity is treated as the formation of the quaternary phase $Al_{15}(FeMn)_3Si_2$. Invariant transformations involving (Al), which are possible in the Al–Fe–Mn–Si system within the framework of the accepted version, are given in Table 1.15; and the corresponding bi- and monovariant reactions, in Table 1.16.

On the other hand, Zakharov et al. (1988, 1989a, b, 1992) studied alloys containing 10–14% Si, 0–3% Fe, 0–4% Mn, and found the existence of the quaternary compound $Al_{16}(FeMn)_4Si_3$, the formation of which led to the development of the quasi-ternary section Al–$Al_{16}(FeMn)_4Si_3$–Si. Two secondary systems are then formed on both sides of this section: Al–$Al_{16}(FeMn)_4Si_3$–Al_5FeSi–Si (adjacent to the Al–Fe–Si face) and Al–$Al_{16}(FeMn)_4Si_3$–$Al_{12}Mn_3Si_2$–Si (adjacent to the Al–Mn–Si face). In addition, yet another secondary system, Al_5FeSi–Al_4FeSi_2–$Al_{16}(FeMn)_4Si_3$–Si, which occurs below 596°C, can be singled out. Alloys containing

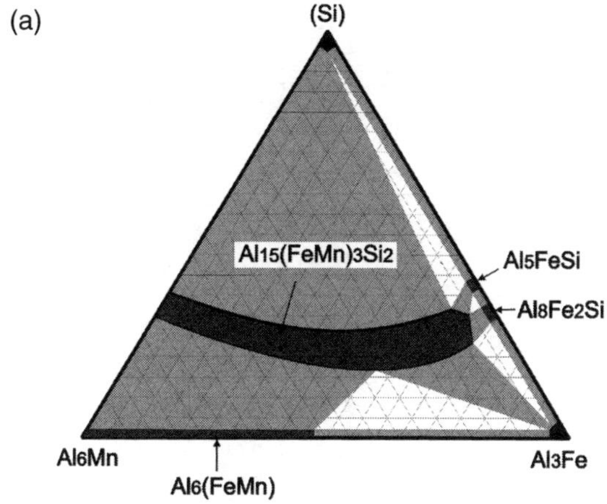

Figure 1.5. Phase diagram of Al–Fe–Mn–Si system: (a) distribution of phase fields in the solid state and (b) polythermal projection of solidification surfaces (Mondolfo, 1976).

Table 1.15. Invariant reactions in quaternary alloys of Al–Fe–Mn–Si system (Mondolfo, 1976)*

Reaction	Point in Figure 1.5b	T, °C	Concentrations in liquid phase		
			Si, %	Fe, %	Mn, %
$L + Al_6(FeMn) + Al_3Fe \Rightarrow (Al) + Al_{15}(FeMn)_3Si_2$	P_4	648	1.75	2.0	0.35
$L + Al_3Fe \Rightarrow (Al) + Al_8Fe_2Si + Al_{15}(FeMn)_3Si_2$	P_3	~628**	3–5	2–2.5	<0.2
$L + Al_8Fe_2Si \Rightarrow (Al) + Al_5FeSi + Al_{15}(FeMn)_3Si_2$	P_2	~607**	5–10	1–2	0.1–0.5
$L + Al_5FeSi \Rightarrow (Al) + (Si) + Al_{15}(FeMn)_3Si_2$	P_1	575	11.7	0.6	0.2

* Reaction $L + Al_6(FeMn) \Rightarrow (Al) + Al_{15}(FeMn)_3Si_2 + Al_3Fe$ is possible

** Our estimate

Table 1.16. Bivariant and monovariant reactions in quaternary alloys of Al–Fe–Mn–Si system

Bivariant reaction	Field in Figure 1.5b	*T*, °C	Monovariant reaction	Line in Figure 1.5b	*T*, °C
$L \Rightarrow (Al) + Al_3Fe$	Al_3Fe–e_3–P_4–P_3–p_3–Al_3Fe	655–~628	$L \Rightarrow (Al) + Al_6(FeMn) + Al_3Fe$	e_3–P_4	654–648
$L \Rightarrow (Al) + Al_6(FeMn)$	Al_6Mn–p_1–P_4–e_3–Al_6Mn	658–648	$L + Al_6(FeMn) \Rightarrow (Al) + Al_{15}(FeMn)_3Si_2$	p_1–P_4	649–648
$L \Rightarrow (Al) + Al_{15}(FeMn)_3Si_2$	p_1–P_4–P_3–P_2–P_1–e_1–p_1	648–573	$L \Rightarrow (Al) + Al_3Fe + Al_{15}(FeMn)_3Si_2$	P_4–P_3	648–~628
$L \Rightarrow (Al) + Al_8Fe_2Si$	p_3–P_3–P_2–p_2–p_3	629–~607	$L + Al_3Fe \Rightarrow (Al) + Al_8Fe_2Si$	p_3–P_3	629–~628
$L \Rightarrow (Al) + Al_5FeSi$	p_2–P_2–P_1–e_2–p_2	611–575	$L \Rightarrow (Al) + Al_8Fe_2Si + Al_{15}(FeMn)_3Si_2$*	P_3–P_2	628–~607
$L \Rightarrow (Al) + (Si)$	(Si)–e_2–P_1–e_1–(Si)	577–573	$L + Al_8Fe_2Si \Rightarrow (Al) + Al_5FeSi$	p_2–P_2	611–~607
			$L \Rightarrow (Al) + Al_{15}(FeMn)_3Si_2 + Al_5FeSi$*	P_2–P_1	~607–575
			$L \Rightarrow (Al) + Al_5FeSi + (Si)$	e_2–P_1	576–575
			$L \Rightarrow (Al) + (Si) + Al_{15}(FeMn)_3Si_2$	P_1–e_1	575–573

* Peritectic reaction is a possible alternative

iron and manganese in the ratio Mn:Fe = 1:1 complete crystallization at 576°C and in the solid state contain the phases (Al), $Al_{16}(FeMn)_4Si_3$, and (Si). Alloys with the ratio of Mn:Fe < 1:1 complete crystallization at 574°C and in the solid state contain the phases (Al), $Al_{16}(FeMn)_4Si_3$, Al_5FeSi and (Si). Finally, alloys with the ratio Mn:Fe > 1:1 complete crystallization at 575°, and in the solid state contain the phases (Al), $Al_{16}(FeMn)_4Si_3$, $Al_{15}Mn_3Si_2$, and (Si).

According to Mondolfo, the solid solution of iron in the $Al_{15}Mn_3Si_2$ phase has a cubic structure with the lattice parameter *a* decreasing with an increase in the Fe content from 1.265 nm (0% Fe) up to 1.25 nm (31.1% Fe) (1976). The quaternary phase found by Zakharov et al. has a face-centered cubic structure with parameter $a = 1.252 \pm 0.04$ nm (Zakharov et al., 1989b). The close lattice parameters of these phases do not enable us to say equivocally if this or that variant of the Al–Fe–Mn–Si phase diagram is true. We may note that one of the recent studies on this quaternary system supports the variant given by Mondolfo, in which no quaternary compound is present but there is a wide homogeneity range of the $Al_{15}(FeMn)_3Si_2$ phase (Davignon et al., 1996). This study used the electron microprobe analysis to investigate 24 Al–Fe–Mn–Si alloys annealed at 550°C for 12 weeks.

The solubilities of the elements in (Al) in the four-phase regions adjacent to the ternary systems are, probably, the same as in the respective three-phase regions of the ternary systems. In other words, in 4XX.0-series alloys (Table 1.14) that fall into the (Al) + (Si) + Al_5FeSi + $Al_{15}(FeMn)_3Si_2$ region, the solubilities of silicon and manganese are approximately the same as those given in Table 1.13 for the (Al) + (Si) + $Al_{15}Mn_3Si_2$ region.

Nonequilibrium crystallization has a significant effect on the phase composition. In particular, in Al–Si alloys, due to the inhibition of peritectic reactions, the Al_5FeSi phase occurs at a much greater Mn:Fe ratio. The solubility of manganese in (Al) in the as-cast state decreases with the increasing Fe and Si concentrations, because a considerable part of it is bound in the $Al_{15}(FeMn)_3Si_2$ phase during solidification.

1.5. COMMERCIAL ALUMINUM AND 8111-TYPE ALLOYS

According to the Al–Fe–Si diagram (Figure 1.1), commercial aluminum and 8111-type alloys (Table 1.1) can contain in the solid state all the phases, which are in equilibrium with (Al). Therefore, due to the variable solubility of Si in (Al) the phase composition of an alloy can strongly vary depending on the heat treatment temperature, as reflected in the isothermal sections shown in Figure 1.6. The isothermal sections below 576°C (Figure 1.6a–c) comprise the same three-phase fields: (Al) + Al_3Fe + Al_8Fe_2Si, (Al) + Al_8Fe_2Si + Al_5FeSi, and (Al) + Al_5FeSi + (Si), the last region widening as temperature decreases, due to the decreasing solubility of

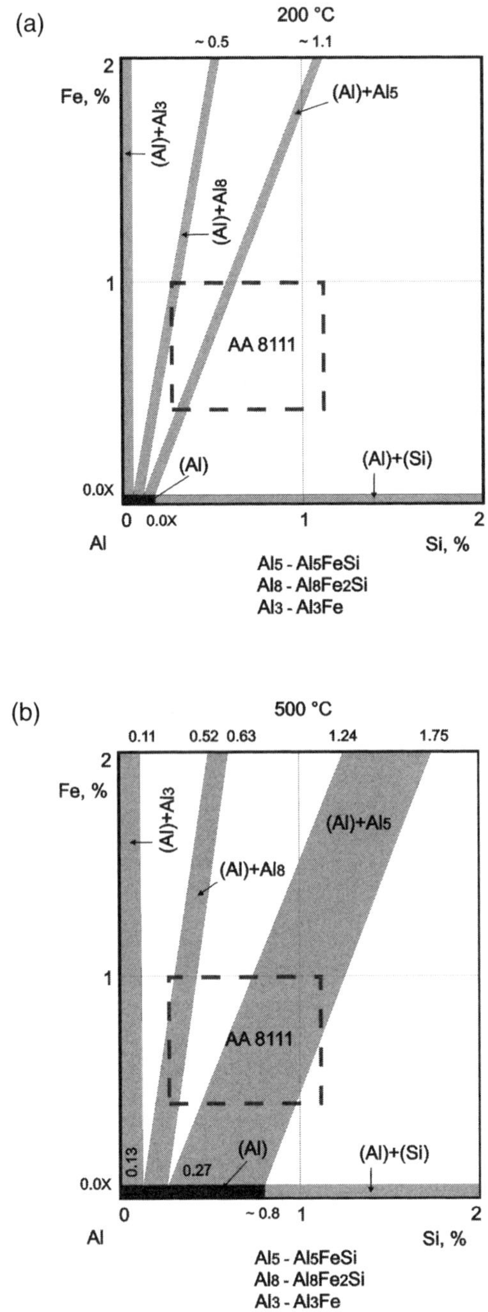

Figure 1.6. Isothermal sections of Al–Fe–Si phase diagram: (a) 200°C; (b) 500°C; (c) 570°C; (d) 600°C; (e) 620°C; and (f) 640°C.

(c)

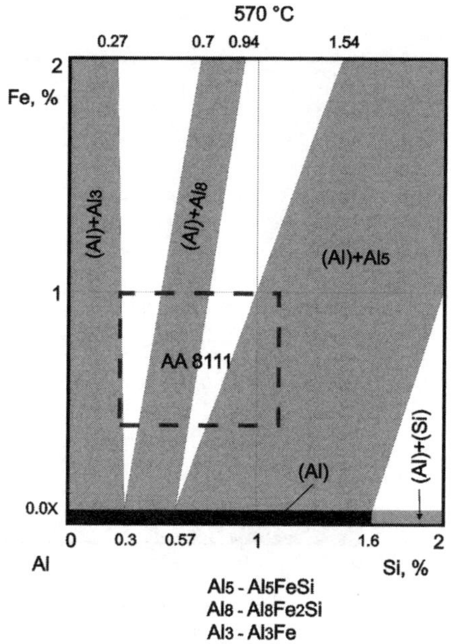

(d)

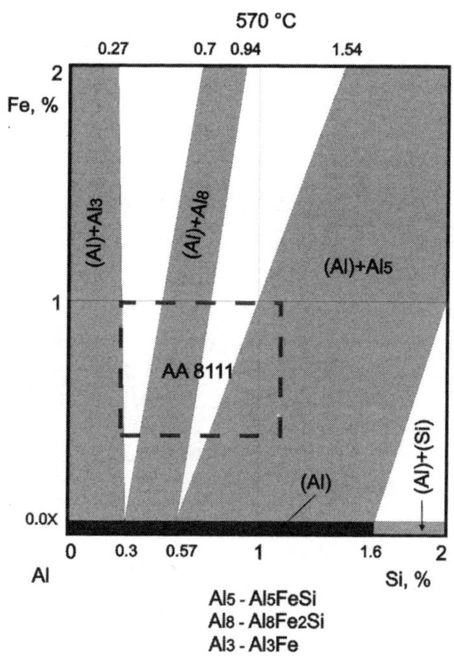

Figure 1.6 (*continued*)

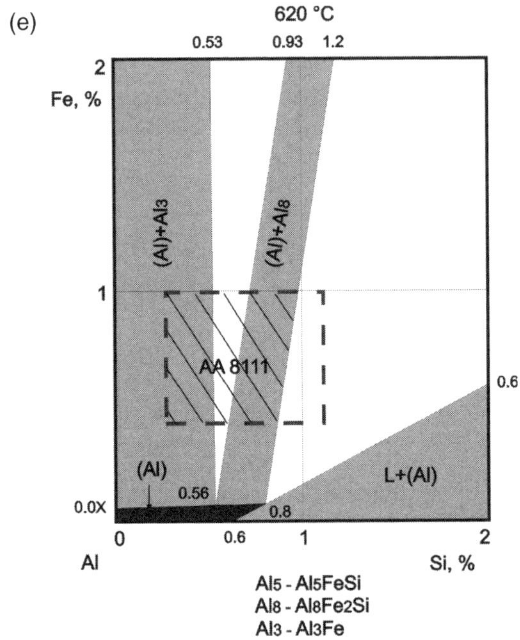

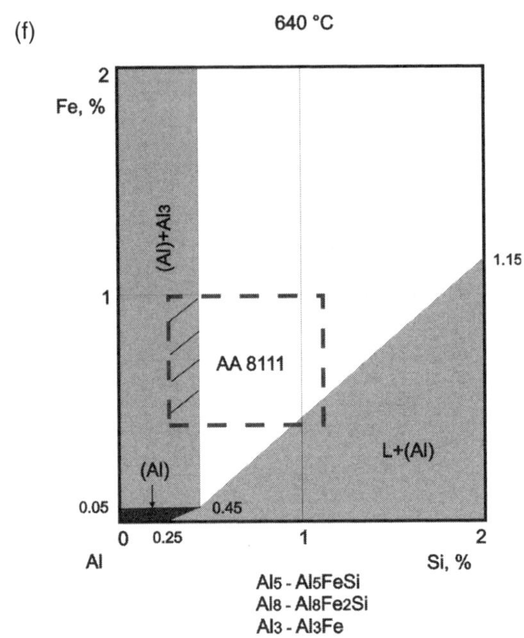

Figure 1.6 *(continued)*

silicon in (Al). At higher temperatures, regions with the liquid phase appear on isothermal sections, which enables one to determine the ranges of allowable heating (Figure 1.6d–f). This is topical for 8111-type alloys, in which the concentration of silicon can reach 1.1%, and, following Figure 1.6e, only alloys within the hatched compositional range can be annealed at 620°C.

Polythermal sections given in Figure 1.7 can be used to analyze the phase composition of as-cast alloys belonging to the Al–Fe–Si system. As the presence of free silicon implies the widening of the solidification range (due to the eutectic reaction given in Table 1.2), commercial alloys usually contain more iron than silicon, e.g. an 8111 alloy with the compositional ranges of Fe and Si considerably larger than those of 1XXX-series alloys (Table 1.1).

Let us consider the microstructure of a cast 8111-alloy strip and the products of downstream process stages (semifinished rolled products and foil). The change in the phase composition of an 8111 alloy with respect to specific concentrations of Fe and Si and to the annealing temperature can be traced in polythermal sections in Figure 1.7. These sections are characterized by a rather complex constitution, especially at small concentrations of silicon, which is due to the peritectic reactions and variable solubility of Si in (Al). Some points on the invariant horizontals are so close to one another that they can be separated only by calculation. Compositional ranges of an 8111 alloy are given on the polythermal sections at 1% Fe, 0.5% Si and 1% Si in Figure 1.7c, e, and f, respectively.

At a Si content close to the upper limit, considerably large particles of the (Si) phase are formed and can be retained in the subsequent process stages up to the final foil. This can result in spoilage and perforation, especially in thin foils (6–14 μm) when hard and brittle (Si) particles lead to scratches and even to ruptures (Figure 1.8a, b). A negative effect can also be caused by needle-like inclusions of the Al_3Fe and Al_5FeSi phases (Figure 1.8c), incapable of spheroidization even at large holding times during annealing (Belov and Zolotorevskii, 2001). At a Si content close to the lower limit (and at 1% Fe), only one excess phase, Al_8Fe_2Si, is formed during solidification (Figure 1.7c, e). This phase is stable within a relatively broad range of annealing temperatures and is capable of spheroidization upon annealing (Figure 1.8d).

A more reliable selection of the optimal Si concentration and the annealing temperature is possible using the calculated dependences of phase volume fractions on the alloy composition. In particular, the plots shown in Figure 1.9 suggest that when the concentration of Si is at the upper limit in an 8111 alloy, the annealing temperature should be equal to or higher than 500°C (Belov and Matveeva, 2001). This ensures the complete dissolution of eutectic silicon.

The as-cast structure of Al–Fe–Si alloys is typically quite different from the structure (phase composition) predicted by the equilibrium phase diagram

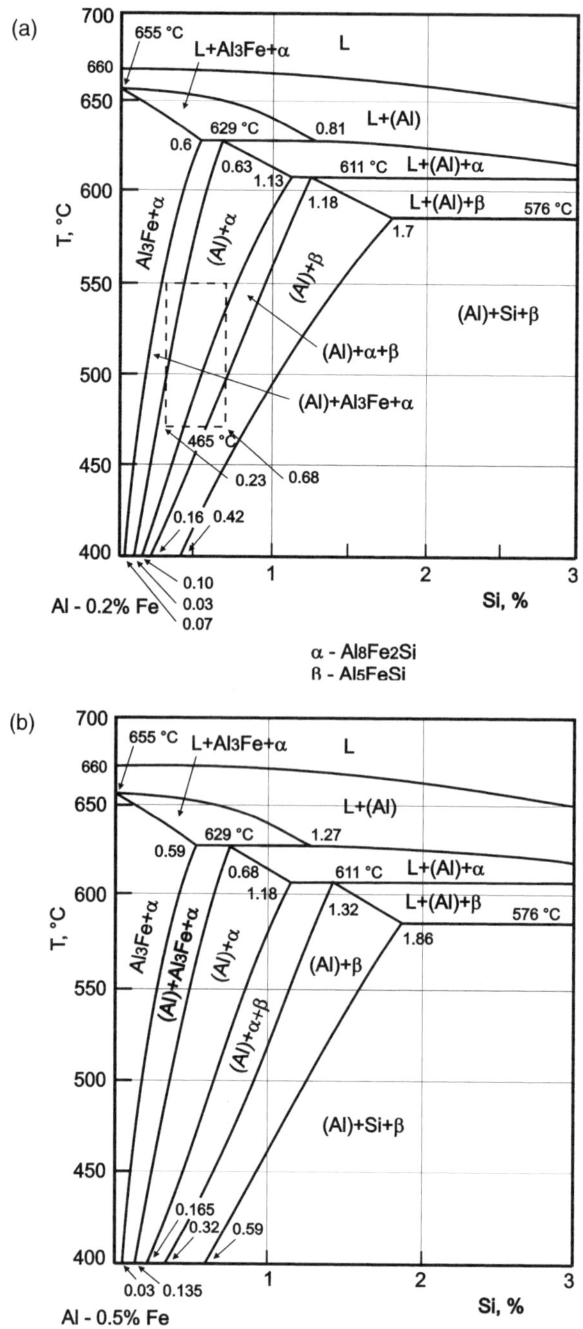

Figure 1.7. Polythermal sections of Al–Fe–Si phase diagram: (a) 0.2% Fe; (b) 0.5% Fe; (c) 1% Fe; (d) 0.2% Si; (e) 0.5% Si; and (f) 1% Si. Compositional range of an 8111 alloy is marked on some sections.

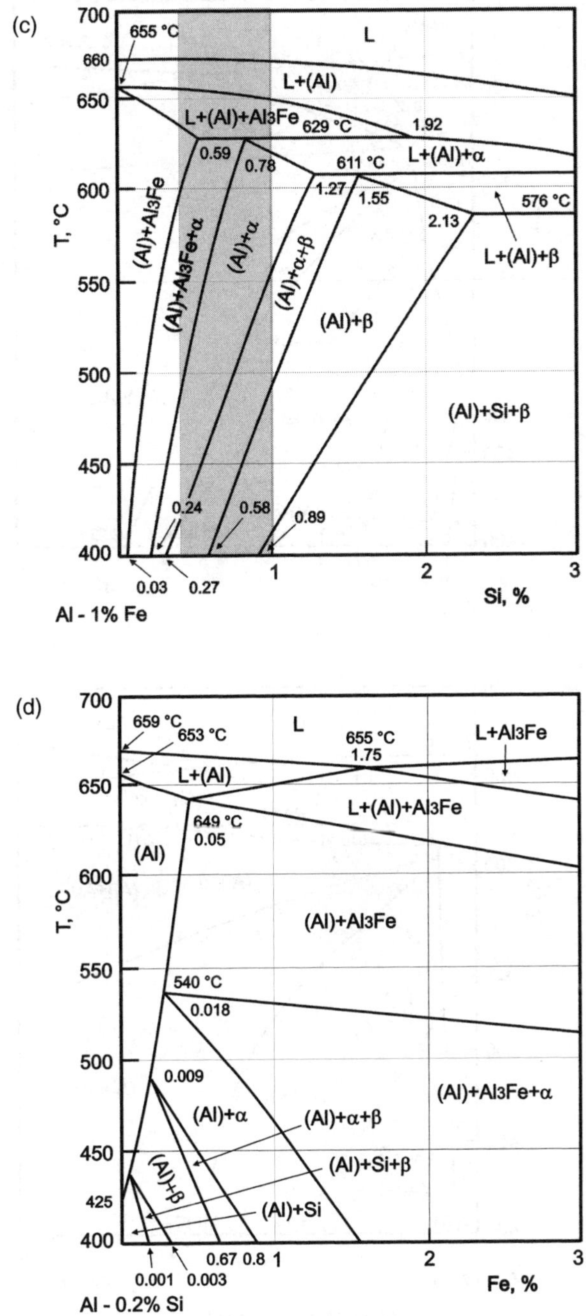

Figure 1.7 *(continued)*

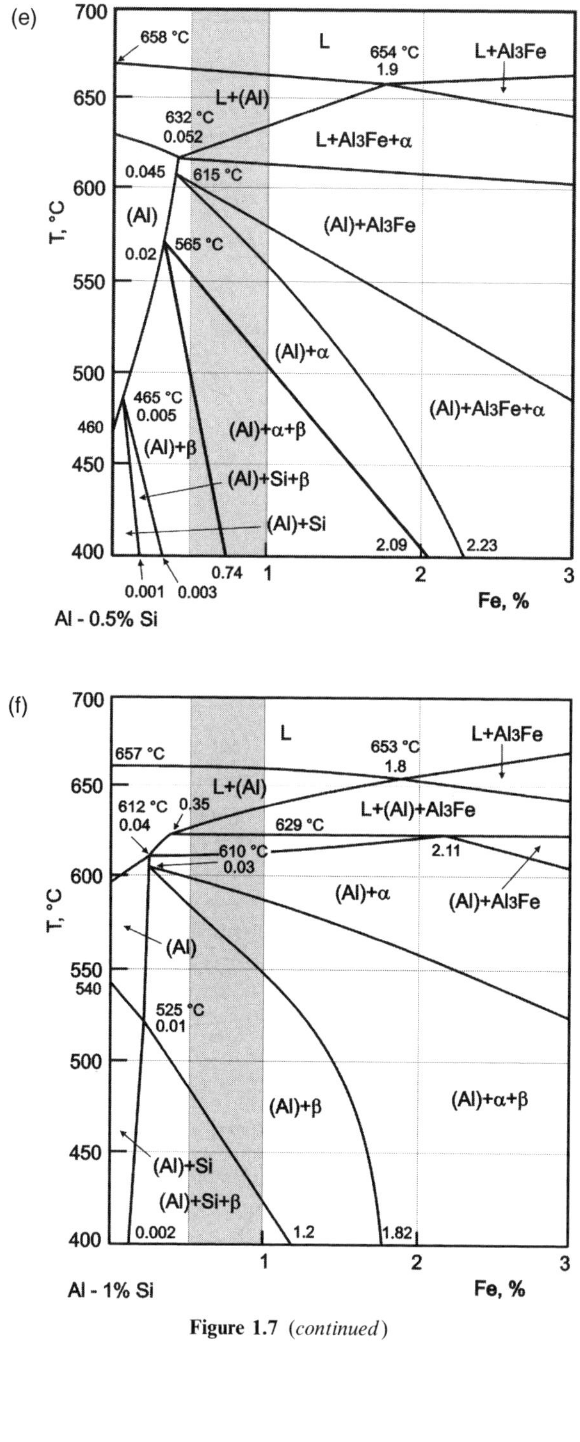

Figure 1.7 *(continued)*

(a)

(b)

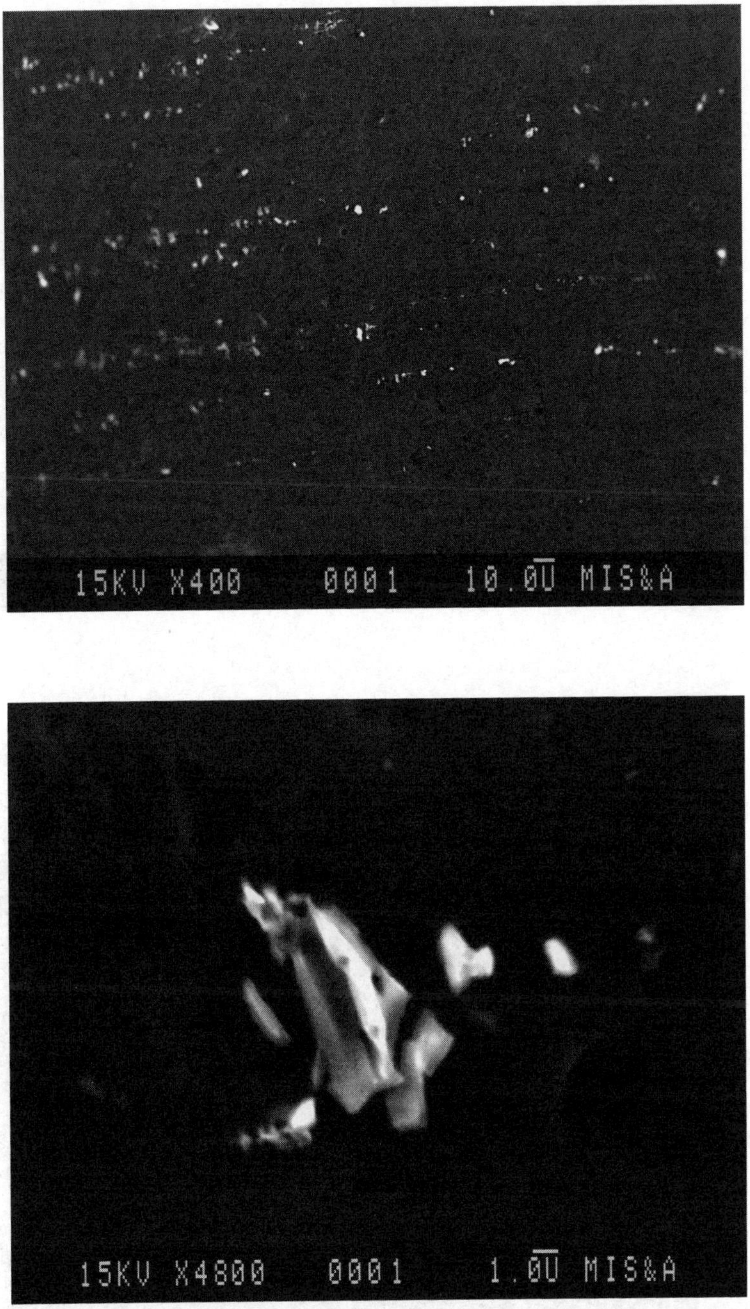

Figure 1.8. Microstructure of an 8111 alloy, SEM: (a, b) eutectic silicon in 10 μm foil; (c) needle particles of Al_5FeSi phase in a twin-roll cast strip annealed at 550°C, 10 h; (d) globular particles of Al_8Fe_2Si phase in a twin-roll cast strip annealed at 550°C, 10 h: (a–c) 1% Fe, 0.8% Si; and (d) 1% Fe, 0.4% Si.

(c)

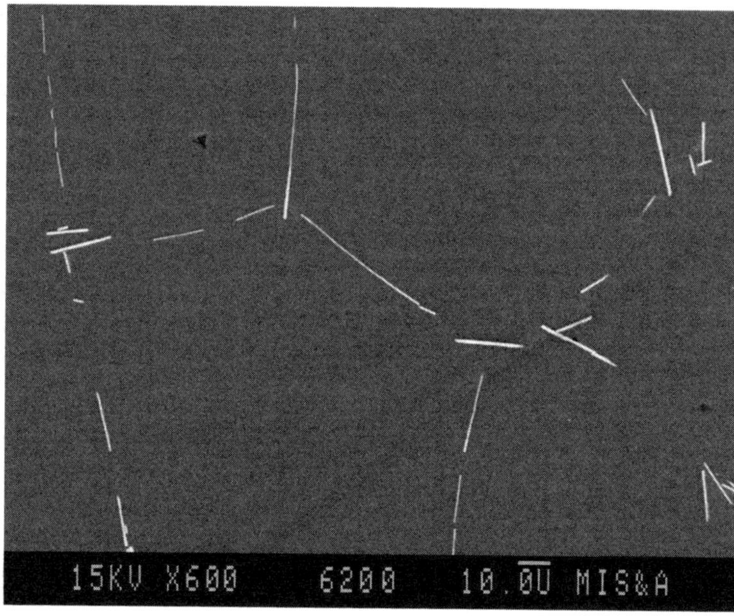

(d)

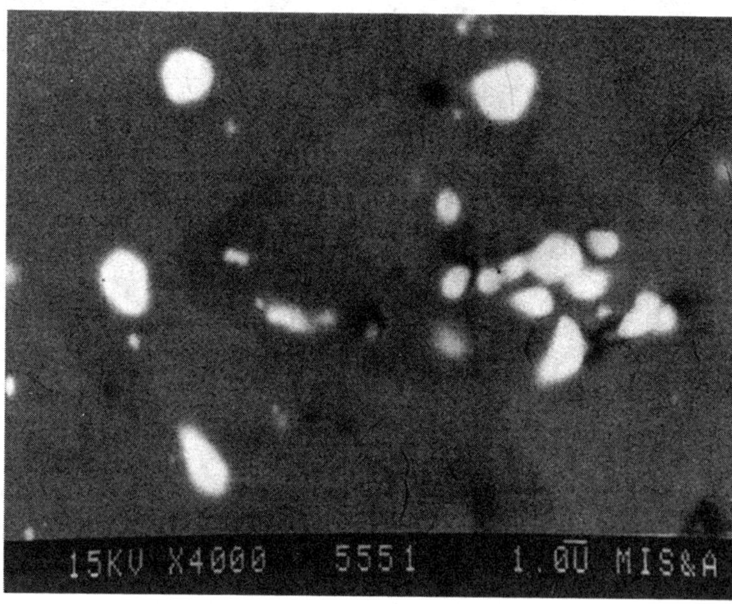

Figure 1.8 (*continued*)

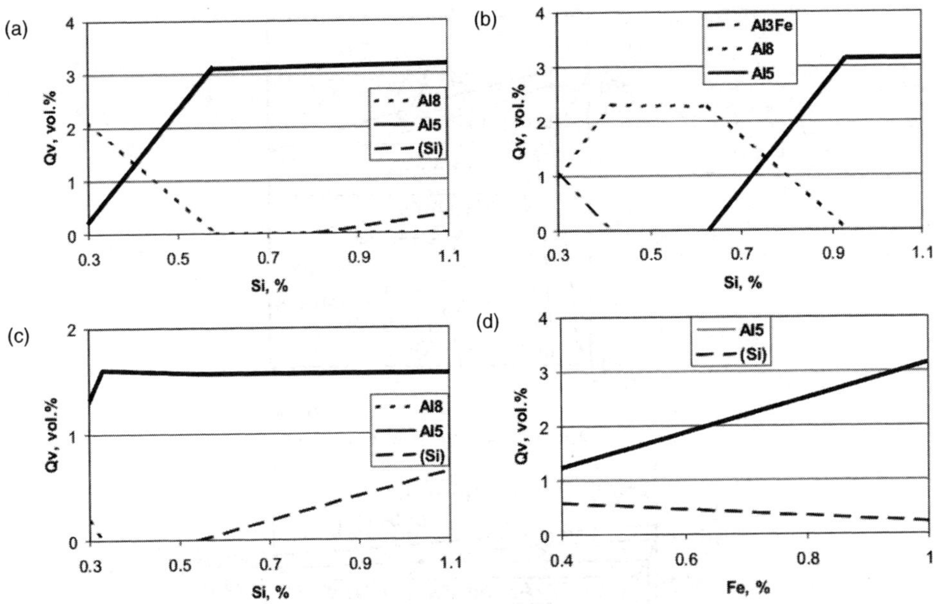

Figure 1.9. Calculated dependence of volume fractions of phases on Fe and Si concentrations in an 8111 alloy: (a) 1% Fe, 400°C; (b) 1% Fe, 550°C; (c) 0.5% Fe, 400°C; and (d) 1% Si, 400°C. Al_8 – Al_8Fe_2Si and Al_5 – Al_5FeSi.

(Figure 1.1). The nonequilibrium soldification can be analyzed from polythermal sections constructed for different cooling rates (Belov et al., 2002a). Figure 1.10 demonstrates how the cooling rate affects the solidification path of Al–1.7%Fe–Si alloys. One can see that the probability of Al_3Fe formation during solidification first increases and then decreases on increasing the cooling rate. At relatively high cooling rates, this phase is formed only in alloys containing Fe:Si > 2.

The as-cast structure of commercial aluminum (1XXX series in Table 1.1) can also be analyzed using isothermal (Figure 1.6) and polythermal sections (Figure 1.7 a, d). The main phases in cast aluminum would be Al_3Fe, Al_8Fe_2Si, and Al_5FeSi as well as different metastable phases listed in Table 1.5. Free (Si) is rare as the Fe:Si ratio is usually maintained above unity in order to prevent hot tearing during casting, through avoiding the low-temperature eutectic reaction $L \Rightarrow (Al) + Al_5FeSi + (Si)$ (Table 1.2). The phase selection in as-cast commercial aluminum is a function of the Fe:Si ratio and the cooling rate. Table 1.17 shows experimentally observed temperatures during solidification of a 1050 alloy containing 0.37% Fe and 0.05% Si (Bäckerud et al., 1986). The solidus temperature decreases with increasing the cooling rate, reaching 630°C (close to point P_1 in Figure 1.1b and Table 1.2) at which the invariant peritectic reaction specified by Bäckerud et al. (1986) in

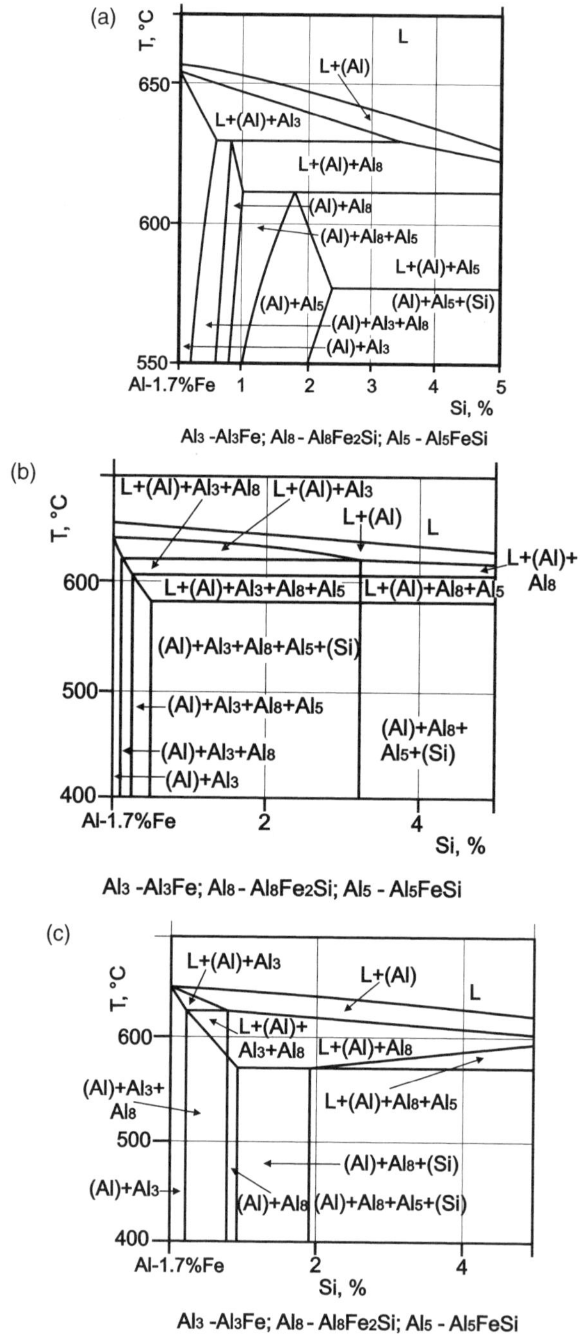

Figure 1.10. Effect of cooling rate (V_c) on the polythermal section of Al–Fe–Si phase diagram at 1.7% Fe: (a) equilibrium; (b) $V_c \sim 10^{-2}$ K/s; and (c) $V_c \sim 10$ K/s.

Table 1.17. Solidification reactions under nonequilibrium conditions in commercial aluminium containing 0.37% Fe and 0.05% Si (Bäckerud et al., 1986)

Reaction	Temperatures (°C) at a cooling rate		
	0.4 K/s	1.2 K/s	18 K/s
$L \Rightarrow (Al)$	659	659	659
$L \Rightarrow (Al) + Al_3Fe$	650	649	647
$L + Al_3Fe \Rightarrow (Al) + Al_8Fe_2Si$	–	642–638	630
Solidus	642	638	630

Table 1.17 should occur under equilibrium conditions. The formation of the metastable Al_6Fe phase is possible in a 1050 alloy cast at cooling rates above 1 K/s.

1.6. WROUGHT ALLOYS WITH MANGANESE (3XXX SERIES AND 8006 TYPE)

Analysis of Mn-containing alloys is more complex than that of 8111-type alloys and can be performed using the ternary diagrams in some cases only. For example, the phase composition of an 8006 alloy (Table 1.6) at a low content of Si impurity can be considered using the Al–Fe–Mn phase diagram. The isothermal sections of this diagram in the solid state are simple and have only one three-phase region $(Al) + Al_6(FeMn) + Al_3Fe$ as it follows from Figure 1.3b. However, we should note relatively wide two-phase regions $(Al) + Al_6(FeMn)$ and $(Al) + Al_3Fe$. The appearance of the former phase field is due to a considerable solubility of iron in the Mn-containing aluminide; and the latter phase field results from a high solubility of manganese in (Al) at sub-solidus temperatures. Polythermal sections within the compositional range of commercial alloys are also rather simple, as they have only one invariant horizontal (Figure 1.11). The sections at 0.7% Mn (Figure 1.11a) and at 1.6% Mn (Figure 1.11b) show that primary crystals of the Al_3Fe and $Al_6(FeMn)$ phases (which, as a rule, are undesirable) can form when the iron concentration is at the upper limit of an 8006 alloy. However, upon faster solidification, the boundary of the occurrence of these primary crystals shifts towards higher concentrations of Fe and Mn.

The effect of temperature on the structure of an 8006 alloy is determined by the ratio between the equilibrium and nonequilibrium solubilities of manganese in solid aluminum. The latter depends on the alloy composition and, to a significant extent, on the cooling rate (see Table 9.6 and Figure 9.18 in Chapter 9). The nonequilibrium concentration of Mn in (Al) in the as-cast state determines the amount of dispersoids formed during annealing at temperatures above 300–350°C. As the solubility of

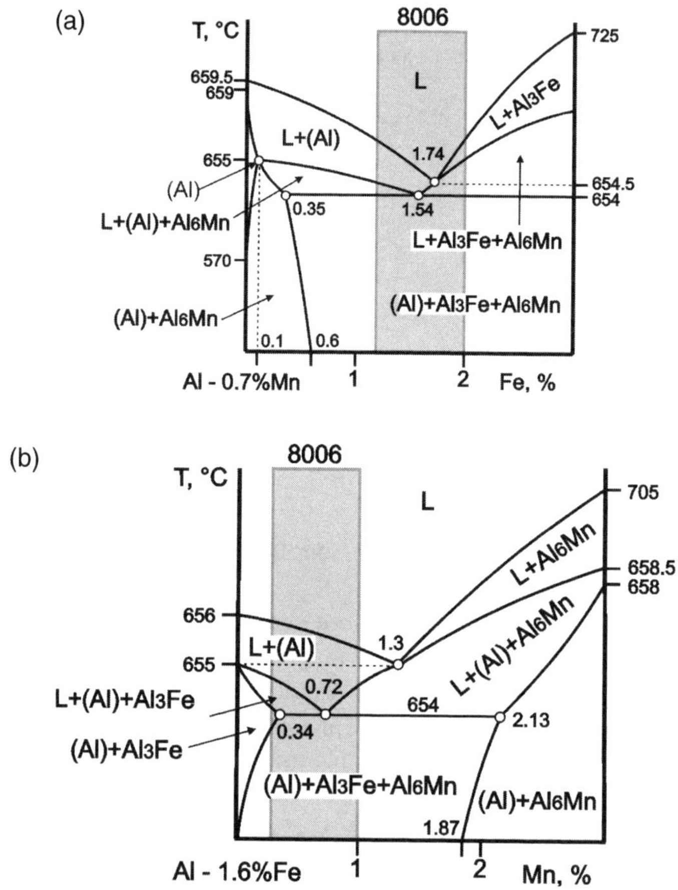

Figure 1.11. Polythermal sections of Al–Fe–Mn phase diagram: (a) 0.7% Mn and (b) 1.6% Fe.

iron in (Al) is negligible, most of these dispersoids are represented by Al_6Mn but not $Al_6(FeMn)$.

The phase composition of a 3009 alloy containing silicon as a main alloying component (Tables 1.10 and 1.14), and the effect of Si impurity on the phase composition of a 3003 alloy (at a low Fe concentration in these alloys) can be analyzed using isothermal and polythermal sections of the Al–Mn–Si phase diagram (Figure 1.12). This ternary diagram is more complicated than the previous one because of the presence of the ternary $Al_{15}Mn_3Si_2$ compound. Due to a relatively high solubility of Si in (Al), this compound can form not only in solidification but also during annealing (forming dispersoids).

A 3003 alloy can be obtained in the single-phase state providing low Mn (<1%) and Si (<0.1%) concentrations, Figure 1.12d. This suggests the possibility of

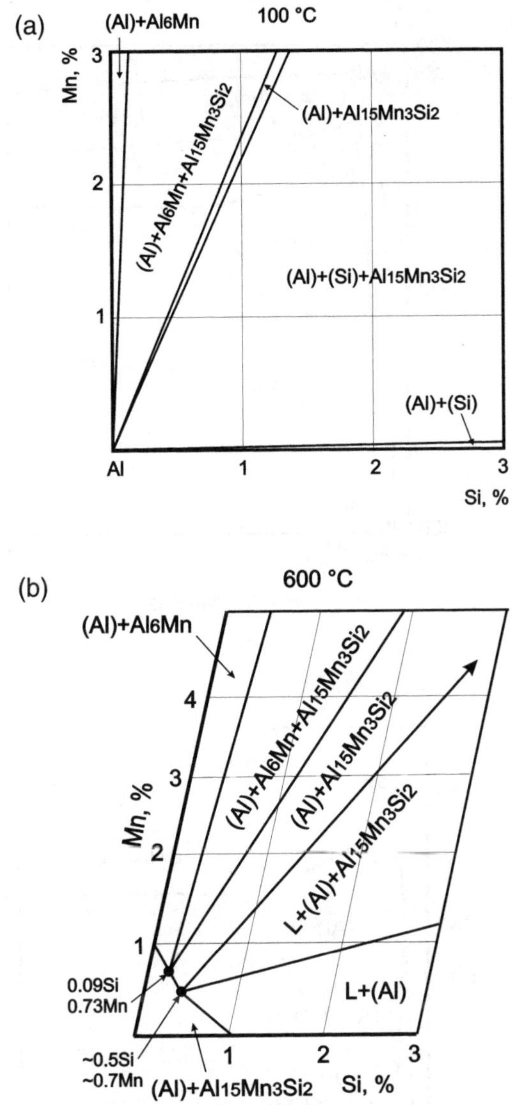

Figure 1.12. Isothermal (a, b) and polythermal (c, d) sections of Al–Mn–Si phase diagram: (a) 100°C; (b) 600°C; (c) 1.5% Mn; and (d) 0.5% Mn.

eliminating the microsegregation with respect to manganese, which is characteristic of most commercial alloys, by annealing at temperatures above 540–550°C. However, Figures 1.12b and 1.12d show that a 3003 alloy containing more Si at the same annealing temperature falls into phase regions containing one or two excess phases. The isothermal section at 600°C in Figure 1.12b also suggests a danger

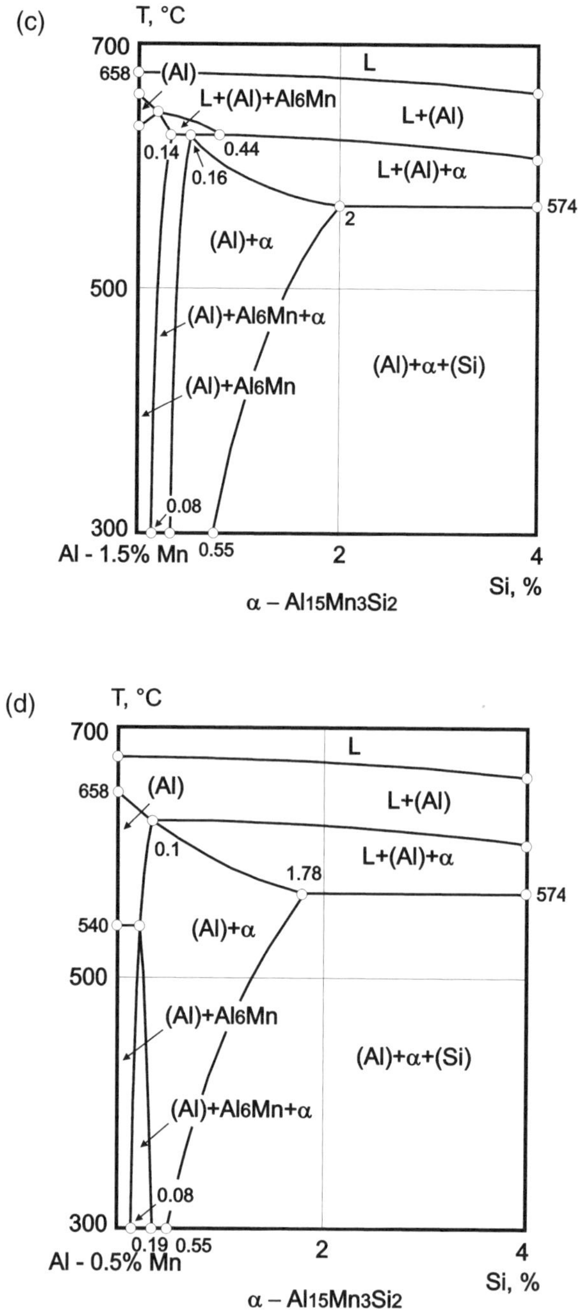

Figure 1.12 (*continued*)

of partial melting of 3XXX alloys with silicon, even at a temperature that is almost 60°C lower than the solidus of binary Al–Mn alloys. At lower temperatures, a 3003 alloy can fall into the two- and three-phase regions involving the phases Al_6Mn, $Al_{15}Mn_3Si_2$, and (Si) (Figure 1.12a).

The polythermal section of the Al–Mn–Si phase diagram at 1.5% Mn in Figure 1.12c shows that the Al_6Mn phase does not form in a 3009 alloy, either during solidification or during annealing.

The combined effect of all three elements (e.g. Mn, Fe, and Si) on the phase composition of 8006, 3003, and 3009 alloys can be analyzed only by the quaternary phase diagram. The isothermal and polythermal sections of the Al–Fe–Mn–Si phase diagram show that the presence of Fe and Si leads generally to the formation of the $Al_{15}(FeMn)_3Si_2$ phase, though other phases can be formed as well (Figure 1.13).

The calculated dependence of the volume fraction of phases in an 8006 alloy (containing 1.6% Fe and 0.7% Mn) versus the concentration of silicon impurity are shown in Figure 1.14.

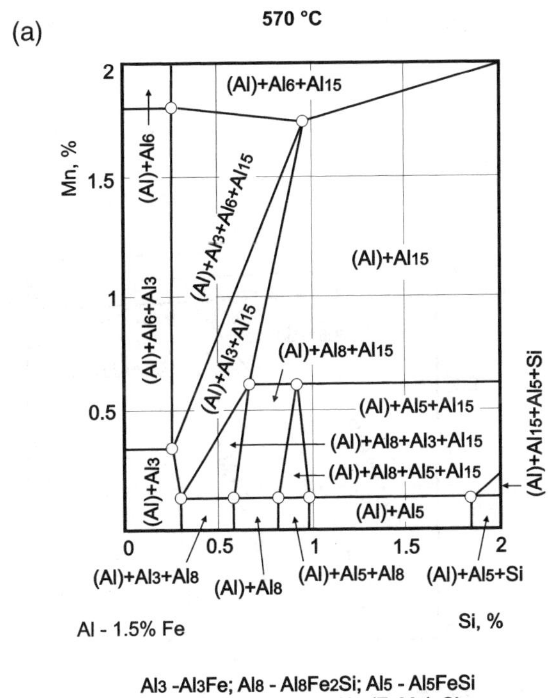

Figure 1.13. Isothermal (a, b) and polythermal (c) sections of Al–Fe–Mn–Si phase diagram: (a) 570°C and 1.5% Fe; (b) 570°C and 1% Mn; (c) 1.5% Fe and 1% Mn.

(b)

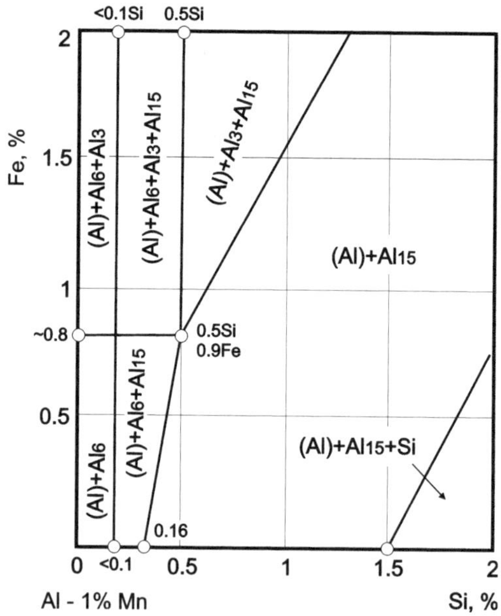

(c)

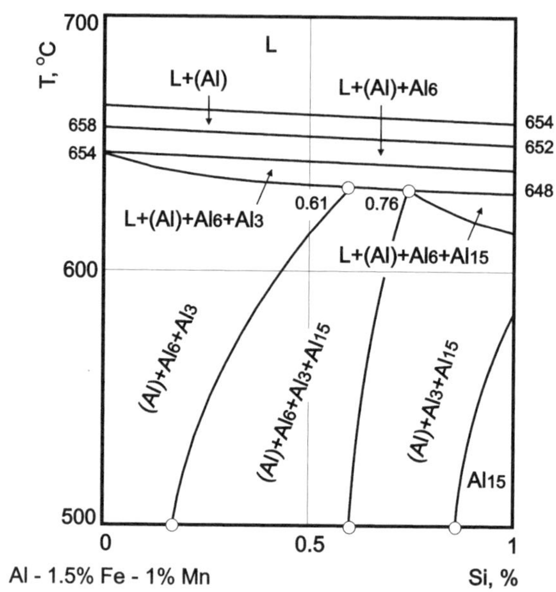

Figure 1.13 *(continued)*

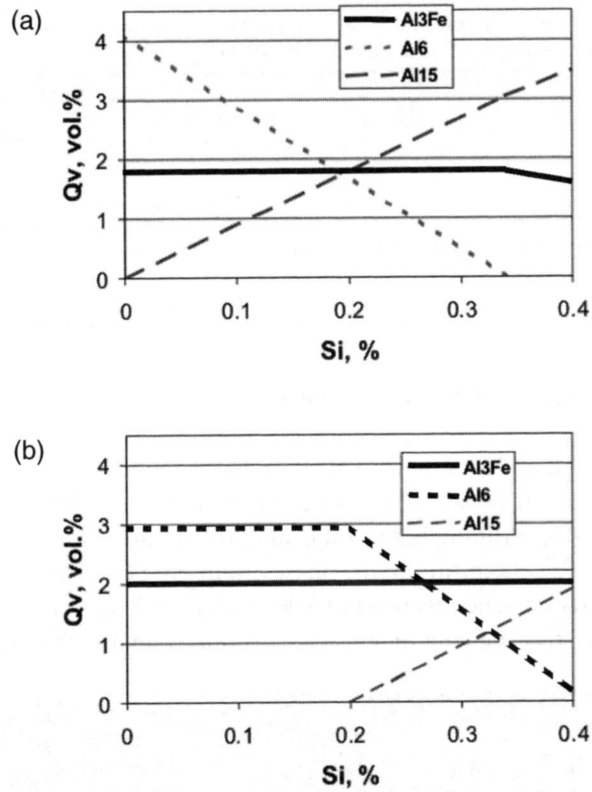

Figure 1.14. Calculated dependence of volume fractions of phases on the Si concentration in an 8006 alloy (1.6% Fe and 0.7% Mn): (a) 400°C and (b) 550°C. Al_6 – $Al_6(FeMn)$; Al_{15} – $Al_{15}(FeMn)_3Si_2$.

Nonequilibrium solidification can complicate the phase composition, which is due to the inhibition of peritectic reactions listed in Tables 1.15 and 1.16. Bäckerud et al. (1986) experimentally observed the solidification reactions given in Table 1.18 during nonequilibrium solidification of a 3003 alloy. Note the considerable decrease in the solidus temperature. The as-cast structure contains $Al_6(FeMn)$ and $Al_{15}(FeMn)_3Si_2$ phases. The amount of the latter phase increases with the silicon concentration. If the concentration of silicon is at the upper level (0.6%) and that of iron is at the lower level, the formation of free (Si) is possible with corresponding decrease of the solidus to 573°C (Table 1.16). The typical microstructures of 3003 and 8006 alloys are shown in Figure 1.15.

The situation becomes simpler with the complete binding of iron and silicon in the $Al_{15}(FeMn)_3Si_2$ phase. The major issue is then the concentration of manganese and silicon in (Al), which determines the amount of dispersoids. In this case, one should use the Al–Mn–Si diagram.

Table 1.18. Solidification reactions under nonequilibrium conditions in a 3003 alloy (1.19% Mn, 0.55% Fe, and 0.18% Si) (Bäckerud et al., 1986)

Reaction	Temperatures (°C) at a cooling rate	
	0.5 K/s	17 K/s
L ⇒ (Al)	655	655
L ⇒ (Al) + $Al_6(FeMn)$	653	646–615
L + $Al_6(FeMn)$ ⇒ (Al) + $Al_{15}(FeMn)_3Si_2$ and/or L ⇒ (Al) + $Al_{15}(FeMn)_3Si_2$	641–634	589
Solidus	634	589

1.7. Al–Si CASTING ALLOYS (4XX.0 SERIES)

Casting alloys of the 4XX.0 series with low Mn content, irrespective of the concentrations of Fe and Si fall at temperatures below 576°C into the phase region (Al) + Al_5FeSi + (Si). Although the total amount of the Al_5FeSi phase increases linearly with the Fe concentration, the phase origin can be different, i.e. the result of binary and ternary eutectic reactions (Tables 1.2 and 1.3) or primary solidification. The particles of this Fe-containing phase (both primary and eutectic) have a

(a)

Figure 1.15. Microstructure of 8006 (a, b) and 3003 (c) alloys, SEM: 8006 (0.6% Mn, 1% Fe, 0.1% Si), 100 μm foil (a – height–width plane, b – length–width plane); main phases $Al_3(FeMn)$ and $Al_6(FeMn)$ and 3003 (1.3%Mn, 0.5%Fe, 0.3%Si), extruded pipe, main phase $Al_{15}(FeMn)_3Si_2$.

(b)

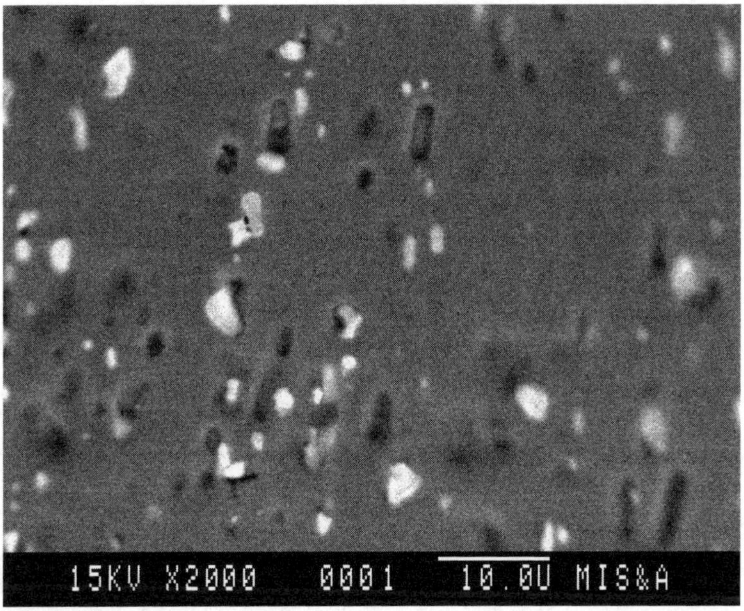

(c)

Figure 1.15 (*continued*)

plate-like morphology (needles in cross sections as shown in Figure 1.16a, b), but their size depends on how the phase has been formed. The finest plates are characteristic of the ternary eutectic, and the largest particles represent primary crystals (Belov et al., 2002a).

(a)

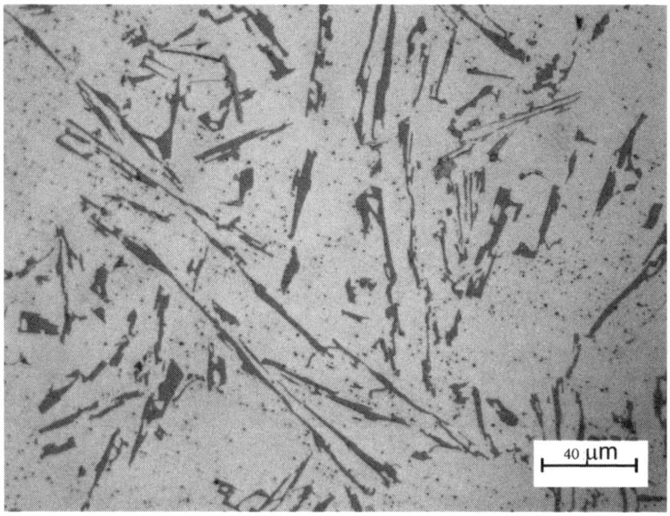

(b)

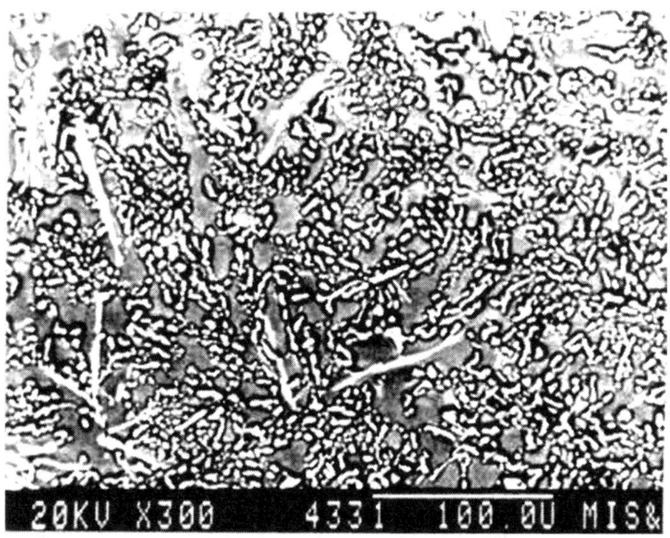

Figure 1.16. Microstructure of a 413.0 alloy cast in a metallic mold: (a) as-cast state (~12% Si, 0.8% Fe, <0.1% Mn), optical microscope; (b) and (c) annealed at 550°C, 10 h (~12% Si, 0.3% Fe, 0.2% Mn), SEM.

(c)

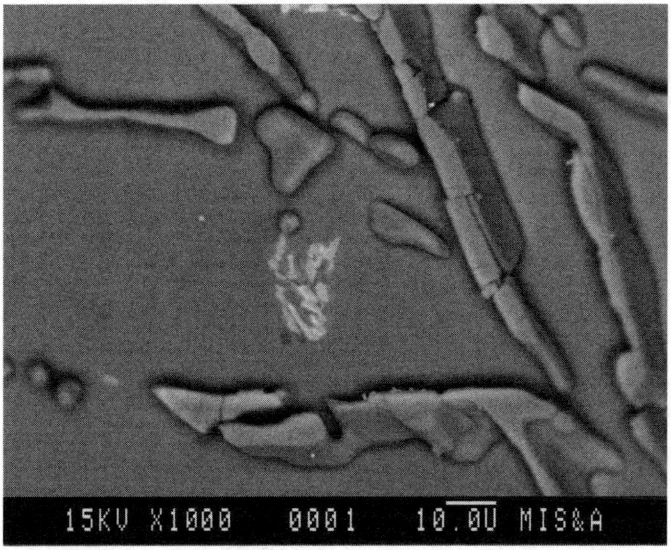

Figure 1.16 *(continued)*

Figure 1.17 shows polythermal sections of the Al–Fe–Si phase diagram at different silicon concentrations. From the polythermal section at 7% Si (Figure 1.17b), it follows that at this Si concentration the primary Al_5FeSi phase is formed at iron concentrations above 1.6% Fe; and the binary eutectic, at >0.37% Fe. At 10%

(a)

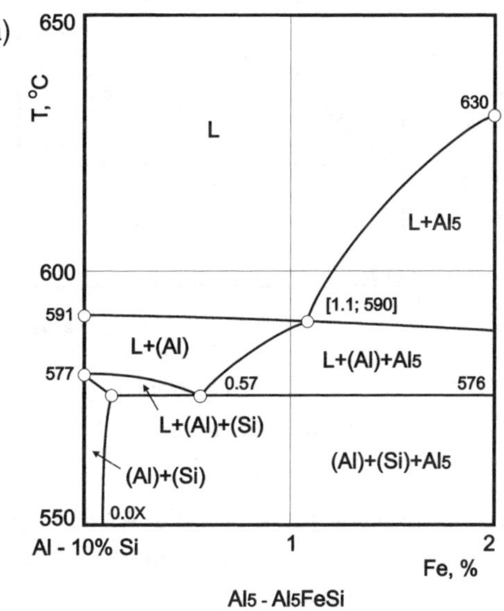

Figure 1.17. Polythermal sections of Al–Fe–Si phase diagram at 10% Si (a); 7% Si (b); and 5% Si (c).

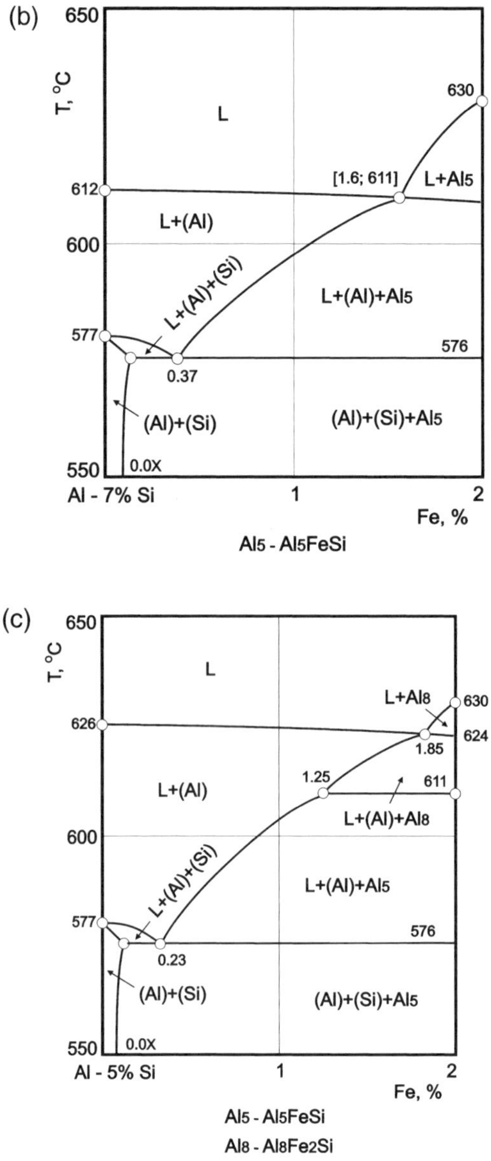

Figure 1.17 *(continued)*

Si (Figure 1.17a), the primary crystals of the Al_5FeSi phase are formed at a lower concentration of iron (1%), but the boundary for the occurrence of the binary eutectic shifts towards higher iron concentrations (0.6% Fe). Obviously, no other Fe-containing phases appear at these Si concentrations under typical industrial solidification conditions. However, at 5% Si the Al_8Fe_2Si phase can appear as a result of nonequilibrium solidification (incomplete peritectic reaction $L + Al_8Fe_2Si \Rightarrow (Al) + Al_5FeSi$), if the concentration of Fe impurity exceeds 1.25% (Figure 1.17c). Therefore, two Fe-containing phases can occur in the as-cast structure. As the Fe concentration and cooling rate increase, the amount of the Al_8Fe_2Si phase should go up.

Even small manganese additions to 4XX.0 series alloys, e.g. 444.0 (Table 1.14), lead to the formation of the $Al_{15}(FeMn)_3Si_2$ phase as follows from the isothermal section at 9% Si (Figure 1.18a). This phase has a more favorable skeleton morphology (Figure 1.16b, c) as compared with needle-like Al_5FeSi particles; therefore the presence of manganese can be useful. However, under real solidification conditions the complete binding of iron in the Mn-containing phase can be achieved only if the Al_5FeSi phase has not formed before in solidification. This becomes obvious from the analysis of peritectic reactions in this system (Tables 1.15 and 1.16). During these reactions, the Al_5FeSi phase should vanish. But the peritectic reactions are usually incomplete and the Al_5FeSi phase remains in the structure as follows from the polythermal section at 9% Si and 0.15% Mn given in Figure 1.18b. If one assumes the total suppression of the peritectic reactions, then the complete binding of iron in the $Al_{15}(FeMn)_3Si_2$ phase is achieved at ~0.4%Fe but not at >1% Fe, as it follows from the equilibrium diagram. Typically, conglomerates of these phases are formed, joined by silicon particles (Figure 1.16c). This situation is unfavorable not only because of the presence of Al_5FeSi needles, but also because the Mn-containing phase grows on these needles instead of forming isolated dendritic inclusions. By taking this into account, the Mn:Fe ratio required to prevent the formation of needle-like inclusions should be significantly higher than it follows from equlibrium phase diagram (~1:20 as it follows from the composition of the $Al_{15}(FeMn)_3Si_2$ phase – 1.5% Mn and 31% Fe). On the other hand, the increase of the total Fe and Mn concentration above 2.0–2.5% may result in the formation of primary $Al_{15}(FeMn)_3Si_2$ particles that have polygonal shape and often occur as big clusters (Belov et al., 2002a), which is evidently harmful for many properties, in particular for ductility and machinability. At high concentrations of silicon (>8%) and iron (>1%), the use of manganese as a modifier of the Fe-containing phase appears to be inefficient.

The polythermal section of the quaternary Al–Fe–Mn–Si system calculated using Thermocalc software for Al–10%Si–1%Fe–Mn alloys shows that one or another primary iron-containing phases is formed at any Mn concentrations in the range from 0 to 4%, Figure 1.19a (Balitchev et al., 2003). The isothermal section of Al–10%Si–Fe–Mn alloys at 660°C (Figure 1.19b) demonstrates that iron and manganese can considerably increase the liquidus of quaternary alloys, therefore making their casting difficult.

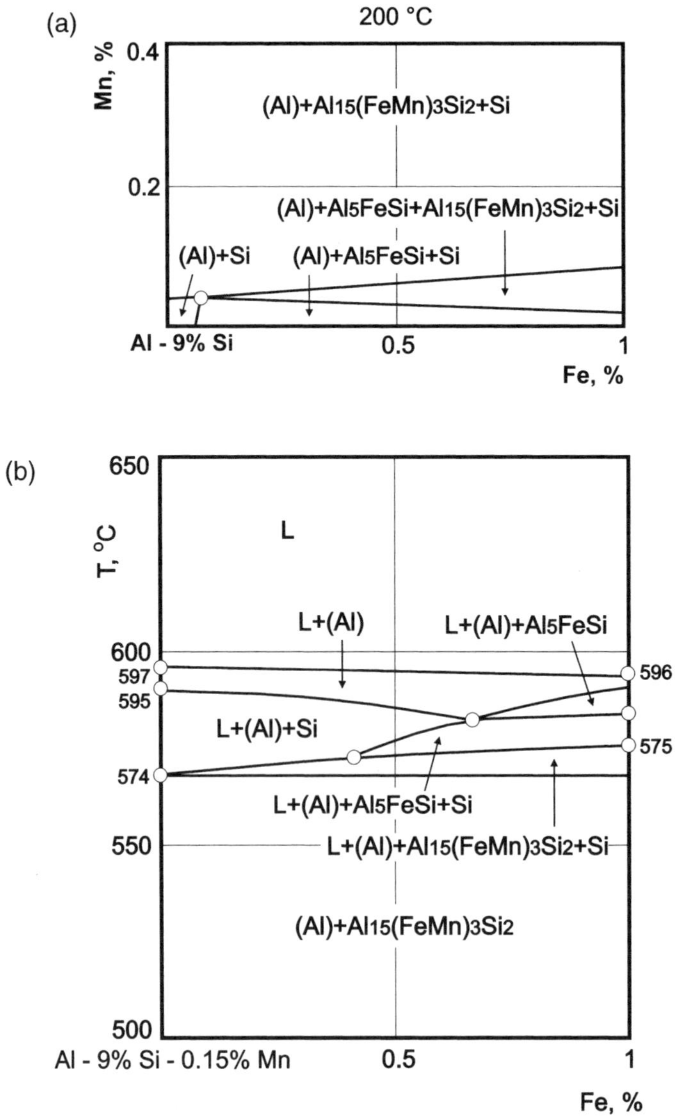

Figure 1.18. Isothermal (a) and polythermal (b) sections of Al–Fe–Mn–Si phase diagram at 9%Si: (a) 200°C and (b) 0.15% Mn.

Bäckerud et al. (1990) examined the solidification of a "eutectic" 413.0 alloy (Table 1.14) under nonequilibrium conditions and revealed the solidification reactions shown in Table 1.19. Primary (Al) grains and primary (Si) crystals can be simultaneously found in the structure, alongside eutectic particles of (Si), Al_5FeSi (needles), and $Al_{15}(FeMn)_3Si$ (skeletons).

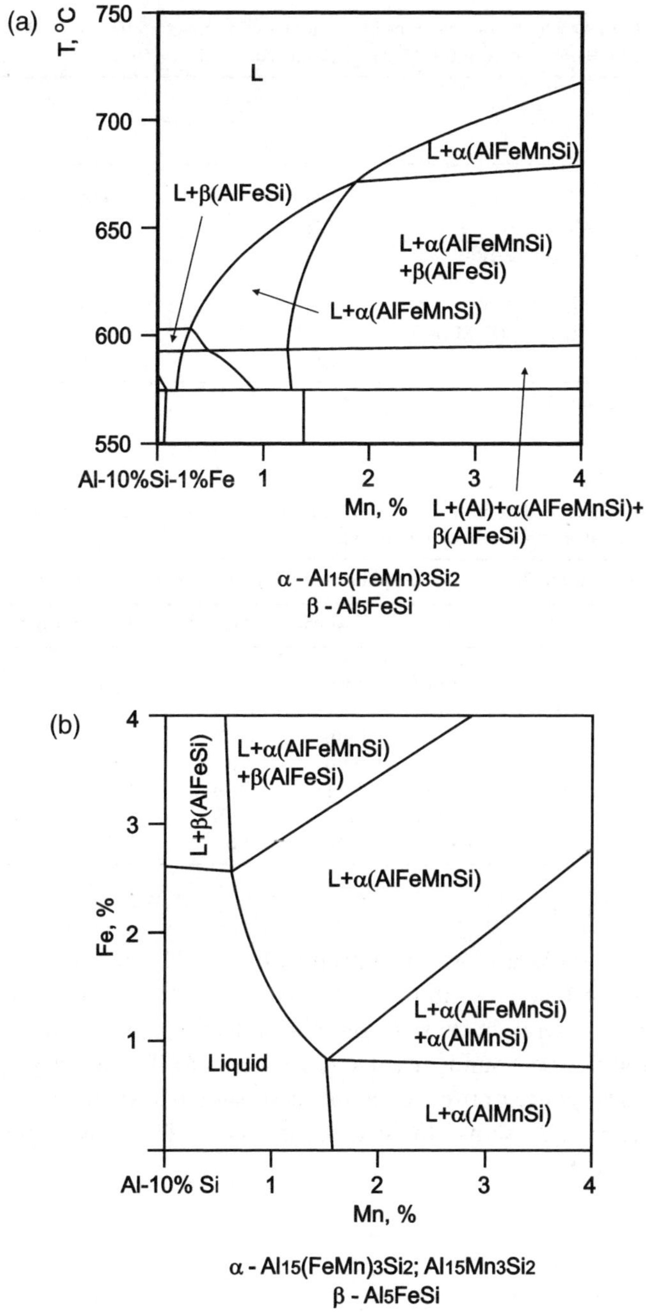

Figure 1.19. Polythermal (a) and isothermal (b) sections of Al–Fe–Mn–Si phase diagram calculated by Themocalc at 10% Si: (a) 1% Fe and (b) 660°C (after Balitchev et al., 2003).

Table 1.19. Solidification reactions under nonequilibrium conditions in a 413.1 alloy (11.4% Si, 0.46% Fe, and 0.18% Mn*) (Bäckerud et al., 1990)

Reaction	Temperatures (°C) at a cooling rate	
	0.3 K/s	5 K/s
L ⇒ (Al)	574–573	574
L ⇒ (Al) + Al_5FeSi	572	574–573
L ⇒ (Al) + $Al_{15}(FeMn)_3Si_2$		
L ⇒ (Si)		
L ⇒ (Al) + (Si) + Al_5FeSi	572–557	573–546
L ⇒ (Al) + (Si) + $Al_{15}(FeMn)_3Si_2$		
Solidus	557	546

* also contains 1.1% Zn

Table 1.20. Calculated volume fractions of eutectic phases in as-cast Al–Si–Fe–Mn alloys (equilibrium values are given in parentheses)

Alloy composition, %			Nonequilibrium (equilibrium) volume fractions, vol. %				
Si	Fe	Mn	(Si)	Al_5FeSi	Al_8Fe_2Si	$Al_{15}(FeMn)_3Si_2$	Σ
4	0.5	0	3.3 (4.3)	1.6 (1.6)	0 (0)	0 (0)	4.9
4	1	0	3.0 (4.1)	3.2 (3.1)	0 (0)	0 (0)	6.2
4	0.5	0.5	3.3 (4.4)	0 (0)	0 (0)	2.3 (2.3)	5.6
4	1	1	3.1 (4.2)	0 (0)	0 (0)	4.6 (4.6)	7.7
5	0.5	0	4.4 (5.5)	1.6 (1.6)	0 (0)	0 (0)	6.0
5	1	0	4.2 (5.2)	3.2 (3.2)	0 (0)	0 (0)	7.4
5	0.5	0.5	4.5 (5.6)	0 (0)	0 (0)	2.3 (2.3)	6.8

Although we showed that the introduction of Mn in high-silicon alloys is useless with respect to preventing the formation of Al_5FeSi particles, manganese addition can be useful in low-silicon alloys like 443.0 (Table 1.14). In such alloys, all iron can be bound in eutectic $Al_{15}(FeMn)_3Si_2$ particles with favorable morphology. The concentration of manganese should then be close to the upper grade limit. Table 1.20 shows the calculated volume fractions of phases in alloys containing 4–5% Si.

Chapter 2
Alloys of the Al–Mg–Si–Fe System

This chapter considers the phase composition of alloys that contain magnesium and silicon in the absence of copper. These are heat treatable, low-alloyed wrought alloys of 6XXX series; heat treatable, casting Al–Si alloys (356/357 type); and some casting and wrought Al–Mg-based alloys that are not strengthened by heat treatment (5XX.0 and 5XXX series). The properties of all these alloys are largely determined by the Mg_2Si phase, so their analysis should be started from the Al–Mg–Si phase diagram that is comparatively simple and has been treated in literature in sufficient detail. However, as most alloys have an iron impurity in the amount appreciably affecting the phase composition, special attention in this chapter is given to the Al–Fe–Mg–Si phase diagram that is fairly complex. This quaternary diagram is actually the basis for most commercial alloys of the given series.

Some commercial alloys contain manganese, which has significant consequences for their phase composition. By taking into account the complexity of multicomponent diagrams with manganese, these alloys alongside 5XX.0- and 5XXX-series alloys are discussed separately, in Chapter 4.

2.1. Al–Mg–Si PHASE DIAGRAM

The Al–Mg–Si phase diagram can be used for the analysis of many wrought alloys of 6XXX series and casting alloys of the 356.0 type, provided the concentration of iron impurity is low (Table 2.1). This diagram is also the basic diagram for casting alloys of the 512.0 type that are considered in Chapter 4. The knowledge of this phase diagram is also required for the analysis of more complex systems involving Mg and Si, in particular, Al–Cu–Mg–Si and Al–Fe–Mg–Si.

In the aluminum corner of the Al–Mg–Si system the following phases are in equilibrium with the aluminum solid solution: Al_8Mg_5, (Si) and Mg_2Si (Figures 2.1a, b) (Mondolfo, 1976; Drits et al., 1977; Phillips, 1959). The Al_8Mg_5 phase (often designated as Al_3Mg_2) has an fcc structure (space group *Fd*3*m*, 1166 atoms in the unit cell) with lattice parameter $a = 2.82$–2.86 nm. The density of this phase is 2.23 g/cm^3; Vickers hardness, 2–3.4 GPa at room temperature and 1.6 GPa at 327°C; Young's modulus, 46–52 GPa; microhardness at 20°C, 2.8 GPa and 1-h microhardness at 300°C, 0.65 GPa (Kolobnev, 1973; Mondolfo, 1976). This compound is not heat resistant. The Mg_2Si phase (63.2% Mg, 36.8% Si) has a cubic structure (space group *Fm*3*m*, 12 atoms in the unit cell) with lattice parameter $a = 0.635$–0.640 nm. The

Table 2.1. Chemical composition of some commercial alloys whose phase composition can be analyzed using Al–Mg–Si phase diagram

Grade	Si, %	Mg, %	Other		
			Fe, %	Mn, %	Cu, %
6160	0.3–0.6	0.35–0.6	0.15	0.05	0.2
6463	0.2–0.6	0.45–0.9	0.15	0.2	0.05
6005	0.6–0.9	0.4–0.6	0.35	0.1	0.1
6105	0.6–1.0	0.45–0.8	0.35	0.10	0.10
356.0	6.5–7.5	0.25–0.45	0.2	0.1	0.2
357.0	6.5–7.5	0.45–0.6	0.15	0.03	0.05
358.0	7.6–8.6	0.4–0.6	0.3	0.2	0.2
359.0	8.5–9.5	0.5–0.7	0.2	0.1	0.2
511.0	0.3–0.7	3.5–4.5	0.5	–	0.15
512.0	1.4–2.2	3.5–4.5	0.6	0.35	0.35
514.0	0.35	3.5–4.5	0.5	–	0.15

melting temperature of this compound is 1087°C; density, 1.88 g/cm^3; Vickers hardness, 4.5 GPa (Mondolfo, 1976). The microhardness of the compound at room temperature is 5.36 GPa, and 1-h microhardness at 300°C, 1.77 GPa (Kolobnev, 1973).

The quasi-binary section between (Al) and Mg_2Si shown in Figure 2.1d corresponds to the concentration ratio Mg:Si = 1.73 (in wt%). This section divides the diagram into two simple systems of eutectic type: Al–Mg–Mg_2Si and Al–Si–Mg_2Si. The invariant eutectic reactions occurring in ternary alloys are given in Table 2.2. In almost all commercial alloys belonging to this system, (Al) is primarily solidified (Figure 2.1a), and then one of the binary eutectics is formed in temperature ranges given in Table 2.3.

The binary and ternary eutectics, involving the Al_8Mg_5 phase, can solidify in commercial alloys given in Table 2.1, only under nonequilibrium conditions. The distribution of the phases in the as-cast state, characterized mainly by the appearance of nonequilibrium eutectics, is shown in Figure 2.2. In as-cast Al–Si alloys (356.0, 357.0 type), the Mg_2Si phase appears only as a result of nonequilibrium ternary eutectic reaction at 555°C (Table 2.2), its amount is small (less than 1 vol.%), which makes its identification difficult in an optical microscope. Figure 2.2 shows that the formation of both magnesium silicide and the silicon phase is possible in as-cast ingots of 6XXX series alloys.

As it follows from the solidus surface boundaries (Figure 2.1c), most alloys of the 6XXX series (Table 2.1) with low iron content can be completely transformed into the single-phase state during homogenization. On the contrary, as-cast and heat-treated 356.0 and 512.0 alloys are always heterophase (Figure 2.1b); the excess phase being (Si) in the former alloy and Mg_2Si, in the latter.

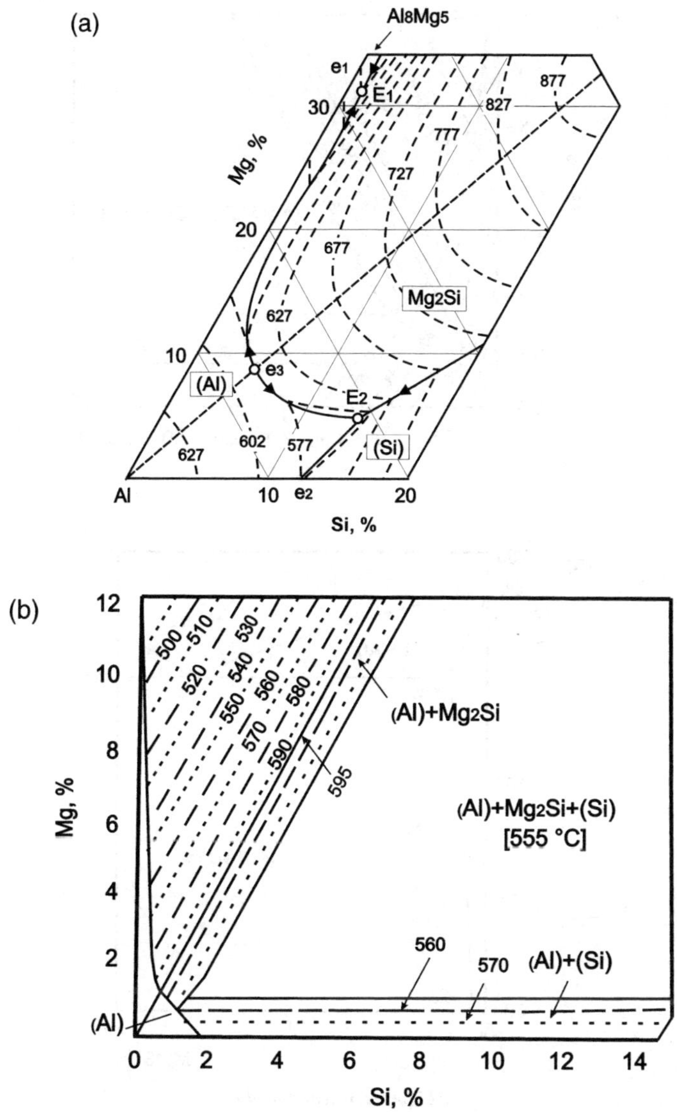

Figure 2.1. Phase diagram of Al–Mg–Si system: (a) liquidus; (b) solidus; (c) solidus detail in the Al corner; and (d) quasi-binary section Al–Mg_2Si.

In spite of the comparatively low mutual solubility of Mg and Si in solid (Al), it enables a significant effect of precipitation hardening due to the formation of metastable coherent and semi-coherent modifications (β'', β') of the Mg_2Si phase during aging. Recent results showed that the composition of metastable precipitates differs from that of the equilibrium Mg_2Si phase. Early precipitates contain

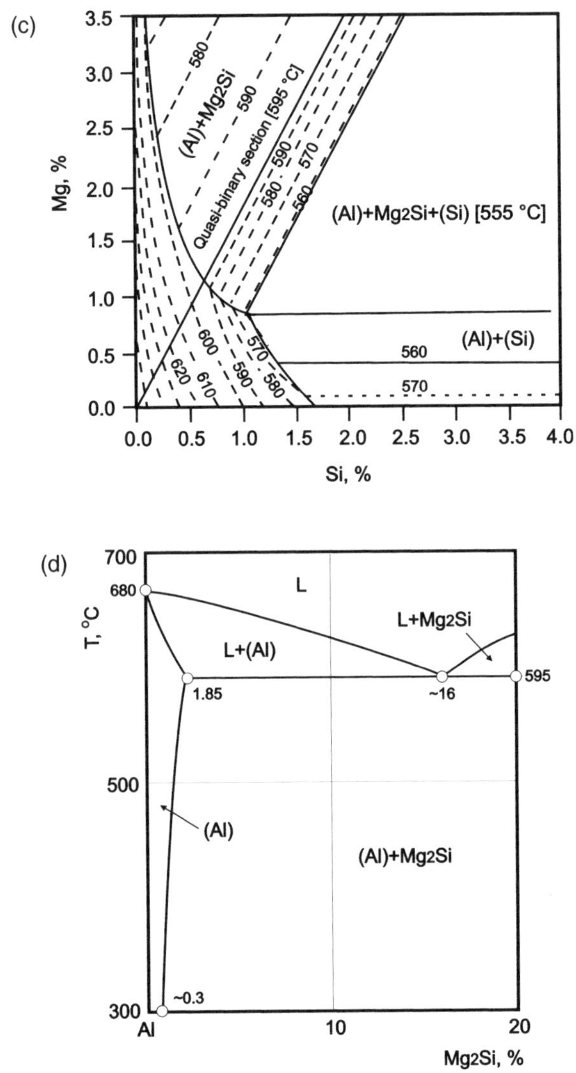

Figure 2.1 (*continued*)

aluminum in addition to Mg and Si, and coherent β″ phase contains an excess of silicon with one of the possible formulae Mg_5Si_6 (Marioara et al., 2001). The precipitation of metastable phases in Al–Mg–Si alloys is considered in greater detail in Section 2.4.

As it follows from Table 2.4, the mutual solid solubility of magnesium and silicon in (Al) strongly depends on temperature, which requires strict observation of a heat treatment regime.

Table 2.2. Invariant reactions in ternary alloys of Al–Mg–Si system (Mondolfo, 1976)

Reaction	Point in Figure 2.1a	*T*, °C	Concentrations in liquid phase	
			Mg, %	Si, %
$L \Rightarrow (Al) + Mg_2Si$ (quasi-binary)	e_3	595	8.15	7.75
$L \Rightarrow (Al) + (Si) + Mg_2Si$	E_2	555	4.96	12.95
$L \Rightarrow (Al) + Mg_2Si + Al_8Mg_5$	E_1	449	32.2	0.37

Table 2.3. Monovariant reactions in ternary alloys of Al–Mg–Si system

Reaction	Lines in Figure 2.1a	*T*, °C
$L \Rightarrow (Al) + Mg_2Si$	e_3–E1 and e_3–E_2	595–555 and 595–449
$L \Rightarrow (Al) + (Si)$	e_2–E_2	577–555
$L \Rightarrow (Al) + Al_8Mg_5$	e_1–E_1	450–449

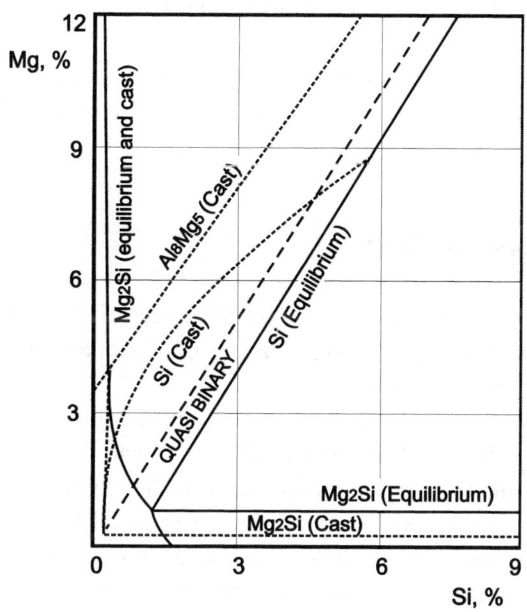

Figure 2.2. Nonequilibrium distribution of phase fields in Al–Mg–Si system in the as-cast state ($V_c \sim 10^{-1}$ K/s) (Phillips, 1961). Lines show the boundaries of the first phase appearance in equilibrium and in as-cast conditions.

Table 2.4. Limit solid solubility of Mg and Si in aluminum in Al–Mg–Si system (Mondolfo, 1976)

T, °C	(Al) + Mg_2Si + Al_8Mg_5		(Al) + Mg_2Si		(Al) + (Si) + Mg_2Si	
	Mg, %	Si, %	Mg, %	Si, %	Mg, %	Si, %
595	–	–	1.17	0.68	–	–
577	–	–	1.10	0.63	–	–
552	–	–	1.00	0.57	0.83	1.06
527	–	–	0.83	0.47	0.6	0.8
502	–	–	0.70	0.40	0.5	0.65
452	15.3	0.1	0.48	0.27	0.3	0.45
402	11	<0.01	0.33	0.19	0.22	0.3
302	5	<0.01	0.19	0.11	0.1	0.15

The aluminum solid solution of Al–Si–Mg casting alloys (356.0 type) always has an excess of silicon with respect to the stoichiometric Mg:Si ratio; therefore, the amount of secondary Mg_2Si precipitates is determined by the concentration of magnesium (the maximum volume fraction Q_V being about 1 vol.%).

In 6XXX series alloys, both elements can be present in excess depending on the Mg:Si ratio, even within the compositional range of one alloy. The volume fraction of secondary Mg_2Si precipitates after aging can be assessed from the dependences shown in Figure 2.6. Their maximal amount is achieved in an alloy lying at the quasi-binary section.

In Al–Mg alloys containing more than 3–4% Mg, no secondary precipitates of the Mg_2Si phase are formed due to the low solubility of Si in (Al) (Table 2.4). Almost all silicon is bound in eutectic Mg_2Si particles as suggested by Figure 2.1b, c.

2.2. Al–Fe–Mg PHASE DIAGRAM

This phase diagram can be used to analyze the effect of iron on the phase composition of Al–Mg alloys with low concentrations of silicon and manganese. Examples of such alloys are given in Table 2.5.

No ternary compounds have been found in the ternary Al–Fe–Mg system (Phillips, 1959; Mondolfo, 1976; Drits et al., 1977; Belov et al., 2002a). The binary phases Al_3Fe and Al_8Mg_5 are in equilibrium with the aluminum solid solution. The solubility of manganese in Al_3Fe and that of iron in Al_8Mg_5 are negligibly small. In the aluminum corner of the Al–Mg–Fe phase diagram (Figure 2.3), invariant and monovariant eutectic transformations take place as shown in Table 2.6. The Al_3Fe phase, in contrast with Al_8Mg_5, is formed within a wide temperature range.

The low solubility of iron in (Al) becomes even lower in the presence of magnesium. In turn, iron noticeably decreases the solubility of magnesium in aluminum,

Table 2.5. Chemical composition of some commercial alloys whose phase composition can be analyzed using Al–Mg–Fe phase diagram

Grade	Mg, %	Fe, %	Other		
			Si, %	Mn, %	Cu, %
514.0	3.5–4.5	0.5	0.35	–	0.15
518.0	7.5–8.5	1.8	0.35	0.35	0.25
585.0	10	0.3	0.25	0.15	0.25
520.0	9.5–10.6	0.3	0.25	0.18	0.25
5005	0.5–1.1	0.7	0.3	0.2	0.2
1530 (rus)					
5050	1.1–1.8	0.7	0.4	0.1	0.2
5151	1.5–2.1	0.35	0.2	0.1	0.15

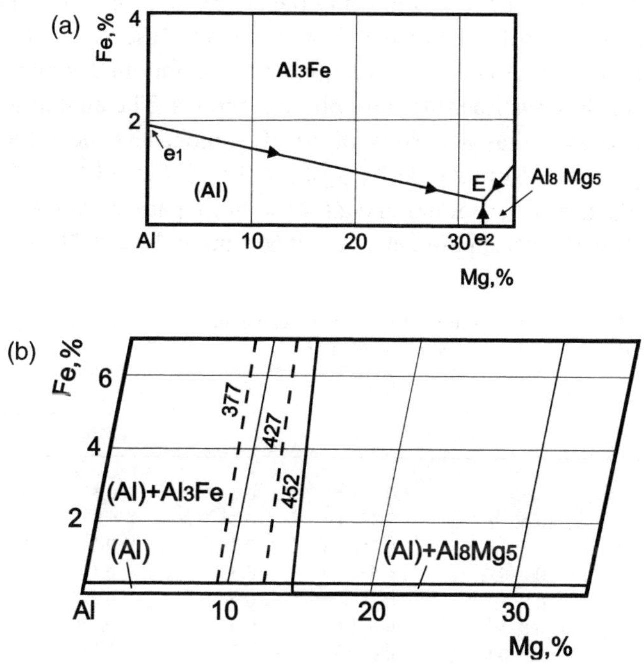

Figure 2.3. Phase diagram of Al–Fe–Mg system: (a) liquidus and (b) solidus.

which becomes 14.1% Mg at 449°C as compared with 17.45 in the binary Al–Mg system.

Nonequilibrium solidification facilitates the formation of the degenerated ternary eutectics with large Al_3Fe particles formed at low Fe concentration and with Al_8Mg_5 compound appearing even at 2–3% Mg.

Table 2.6. Invariant and monovariant reactions in ternary alloys of Al–Mg–Fe system (Mondolfo, 1976)

Reaction	Point/Line in Figure 2.3a	*T*, °C	Concentrations in liquid phase	
			Mg, %	Fe, %
$L \Rightarrow (Al) + Al_3Fe + Al_8Mg_5$	E	449	32.2	0.37
$L \Rightarrow (Al) + Al_3Fe$	e_1–E	655–449	–	–
$L \Rightarrow (Al) + Al_8Mg_5$	e_2–E	450–449	–	–

2.3. Al–Fe–Mg–Si PHASE DIAGRAM

The phase composition of most wrought 6XXX-series alloys and of many casting alloys based on the Al–Si and Al–Mg systems (in particular, 356.0 and 512.0; Tables 2.1, 2.5, 2.7) can be analyzed using the Al–Fe–Mg–Si phase diagram. Alloys containing Mn are considered in Chapter 4. The joint presence of Mg, Si, and Fe in the composition of an alloy produces a quaternary compound that makes inappropriate the use of the constituent ternary phase diagrams. The quaternary compound, often designated as π, has a narrow range of homogeneity near the composition corresponding to the formula $Al_8FeMg_3Si_6$ (10.9% Fe, 14.1% Mg, 32.9% Si). This compound has a hexagonal crystal structure (space group $P\bar{6}2m$, 18 atoms in the unit cell) with lattice parameters $a = 0.663$ nm and $c = 0.794$ nm. Its density is

Table 2.7. Chemical composition of some commercial alloys whose phase composition can be analyzed using Al–Fe–Mg–Si phase diagram

Grade	Si, %	Mg, %	Fe, %	Other	
				Mn, %	Cu, %
6003	0.35–1.0	0.8–1.5	0.6	0.8	0.1
6017	0.55–0.7	0.45–0.6	0.15–0.3	0.1	0.05–0.2
6060	0.3–0.6	0.35–0.6	0.1–0.3	0.1	0.1
6063	0.2–0.6	0.45–0.9	0.35	0.1	0.1
6016	0.9–1.5	0.25–0.6	0.5	0.2	0.2
6081	0.8–1.2	0.6–1.0	0.45	0.15	0.1
6301	0.5–0.9	0.6–0.9	0.7	0.15	0.1
6201	0.5–0.9	0.6–0.9	0.5	0.03	0.1
6162	0.4–0.8	0.7–1.1	0.5	0.1	0.2
356.1	6.5–7.5	0.2–0.45	0.50	0.35	0.25
364.0	7.5–9.5	0.2–0.4	1.5	0.1	0.2
364.2	7.5–9.5	0.25–0.4	0.7–1.1	0.1	0.2
360.2	9.0–10.0	0.45–0.6	0.7–1.1	0.1	0.1
360.0	9.0–10.0	0.4–0.6	1.0	0.35	0.5
369.1	11.0–12.0	0.3–0.45	2.0	0.35	0.6
518.0	0.35	7.5–8.5	1.8	0.35	0.25
AMg11(rus)	0.8–1.2	10.5–13	0.9	–	–

2.82 g/cm^3; microhardness at room temperature, 5.85 GPa; and 1-h microhardness at 300°C, 3.76 GPa (Kolobnev, 1973; Mondolfo, 1976). The quaternary compound is sufficiently heat-resistant.

Apart from the quaternary compound, phases from the binary and ternary systems – Al_3Fe, Al_8Mg_5, Mg_2Si, Al_8Fe_2Si, Al_5FeSi, and (Si) – can be in equilibrium with the aluminum solid solution (Phillips, 1959; Mondolfo, 1976; Drits et al., 1977; Belov et al., 2002a). It should be noted that the compound Mg_2Si is in equilibrium with all the other phases and occurs in most alloys in the solid state.

Figure 2.4 shows the distribution of phase regions in the solid state (a) and the liquidus surface projection (b) in the Al–Fe–Mg–Si system. Invariant five-phase

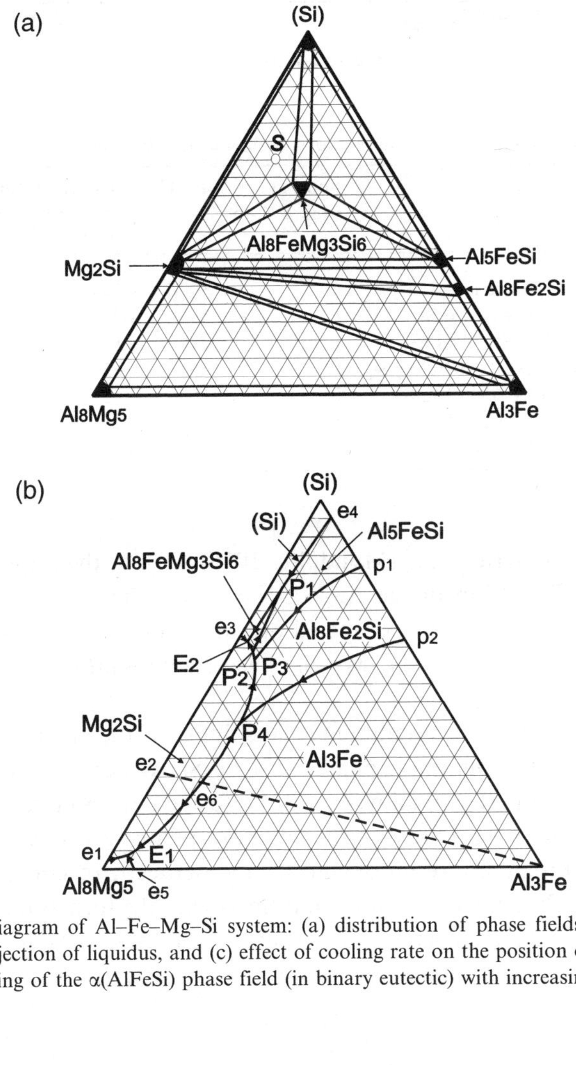

Figure 2.4. Phase diagram of Al–Fe–Mg–Si system: (a) distribution of phase fields in the solid state, (b) polythermal projection of liquidus, and (c) effect of cooling rate on the position of liquidus surfaces, widening of the α(AlFeSi) phase field (in binary eutectic) with increasing V_c.

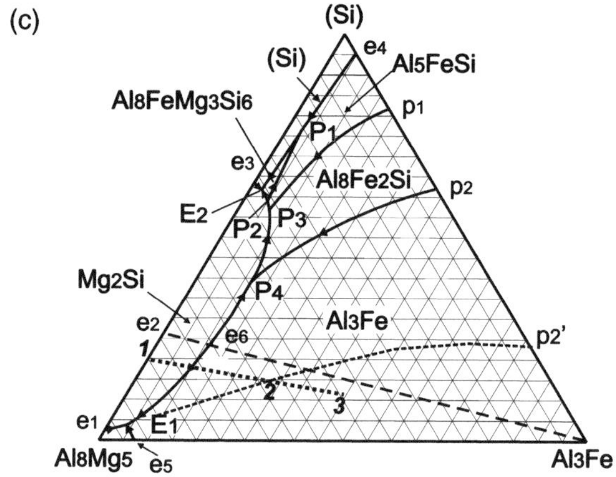

Figure 2.4 (*continued*)

Table 2.8. Invariant reactions in quaternary alloys of Al–Fe–Mg–Si system (Mondolfo, 1976)

Reaction	Point in Figure. 2.4b	*T*, °C	Concentrations in liquid phase		
			Fe, %	Mg, %	Si, %
$L \Rightarrow (Al) + Al_3Fe + Mg_2Si$ (quasi-ternary)	e_6	>587	~1.0	~10.0	~7.0
$L + Al_3Fe \Rightarrow (Al) + Mg_2Si + Al_8Fe_2Si$	P_4	586	1.35	7.25	7.05
$L + Al_8Fe_2Si \Rightarrow (Al) + Mg_2Si + Al_5FeSi$	P_3	576	0.82	6.45	9.50
$L + Al_5FeSi + Mg_2Si \Rightarrow (Al) + Al_8FeMg_3Si_6$	P_2	568	0.55	6.0	11.4
$L + Al_5FeSi \Rightarrow (Al) + (Si) + Al_8FeMg_3Si_6$	P_1	567	0.52	2.9	12.15
$L \Rightarrow (Al) + (Si) + Mg_2Si + Al_8FeMg_3Si_6$	E_2	554	0.15	4.9	12.9
$L \Rightarrow (Al) + Al_3Fe + Al_8Mg_5 + Mg_2Si$	E_1	448	0.11	33.3	0.35

reactions are given in Table 2.8 (Mondolfo, 1976). Due to the presence of the quasi-binary (Al)–Mg_2Si section in the Al–Mg–Si system (Figure 2.1d), a quasi-ternary section (Al)–Mg_2Si–Al_3Fe can be singled out in the quaternary system. This quasi-ternary section divides the Al–Fe–Mg–Si phase diagram into two parts as shown in Figure 2.4b.

In commercial alloys, the aluminum solid solution is the main primary phase, but primary crystals of Fe-containing phases – Al_8Fe_2Si, Al_5FeSi, $Al_8FeMg_3Si_6$ (in Al–Si alloys) and Al_3Fe (in Al–Mg alloys) – can be formed at an increased iron content. In 6XXX-series alloys, primary iron-containing phases are rare, though all these phases may be present as a result of eutectic and peritectic reactions. In the majority of commercial 6XXX alloys, solidification starts with the formation of primary (Al), followed by secondary eutectic and peritectic reactions to form small quantities of intermetallic particles in interdendritic regions. The temperature ranges of all possible mono- and bivariant reactions involving (Al) in the Al–Fe–Mg–Si system are given in Table 2.9.

The low solubility of iron in (Al) makes all alloys of the Al–Fe–Mg–Si system heterophase in any state. The composition of (Al) and the formation of Mg_2Si precipitates during annealing can be analyzed using the Al–Mg–Si phase diagram; however one should take into account the binding of silicon and magnesium to Fe-containing phases. At insufficiently high annealing temperatures (less than 500–550°C) the Fe-containing phases, as a rule, undergo no changes. Therefore, the phase composition of the aluminum matrix after aging or annealing should be analyzed using the actual (not nominal!) concentrations of Si and Mg in (Al) at the temperature of annealing.

The calculated dependences of the volume fraction of various Fe-containing phases on the content of iron in 6XXX-series alloys (shown in Table 2.10) vividly illustrate that the amounts of these phases are different at the same concentration of iron. For example, at 0.2% Fe the volume fractions of π, β(AlFeSi), α(AlFeSi), and Al_3Fe phases are 1.8, 0.6, 0.5, and 0.4 vol.%, respectively. Therefore, in alloys with a higher concentration of silicon (6081, 6016), in which the formation of the π phase is most likely, the total volume fraction of Fe-containing phases will be considerably larger than that in 6063-type alloys where the main Fe-containing phase is α(AlFeSi) (for alloy compositions see Table 2.7).

Under nonequilibrium solidification conditions, most peritectic reactions do not complete, and more phases are present in the alloys than there should be according to the equilibrium phase diagram. An especially complex structure is characteristic of 6XXX series alloys, because different sets of phases can form in different areas of an ingot due to the gradient of the cooling rate (Belov et al., 2002a). At a high Fe:Si ratio and slow cooling the Al_3Fe phase can form, and at an inverse ratio and a high silicon content (>1%) one can expect the formation of the quaternary π compound. In most cases, phases α(AlFeSi) and β(AlFeSi) are formed as well. However, in different sections of the ingot their relative concentration can be quite different, which can be explained by the proximity of the points of the invariant reactions on line e_6–E_2 of the polythermal diagram, especially P_2, P_3, and E_2 (Figure 2.4c).

In Al–Si alloys, nonequilibrium solidification suppresses the peritectic reaction $L + \beta(AlFeSi) \Rightarrow (Al) + (Si) + \pi$, which causes the appearance of π-phase rims on earlier formed needle-like β(AlFeSi) crystals. These conglomerates remain almost unchanged after heat treatment.

In Al–Mg alloys, as it follows from the equilibrium Al–Fe–Mg–Si phase diagram (Figure 2.4a), only one Fe-containing phase – Al_3Fe – can form in the presence of iron and irrespective of the silicon concentration. As in the ternary Al–Fe–Si system, an increase in the cooling rate (V_c) during solidification markedly narrows the region

Table 2.9. Bivariant and monovariant reactions in quaternary alloys of Al–Fe–Mg–Si system

Bivariant reactions*	Field in Figure 2.4b	*T*, °C	Monovariant reactions*	Line in Figure 2.4b	*T*, °C
$L \Rightarrow (Al) + Al_3Fe$	$Al_3Fe–p_2–P_4–e_6–E_1–e_5$	655–586	$L \Rightarrow (Al) + Mg_2Si + Al_3Fe$	$e_6–P_4$	590–586
				$e_6–E_1$	590–448
$L \Rightarrow (Al) + \alpha$	$p_2–p_1–P_3–P_4$	629–576	$L \Rightarrow (Al) + Mg_2Si + \alpha$	$P_4–P_3$	586–576
$L \Rightarrow (Al) + \beta$	$p_1–e_4–P_1–P_2–P_3$	611–567	$L \Rightarrow (Al) + Mg_2Si + \beta$	$P_3–P_2$	576–568
$L \Rightarrow (Al) + \pi$	$P_2–P_1–E_2$	668–554	$L \Rightarrow (Al) + Mg_2Si + \pi$	$P_2–E_2$	568–554
$L \Rightarrow (Al) + (Si)$	$(Si)–e_3–E_2–P_1–e_4$	577–554	$L \Rightarrow (Al) + (Si) + Mg_2Si$	$e_3–E_2$	555–554
$L \Rightarrow (Al) + Mg_2Si$	$e_1–E_1–e_6–P_4–P_3–P_2–E_2–e_3–e_2$	595–554	$L + Al_3Fe \Rightarrow (Al) + \alpha$	$p_2–P_4$	629–586
			$L + \alpha \Rightarrow (Al) + \beta$	$p_1–P_3$	611–576
			$L \Rightarrow (Al) + (Si) + \beta$	$e_4–P_1$	576–567
			$L + \beta \Rightarrow (Al) + \pi$	$P_2–P_1$	568–567
			$L \Rightarrow (Al) + (Si) + \pi$	$P_1–E_2$	567–554
			$L \Rightarrow (Al) + Mg_2Si + Al_8Mg_5$	$e_1–E_1$	449–448
			$L \Rightarrow (Al) + Al_3Fe + Al_8Mg_5$	$e_5–E_1$	449–448

* α is Al_8Fe_2Si; β is Al_5FeSi; and π is $Al_8FeMg_3Si_6$

Table 2.10. Calculated maximal volume fractions of Fe-containing phases in 6XXX alloys

Fe content in alloy, %	Volume fractions of phases, vol.%			
	Al_3Fe	α(AlFeSi)	β(AlFeSi)	π(AlFeMgSi)
0.1	0.19	0.23	0.31	0.88
0.2	0.38	0.46	0.63	1.76
0.3	0.56	0.69	0.94	2.64
0.4	0.75	0.91	1.25	3.51
0.5	0.94	1.14	1.57	4.39
1.0	1.88	2.29	3.13	8.78

of Al_3Fe primary crystallization (Belov et al., 2002a). Therefore, commercial Al–Mg alloys with Fe and Si impurities, containing less than 6% Mg and obtained by casting into metallic molds or by direct-chill casting, frequently contain the α (Al_8Fe_2Si) phase. The higher the cooling rate V_c, the greater the probability of the Al_8Fe_2Si phase to be formed. This can be illustrated by the liquidus projection of the quaternary diagram, on which the dashed line shows the shift of the boundary of the binary eutectic reaction $L \Rightarrow (Al) + Al_8Fe_2Si$ towards the Al–Mg side upon increasing V_c (from line p_2–P_4 towards line p_2'–E_1 in Figure 2.4c). Accordingly, the compositional range of the eutectics $L \Rightarrow (Al) + Al_8Fe_2Si$ should expand, and, as a result, the phase composition of as-cast Al–Mg alloys (line 1–2 in Figure 2.4c) should change. In the alloys falling in the range 1–2, the ternary eutectic reaction $L \Rightarrow (Al) + Mg_2Si + Al_8Fe_2Si$ should proceed after the solidification of (Al) and the binary eutectics $(Al) + Mg_2Si$ or $(Al) + Al_8Fe_2Si$. The solidification of the binary eutectics $(Al) + Al_3Fe$ is possible only within segment 2–3.

2.4. Al–Mg–Si WROUGHT ALLOYS OF 6XXX SERIES

Alloys of the 6XXX series would be easy to analyze if they did not contain other elements, apart from magnesium and silicon, capable of affecting the phase composition. However, this is not always the case (see Tables 2.1 and 2.7).

Nevertheless, the Al–Mg–Si phase diagram gives important information and it is appropriate to start the analysis of commercial 6XXX alloys with this diagram. The isothermal sections at 600, 550, and 200°C appear to be the most characteristic (Figure 2.5a–c). The first temperature is the upper limit for solution heat treatment (homogenization and heating for quenching) and is allowable only for alloys with the minimum content of magnesium and silicon, so as to avoid melting (Figure 2.5a). It is true, though, that such a high temperature is not always necessary, because the solvus of most 6XXX series alloys is sufficiently low to assure the complete

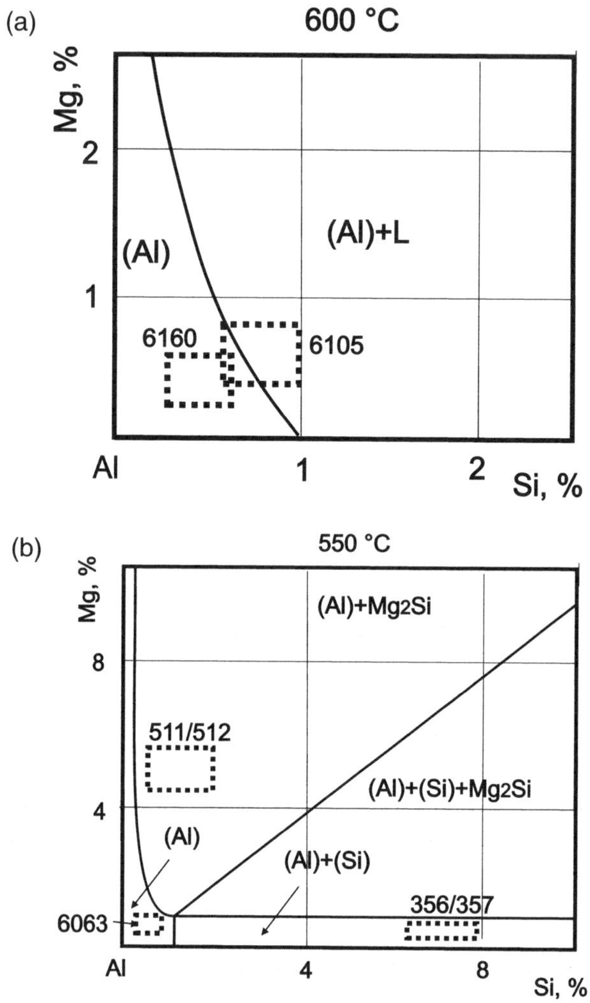

Figure 2.5. Isothermal (a–c) and polythermal (d, e) sections of Al–Mg–Si phase diagram: (a) 600°C; (b) 550°C; (c) 200°C; (d) 0.6% Si; and (e) 0.9% Mg.

dissolution of the Mg_2Si phase in (Al) at much lower temperatures. However, if iron is present (and this is usually the case), the high-temperature homogenization makes it possible to achieve a relatively globular morphology of Fe-containing inclusions (namely, α (Al_8Fe_2Si) phase), which is good for mechanical properties. Alloys containing more than 1% Mg and Si require a more stringent temperature control, because due to the narrow temperature gap between the solidus and the solvus, there is a danger of either melting or incomplete dissolution of magnesium silicide. This follows from the polythermal sections shown in Figure 2.5d, e. If the ternary

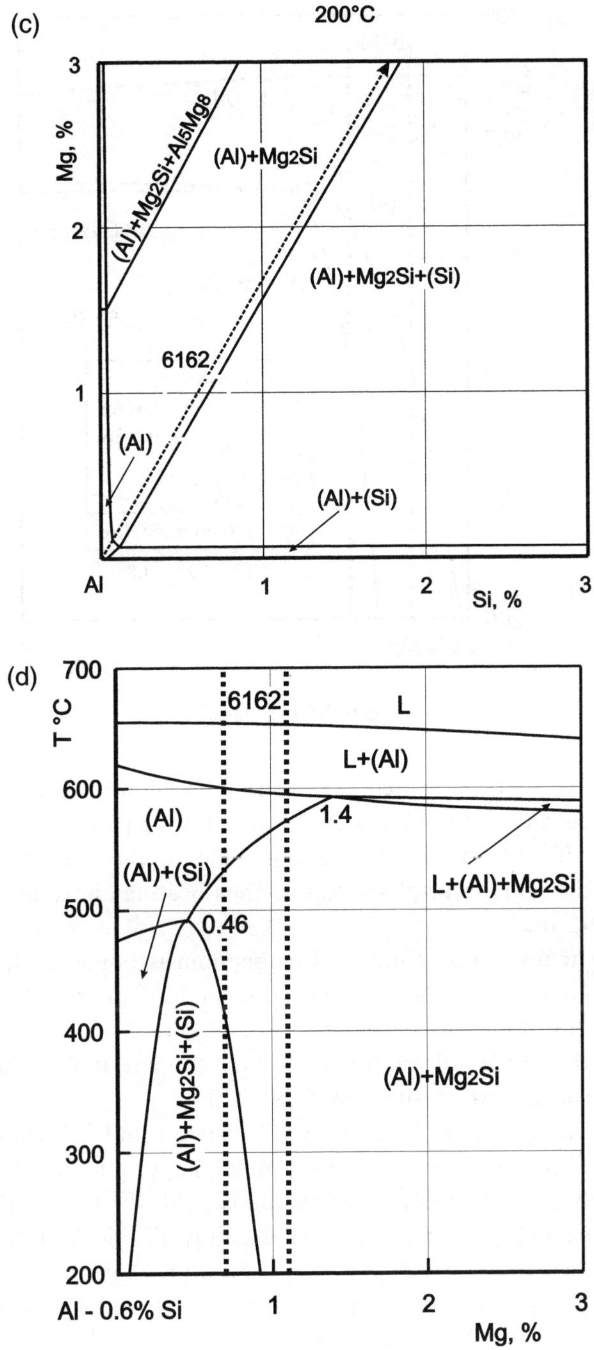

Figure 2.5 *(continued)*

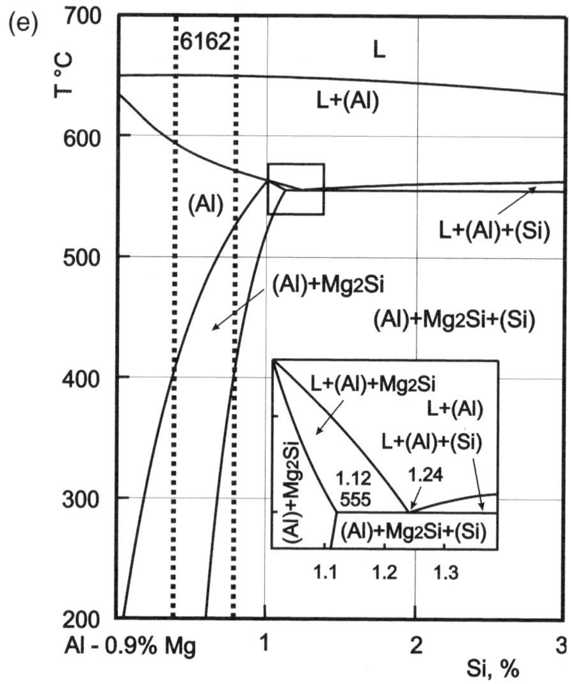

Figure 2.5 (*continued*)

eutectics (Al) + (Si) + Mg_2Si is present (as a rule, the nonequilibrium one), the first stage of homogenization should be carried out at a temperature lower than 555°C (see Table 2.2). If, even at this temperature, the equilibrium phase composition remains within the same three-phase region, then heating above this temperature is not allowed altogether.

At the aging temperature, almost all magnesium is bound to the Mg_2Si phase (Figure 2.5c). To estimate the amount of secondary precipitates of this phase (metastable modifications, to be more exact), Figure 2.6 presents the calculated dependences of the Mg_2Si volume fraction on the content of magnesium and silicon in a 6162 alloy (see the composition in Table 2.7).

The occurrence of iron in 6XXX alloys (Tables 2.1 and 2.7) calls for the use of the Al–Fe–Mg–Si phase diagram and for a much larger number of isothermal and polythermal sections to be analyzed as compared with the ternary diagram (Belov, 2005). As an example, we consider a 6003 alloy (Table 2.7) that has a wide compositional range (0.8–1.5% Mg, 0.35–1% Si, up to 0.6% Fe), so its phase composition can vary rather strongly within the grade limits. The compositional range of this alloy is marked in all sections shown in Figure 2.7. The combined effect of magnesium and silicon can be seen in the isothermal sections at 0.2% Fe, which

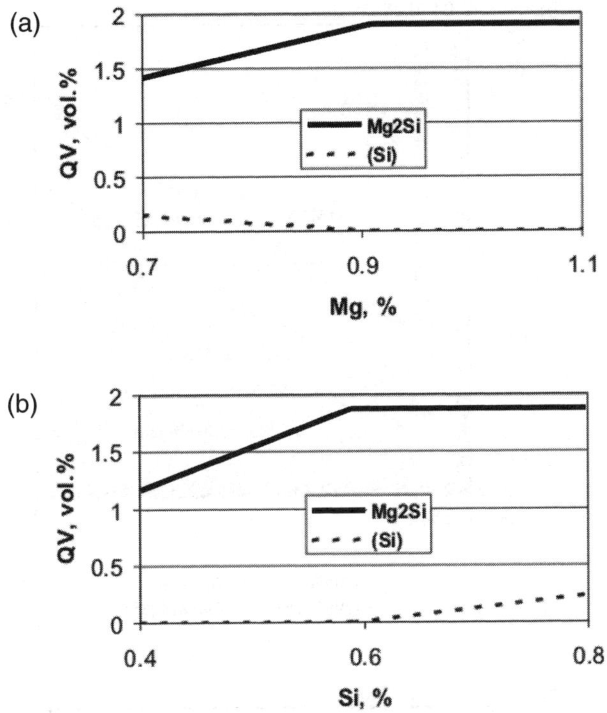

Figure 2.6. Calculated dependence of Mg_2Si and (Si) volume fractions on the concentration of Mg (a) and Si (b) in a 6162 alloy at 200°C.

corresponds to a typical concentration of this element in many alloys of the 6XXX series. At 200°C (Figure 2.7a), all four Fe-containing phases can be present in a 6003 alloy. At a high Mg:Si ratio, all iron should be bound to the Al_3Fe phase, at the inverse ratio of these elements the β(AlFeSi) and π phases become dominant in the equilibrium state. At 550°C (Figure 2.7c), when the solubility of Mg and Si in (Al) is considerable, the occurrence of α(AlFeSi) and β(AlFeSi) phases is most probable. Note that the compositional range where iron is completely bound in the α(AlFeSi) phase, which has the most favorable morphology among all Fe-containing phases, is quite narrow at all three given temperatures, i.e. 200, 450, and 550°C.

The combined effect of iron and magnesium can be analyzed using sections at a constant concentration of silicon. At a Si content close to the lower limit of the 6003 grade, the Al_3Fe phase is present within the entire range of Mg and Fe concentrations (Figure 2.7d). In contrast, when the Si concentration is close to the upper level, the binding of iron to phase β(AlFeSi) becomes most probable (Figure 2.7e).

Figure 2.8 shows calculated volume fractions of excess phases in a 6063 alloy (at 450 and 300°C) with respect to the ratio of Mg, Si, and Fe. These dependences

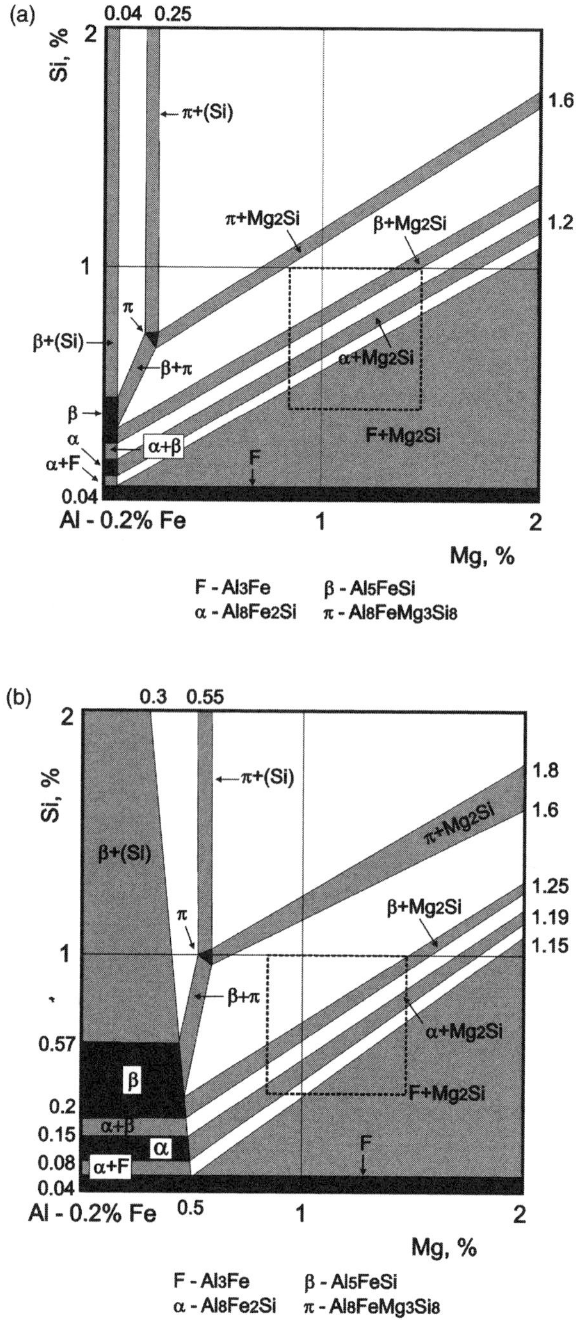

Figure 2.7. Isothermal sections of Al–Fe–Mg–Si phase diagram: (a) 0.2% Fe, 200°C; (b) 0.2% Fe, 450°C; (c) 0.2% Fe, 550°C; (d) 0.5% Si, 400°C; and (e) 1% Si, 550°C. All phase fields also contain (Al).

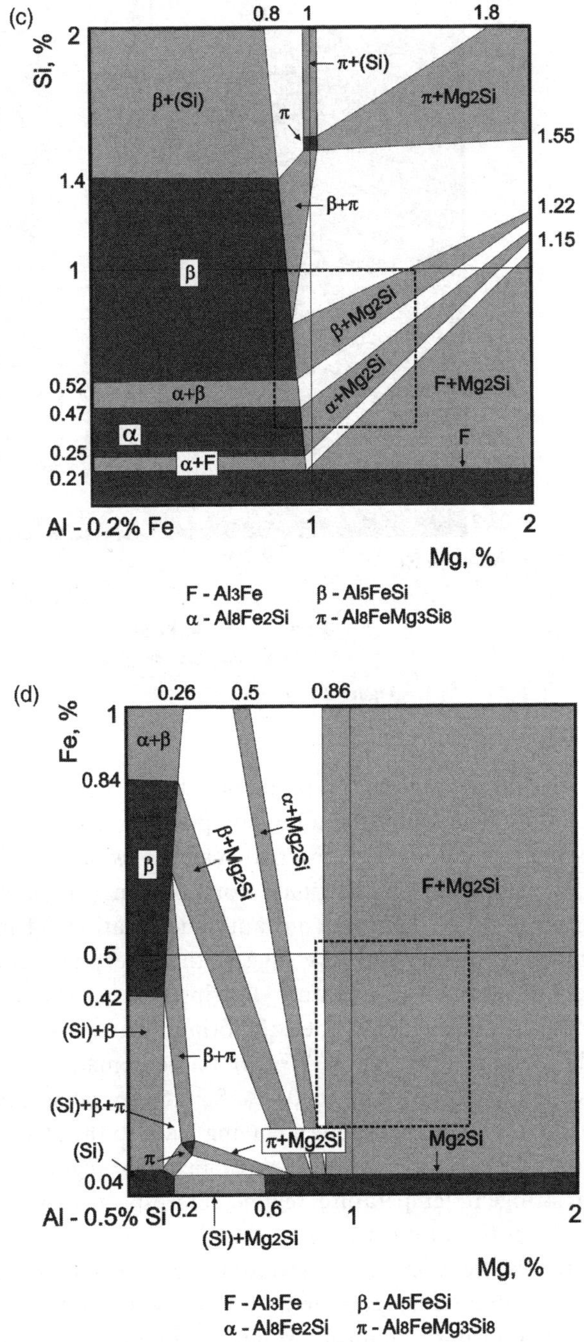

Figure 2.7 (*continued*)

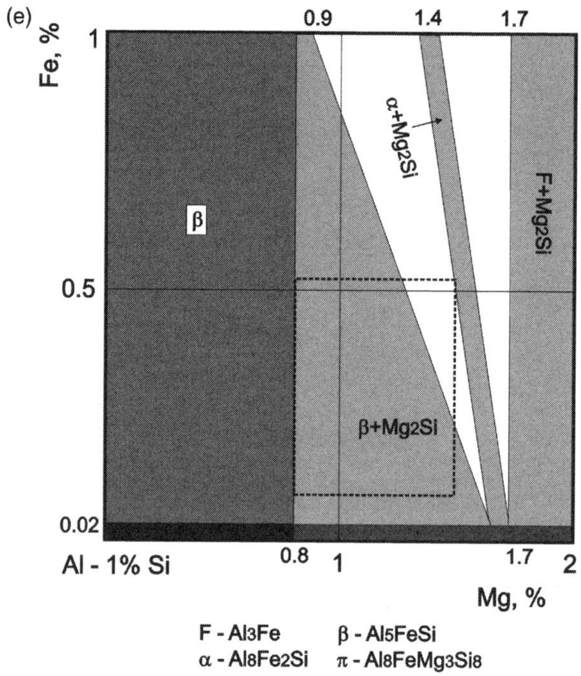

Figure 2.7 (*continued*)

clearly demonstrate that the amount of excess phases can vary strongly even within the grade composition. Therefore, the alloy composition and the alloying element ratio should be strictly maintained in order to achieve desirable properties.

The effects of temperature on the phase composition can be analyzsed using polythermal sections. Two of them, at constant concentrations of iron and silicon, are given in Figure 2.9. These isopleths suggest that the phase composition of a 6003 alloy strongly depends on temperature. For instance, at 1% Si, 0.8% Mg, and 0.2% Fe at temperatures below 500°C the equilibrium phases are (Si), Mg_2Si, and π, and closer to the solidus only the β(AlFeSi) phase remains (Figure 2.9b). On decreasing the silicon concentration to 0.5% Si (at the same concentrations of magnesium and iron) the π phase disappears giving place to the α(AlFeSi) phase and, at higher temperatures to Al_3Fe (Figure 2.9a). Some phase regions are very narrow, which requires a stringent temperature regime for respective operations, e.g. for homogenization of ingots and billets.

Nonequilibrium solidification causes deviation from the equilibrium phase composition. For example in an alloy containing 0.5% Mg, 0.5% Si, and 0.2% Fe, the Al_3Fe phase is formed during equilibrium solidification as follows from the isopleth shown in Figure 2.9a. However, as the formation of this phase requires larger

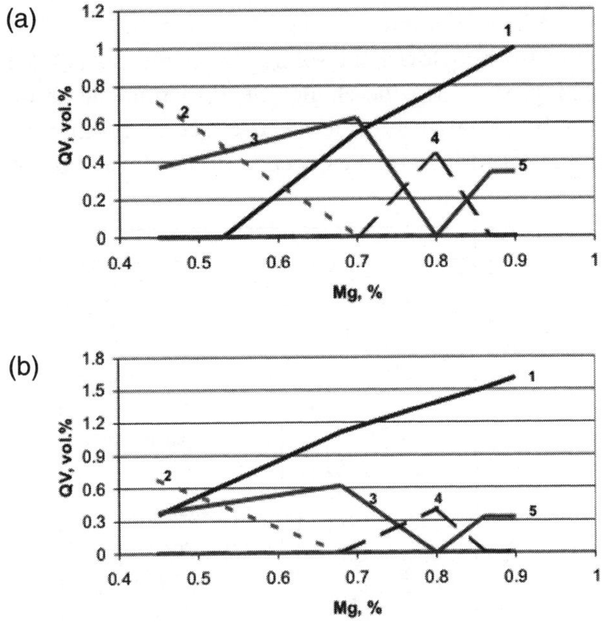

Figure 2.8. Calculated dependence of volume fractions of excess phases in a 6063 alloy (0.5% Si, 0.2% Fe) on the concentration of Mg: (a) 450°C and (b) 300°C. 1 – Mg_2Si, 2 – $Al_8FeMg_3Si_6$, 3 – Al_5FeSi, 4 – Al_8Fe_2Si, and 5 – Al_3Fe.

undercooling as compared to the α(AlFeSi) phase, the latter usually forms under real casting condition (at this alloy composition). At a stoichiometric ratio of Mg and Si (or larger Mg:Si), the Al_3Fe phase should be the only Fe-containing phase formed during equilibrium solidification (Figure 2.4).

Using the polythermal sections, all reactions during solidification can be traced, which enables one to reveal the causes for the formation of nonequilibrium phases. For instance, in an alloy with 1% Si and 0.2% Fe, according to a respective section (Figure 2.9b), the α (Al_8Fe_2Si) phase should be formed early during the solidification. Therefore, due to the incomplete peritectic reaction $L + Al_8Fe_2Si \Rightarrow (Al) + Mg_2Si + Al_5FeSi$ (Table 2.8) it can be retained in the as-cast structure, though the equilibrium phase diagram forbids its occurrence at Mg concentrations lower than ~1.6%. The comparison of polythermal sections in Figure 2.9a, b also suggests that at 1% Si the probability of undesirable β (Al_5FeSi) formation is significantly larger than at 0.5% Si.

When dealing with polythermal sections of multicomponent phase diagrams, one should bear in mind that only qualitative and semi-quantitative information can be obtained. For quantitative data, calculations are required. However, even semiquantitative data can be very useful. Figure 2.5e shows that even a small increase in Si concentration can strongly lower the solidus temperature (T_S). For example, in a 6162 alloy containing 0.9% Mg the change of T_S within the grade compositional limits (for Si) can be as large as 25°C. The effect of magnesium is less strong (Figure 2.5d). The effect of iron on T_S depends on the formation of α(AlFeSi) and

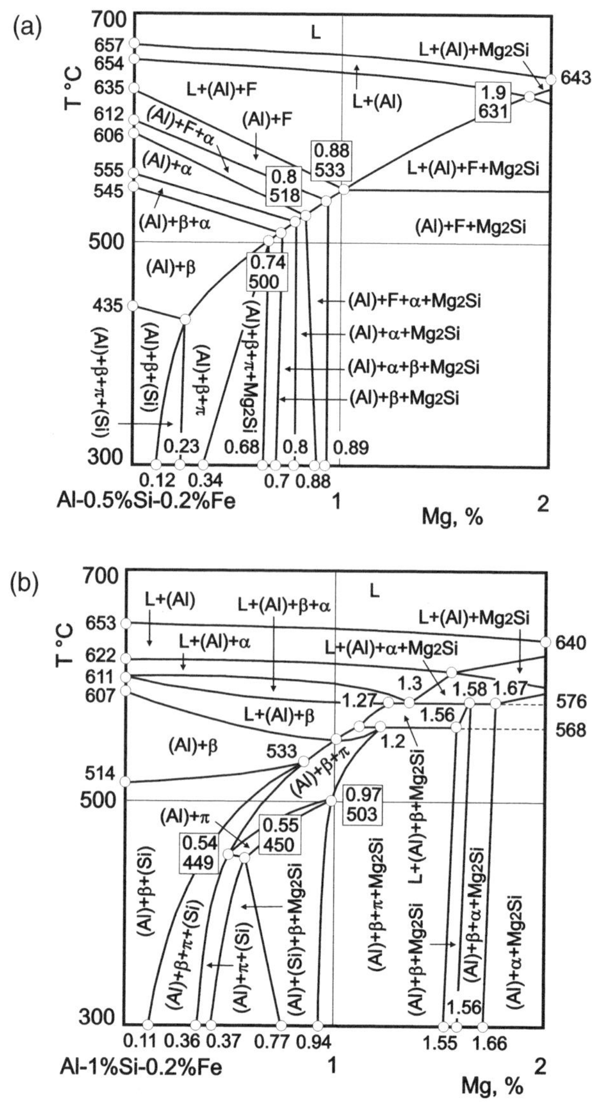

Figure 2.9. Polythermal sections of Al–Fe–Mg–Si phase diagram: (a) 0.5% Si, 0.2% Fe; (b) 1% Si, 0.2% Fe; and (c) 1.5% Si, 0.2% Fe. F – Al_3Fe, α – Al_8Fe_2Si, β – Al_5FeSi, and π – $Al_8FeMg_3Si_6$.

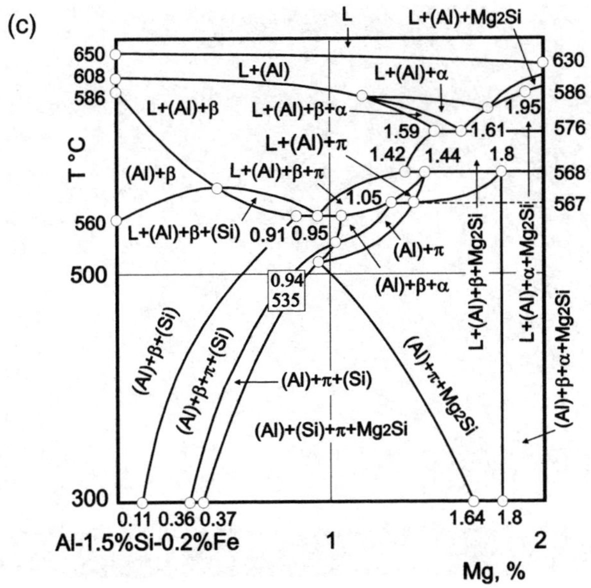

Figure 2.9 *(continued)*

β(AlFeSi) that contain silicon and, therefore decrease the amount of free silicon, resulting in the increased solidus temperature. In our estimate, up to 0.1% Si can be bound in the Fe-containing phases in most 6XXX alloys containing 0.2% Fe.

Experimental studies of nonequilibrium solidification of a 6063 alloy (low-alloyed) were performed by Bäckerud et al. (1986). The results given in Table 2.11 show the simultaneous presence of α(AlFeSi) and β(AlFeSi) phases in the as-cast structure. This agrees well with the casting practice. The amount of Mg_2Si formed at the end of solidification is small, and it is difficult to distinguish its particles in a microscope (Figure 2.10a). Frequently, particles of Mg_2Si form conglomerates with

Table 2.11. Solidification reactions under nonequilibrium conditions in a 6063 alloy (0.43%Mg, 0.39%Si, and 0.2 %Fe) (Bäckerud et al., 1986)

Reaction	Temperatures (°C) at a cooling rate	
	0.5 K/s	15 K/s
L ⇒ (Al)	655–653	654
L ⇒ (Al) + α(AlFeSi)*	618–615	617
L + α(AlFeSi)* ⇒ (Al) + Al_5FeSi	613	610
L + α(AlFeSi)* ⇒ (Al) + Al_5FeSi + Mg_2Si	576**	576**
Solidus	576**	576**

* The crystal structure of α(AlFeSi) is cubic, hence this is a metastable phase (see Table 1.5)
** Estimated value from Mondolfo (1976)

(a)

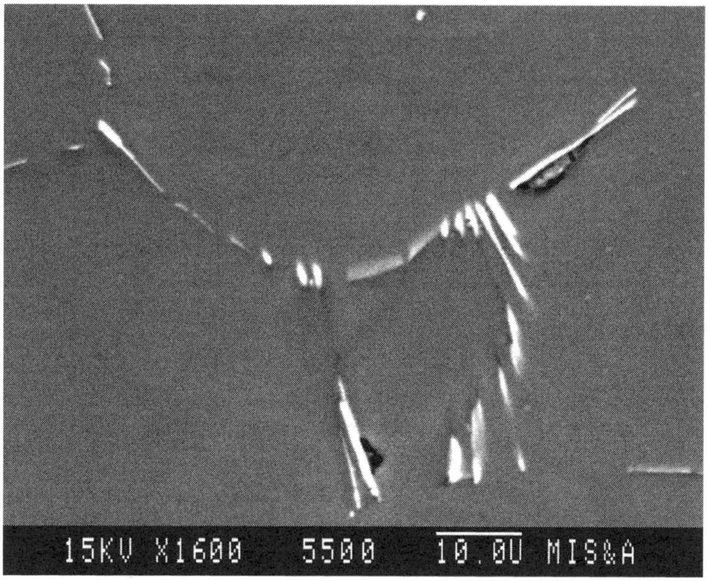

(b)

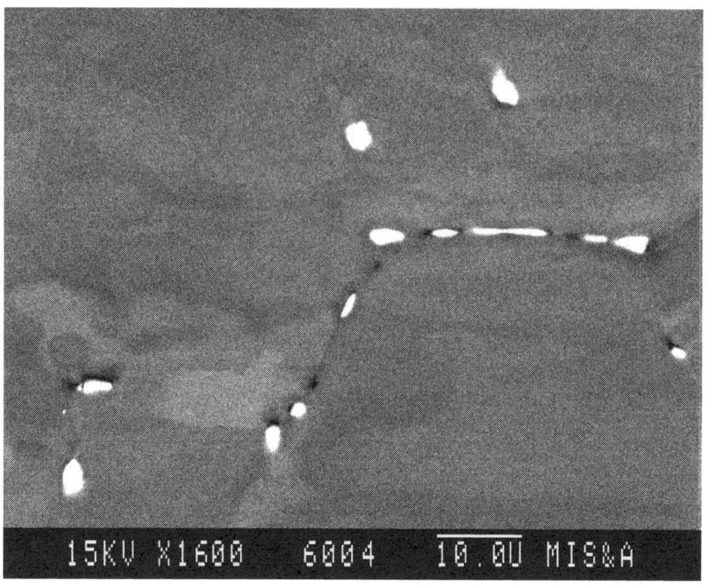

Figure 2.10. Microstructures of 6XXX alloys: (a) as-cast 6063 alloy (Al–0.5%Mg–0.5%Si–0.2%Fe), eutectic phases Al_5FeSi (needles) and Mg_2Si (black), SEM; (b) a 6063 alloy annealed at 600°C, 4 h, fragmented eutectic particles of Al_8Fe_2Si, Mg in (Al), SEM; (c) as-cast Al–0.5%Mg–1.5%Si–0.2%Fe alloy, agglomeration of eutectic phases, i.e. Al_5FeSi (white needles), (Si) (gray), π (gray), and Mg_2Si (black), SEM; (d) an Al–0.5%Mg–1.5%Si–0.2%Fe alloy annealed at 580°C, 4 h, Al_5FeSi (white needles), Mg in (Al), SEM; and (e) precipitates of Mg_2Si in a 6063 alloy, TEM.

(c)

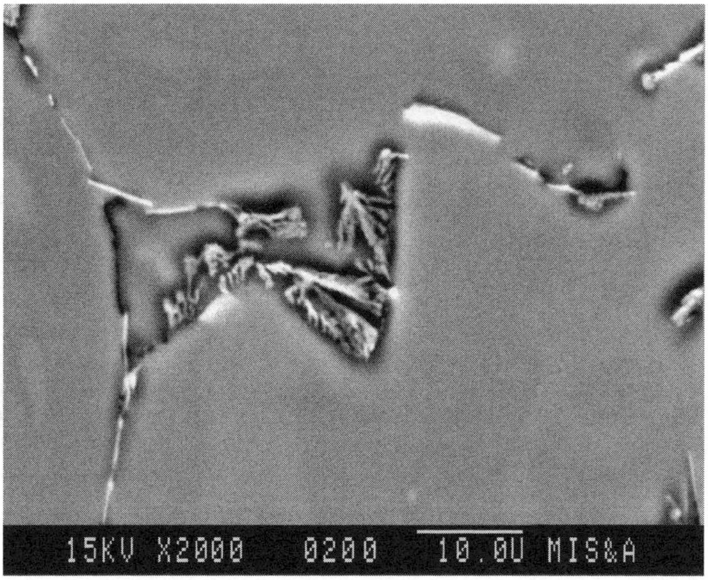

(d)

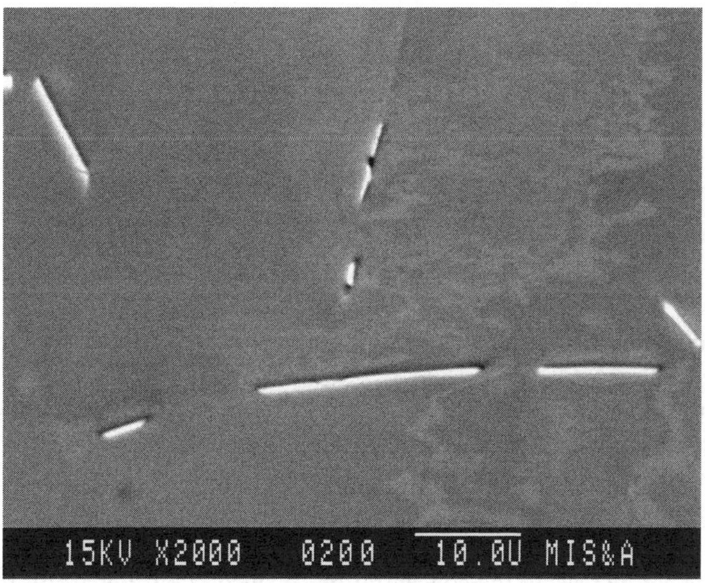

Figure 2.10 (*continued*)

(e)

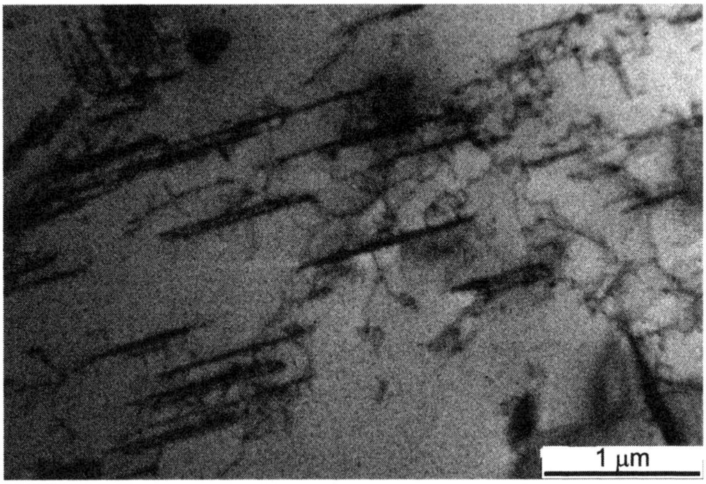

Figure 2.10 (*continued*)

Table 2.12. Solidification reactions under nonequilibrium conditions in a 6063 alloy (0.8% Mg, 0.6% Si, and 0.3% Fe) (Hsu et al., 2001)

Reaction	Temperatures (°C) at a cooling rate of ~0.1 K/s
$L \Rightarrow (Al)$	651
$L \Rightarrow (Al) + Al_3Fe$	625
$L + Al_3Fe \Rightarrow (Al) + \alpha(AlFeSi)$*	617
$L \Rightarrow (Al) + \alpha(AlFeSi)$*	593
$L \Rightarrow (Al) + \alpha(AlFeSi)$* $+ Mg_2Si$	586**
Solidus	586

* The crystal structure of α(AlFeSi) is cubic, hence this is a metastable phase (see Table 1.5)
** This reaction was observed only on partially remelted samples

iron-containing particles that testifies for the occurrence of the last, peritectic reaction in Table 2.11.

On increasing the concentration of Mg and Fe in 6XXX series alloys, the probability of Al_3Fe formation increases, especially at moderate cooling rates. Hsu et al. examined the phase composition of an Al–0.8%Mg–0.6%Si–0.3%Fe alloy (6063 type, high-alloyed) after nonequilibrium solidification at ~0.1 K/s and revealed the solidification reactions listed in Table 2.12 (Hsu et al., 2001).

On further increasing the concentration of silicon (and low iron), the formation of Al_3Fe is unlikely even upon slow cooling, and the probability of the quaternary π phase formation increases. For example, simultaneous presence of Mg_2Si, (Si), β, and π crystals within one conglomerate is observed in an Al–0.86%Mg–1.61%Si–0.072%Fe alloy cast at 0.03 K/s (Liu et al., 1999). Figure 2.10b shows an example of such a structure. The occurrence of two iron-containing phases at a low concentration of iron agrees with the polythermal section shown in Figure 2.9c (of course, after adjustments to the nonequilibrium solidification). The volume fractions of Fe-containing phases found in the as-cast alloy distribute as follows (Liu et al., 1999): 0.21 vol.% Al_5FeSi, and 0.72 vol.% π(AlFeMgSi), which agrees well to the calculated values given in Table 2.10 (0.31 vol.% Al_5FeSi and 0.88 vol.% π(AlFeMgSi)).

Another important phenomenon that occurs under nonequilibrium and/or metastable conditions is the decomposition of a supersaturated solid solution and the corresponding precipitation of metastable phases. The precipitation in Al–Mg–Si alloys and the resultant hardening effect depend very much of the Mg:Si ratio. This ratio is conventionally related to the stoichiometric composition of Mg_2Si (Mg:Si = 1.73 in wt%). Hence, alloys are conditionally divided into alloys with excess of Mg, balanced alloys, and alloys with excess of Si. In the balanced alloy and alloys with excess Mg, the precipitation sequence is typical of aluminum alloys: zone formation, coherent needle-like β'' (Mg_xSi_y) precipitates, semi-coherent rod-shaped β' (Mg_xSi_y) precipitates, and formation of the equilibrium Mg_2Si phase. The excess of silicon can considerably change the kinetics of precipitation and the phase composition. It has been found that in Al–Mg–Si alloys with an excess of silicon, the semicoherent β' phase has several modifications (Matsuda et al., 2000).

The information on metastable phases typical of Al–Mg–Si alloys is listed in Table 2.13.

The composition of metastable phases, i.e. Mg:Si ratio, is different from that of Mg_2Si (Mg:Si = 2 [at.%]). The Mg:Si ratio continuously increases in the series GPZ, β'', β', β (Maruyama et al., 1997), especially in alloys with an excess of silicon. In other words, metastable phases are enriched in silicon. This means that the silicon must eventually form own precipitates. In alloys with the excess of silicon, Si also precipitates independently of and competitively to Mg_2Si. Silicon precipitates have no hardening effect but their formation should always be taken into account when considering the composition of the supersaturated solid solution, sequence of precipitation and mass balance. Figure 2.11 demonstrates how the neglecting of silicon precipitation can lead to the wrong conclusion on the residual composition of the supersaturated solid solution, which eventually determines the phase composition.

According to recent studies by Matsuda et al. (2001), Edwards et al. (1998), and Gupta et al. (2001) the decomposition of the supersaturated solid solution in Al–Mg–Si alloys with an excess of silicon occurs as follows: clusters of Si and clusters of Mg → dissolution of Mg clusters → formation of Mg/Si clusters →"random"

Table 2.13. Characteristics of phases occurring in Al–Mg–Si alloys as compiled in Eskin (2003)

Phase	Crystal structure/morphology	Lattice parameters, nm			Mg:Si ratio [at.%] in balanced (excess Si) alloy
		a	*b*	*c*	
GPZ	–/spherical	–	–	–	~2 (~1)
β''	Monoclinic/needles	0.616	0.616	0.710 $\gamma = 82^o$	
		0.714	0.658	0.405 $\gamma = 75^o$	1.74 (1.00)
		1.534	0.405	0.683 $\gamma = 106^o$	(1.2)
		1.516	0.405	0.674 $\gamma = 105.3^o$	(~1)
β'	Hexagonal/needles, rods	0.705	–	0.405	0.44
β'_A	Hexagonal	0.405	–	0.67	(0.28, 0.2), 40 at.% Al
	Hexagonal	0.407	–	0.405	1.68
β'_C, B', λ	Hexagonal/rods, laths	1.04	–	0.405	(0.83, 1.1)
β'_B	Orthorhombic	0.684	0.793	0.405	(0.4), 40 at.% Al
	Orthorhombic	0.672	0.787	0.405	1.75 (1.21)
β Mg_2Si	Cubic/rods, plates, cubes	0.639 or 0.6351	–	–	2.10 (2.13)
Si	Cubic/faceted particles, plates	0.543	–	–	–

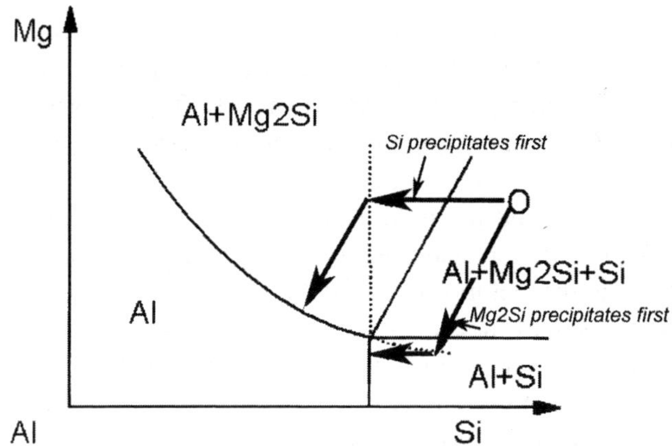

Figure 2.11. A diagram showing the change in the composition of the supersaturated solid solution (thick arrows) when either Si or Mg_2Si precipitates first (after Dons, 2002).

and "parallelogram"-type coherent needle-shaped precipitates (GPZ) → coherent needles β″; fine Si particles → semi-coherent rods β′; rods β'_B; rods β'_A; rods and laths of β'_C (B′), and plates and faceted particles of Si → plate- and cube-shaped β particles.

Depending on the time–temperature conditions (isothermal annealing, temperature of annealing, precipitation upon heating etc.) the precipitation can go through this sequence or start at a certain stage. The decomposition starts directly with the formation of β'_C or β'_A particles at temperatures above 300°C, and the equilibrium β phase directly precipitates upon annealing above 400°C. It should be noted that during high temperature annealing (at 300–350°C) the β′ and equilibrium β phases may coexist for a long time, large incoherent precipitates with the structure of β′ existing in the saturated solid solution (Eskin et al., 1999).

The coherent GP (Mg, Si) zones and β″ phase are efficient hardeners and participate in processes of natural and artificial aging. In the stage of softening they give rise to various modifications of the β′ phase which are considerably stable. According to most references, there is no significant hardening associated with the precipitation of β′-modifications.

2.5. Al–Si CASTING ALLOYS OF 356.0 TYPE

Commercial casting 356.0-type alloys usually contain only silicon and magnesium (Table 2.1), which reduces the analysis of phase composition to the ternary Al–Si–Mg phase diagram. In particular, the solubility values (Table 2.4) show that in the T4 state (after solution treatment at 530–550°C) 356.0-type alloys fall into the

(Al) + (Si) phase region and, after artificial aging, into the (Al) + (Si) + Mg_2Si region. The binary (Al) + (Si) eutectics always forms in the temperature range 577–550°C (Table 2.3), after the primary crystallization of (Al). The ternary eutectic ((Al) + (Si) + Mg_2Si at 550°C) forms in commercial compositions only as a result of non-equilibrium solidification.

The presence of iron impurity in Al–Si alloys (Table 2.1) demands for the use of the quaternary Al–Fe–Mg–Si phase diagram for the correct analysis of the phase composition. Within the compositional range of 356.0-type alloys, variation of silicon concentration does not affect the phase composition, which makes it convenient to use sections at a constant Si concentration. The isothermal sections plotted for equilibrium conditions (Figure 2.12) show that 356.0-type alloys can fall only into two four-phase regions, i.e. (Al) + (Si) + Mg_2Si + π or (Al) + (Si) + β(AlFeSi) + π. As a result, starting from 0.3–0.4% Fe, the iron impurity can

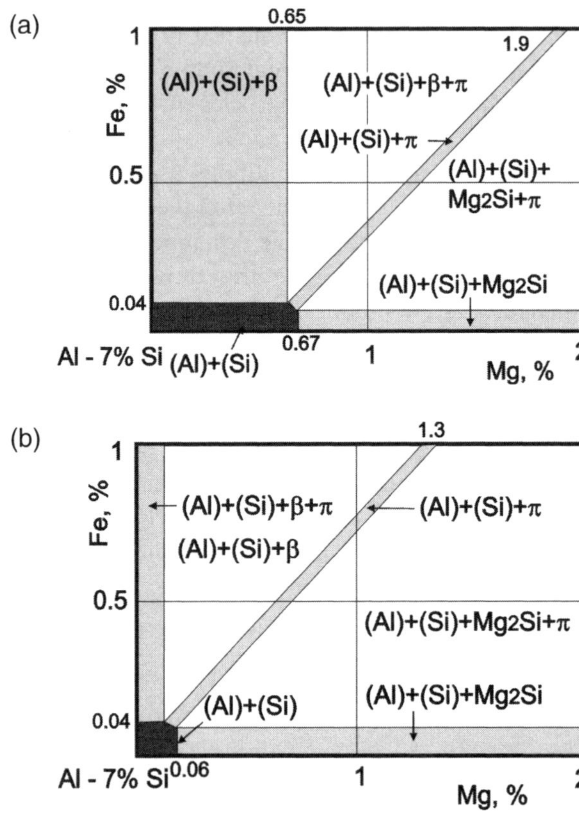

Figure 2.12. Isothermal sections of Al–Fe–Mg–Si phase diagram at 7% Si: (a) 540°C and (b) 200°C. β – Al_5FeSi, and π – $Al_8FeMg_3Si_6$.

completely bind magnesium to the π phase, thus excluding the formation of Mg_2Si precipitates. However, this does not occur in reality, because at early stages of solidification iron mostly enters into the β(AlFeSi) phase that, due to the suppressed peritectic reaction $L + \beta(AlFeSi) \Rightarrow (Al) + \pi$ and low diffusion of Fe in (Al), is retained in the final structure. As a result, magnesium remains in the solid solution after quenching (Figure 2.12a) and can precipitate upon aging. The equilibrium phase composition at a temperature of aging shall be as it is shown in Figure 2.12b. Yet, due to extremely low diffusion of iron in solid (Al) and the preferential precipitation of Mg_2Si (metastable modifications, see in Table 2.13) upon decomposition of supersaturated solid solution, the Al–Fe–Mg–Si phase diagram cannot be directly used for the analysis of a nonequilibrium phase composition formed during aging. Rather the composition of a supersaturated (in Si and Mg) solid solution should be put on the relevant isothermal section of the Al–Mg–Si phase diagram as we show later in Section 3.9, Figure 3.21a.

The polythermal sections at 7% Si and 0.2% Fe (Figure 2.13a) and 0.5% Fe (Figure 2.13b) can be used to follow the reactions during solidification and cooling in the solid state of a 356/357-type alloy at a typical concentration of iron. After primary solidification of (Al), the (Al) + (Si) eutectics is formed, and the remaining liquid reacts through the ternary eutectic reaction involving the Al_5FeSi phase. Under real casting conditions, the quaternary π compound and the Mg_2Si phase are found in as-cast alloys containing over 0.4% Mg (alloys of the 357.0 type) (Wang, 2001). One may notice that the peritectic reaction (point P_1 in Table 2.8) with the formation of the quaternary π phase occurs at a higher magnesium content in the equilibrium phase diagram (0.75–0.77% Mg in Figure 2.13a, b). This discrepancy is an obvious result of nonequilibrium solidification. According to the equilibrium phase diagram, at low magnesium concentrations and at a typical Fe impurity level (356.0-type alloys), iron is bound mainly in the β(AlFeSi) phase (Figure 2.13c). On the other hand, the isopleth in Figure 2.13d shows that at 1% Mg (a concentration much higher than that in 357.0/357.0-type alloys) and at a relatively low iron concentration of less than 0.2%, the β(AlFeSi) phase is completely replaced by the quaternary compound that binds almost all iron.

Under real, nonequilibrium conditions, solidification is completed by the invariant eutectic reaction $L \Rightarrow (Al) + (Si) + Mg_2Si + Al_8FeMg_3Si_6$ at 554°C (Belov et al., 2002a; Wang, 2001). Due to the low concentrations of Fe and Mg, this eutectics usually degenerates into isolated inclusions of phases or their conglomerates. Bäckerud et al. report that the solidus of 356-type alloys can be as low as 505–519°C at a cooling rate of 5 K/s (Bäckerud et al., 1990). Figure 2.14 shows the distribution of phase fields in the solid state after nonequilibrium solidification, and Figure 2.15 demonstrates corresponding as-cast structures with participation of (Si), Al_5FeSi, $Al_8FeMg_3Si_6$, and Mg_2Si.

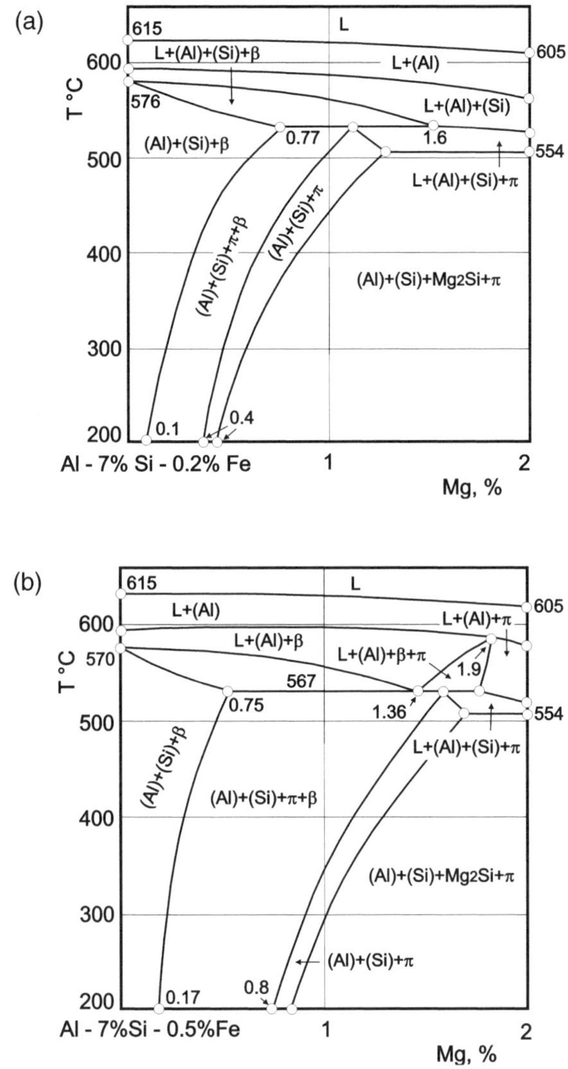

Figure 2.13. Polythermal sections of Al–Fe–Mg–Si phase diagram at 7% Si: (a) 0.2% Fe; (b) 0.5% Fe; (c) 0.3% Mg; and (d) 1% Mg. β – Al_5FeSi, and π – $Al_8FeMg_3Si_6$.

2.6. Al–Mg–Si CASTING ALLOYS (5XX.0 SERIES)

Alloys of 5XX.0 and 5XXX series that contain, besides magnesium, manganese as an alloying element are considered in Chapter 4. Without manganese, the phase composition of such alloys (Tables 2.1, 2.5, 2.7) can be analyzed using the Al–Fe–Mg–Si phase diagram. In the range of high-magnesium alloys (>5% Mg), this phase

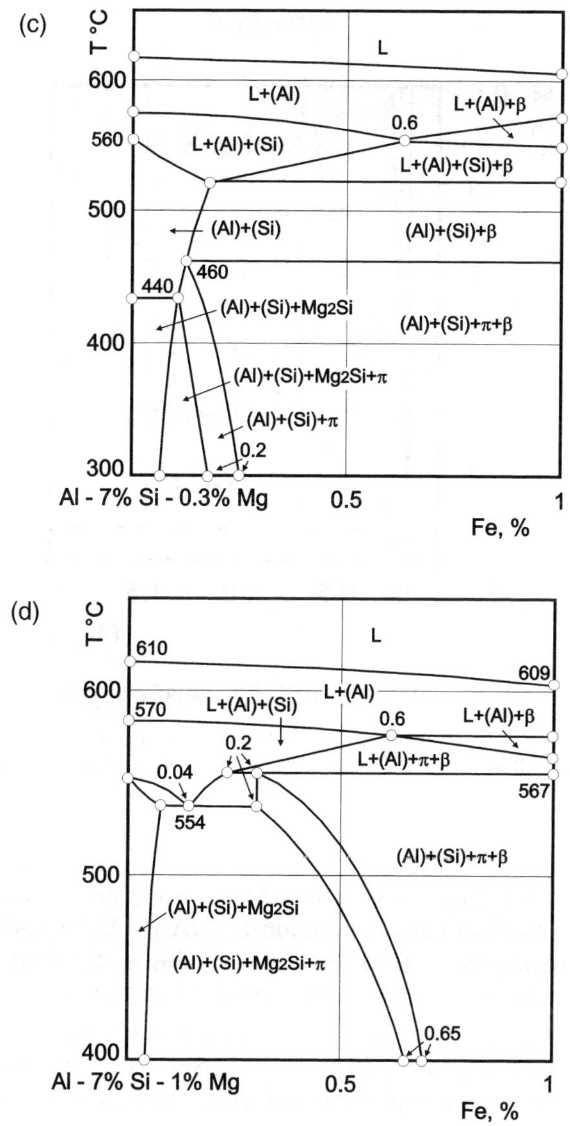

Figure 2.13 (*continued*)

diagram has a relatively simple constitution, with most commercial alloys (except those containing 3–4% Mg) falling at room temperature into the phase region (Al) + Al_3Fe + Al_8Mg_5 + Mg_2Si (Figure 2.4a). According to the equilibrium phase diagram, the Al_8Mg_5 phase is formed in casting 5XX.0-series alloys only in the solid state, by precipitation from the aluminum solid solution. However, under real

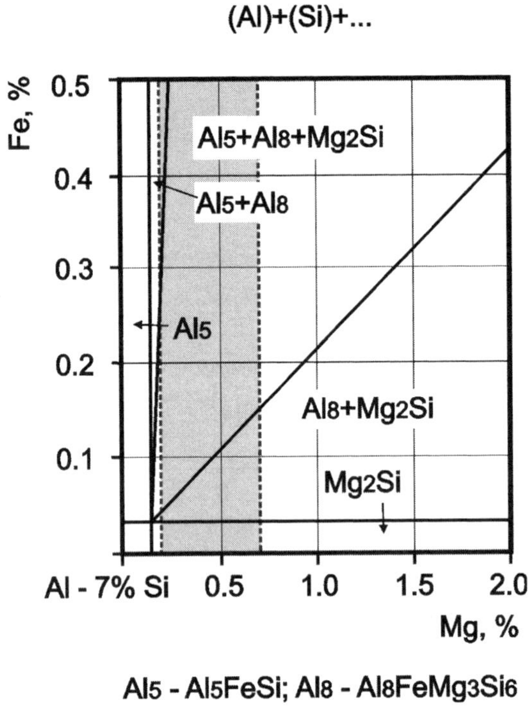

Figure 2.14. Nonequilibrium distribution of phase fields in Al–Fe–Mg–Si system at 7% Si in the as-cast state ($V_c \sim 10^{-1}$ K/s). All phase fields contain (Al) and (Si). Composition range of 356- and 357-type alloys is marked.

casting conditions the majority of commercial casting alloys complete the solidification with the invariant eutectic reaction $L \Rightarrow (Al) + Al_3Fe + Al_8Mg_5 + Mg_2Si$ at 447°C. And the nonequilibrium solidus can be as low as 428°C at a cooling rate of 6 K/s as measured by Bäckerud et al. (1990) for a 518.2 alloy.

As the concentrations of Fe and Si in the eutectic liquid are rather small (point E_1 in Figure 2.4b and Table 2.8), the Al_3Fe and Mg_2Si phases are formed in commercial alloys (except those that are high-pure with respect to Fe and Si impurities) through bi- and monovariant reactions in a wide range of temperatures as shown in Table 2.9. Note, however, that there are some 5XX.0-series alloys that contain iron or silicon as alloying components, e.g. up to 2.2% Si in a 512.2 alloy and up to 1% Fe in a 516.0 alloy. The solidification of such alloys can be traced using the polythermal section at 10% Mg and 0.5% Fe shown in Figure 2.16. After the solidification of primary (Al) grains, either $L \Rightarrow (Al) + Al_3Fe$ (Si < 1%) or $L \Rightarrow (Al) + Mg_2Si$ (Si > 1%) eutectics is formed. Under equilibrium conditions the alloys become solid after the formation of the ternary $L \Rightarrow (Al) + Al_3Fe + Mg_2Si$ eutectics. During

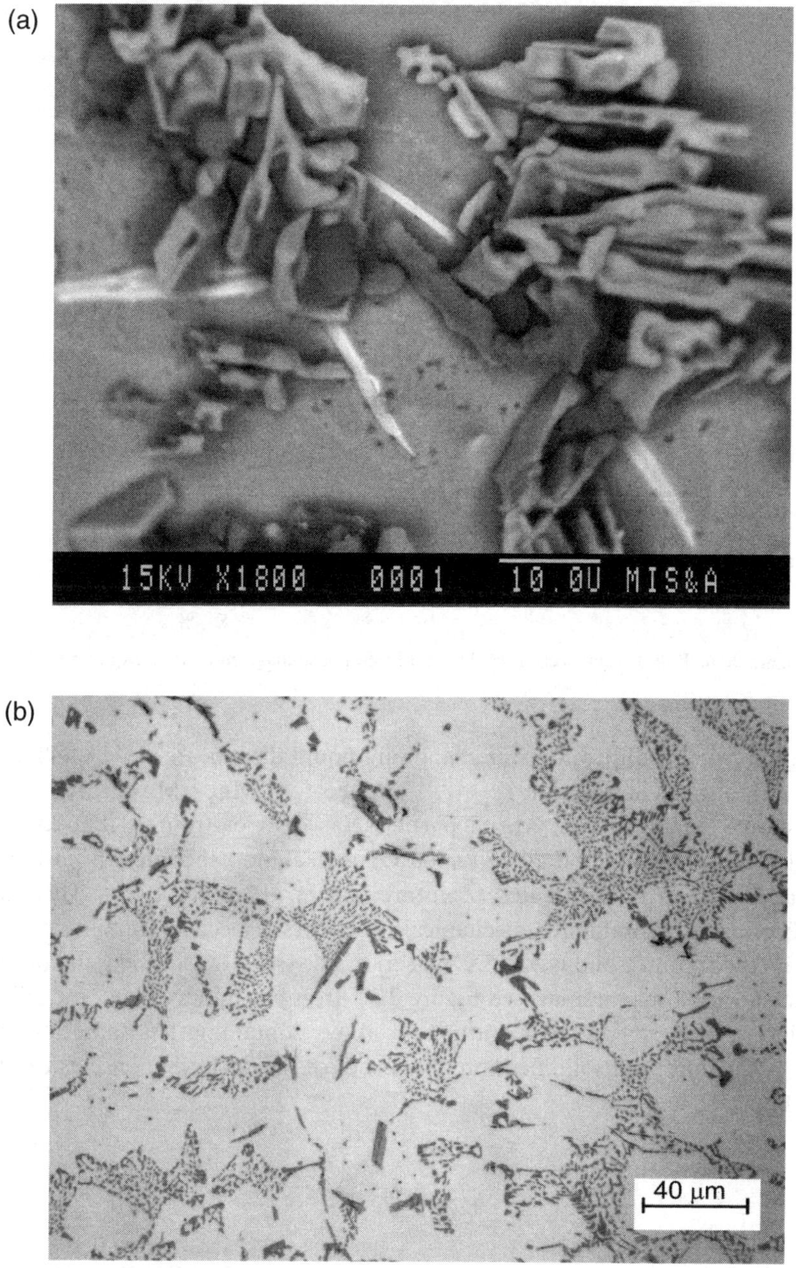

Figure 2.15. Microstructures of as-cast 356 (a) and 357 (b) alloys: (a) ~7% Si, 0.3% Mg, 0.5% Fe, gray particles of (Si) and white needles Al_5FeSi (SEM) and (b) ~8% Si, 0.5% Mg, 0.6% Fe, primary dendrites of (Al), colonies of (Al) + (Si) eutectics, needles of Al_5FeSi phase with inclusions of $Al_8FeMg_3Si_6$, small Mg_2Si particles are mainly in the (Al) + (Si) eutectics (optical microscope).

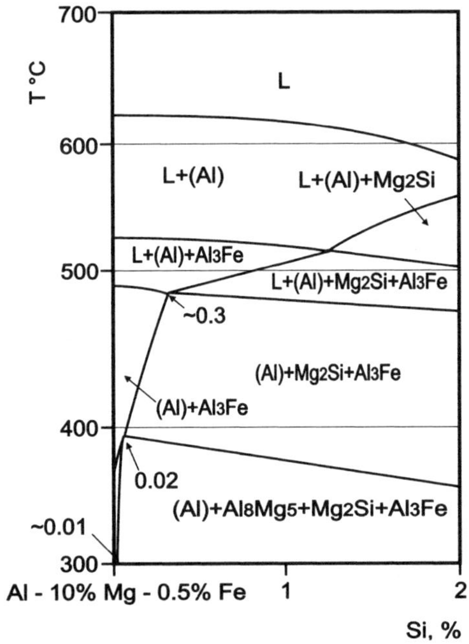

Figure 2.16. Polythermal section of Al–Fe–Mg–Si phase diagram at 10% Mg and 0.5% Fe.

nonequilibrium solidification, the remaining liquid disappears at 447–448°C during the invariant eutectic reaction $L \Rightarrow (Al) + Al_3Fe + Al_8Mg_5 + Mg_2Si$.

It should be noted that Mg_2Si particles (as distinct from Al_3Fe) can become globular upon high-temperature (>500°C) annealing, especially in cast products produced at high cooling rates (Zolotorevskii et al., 1986, 1988). This structure modification is favorable for mechanical properties, especially ductility.

The equilibrium solidus of 5XX.0-series alloys is determined mainly by the concentration of magnesium (see Figure 2.3). Iron has minor effect, and silicon can even increase the solidus temperature. In alloys containing less than 5% Mg, e.g. 512.2, nonequilibrium solidification may produce the Al_8Fe_2Si phase as can be seen from Figure 2.4c.

Chapter 3
Alloys of the Al–Cu–Si–(Mg, Fe) System

This chapter discusses the phase composition of alloys containing silicon and copper as main components. First of all, there are numerous casting alloys of 3XX.0 series (except those that contain nickel and manganese additions, or do not contain copper), and also some casting alloys of 2XX.0 series with a silicon addition (242.0 type). These alloys often contain magnesium (usually as an alloying element) and iron (as a rule, as an impurity), so the analysis of at least quaternary phase diagrams is usually required. Yet in reality, most of the alloys belong to the Al–Cu–Fe–Mg–Si system, and this five-component phase diagram will be considered in detail. The equilibrium and nonequilibrium phase composition of wrought alloys of the 2XXX series and of the 6XXX-series alloys containing copper can be properly analyzed only using the Al–Cu–Mg–Si–(Fe) phase diagram.

3.1. Al–Cu–Si PHASE DIAGRAM

The Al–Cu–Si phase diagram can be used to correctly analyze the phase composition of casting alloys of 3XX.0 and 2XX.0 (242.0 type) series with low concentrations of iron and magnesium impurities (Table 3.1). It is also required for the analysis of more complex phase diagrams involving Cu and Si (Sections 3.4, 3.5, and 3.7).

No ternary compounds are formed in the aluminum corner of the Al–Cu–Si system. The phases Al_2Cu and (Si) are in equilibrium with the aluminum solid solution. The Al_2Cu phase (θ) has a tetragonal structure (space group *I*4/*mmm*, 12 atoms per unit cell) with lattice parameters $a = 0.6063$ nm, $c = 0.4872$ nm (Mondolfo, 1976). This phase exists in a homogeneity range of 52.5–53.9% Cu which does not reach the stoichiometric concentration of copper (54.2%). The density of this phase in binary alloys is 4.34 g/cm^3.

Data on invariant and monovariant reactions occurring in the aluminum corner of the system are given in Table 3.2. The only one invariant (eutectic) reaction occurs in aluminum-rich alloys of this system. Figure 3.1 shows liquidus and solidus surfaces of the Al–Cu–Si system. Once can see that the liquidus (Figure 3.1a) and, especially, the solidus (Figure 3.1b) temperatures strongly decrease with increasing copper and silicon concentrations.

The solubilities of copper in (Si) and silicon in Al_2Cu are negligibly small. The maximal mutual solubility of copper and silicon in (Al) at the eutectic temperature

Table 3.1. Chemical composition of some commercial alloys whose phase composition can be analyzed using Al–Cu–Si phase diagram

Grade	Si, %	Cu, %	Other		
			Mg, %	Fe, %	Mn, %
305.0	4.5–5.5	1.0–1.5	0.1	0.2	0.1
308.2	5.0–6.0	4.0–5.0	0.1	0.8	0.3
319.2	5.5–6.5	3.0–4.0	0.1	0.6	0.1
380.2	7.5–9.5	3.0–4.0	0.1	0.6	0.1
384.1	10.5–12.0	3.0–4.5	0.1	1	0.5
295.1	0.7–1.5	4.0–5.0	0.03	0.8	0.35
296.2	2.0–3.0	4.0–5.0	0.03	0.8	0.30

Table 3.2. Invariant and monovariant reactions in ternary alloys of Al–Cu–Si system (Mondolfo, 1976; Drits et al., 1977)

Reaction	Point or line in Figure 3.1a	Concentrations in phases				*T*, °C
		L		(Al)		
		Cu, %	Si, %	Cu, %	Si, %	
$L \Rightarrow (Al) + Al_2Cu + (Si)$	E	27	5	4.5	1.1	525
$L \Rightarrow (Al) + Al_2Cu$	e_1–E	–	–	–	–	547–525
$L \Rightarrow (Al) + (Si)$	e_2–E	–	–	–	–	577–525

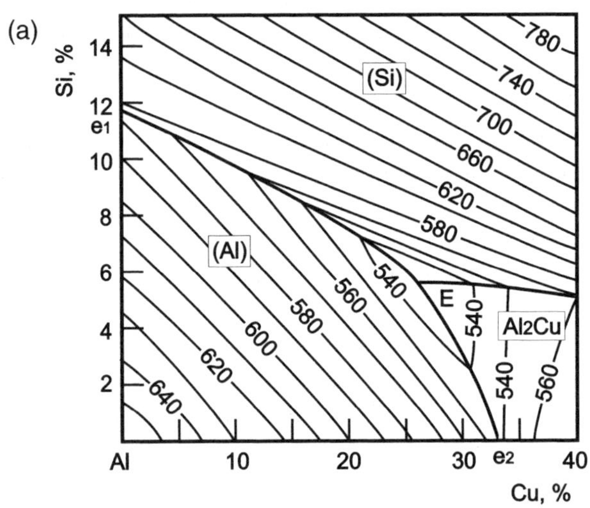

Figure 3.1. Phase diagram of Al–Cu–Si system (Drits et al., 1977): (a) liquidus; (b) solidus; and (c) solvus.

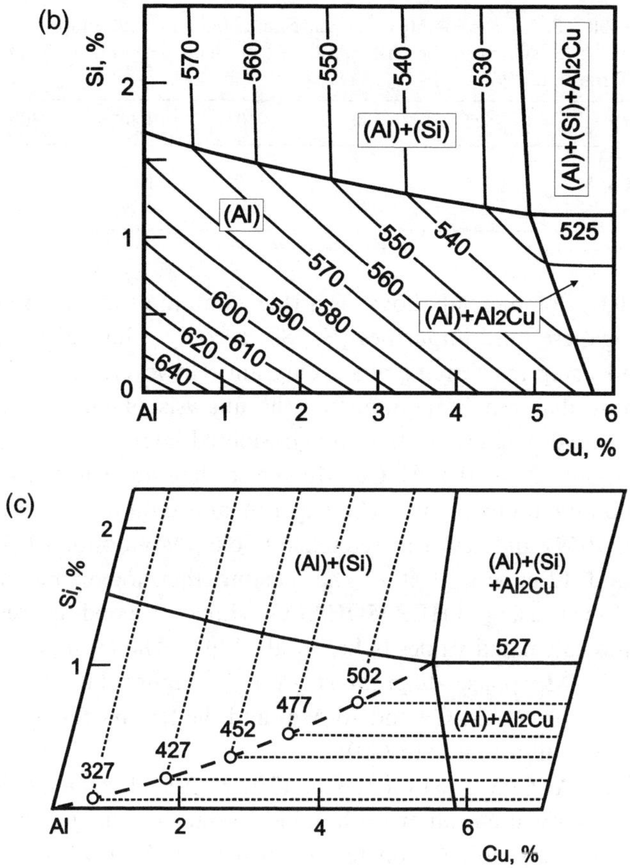

Figure 3.1 (*continued*)

(525°C) is 4.5% Cu and 1.1% Si. As the temperature lowers, the solid solubility of these elements in (Al) decreases as shown in Table 3.3 and Figure 3.1c.

Under real solidification conditions, eutectic particles of the Al_2Cu and (Si) phases are formed at smaller concentrations of copper and silicon than what follows from Table 3.3. The distribution of phase regions in the as-cast state depends on the cooling rate.

3.2. Al–Cu–Mg PHASE DIAGRAM

This system is the basis of so-called duralumins which are more often used as wrought alloys (2XXX series), though there are some casting alloys of the same

Table 3.3. Limit solubility of Cu and Si in aluminum solid solution of Al–Cu–Si system in the (Al) + Al_2Cu + (Si) phase field (Figure 3.1c) (Drits et al., 1977)

T, °C	525	500	460	400	300
Cu, %	4.5	4.1	3.6	1.5	0.4
Si, %	1.1	0.85	0.6	0.25	0.1

system, e.g. 206.0. Commercial alloys, as a rule, contain other alloying elements (in particular, manganese) and impurities (Fe, Si), so only a limited number of alloys can be analyzed using the ternary phase diagram. However, the knowledge of the Al–Cu–Mg phase diagram is required for the analysis of more complex systems involving copper and magnesium that are considered later in this chapter.

Due to its significance, the Al–Cu–Mg phase diagram has been well studied, especially in the aluminum corner. The experimental data are generalized in great detail (Phillips, 1959) and given in reference books (Mondolfo, 1976; Drits et al., 1977; Prince and Effenberg, 1991). The existing thermodynamic model of this diagram (developed using THERMOCALC) shows a good correspondence of experimental and calculated values (Chen et al., 1997). The most generally accepted variant of Al–Cu–Mg phase diagram is given in Figure 3.2. According to this version, the binary phases Al_2Cu and Al_8Mg_5 and the ternary phases Al_2CuMg and Al_6CuMg_4 are in equilibrium with (Al).

The compound Al_2CuMg (S) (46% Cu, 17% Mg) is characterized by a narrow region of homogeneity; it has an orthorhombic crystal structure (space group *Cmcm*, 16 atoms per unit cell) with parameters $a = 0.401$ nm, $b = 0.925$ nm, $c = 0.715$ nm, its density is 3.55 g/cm^3 (Mondolfo, 1976). Metastable modifications of the phase Al_2CuMg (S′) ensure high effect of precipitation hardening during the decomposition of a supersaturated solid solution. The compound Al_6CuMg_4 (22–27%Cu, 27.5–30%Mg) has a defected body-centered cubic structure (space group *Im*3, 162 atoms per unit cell) with the lattice parameter $a = 1.428$–1.431 nm (Hatch, 1984). The density of the phase is 2.69 g/cm^3 (Mondolfo, 1976). This compound is usually designated as T and is isomorphic to the $Al_2Mg_3Zn_3$ phase from the Al–Mg–Zn system. Two other ternary phases – AlCuMg and $Al_5Cu_6Mg_2$ – are not in equilibrium with (Al), but (as does Al_6CuMg_4) they form continuous series of solid solutions with the $MgZn_2$ and Mg_2Zn_{11} phases from the Al–Mg–Zn system (see Section 6.1). This should be taken into account in the analysis of the phase composition of the Al–Cu–Mg–Zn quaternary system (see Section 6.3).

Four invariant reactions occur in the aluminum corner of the Al–Cu–Mg system as shown in Table 3.4. Monovariant reactions are given in Table 3.5. The liquidus and solidus isotherms in aluminum-rich alloys are shown in Figure 3.2. One can see

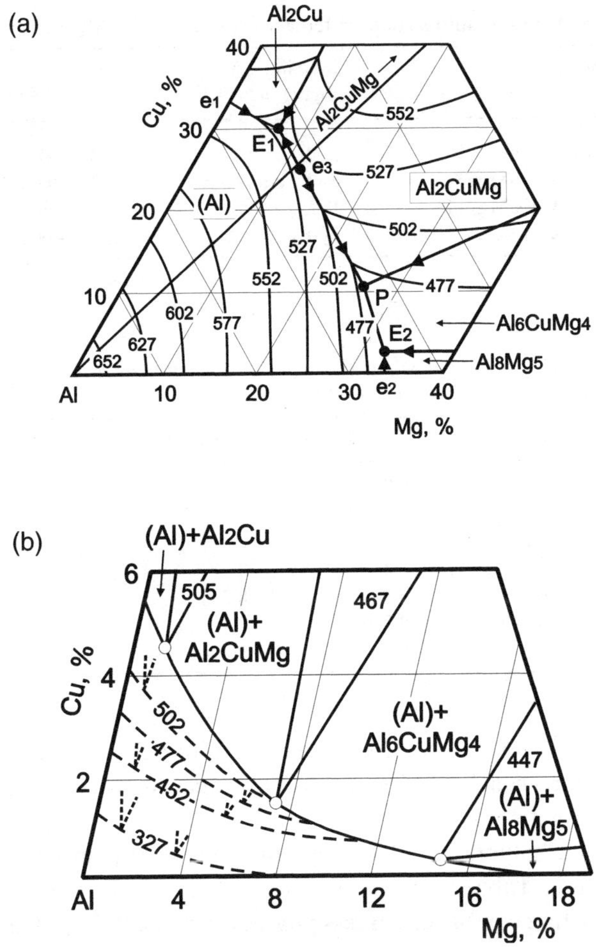

Figure 3.2. Phase diagram of Al–Cu–Mg system (Mondolfo, 1976): (a) liquidus; (b) solidus.

Table 3.4. Invariant reactions in ternary alloys of Al–Cu–Mg system (Mondolfo, 1976; Drits et al., 1977)

Reaction	Point in Figure 3.2a	*T*, °C	Concentrations in the liquid phase	
			Cu, %	Mg, %
L ⇒ (Al) + Al_2Cu + Al_2CuMg (S)	E_1	507	30	6
L ⇒ (Al) + Al_2CuMg (quasi-binary)	e_3	518	24.5	10.1
L + Al_2CuMg ⇒ (Al) + Al_6CuMg_4 (T)	P	467	10	26
L ⇒ (Al) + Al_8Mg_5 + Al_6CuMg_4	E_2	449	2.7	32

Table 3.5. Monovariant reactions in ternary alloys of Al–Cu–Mg system

Reaction	Line in Figure 3.2a	*T*, °C
$L \Rightarrow (Al) + Al_2Cu$	e_1–E_1	547–507
$L \Rightarrow (Al) + Al_2CuMg$ (S)	e_3–E_1 and e_3–P	518–507 and 518–467
$L \Rightarrow (Al) + Al_6CuMg_4$ (T)	P–E_2	467–449
$L \Rightarrow (Al) + Al_8Mg_5$	e_2–E_2	450–449

Table 3.6. Limit solubility of Cu and Mg in aluminum solid solution in Al–Cu–Mg system (Figure 3.2b) (Phillips, 1959; Mondolfo, 1976; Drits et al., 1977)

T, °C	$(Al) + Al_2Cu + Al_2CuMg$		$(Al) + Al_2CuMg + Al_6CuMg_4$		$(Al) + Al_8Mg_5 + Al_6CuMg_4$	
	Cu, %	Mg, %	Cu, %	Mg, %	Cu, %	Mg, %
507	4.1	1.6	–	–	–	–
467	–	–	0.4	9	–	–
450	2.0–2.6	0.6–1.1	0.3–0.35	8.5	0.3	10.5
400	1.4–1.8	0.4–0.8	0.2–0.3	7.4	0.2	9.2–9.5
350	0.9	0.5	0.1	6.2	0.1	7.6

that in commercial alloys of the 2XXX series (up to ~5% Cu, up to ~2% Mg) the solidus temperature can be as low as 507°C, which is often the cause of melting during solution treatment.

The limit solid solubilities of copper and magnesium in (Al) for three-phase regions are given in Table 3.6.

Under real casting conditions, excess phases (in particular, Al_2Cu, and Al_2CuMg) are formed at lower concentrations of copper and silicon than what follows from Figure 3.2 and Table 3.6.

3.3. Al–Cu–Fe PHASE DIAGRAM

Analysis of this system makes it possible to follow the effect of iron impurity on the phase composition of Al–Cu commercial alloys, if the concentrations of the other elements, especially those that may form own phases with iron, are low. This phase diagram is also required for the analysis of more complex phase diagrams involving Cu and Fe and are considered later in this chapter (see Sections 3.5–3.7).

Besides the phases from the binary systems (Al_3Fe and Al_2Cu), two ternary compounds – Al_7Cu_2Fe and $Al_6(FeCu)$ – can be in equilibrium with (Al). The Al_3Fe

phase can dissolve up to 0.5% Cu. The phase $Al_6(FeCu)$ (7% Cu, 24.6% Fe), which is also designated as $Al_{23}CuFe_4$ and α(FeCu), is a modification of the metastable phase Al_6Fe, which becomes stable at 7–8% Cu and 22–25% Fe. This compound has an orthorhombic crystal structure of the Al_6Mn type (space group $Ccm2_1$, 28 atoms per unit cell) with parameters $a = 0.64343$ nm, $b = 0.74604$ nm, $c = 0.87769$ nm (Legendre and Harmelin, 1991). The density of the phase is 3.45 g/cm^3 (Mondolfo, 1976). The Al_7Cu_2Fe phase (36.9% Cu, 16.2% Fe), also designated as β(FeCu) and N, has a broad homogeneity range of 29–39% Cu and 12–20% Fe. The structure of this phase belongs to the tetragonal crystal system (space group $P4/mnc$, 40 atoms per unit cell) with lattice parameters $a = 0.6336$ nm, $c = 1.4879$ nm (Legendre and Harmelin, 1991). Its density is 4.3 g/cm^3 (Mondolfo, 1976). Depending on the alloy composition, ternary phases can crystallize primarily or form by peritectic reactions.

The liquidus and solidus projections in the aluminum corner of the system are shown in Figure 3.3. The invariant reactions in the Al–Cu–Fe system are given in Table 3.7, and the monovariant reactions, in Table 3.8.

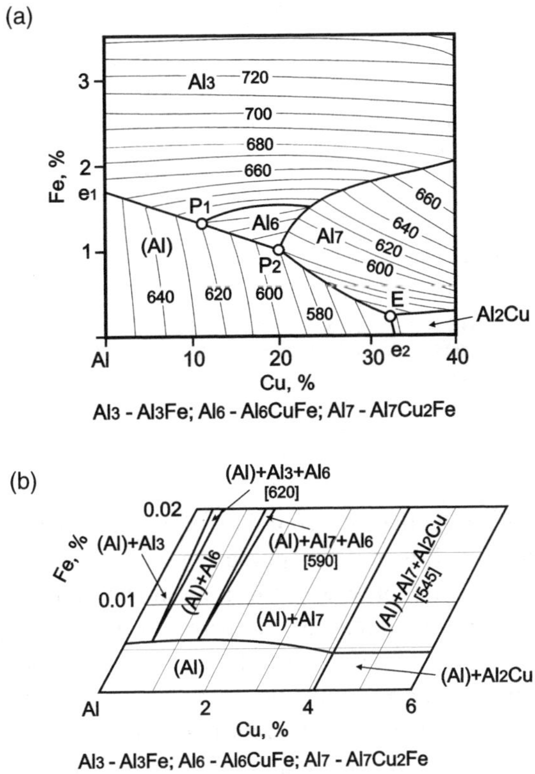

Figure 3.3. Phase diagram of Al–Cu–Fe system (Mondolfo, 1976): (a) liquidus; (b) solidus.

Table 3.7. Invariant reactions in ternary alloys of Al–Cu–Fe system (Mondolfo, 1976; Belov et al., 2002a)

Reaction	Point in Figure 3.3a	*T*, °C	Concentration in phases			
			Liquid		(Al)	
			Cu, %	Fe, %	Cu, %	Fe, %
$L + Al_3Fe \Rightarrow (Al) + Al_6(FeCu)$	P_1	620	10.8	1.4	1.5	0.03
$L + Al_6(FeCu) \Rightarrow (Al) + Al_7Cu_2Fe$	P_2	590	20.0	1.0	3.0	0.02
$L \Rightarrow (Al) + Al_2Cu + Al_7Cu_2Fe$	E	545	32.5	0.3	5.3	0.03

Table 3.8. Monovariant reactions in ternary alloys of Al–Cu–Fe system

Reaction	Line in Figure 3.3a	*T*, °C
$L \Rightarrow (Al) + Al_3Fe$	e_1–P_1	655–622
$L \Rightarrow (Al) + Al_6(FeCu)$	P_1–P_2	622–590
$L \Rightarrow (Al) + Al_7Cu_2Fe$	P_2–E	590–545
$L \Rightarrow (Al) + Al_2Cu$	e_2–E	547–545

Table 3.9. Limit solubility of Cu and Fe in aluminum solid solution in Al–Cu–Fe system (Figure 3.3b) (Mondolfo, 1976)

T, °C	$(Al) + Al_2Cu + Al_7Cu_2Fe$		$(Al) + Al_6(FeCu) + Al_7Cu_2Fe$		$(Al) + Al_3Fe + Al_6(FeCu)$	
	Cu, %	Fe, %	Cu, %	Fe, %	Cu, %	Fe, %
552	5.65	0.018	2.00	0.015	0.60	0.013
527	5.00	0.012	1.75	0.010	0.50	0.009
502	4.00	0.005	1.50	0.005	0.40	0.006
477	3.30	0.003	1.00	0.003	0.30	0.003
452	2.56	0.002	0.80	0.002	0.23	0.002
427	1.50	0.001	0.58	0.001	0.19	0.001

Iron only slightly affects the solid solubility of copper in (Al) in the phase regions where Al_2Cu is present. But in other phase regions the solid solubility of copper decreases significantly in the presence of iron as shown in Table 3.9.

Under real solidification conditions the peritectic reactions shown in Table 3.7 do not, as a rule, complete; therefore, the phase composition of as-cast alloys differs from the equilibrium composition. For example, in as-cast Al–Cu alloys iron impurity can form, in addition to the equilibrium Al_7Cu_2Fe phase, two other phases, i.e. Al_3Fe and $Al_6(FeCu)$, that may remain in the structure after the end of solidification. In most cases, heat treatment has no noticeable effect on the composition and morphology of Fe-containing phases, so the nonequilibrium phases are retained in the final structure of a product.

3.4. Al–Cu–Mg–Si PHASE DIAGRAM

This system is exceptionally important for most casting Cu-containing alloys of the 3XX.0 series, and also for some wrought alloys of the 2XXX series (e.g. 2008, 2014, 2037, and 2024) and the 6XXX series with composition given in Table 3.10 (see also Table 5.5 in Section 5.2). These alloys cannot be satisfactorily analyzed by the ternary phase diagrams, primarily due to the formation of the quaternary $Al_5Cu_2Mg_8Si_6$ compound that is usually designated as the Q phase. This compound, together with the phases from the binary and ternary systems, can be in equilibrium with (Al) in alloys containing simultaneously copper, magnesium, and silicon, and is present in almost all phase fields, which is reflected in Figure 3.4a.

Analysis of the primary crystals of the quaternary compound in Al-rich alloys shows the following compositional range: 14–17% Cu, 28–30% Mg, and 27–29% Si (Mondolfo, 1976). This composition conforms to the formula $Al_4CuMg_{4-5}Si_4$. Crystals formed in more alloyed material fall into a compositional range of 19.2–20.6% Cu, 31.8–33% Mg, and 31.4–32.1% Si, and can be adequately described by the formula $Al_5Cu_2Mg_8Si_6$ (20.3% Cu, 31.1% Mg, 27% Si) (Mondolfo, 1976). The quaternary phase has a hexagonal structure with parameters $a = 1.032$ nm, $c = 0.405$ nm (Mondolfo, 1976; Drits et al., 1977); its density is 2.79 g/cm^3 (Hatch, 1984). The presence of the quaternary phase is very important for the analysis of

Table 3.10. Chemical composition of some commercial alloys whose phase composition can be analyzed using Al–Cu–Mg–Si phase diagram

Grade	Cu, %	Mg, %	Si, %	Other	
				Fe, %	Mn, %
355.2	1.0–1.5	0.5–0.6	4.5–5.5	0.06	0.03
354.1	1.6–2.0	0.45–0.6	8.6–9.4	0.15	0.1
390.0	4.0–5.0	0.5–0.65	16–18	0.4	0.1
204.0	4.2–5.0	0.15–0.35	0.2	0.35	0.1
222.1	9.2–10.7	0.2–0.35	2	1.2	0.5
2009	3.2–4.4	1.0–1.6	0.25	0.05	–
2008	0.7–1.1	0.25–0.5	0.5–0.8	0.4	0.3
2002	1.5–2.5	0.5–1.0	0.35–0.8	0.3	0.2
6951	0.15–0.4	0.4–0.8	0.2–0.5	0.8	0.1
6763	0.04–0.16	0.45–0.9	0.2–0.6	0.08	0.03
6111	0.5–0.9	0.5–1.0	0.7–1.1	0.4	0.15–0.45

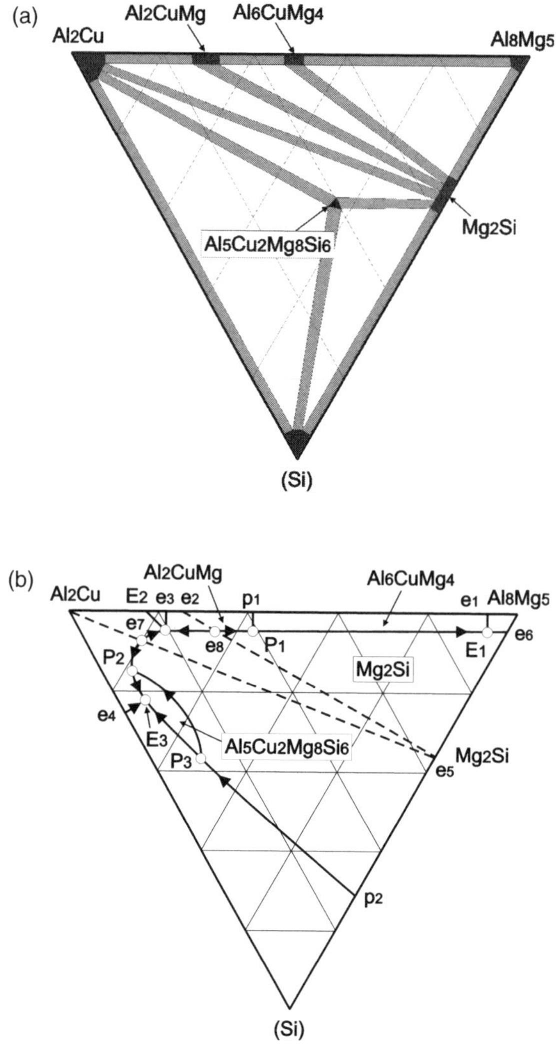

Figure 3.4. Phase diagram of Al–Cu–Mg–Si system: (a) distribution of phase fields in the solid state (Mondolfo, 1976); (b) polythermal projection of liquidus (Mondolfo, 1976); (c) polythermal projection of (Al) single-phase region.

both the phase composition after solidification and the phase composition after decomposition of a supersaturated solid solution (see also Section 3.9).

According to the most reliable variant of the Al–Cu–Mg–Si phase diagram (Mondolfo, 1976) (Figure 3.4), the Mg_2Si phase is formed after (Al) within a wide concentration range, (Si) or Q phase being dominant only in Si- and Cu-rich alloys. Invariant reactions involving (Al) in quaternary alloys of this system are given in

(c)

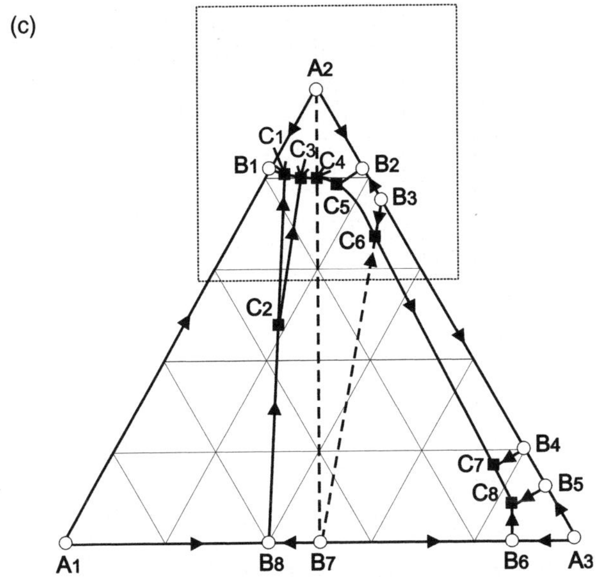

Figure 3.4 *(continued)*

Table 3.11. Invariant reactions in quaternary alloys of Al–Cu–Mg–Si system (Mondolfo, 1976; Chakrabarti and Murray, 1996)

Reaction	Point in Figure 3.4b	Concentration in liquid phase			T, °C
		Cu, %	Mg, %	Si, %	
$L \Rightarrow (Al) + Al_8Mg_5 + Mg_2Si + Al_6CuMg_4$	E_1	1.5	32.9–33	0.3	444–448
$L + Al_2CuMg \Rightarrow (Al) + Mg_2Si + Al_6CuMg_4$	P_1	10	25.5	0.3	467
$L \Rightarrow (Al) + Mg_2Si + Al_2CuMg$ (quasi-ternary)	e_8	23	10.5	0.3	516
$L \Rightarrow (Al) + CuAl_2 + Al_2CuMg + Mg_2Si$	E_2	30–33	6.1–7.1	0.3–0.4	500
$L \Rightarrow (Al) + Al_2Cu + Mg_2Si$ (quasi-ternary)	e_7	31.5	3.9	2.3	515
$L + Mg_2Si \Rightarrow (Al) + Al_2Cu + Al_5Cu_2Mg_8Si_6$	P_2	31	3.3	3.3	512
$L \Rightarrow (Al) + (Si) + Al_2Cu + Al_5Cu_2Mg_8Si_6$	E_3	28	2.2	6	507
$L + Mg_2Si + (Si) \Rightarrow (Al) + Al_5Cu_2Mg_8Si_6$	P_3	13.8	3.3	9.6	529

Table 3.11; and the bi- and monovariant reactions, in Table 3.12. Taking into account that Si, Cu, and Mg have considerably high solubilities in solid (Al) (Table 3.13), their combined effect plays an exceptionally important role in choosing optimal alloy concentrations and heat treatment temperatures. This effect is illustrated by the solidus projection of the single-phase region that is based on our own estimates of the available data (Figure 3.4c). The projection is plotted using the

Table 3.12. Bivariant and monovariant reactions in quaternary alloys of Al–Cu–Mg–Si system (Mondolfo, 1976)

Bivariant reaction*	Field in Figure 3.4b	*T*, °C	Monovariant reaction*	Line in Figure 3.4b	*T*, °C
$L \Rightarrow (Al) + Al_8Mg_5$	e_1–E_1–e_6–Al_8Mg_5	450–444	$L \Rightarrow (Al) + Al_8Mg_5 + Mg_2Si$	e_6–E_1	450–444
$L \Rightarrow (Al) + T$	e_1–E_1–P_1–p_1	467–444	$L \Rightarrow (Al) + Al_8Mg_5 + T$	e_1–E_1	449–444
$L \Rightarrow (Al) + S$	p_1–P_1–E_2–e_3	518–467	$L \Rightarrow (Al) + Mg_2Si + T$	P_1–E_1	467–444
$L \Rightarrow (Al) + Al_2Cu$	Al_2Cu–e_4–E_3–P_2–E_2–e_3	448–500	$L \Rightarrow (Al) + Mg_2Si + S$	e_8–E_1	516–467
				e_8–P_1	
$L \Rightarrow (Al) + (Si)$	(Si)–e_4–E_3–P_3–p_2	577–507	$L \Rightarrow (Al) + Mg_2Si + Al_2Cu$	e_3–E_1	515–500
$L \Rightarrow (Al) + Q$	P_3–E_3–P_2	529–507	$L + Mg_2Si \Rightarrow (Al) + Q$	P_3–P_2	529–512
$L \Rightarrow (Al) + Mg_2Si$	p_2–P_3–P_2–E_2–P_1–E_1–e_6	595–444	$L \Rightarrow (Al) + Mg_2Si + (Si)$	p_2–P_3	555–529
			$L \Rightarrow (Al) + (Si) + Q$	P_3–E_3	529–507
			$L \Rightarrow (Al) + Al_2Cu + (Si)$	e_4–E_3	525–507
			$L \Rightarrow (Al) + Al_2Cu + Q$	P_2–E_3	512–507

* S – Al_2CuMg, T – Al_6CuMg_4, Q – $Al_5Cu_2Mg_8Si_6$

Table 3.13. Limit solid solubilities of Cu, Mg, and Si in aluminum in Al–Cu–Mg–Si system (Figure 3.4c)

Point in Figure 3.4c	Si, %	Cu, %	Mg, %	*T*, °C	Phases in equilibria with (Al)*
A_1	1.65	–	–	577	(Si)
A_2	–	5.7	–	547	Al_2Cu
A_3	–	–	17.4	450	Al_8Mg_5
B_1	1.1	4.9	–	525	(Si) + Al_2Cu
B_2	–	3.9	1.5	507	Al_2Cu + S
B_3	–	3	2	518	S
B_4	–	0.7	9.8	467	S + T
B_5	–	0.2	12	449	T + Al_8Mg_5
B_6	0.1	–	15.3	449	Mg_2Si + Al_8Mg_5
B_7	0.68	–	1.17	595	Mg_2Si,
B_8	1.1	–	0.85	555	Mg_2Si + (Si)
C_1	0.77	4	0.3	507	(Si) + Al_2Cu + Q
C_2	0.77	1.05	0.4	529	Mg_2Si + (Si) + Q
C_3	0.35	3.85	0.6	512	Mg_2Si + Al_2Cu + Q
C_4	n/a	n/a	n/a	515	Mg_2Si + Al_2Cu
C_5	0.05	3.7	1.4	500	Al_2Cu + S + Mg_2Si
C_6	>0.1	n/a	n/a	516	S + Mg_2Si
C_7	>0.1	0.7	9.8	n/a	S + T + Mg_2Si
C_8	>0.1	0.2	12	n/a	T + Mg_2Si + Al_8Mg_5

* S – Al_2CuMg, T – Al_6CuMg_4, Q – $Al_5Cu_2Mg_8Si_6$

values of limit solid solubility given in Table 3.13 and can be used to assess the solidus of quaternary alloys.

The considered version of the quaternary phase diagram makes in possible to separate two quasi-ternary sections, i.e. Al–Al_2Cu–Mg_2Si and Al–Al_2CuMg–Mg_2Si, as shown in Figure 3.5. This peculiar feature of the quaternary phase diagram enables the choice of alloys with a required phase composition, e.g. alloys containing Al_2CuMg and Mg_2Si phases but without Al_2Cu and Q.

The phase composition of cast alloys significantly differs, as a rule, from the equilibrium selection of phases, which is due to incomplete peritectic reactions and hindered diffusion of copper, magnesium, and silicon in (Al) during solidification. The former factor explains the appearance of "extra" Mg_2Si particles in some alloys of the 3XX.0 series that, apart from (Al) and (Si), can contain only Al_2Cu and Q in the equilibrium solid state. The latter factor explains the formation of eutectic nonequilibrium phases in alloys with magnesium and copper concentrations within the limits of their solubility in (Al) by the time the solidification is completed, i.e. within the limits of the single-phase region (Figure 3.4c). Almost all phases from the aluminum corner of the Al–Cu–Mg–Si system – (Si), Mg_2Si, Al_2Cu, Al_2CuMg, and Q – can simultaneously be found in as-cast ingots and billets of wrought alloys of the 2XXX series.

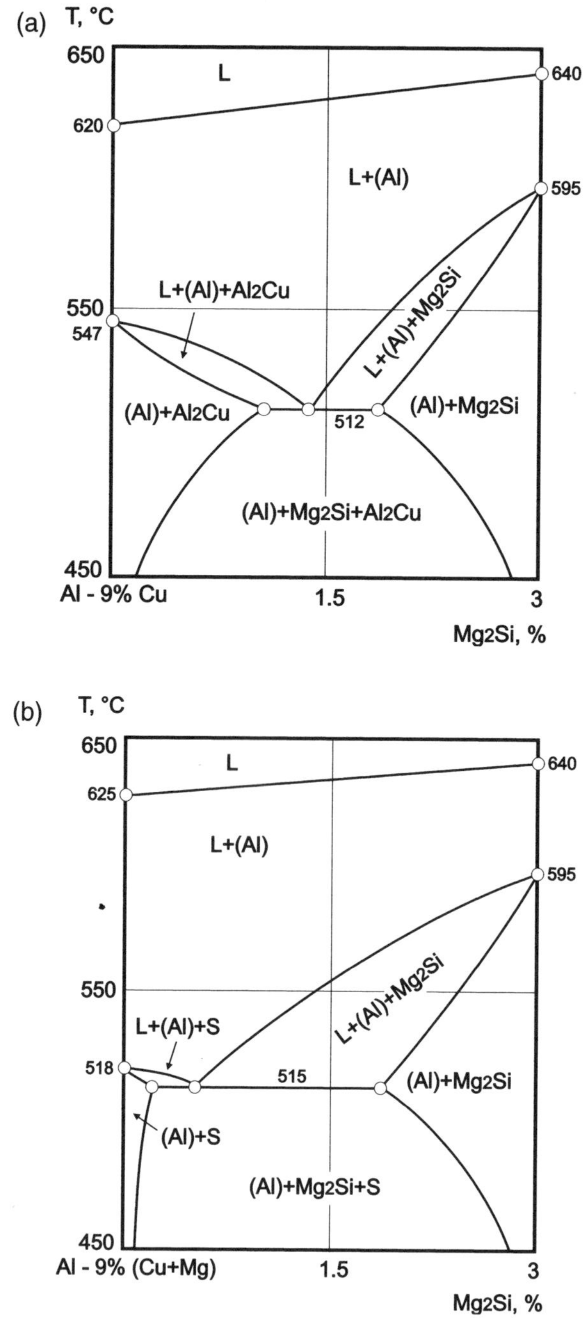

Figure 3.5. Quasi-ternary sections of Al–Cu–Mg–Si phase diagram: (a) (Al)–Al_2Cu–Mg_2Si; (b) (Al)–Al_2CuMg–Mg_2Si.

3.5. Al–Cu–Fe–Si PHASE DIAGRAM

This quaternary phase diagram can be used to analyze the effect of iron impurity on the phase composition of some casting alloys of 3XX.0 and 2XX.0 series containing silicon and copper at low concentrations of magnesium, manganese and nickel impurities (Table 3.14). By taking into account that four (two in each system) different Fe-containing ternary compounds from the Al–Fe–Si and Al–Cu–Fe systems (Sections 1.1 and 3.3) can be in equilibrium with (Al), the necessity of the quaternary phase diagram for the analysis of four-component alloys becomes evident. This phase diagram is also required for the consideration of quinary systems, in particular, Al–Cu–Fe–Mg–Si.

No quaternary phases have been found in Al–Cu–Fe–Si alloys. The phases Al_2Cu, Al_3Fe, Al_7Cu_2Fe, $Al_6(CuFe)$, Al_8Fe_2Si, Al_5FeSi, and (Si) from the constitutive systems can be in equilibrium with the aluminum solid solution. According to Mondolfo the solubility of copper in the (AlFeSi) phases and the solubility of silicon in the (AlCuFe) phases do not exceed 1% (Mondolfo, 1976).

Five-phase invariant reactions that occur in the aluminum corner of the Al–Cu–Fe–Si system are given in Table 3.15; and bi- and monovariant reactions, in Table 3.16. Figure 3.6 shows the liquidus projection and the distribution of phase regions in the solid state.

The solubilities of the components (first of all, copper and silicon) in (Al) are probably the same as in the respective phase regions of the ternary systems.

Table 3.14. Chemical composition of some commercial alloys whose phase composition can be analyzed using Al–Cu–Fe–Si phase diagram

Grade	Si, %	Cu, %	Fe, %	Other		
				Mg, %	Ni, %	Mn, %
305.2	4.5–5.5	1.0–1.5	0.14–0.25	–	–	0.05
308.0	5.0–6.0	4.0–5.0	1.0	0.1	–	0.5
319.0	5.5–6.5	3.0–4.0	1.0	0.1	0.35	0.5
343.0	6.7–7.7	0.5–0.9	1.2	0.1	–	0.5
380.2	7.5–9.5	3.0–4.0	0.7–1.1	0.1	0.1	0.1
383.0	9.5–11.5	2.0–3.0	1.3	0.1	0.3	0.5
385.0	11.0–13.0	2.0–4.0	2.0	0.3	0.5	0.5
208.2	2.5–3.4	3.5–4.5	0.8	0.3	–	0.03
295.2	0.7–1.2	4.0–5.0	0.8	0.3	–	0.03
296.0	2.0–3.0	4.0–5.0	1.2	0.35	0.35	0.05
213.0	1.0–3.0	6.0–8.0	1.2	0.6	0.35	0.1
238.0	4.0	10.0–11.0	1.5	–	–	0.25

Table 3.15. Invariant reactions in quaternary alloys of Al–Cu–Fe–Si system (Mondolfo, 1976; Belov et al., 2002a)

Reaction	Point in Figure 3.6b	T, °C	Concentrations in the liquid phase		
			Cu, %	Fe, %	Si, %
$L + Al_3Fe \Rightarrow (Al) + Al_6(CuFe) + Al_8Fe_2Si$	P_1	612	13.7	1.4	0.65
$L + Al_6(CuFe) \Rightarrow (Al) + Al_8Fe_2Si + Al_7Cu_2Fe$	P_2	607	15.2	1.3	0.7
$L + Al_8Fe_2Si \Rightarrow (Al) + Al_5FeSi + Al_7Cu_2Fe$	P_3	579	16.7	0.8	3.2
$L + Al_7Cu_2Fe \Rightarrow (Al) + Al_5FeSi + Al_2Cu$	P_4	534	27.5	0.35	4.1
$L \Rightarrow (Al) + (Si) + Al_2Cu + Al_5FeSi$	E	525	26.2	0.4	5.5

In particular, in 3XX.0-series alloys, almost all of which fall into the $(Al) + (Si) + Al_2Cu + Al_5FeSi$ phase region, the solid solubilities of the constitutive elements in (Al) can be assessed by the Al–Cu–Si diagram (Section 3.1). On the other hand, in 2XX.0-series alloys with a relatively low content of silicon the solubility of copper should be assessed by the Al–Cu–Fe phase diagram (Section 3.3).

The phase composition of cast alloys can significantly differ from the equilibrium phase selection as a result of incomplete peritectic reactions and hindered diffusion of copper and silicon in (Al) during solidification. The former factor explains the appearance of "extra" Fe-containing phases. For instance, in 3XX.0-series alloys, these can be Al_8Fe_2Si and Al_7Cu_2Fe. The latter factor explains the appearance of nonequilibrium phases of eutectic origin – (Si) and Al_2Cu – at silicon and copper concentrations within the limits of their maximal solubility in solid (Al). In 3XX.0-series alloys (usually containing less than 4% Cu), the solidification will be completed by the eutectic reaction $L \Rightarrow (Al) + (Si) + Al_2Cu + Al_5FeSi$ at ~525°C. Bäckerud et al. (1990) report that even a small amount of Mg (0.1%) can decrease the solidus temperature of a 319.0-type alloy under nonequilibrium solidification conditions down to 507°C (invariant eutectic reaction with the Q phase, E_3 in Table 3.11).

Although, in our opinion, the version of the phase diagram given by Mondolfo (1976) is the most reliable one, we believe it necessary to present an alternative variant.

Zakharov et al. (1992), investigating the phase equilibria in Al–Cu–Fe–Si alloys with copper concentrations of 25 and 40%, observed the following invariant reactions:

$$L + Al_3FeSi \Rightarrow (Si) + Al_7Cu_2Fe + Al_8Fe_2Si\ (620°C);$$

$$L + Al_4FeSi_2 \Rightarrow (Si) + Al_2Cu + Al_7Cu_2Fe\ (567°C);\ \text{and}$$

$$L \Rightarrow (Al) + (Si) + Al_2Cu + Al_7Cu_2Fe\ (520°C).$$

Table 3.16. Bivariant and monovariant reactions in quaternary alloys of Al–Cu–Fe–Si system

Bivariant reactions*	Field in Figure 3.6b	T, °C	Monovariant reactions*	Line in Figure 3.6b	T, °C
$L \Rightarrow (Al) + Al_2Cu$	Al_2Cu–e_1–E–P_4	548–525	$L \Rightarrow (Al) + Al_2Cu + (Si)$	e_1–E	525–525**
$L \Rightarrow (Al) + (Si)$	(Si)–e_1–E–e_3	577–525	$L \Rightarrow (Al) + \beta(AlFeSi) + (Si)$	e_3–E	576–525
$L \Rightarrow (Al) + Al_3Fe$	Al_3Fe–p_3–P_1–p_2	655–612	$L \Rightarrow (Al) + Al_2Cu + N$	P_4–E	545–534
$L \Rightarrow (Al) + \alpha(AlFeSi)$	p_4–P_3–P_2–P_1	629–579	$L \Rightarrow (Al) + Al_2Cu + \beta(AlFeSi)$	e_2–P_4	534–525
$L \Rightarrow (Al) + \beta(AlFeSi)$	e_3–E–P_4–P_3–p_4	611–525	$L + N \Rightarrow (Al) + \beta(AlFeSi)$	P_3–P_4	579–534
$L \Rightarrow (Al) + N$	p_1–P_2–P_3–P_4	590–534	$L + D \Rightarrow (Al) + N$	p_1–P_2	590–579
$L \Rightarrow (Al) + D$	p_2–P_1–P_2–p_1	620–607	$L + Al_3Fe \Rightarrow (Al) + D$	p_2–P_1	620–612
			$L + Al_3Fe \Rightarrow (Al) + \alpha(AlFeSi)$	p_3–P_1	629–612
			$L + \alpha \Rightarrow (Al) + \beta(AlFeSi)$	p_4–P_3	611–579
			$L \Rightarrow (Al) + \alpha(AlFeSi) + D$	P_1–P_2	612–579

* $\alpha(AlFeSi)$ – Al_8Fe_2Si; $\beta(AlFeSi)$ – Al_5FeSi; N – Al_7Cu_2Fe; D – $Al_6(CuFe)$
** Temperature range <1°C

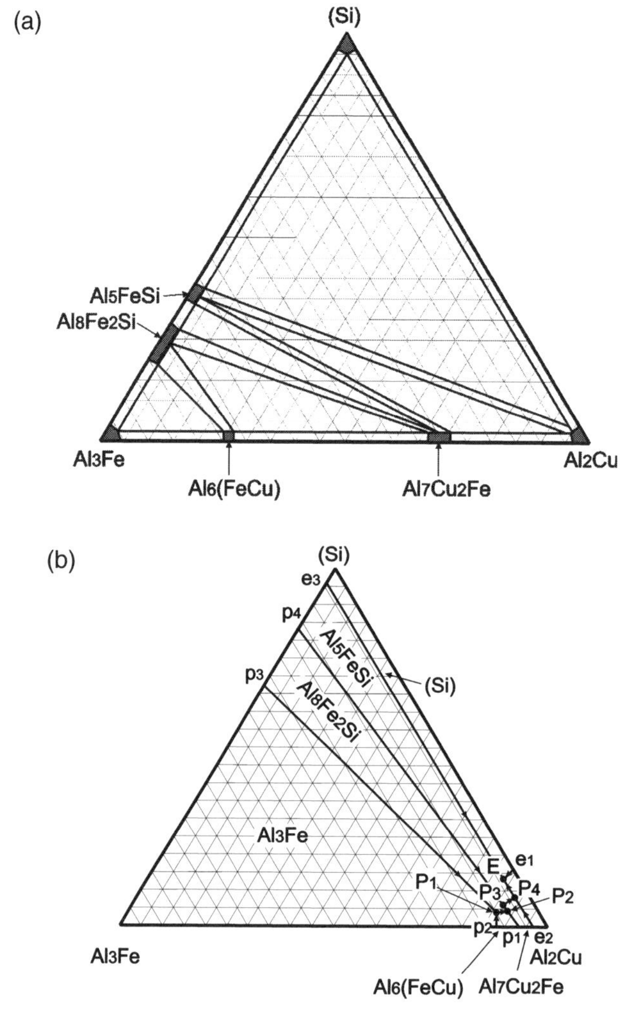

Figure 3.6. Phase diagram of Al–Cu–Fe–Si system (Mondolfo, 1976): (a) distribution of phase fields in the solid state; (b) polythermal projection of liquidus.

These authors give the following composition of the primary crystals formed in the examined alloys:

	Al_3Fe	Al_8Fe_2Si	Al_7Cu_2Fe
Fe, %	29–35	19.5–32.5	15–18
Cu, %	5–11	10–22	31–38
Si, %	0.01–0.9	4.5–10.5	0.06–0.9

These data indicate a significant solubility of copper in the Al_8Fe_2Si phase. The Al_8Fe_2Si phase, in which copper is dissolved, has a hexagonal structure with lattice parameters $a = 1.245$ nm, $c = 2.459$ nm (Zakharov et al., 1989a). The Al_7Cu_2Fe phase in quaternary alloys has a tetragonal crystal structure with parameters $a = 0.643$ nm, $c = 1.483$ nm (Zakharov et al., 1989a).

3.6. Al–Cu–Fe–Mg PHASE DIAGRAM

The phase diagram of this system enables the analysis of the effect of iron impurity on the phase composition of Al–Cu–Mg alloys at low concentrations of silicon, manganese, and nickel. This diagram is also required for the analysis of quinary systems, in particular, Al–Cu–Fe–Mg–Si.

No quaternary phases are formed in the Al–Cu–Fe–Mg system. The phases Al_2Cu, Al_3Fe, Al_8Mg_5, Al_7Cu_2Fe, Al_2CuMg, Al_6CuMg_4, and $Al_6(CuFe)$ from the constitutive systems are in equilibrium with (Al). The liquidus projection and the distribution of phase regions in solid state are shown in Figure 3.7, the invariant reactions in the Al-corner of the systems are given in Table 3.17, and mono- and bivariant reactions, in Table 3.18.

All monovariant lines of the quaternary phase diagram are concentrated near the Al–Cu–Mg face of the concentration tetrahedron, and respective invariant points are close to the invariant points of the Al–Cu–Mg ternary system (Figure 3.2a).

The solubilities of copper and magnesium in (Al) in the quaternary system in the presence of the Al_2Cu phase are, probably, the same as in the phase region (Al) + $Al_2Cu + Al_2CuMg$ of the ternary Al–Cu–Mg diagram. In the absence of this phase, iron should reduce the solubility of copper similar to how it occurs in the Al–Cu–Fe system (see Table 3.9).

3.7. Al–Cu–Fe–Mg–Si PHASE DIAGRAM

This system is sufficiently complex to adequately describe the phase composition and phase transformations in a large number of commercial alloys, e.g. of 3XX.0, 2XX.0, and 2XXX series, in which concentrations of manganese, nickel, and chromium are at a low level.

The procedure that we use for the analysis of five-component phase diagrams is considered in detail in Appendix 3.

The quinary system can be conventionally subdivided into three subsystems, i.e. in the range of Si-rich alloys (3XX.0 series), in the range of Cu-rich alloys (2XX.0 and

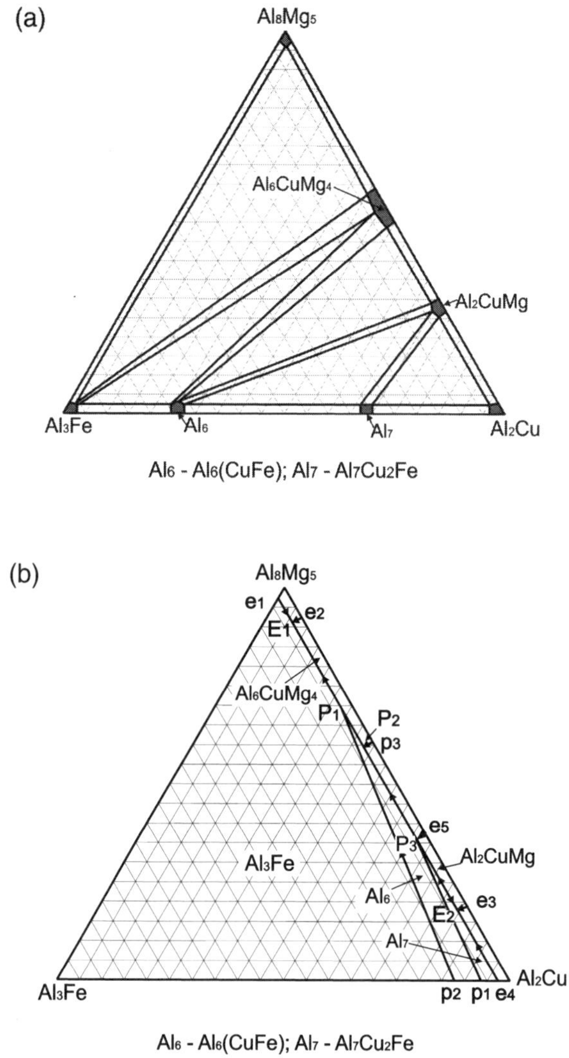

Figure 3.7. Phase diagram of Al–Cu–Fe–Mg system (Mondolfo, 1976): (a) distribution of phase fields in the solid state; (b) polythermal projection of liquidus.

2XXX series), and in the range of Mg-rich alloys (5XX.0 and 5XXX series). In this section we consider these three compositional ranges separately.

Silicon-rich alloys. According to the literature data, the following phases from the ternary and quaternary systems – θ (Al_2Cu), M (Mg_2Si), β (Al_5FeSi), Q ($Al_5Cu_2Mg_8Si_6$), and π ($Al_8FeMg_3Si_6$) – can be expected to occur in Al-rich alloys of the Al–Cu–Fe–Mg–Si system, in addition to (Al) and (Si) that are present in all

Table 3.17. Invariant reactions in quaternary alloys of Al–Cu–Fe–Mg system (Mondolfo, 1976)

Reaction*	Point in Figure 3.7b	*T*, °C	Concentrations in liquid phase		
			Cu, %	Fe, %	Mg, %
L + N ⇒ (Al) + D + S	P_3	<517	25	<1	<5
L ⇒ (Al) + Al_2Cu + N + S	E_2	505	33	0.3	5
L + S ⇒ (Al) + D + T	P_2	465	10	<1	20–25
L + D ⇒ (Al) + Al_3Fe + T	P_1	450	5	<1	25–30
L ⇒ (Al) + Al_3Fe + Al_8Mg_5 + T	E_1	445	2	0.1	33
L ⇒ (Al) + N + S (quasi-ternary?)	E_5	<517	~24	<1	~10

* S – Al_2CuMg; T – Al_6CuMg_4; N – Al_7Cu_2Fe; D – $Al_6(CuFe)$

Table 3.18. Bivariant and monovariant reactions in quaternary alloys of Al–Cu–Fe–Mg system

Bivariant reactions*	Field in Figure 3.7b	*T*, °C	Monovariant reactions*	Line in Figure 3.7b	*T*, °C
L ⇒ (Al) + Al_8Mg_5	Al_8Mg_5–e_2–E_1–e_1	450–445	L ⇒ (Al) + Al_8Mg_5 + T	e_2–E_1	449–445
L ⇒ (Al) + T	e_2–p_3–P_2–P_1–E_1	467–445	L + S ⇒ (Al) + T	p_3–P_2	467–465
L ⇒ (Al) + S	p_3–e_3–E_2–e_5–P_3–P_2	515–465	L ⇒ (Al) + Al_2Cu + S	e_3–E_2	507–500
L ⇒ (Al) + Al_2Cu	Al_2Cu–e_4–E_2–e_3	448–500	L ⇒ (Al) + Al_2Cu + N	e_4–E_2	545–500
L ⇒ (Al) + D	p_2–P_1–P_2–P_3–p_1	620–450	L + D ⇒ (Al) + N	p_1–P_3	590–517
L ⇒ (Al) + N	p_1–P_3–e_5–E_2–e_4	590–500	L + Al_3Fe ⇒ (Al) + D**	p_2–P_1	620–450
L ⇒ (Al) + Al_3Fe	Al_3Fe–e_1–E_1–P_1–p_2	655–445	L ⇒ (Al) + Al_8Mg_5 + Al_3Fe	e_1–E_1	450–445
			L ⇒ (Al) + T + Al_3Fe	P_1–E_1	450–445
			L ⇒ (Al) + T + D	P_2–P_1	465–450
			L ⇒ (Al) + S + D	P_3–P_2	587–465
			L ⇒ (Al) + S + N	e_5–P_3	518–517
				c_5 E_2	517 500

* S – Al_2CuMg; T – Al_6CuMg_4; N – Al_7Cu_2Fe; and D – $Al_6(CuFe)$
** The reaction type possibly changes to the eutectic reaction

phase fields. This assumption has been supported by the results of metallografic studies discussed elsewhere (Belov et al., 1998, 2002a).

Based on our method of a priori analysis (Appendix 3) and on the experimental data, we suggest a version of the Al–Cu–Fe–Mg–Si phase diagram at 10% Si. Figure 3.8a shows the distribution of phase regions in the solid state, from which follows that the β (AlFeSi) and Q phases cannot be in equilibrium with each other. This distribution is valid for any concentration of silicon, at which the (Al) and (Si) phases are present in all phase fields.

The vertices of the triangle in Figure 3.8a correspond to three-phase regions (Al) + (Si) + Y_i, where Y_i is a phase from the respective ternary system: Al_2Cu, Mg_2Si, or Al_5FeSi. All ternary alloys containing the (Al) and (Si) phases are confined

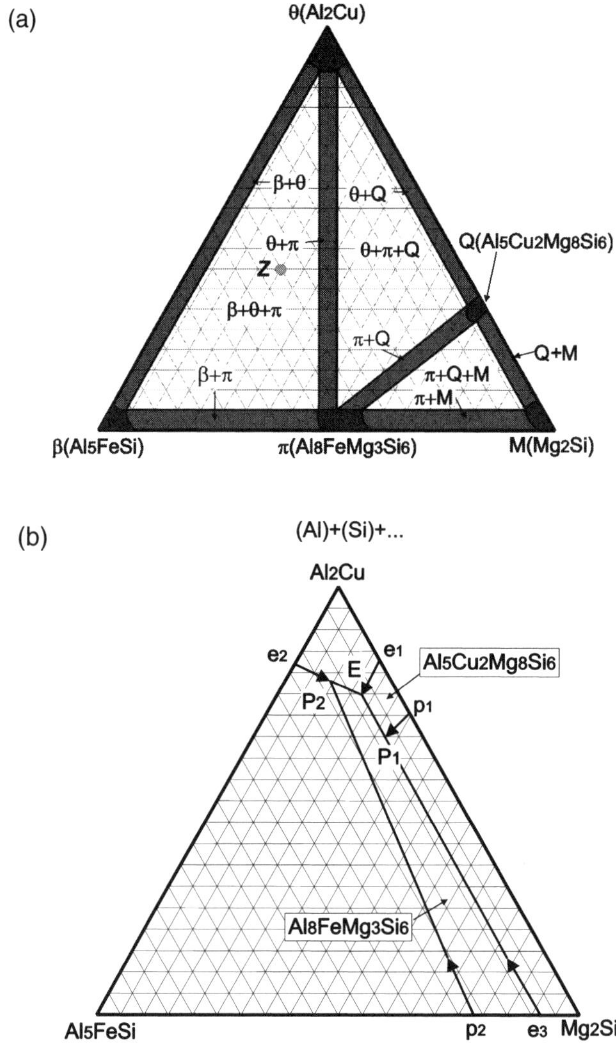

Figure 3.8. Phase diagram of Al–Cu–Fe–Mg–Si system in the range of Al–Si alloys (Belov et al., 2002a): (a) distribution of phase fields in the solid state; (b) polythermal projection of solidification surfaces.

to these vertices. The sides characterize the phase compositions of quaternary alloys, and five-component alloys fall inside the triangle. As all given phases have comparatively narrow homogeneity ranges, for simplification the respective regions are shown as points.

All bi-, mono-, and invariant reactions, and also their respective lines and points in Figure 3.8b are given in Tables 3.19–3.21. The proposed constitution of the Al–Cu–Fe–Mg–Si diagram suggests that under real casting conditions 3XX.0-series

Table 3.19. Invariant reactions with participation of (Al) and (Si) in Al–Cu–Fe–Mg–Si system (Belov et al., 2002a)

Reaction*	Point in Figure 3.8b	Concentrations in liquid phase				*T*, °C
		Si, %	Fe, %	Cu, %	Mg, %	
$L \Rightarrow (Al) + (Si) + Al_2Cu + Q + \pi$	E	5–6	0.1–0.2	26–28	2–3	503–507
$L + Mg_2Si \Rightarrow (Al) + (Si) + Q + \pi$	P_1	7–10	0.1–0.2	14–17	3–6	516–520
$L + Al_5FeSi \Rightarrow (Al) + (Si) + Al_2Cu + \pi$	P_2	5–6	0.2–0.4	26–28	1–2	550–554

* Q – $Al_5Cu_2Mg_8Si_6$ and π – $Al_8FeMg_3Si_6$

Table 3.20. Monovariant reactions with participation of (Al) and (Si) in Al–Cu–Fe–Mg–Si system

Reaction	Line in Figure 3.8b
$L + Al_5FeSi \Rightarrow (Al) + (Si) + Al_8FeMg_3Si_6$	p_2–P_2
$L \Rightarrow (Al) + (Si) + Mg_2Si + Al_8FeMg_3Si_6$	e_3–P_1
$L + Mg_2Si \Rightarrow (Al) + (Si) + Al_5Cu_2Mg_8Si_6$	p_1–P_1
$L \Rightarrow (Al) + (Si) + Al_2Cu + Al_5FeSi$	e_2–P_2
$L \Rightarrow (Al) + (Si) + Al_5Cu_2Mg_8Si_6 + Al_8FeMg_3Si_6$	P_1–E
$L \Rightarrow (Al) + (Si) + Al_2Cu + Al_5Cu_2Mg_8Si_6$	e_1–E
$L \Rightarrow (Al) + (Si) + Al_2Cu + Al_8FeMg_3Si_6$	P_2–E

Table 3.21. Bivariant reactions with participation of (Al) and (Si) in Al–Cu–Fe–Mg–Si system

Reaction	Region in Figure 3.8b
$L \Rightarrow (Al) + (Si) + Al_5FeSi$	β–e_2–P_2–p_2–β
$L \Rightarrow (Al) + (Si) + Al_8FeMg_3Si_6$	p_2–P_2–E–P_1–e_3–p_2
$L \Rightarrow (Al) + (Si) + Mg_2Si$	M–e_3–P_1–p_1–M
$L \Rightarrow (Al) + (Si) + Al_5Cu_2Mg_8Si_6$	P_1–P_1–E–e_1–p_1
$L \Rightarrow (Al) + (Si) + Al_2Cu$	θ–e_1–E–P_2–e_2–θ

alloys complete solidification by the eutectic reaction involving the π ($Al_8FeMg_3Si_6$) phase.

It should be noted that the proposed version of invariant reactions differs from that given by Phragmen (1950). The major difference is that, according to Phragmen, the Al_7FeCu_2 phase but not Al_5FeSi phase is in equilibrium with (Al) and (Si), which leads to the invariant reaction $L + Al_5FeSi \Rightarrow (Al) + (Si) + Al_7FeCu_2 + Al_5Cu_2Mg_8Si_6$. However, this is in contradiction to the available experimental data (Belov et al., 2002a) and the most reliable version of the Al–Cu–Fe–Si quaternary phase diagram (see Section 3.5).

Copper-rich alloys. According to the Al–Cu–Fe–Si, Al–Cu–Fe–Mg, and Al–Cu–Mg–Si quaternary phase diagrams, the following phases can be in equilibrium with (Al) and Al_2Cu: Al_7Cu_2Fe, Al_2CuMg, Al_5FeSi, Mg_2Si, $Al_5Cu_2Mg_8Si_6$, and (Si). The $Al_8FeMg_3Si_6$ phase is in equilibrium with the last four phases, and that makes the quinary phase diagram rather complex. A version of the quinary phase diagram and invariant reactions involving the phase Al_2Cu are shown in Figure 3.9 and Table 3.22.

The vertices of the triangle (Figure 3.9a) correspond to three-phase regions $(Al)+\theta+Y_i$, where Y_i is a phase from the respective ternary system: S (Al_2CuMg), (Si), or N (Al_7Cu_2Fe). All ternary alloys containing both (Al) and θ fall onto these vertices. The sides of the triangle characterize the phase composition of quaternary alloys, and five-component alloys are located inside the triangle. No intermediate phases exist at the S–N side, which follows from the analysis of the Al–Cu–Fe–Mg phase diagram, whereas the phases M (Mg_2Si) and Q ($Al_5Cu_2Mg_8Si_6$) and β (Al_5FeSi) appear at the sides S–(Si) and (Si)–N, respectively (Figure 3.9a). As in the case of the vertices of the triangle, the points at the sides correspond to three-phase regions. In addition, a point reflecting the three-phase region $(Al)+\theta+\pi$ appears inside the triangle. This is due to the fact that quaternary compound π can be in equilibrium with the phases (Al) and θ as it follows from the Al–Cu–Fe–Mg–Si diagram plotted for the region Al–Si of the alloys (Figure 3.8a).

Positions of the phases in the concentration space suggest several variants of "triangulation", i.e. determination of five-phase regions $(Al)+\theta+Y_1+Y_2+Y_3$, where Y_i are the phases on the concentration triangle. Figure 3.9a shows a case of equilibrium between phases M and N, at which the triangle is split into two large parts: a simple one with only one five-phase region $(Al)+\theta+S+N+M$, and a complex one, which requires "triangulation" itself. It follows from the Al–Cu–Fe–Mg–Si phase diagram plotted for Al–Si alloys (Figure 3.8) that the β (Al_5FeSi) phase can be in equilibrium with M (Mg_2Si), but not with Q ($Al_5Cu_2Mg_8Si_6$). This fact enables us to suggest the complete "triangulation" of the Al–Cu–Fe–Mg–Si phase diagram in solid state as shown in Figure 3.9a.

A solidification path includes multiphase eutectic and peritectic reactions. The solidification starts with the aluminum solid solution (Al), and then the eutectics $(Al)+Al_2Cu$ is formed. In alloys containing more than 0.1–0.25% Fe, the second stage of solidification can be the formation of either eutectics (Al) + N (Al_7FeCu) or (Al) + β (Al_5FeSi), but this has no consequence for the analysis of the subsequent reactions which are shown by points, lines, and regions on the concentration triangle (Figure 3.9b).

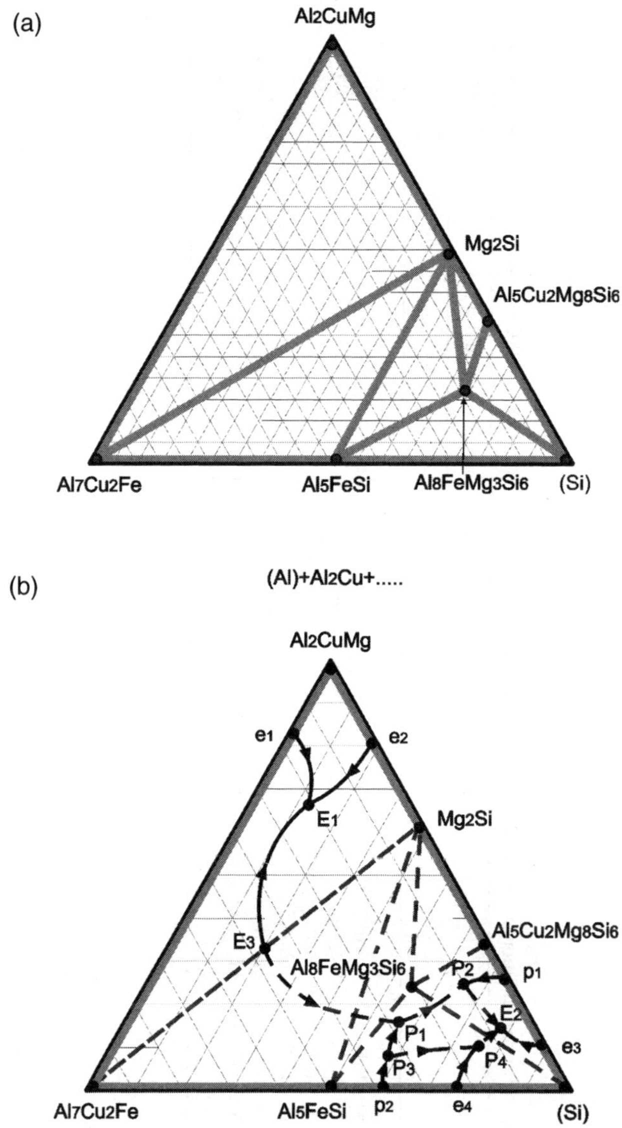

Figure 3.9. Phase diagram of Al–Cu–Fe–Mg–Si system in the range of Al–Cu alloys (Belov et al., 2002b): (a) distribution of phase fields in the solid state; (b) polythermal projection of solidification surfaces.

The occurrence of seven invariant reactions is assumed in the aluminum corner of the system: three eutectic and four peritectic reactions, as given in Table 3.22. Identification of the first eutectic reaction (point E_1 in Figure 3.9b) appears to be the most evident one: $L \Rightarrow (Al) + Al_2Cu + Mg_2Si + Al_7Cu_2Fe + Al_2CuMg$.

Table 3.22. Invariant reactions with participation of (Al) and Al_2Cu in Al–Cu–Fe–Mg–Si system (Belov et al., 2002a, b; Belov and Koltsov, 2002)

Reaction	Point in Figure 3.9b	Concentrations in liquid phase				T, °C
		Cu, %	Fe, %	Mg, %	Si, %	
$L \Rightarrow (Al) + Al_2Cu + Mg_2Si + Al_7Cu_2Fe + Al_2CuMg$	E_1	~33	<0.3	~5	<0.3	~500
$L \Rightarrow (Al) + Al_2Cu + (Si) + Al_8FeMg_3Si_6 + Al_5Cu_2Mg_8Si_6$	E_2	~28	<0.2	~2	~6	~505
$L \Rightarrow (Al) + Al_2Cu + Mg_2Si + Al_7Cu_2Fe$ (quasi-quaternary)	E_3	~23	<0.3	~10	~2	~515
$L + Al_7Cu_2Fe \Rightarrow (Al) + Al_2Cu + Mg_2Si + Al_8FeMg_3Si_6$	P_1	–	–	–	–	–
$L + Al_7Cu_2Fe \Rightarrow (Al) + Al_2Cu + Al_5FeSi + Al_8FeMg_3Si_6$	P_3	~27	<0.3	<1	~4	~530
$L + Mg_2Si \Rightarrow (Al) + Al_2Cu + Al_8FeMg_3Si_6 + Al_5Cu_2Mg_8Si_6$	P_2	~30	<0.3	~3	~3	~510
$L + Al_5FeSi \Rightarrow (Al) + Al_2Cu + (Si) + Al_8FeMg_3Si_6$	P_4	~27	<0.3	<2	<5	~520

From the temperature and composition of the liquid phase, this reaction shall be close to the corresponding quaternary eutectic reaction in the Al–Cu–Mg–Si system (point e_2 in Figure 3.9b corresponds to E_2 in Figure 3.4b and in Table 3.11). The positions of E_2 and P_4 are suggested by Figure 3.8b. Point E_3 is plotted based on the assumption of a "quasi-quaternary" section (Al)–Al_2Cu–Mg_2Si–Al_7Cu_2Fe. The positions of other points (P_1, P_2, and P_3) follow from the assumed variant of "triangulation".

It should be noted that the suggested version does not correspond to the variant by Phragmen (1950) who assumes the eutectic reaction L$\Rightarrow$(Al) + (Si) + Al_2Cu + Al_7Cu_2Fe + Al_2CuMg. However, as already mentioned, Phragmen used a doubtful version of the Al–Cu–Fe–Si quaternary phase diagram.

Magnesium-rich alloys. According to the constitutive phase diagrams, only three phases – Al_3Fe, Mg_2Si, and Al_6CuMg_4 – can be in equilibrium with (Al) and Al_8Mg_5. This corresponds to the invariant eutectic transformation L$\Rightarrow$(Al) + Al_8Mg_5 + Al_3Fe + Mg_2Si + Al_6CuMg_4 and a very simple distribution of phase regions in the solid state with all the phases present (Belov et al., 2002a).

3.8. COMMERCIAL Al–Si–Cu CASTING ALLOYS (3XX.0 AND 2XX.0 SERIES)

Isothermal sections of the Al–Cu–Si phase diagram at 520°C and 200°C (Figure 3.10a) show that in alloys like 319.2 and 305.0 (Table 3.1) with a low concentration of iron impurity, copper totally dissolves in (Al) during solution treatment, and precipitates as metastable modifications of Al_2Cu during aging. From the polythermal section at 6% Si in Figure 3.10b, it follows that after primary solidification of (Al) the binary (Al) + (Si) eutectics is formed, and the Al_2Cu phase can appear as a result of the ternary eutectic reaction at 525°C (Table 3.2), which is nonequilibrium in 319.2 and 305.0 alloys. Figure 3.10a also shows that alloys like 295.9 and 296.0 (Table 3.1) with low concentrations of iron and magnesium fall into the phase region (Al) + (Si) + Al_2Cu at any temperature. In these alloys, the solidification (after formation of primary (Al) grains) proceeds through one of the two possible binary eutectics (Table 3.2, Figure 3.10c), and ceases at 525°C by the invariant eutectic reaction L$\Rightarrow$(Al) + (Si) + Al_2Cu.

The phase composition of 3XX.0-series alloys containing simultaneously magnesium and copper additions at a low concentration of iron impurity is reflected in isothermal sections of the Al–Si–Cu–Mg phase diagram, as shown in Figure 3.11. The phase boundaries in Figure 3.11a are also valid at other concentrations of silicon, because in 3XX.0-series alloys this element is always present in excess. At subsolidus temperatures these alloys, as a rule, fall into the phase region (Al) + (Si), i.e. all copper and magnesium are dissolved in (Al). At 200°C, various combinations

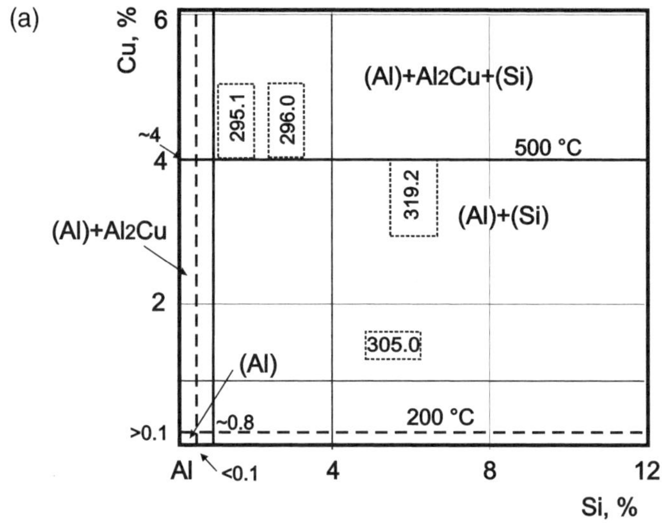

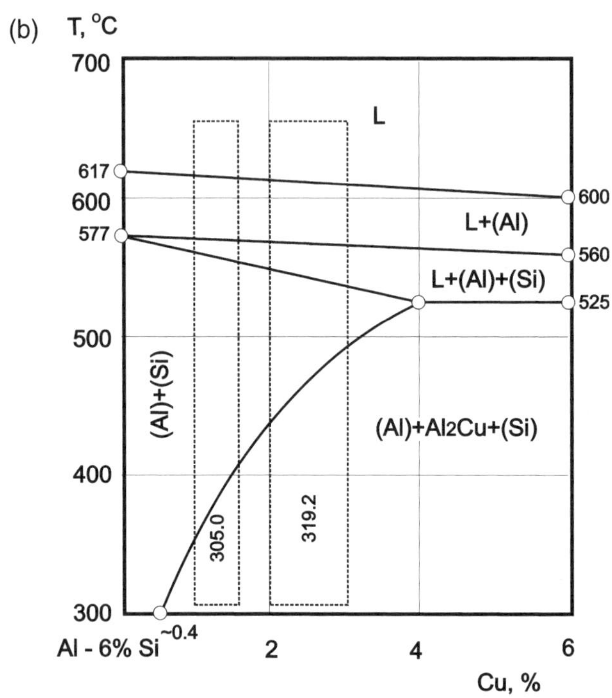

Figure 3.10. Isothermal (a) and polythermal (b, c) sections of Al–Cu–Si phase diagram: (a) 500°C (solid lines) and 200°C (dashed lines); (b) 6% Si; (c) 10% Cu.

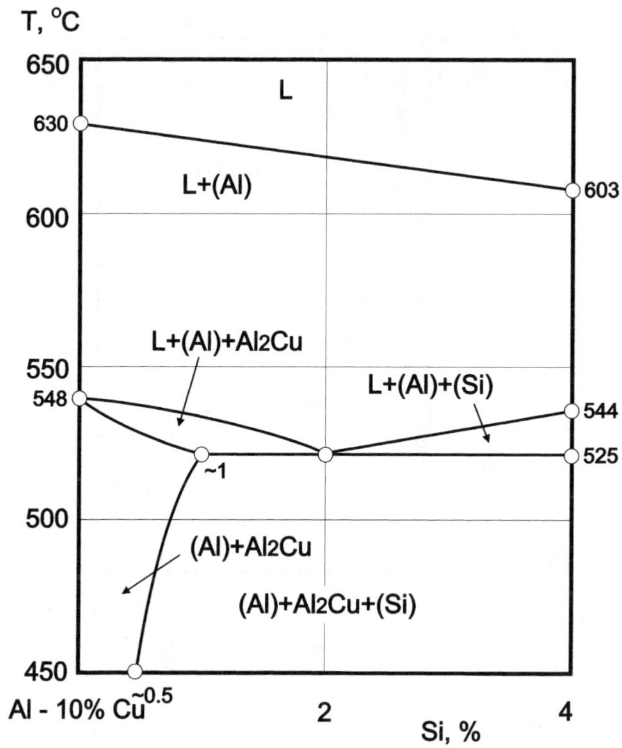

Figure 3.10 (*continued*)

of metastable modifications of the Mg_2Si, Al_2Cu, and Q phases can form depending on the ratio between Cu and Mg and the aging regime. These combinations determine the strengthening after aging.

It should also be noted that the phase composition of the products of aging can be (with caution) analyzed using the isothermal sections only if one knew the chemical composition of the supersaturated solid solution. This composition generally is very different from the composition of the entire alloy, due to the formation of excess phases and dispersoids and because of microsegregation. For instance, for some wrought alloys of 6XXX (6066 type) and 2XXX (2014 type) series the most suitable sections are those at 0.5% Si (Figure 3.11b) and 4% Cu (Figure 3.11c), respectively. However, the real selection of metastable phases that precipitate during decomposition of a supersaturated solid solution and participate in hardening of an alloy cannot be easily derived from the equilibrium phase diagram, as is illustrated in Section 3.9 of this chapter.

The polythermal sections at 10% Si (Figure 3.12) show that in 3XX.0-series alloys the Mg_2Si, Al_2Cu, and Q phases are formed at lower temperatures than (Si), i.e.

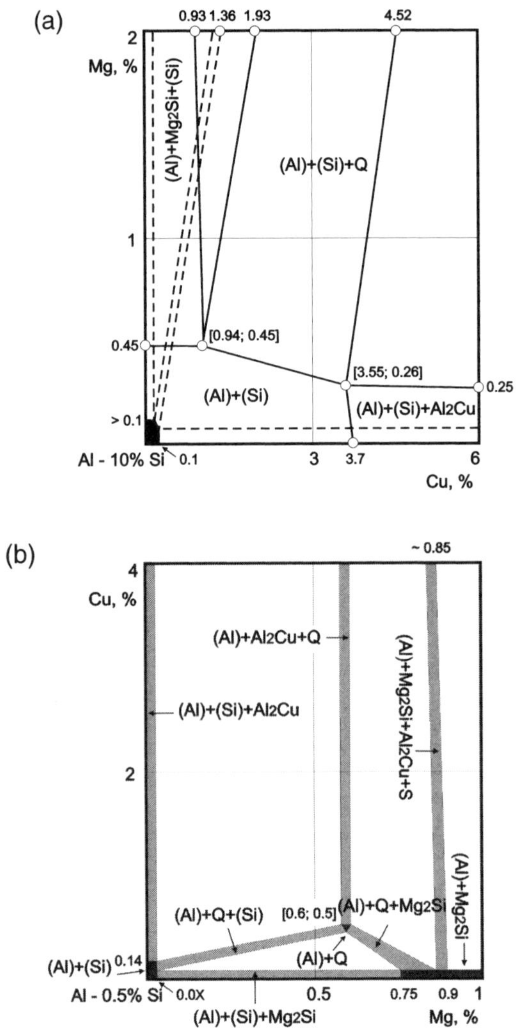

Figure 3.11. Isothermal sections of Al–Cu–Mg–Si phase diagram: (a) 500°C (solid lines) and 200°C (dashed lines), 10% Si; (b) 200°C, 0.5% Si; (c) 200°C, 4% Cu; (d) 500°C, 1% Si.

always after the (Al) + (Si) eutectics. In as-cast structure, the phases containing copper and magnesium appear either as components of complex eutectics involving (Al) and (Si), or as separate inclusions (i.e. as divorced eutectics).

Figure 3.13a shows polythermal sections of the Al–Cu–Fe–Si phase diagram at 6% Si and 1% Fe. Obviously, an iron impurity in 3XX.0-series alloys with low magnesium content can lead to the formation of only one phase – Al_5FeSi.

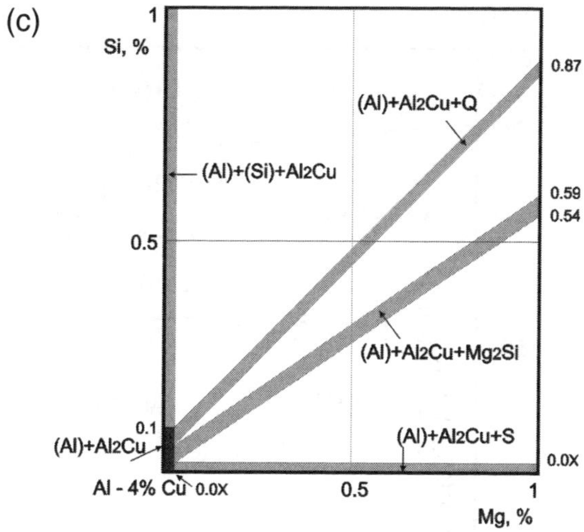

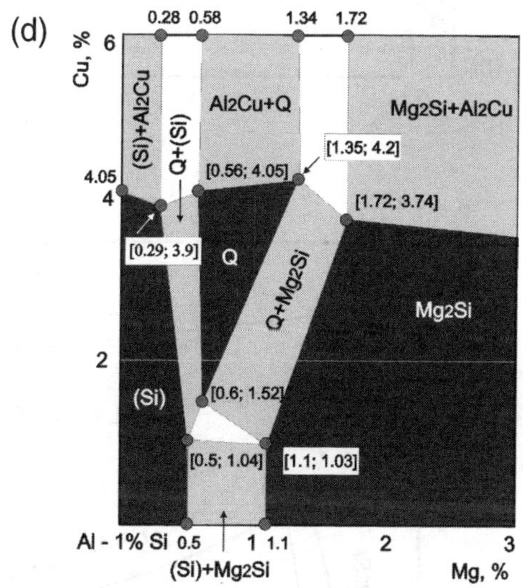

Figure 3.11 (*continued*)

In 2XX.0-series alloys, two Fe-containing phases can form (Al_7Cu_2Fe and Al_5FeSi) as shown in the isopleth at 10% Cu and 0.4% Fe (Figure 3.13b). The probability of Al_5FeSi formation increases with an increasing concentration of Si. Other Fe-containing phases, e.g. $Al_6(FeCu)$ and Al_3Fe, can form upon nonequilibrium solidification in alloys with high content of iron (>0.5%) due to incomplete peritectic reactions (Tables 3.15 and 3.16).

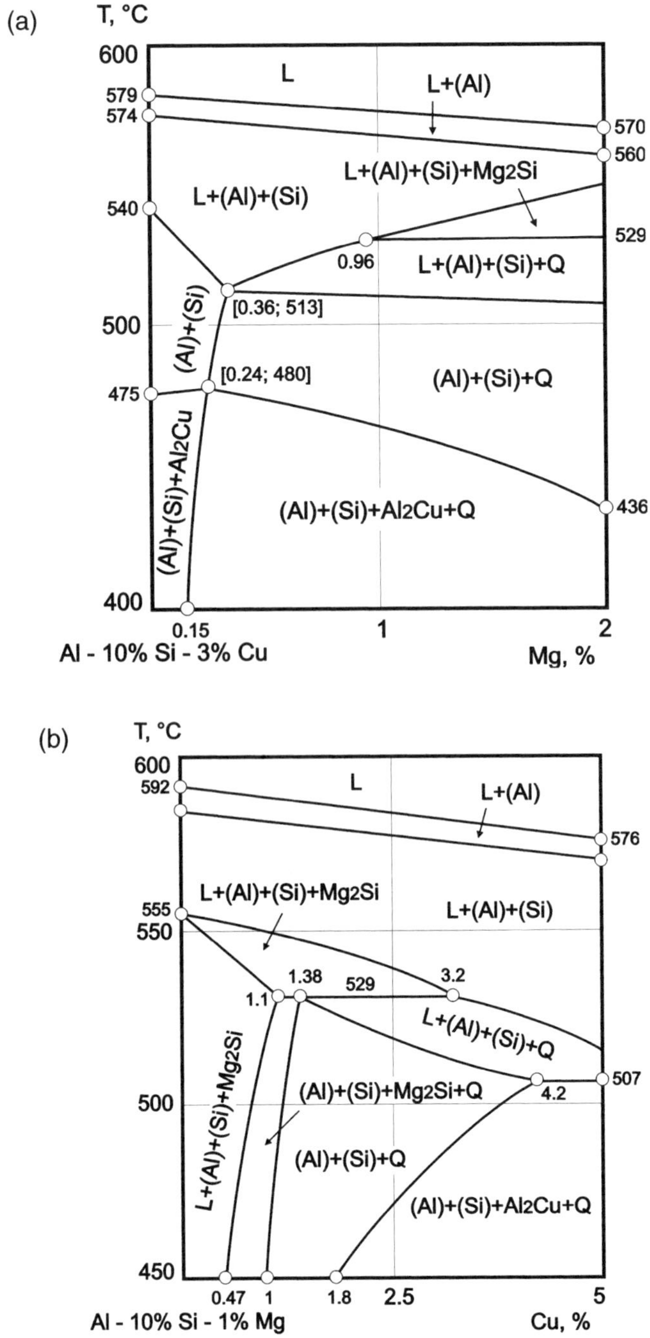

Figure 3.12. Polythermal sections of Al–Cu–Mg–Si phase diagram at 10% Si: (a) 3% Cu; (b) 1% Mg.

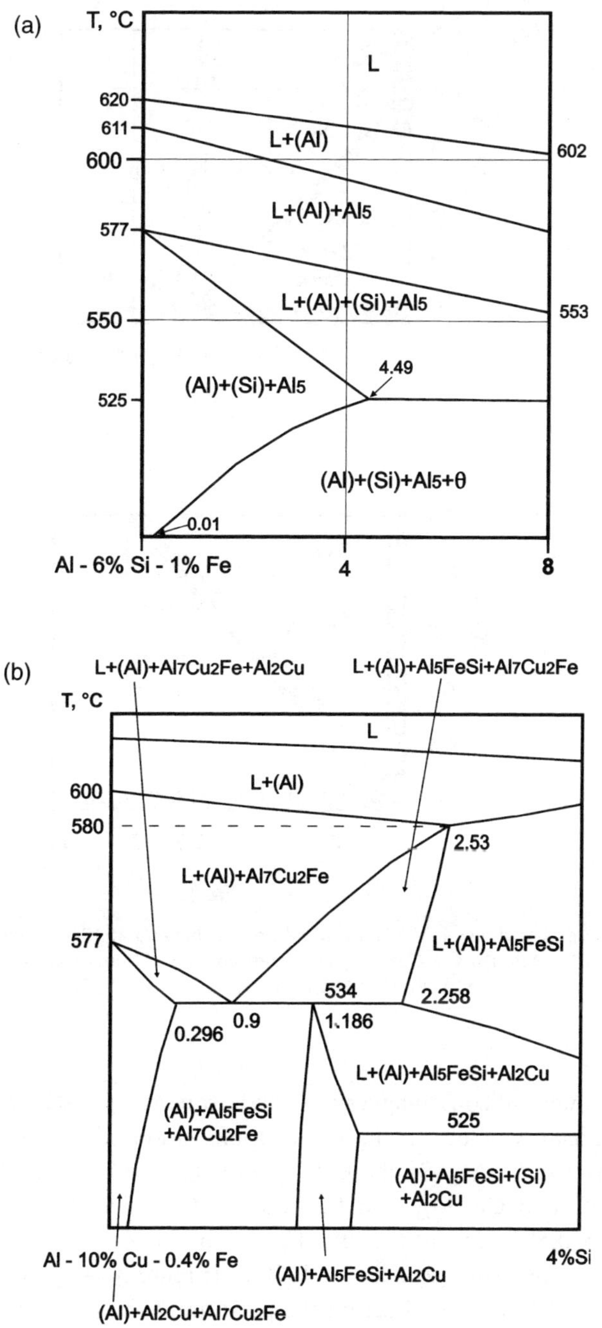

Figure 3.13. Polythermal sections of Al–Cu–Fe–Si phase diagram: (a) 6% Si, 1% Fe; (b) 10% Cu, 0.4% Fe. Al_5 – Al_5FeSi.

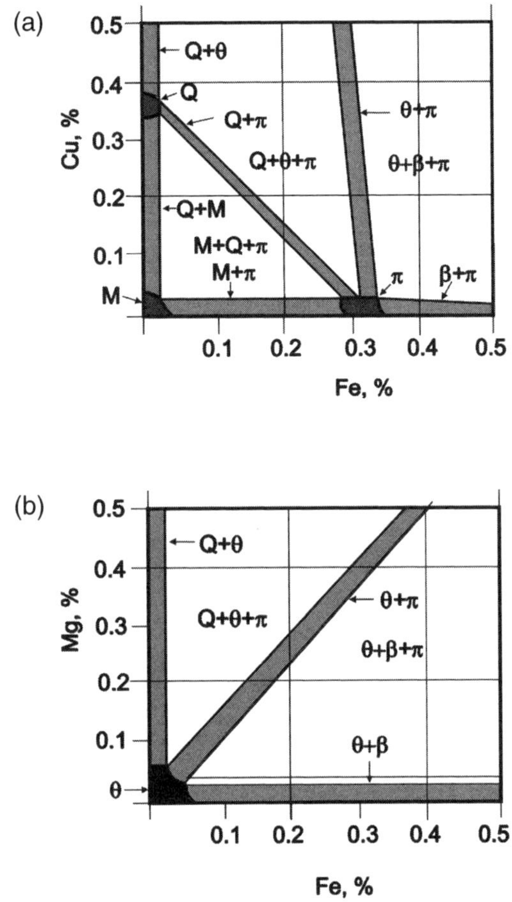

Figure 3.14. Isothermal sections of Al–Cu–Fe–Mg–Si phase diagram at 10% Si and 20°C (Belov et al., 1998, 2002a): (a) 0.3% Mg, (b) 4% Cu. All phase fields contain (Al) and (Si). M – Mg_2Si, θ – Al_2Cu, β – Al_5FeSi, Q – $Al_5Cu_2Mg_8Si_6$, π – $Al_8FeMg_3Si_6$.

Alloys in which silicon, magnesium, and iron are present simultaneously in amounts significantly affecting the phase composition, should be analyzed using the quinary Al–Cu–Fe–Mg–Si phase diagram (Figure 3.8). In 3XX.0-series alloys, the coupled effect of Cu, Mg, and Fe as applied to 333.0-type alloys (8–10% Si, 3–4% Cu, 0.05–0.5% Mg, up to 1.0% Fe, up to 1% Zn, up to 0.5% Mn) can be followed by the isothermal sections at 10% Si (Figure 3.14). The concentration of silicon (above 2–3%) has no effect on the position of the phase boundaries.

A polythermal section of the Al–Cu–Fe–Mg–Si phase diagram at 10% Si, 5% Cu, and 0.5% Mg in Figure 3.15 shows that, at comparatively low concentrations of iron

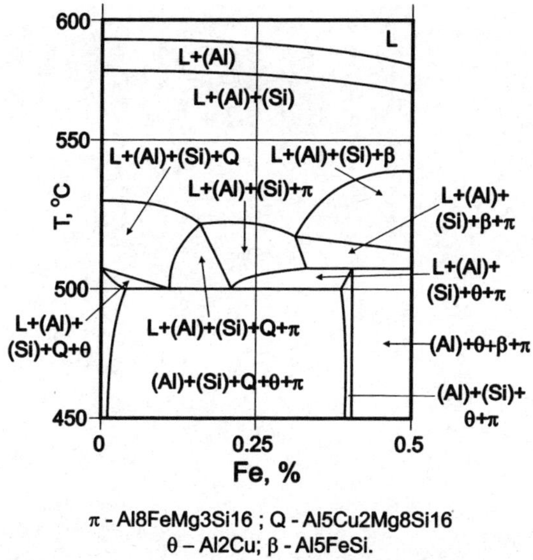

Figure 3.15. Polythermal sections of Al–Cu–Fe–Mg–Si phase diagram at 10% Si, 5% Cu, and 0.5% Mg (Belov et al., 1998, 2002b).

(<0.5%), the (Al) + (Si) eutectic forms in 3XX.0-type alloys after primary solidification of (Al), and only after that the Al_5FeSi phase is formed. The presence of magnesium can lead to the formation of the $Al_8FeMg_3Si_6$ phase in the final stages of solidification. The as-cast structure of 2XX.0- and 3XX.0-type alloys (see compositions in Tables 3.10 and 3.14) can contain more than five phases as a result of nonequilibrium solidification. These phases form complex conglomerates of crystals as shown in Figure 3.16, which are formed during multi-phase solidification reactions listed in Tables 3.19 and 3.20.

For example, Table 3.23 gives solidification reactions experimentally observed in a C355.2 alloy solidified under nonequilibrium conditions (Bäckerud et al., 1990). The as-cast structure of this alloy contains seven excess phases. First five reactions in Table 3.23 agree well with the Al–Cu–Fe–Mg–Si phase diagram (Figure 3.8b), and the last two are different as they do not include the $Al_8FeMg_3Si_6$ phase. However, the as-cast structure is so complex that the reliable experimental identification of all solidification reactions is doubtful.

In hypereutectic Al–Si alloys containing more than 11% Si (390.0 type), (Si) solidifies as a primary phase, and then (Al) is formed through eutectic reactions. The rest of solidification sequence is very much the same as in hypoeutectic Al–Si alloys, as shown in Table 3.24. Note that the quaternary π(AlFeMgSi) phase that has to form in these alloys according to the equilibrium phase diagram (Figure 3.8b,

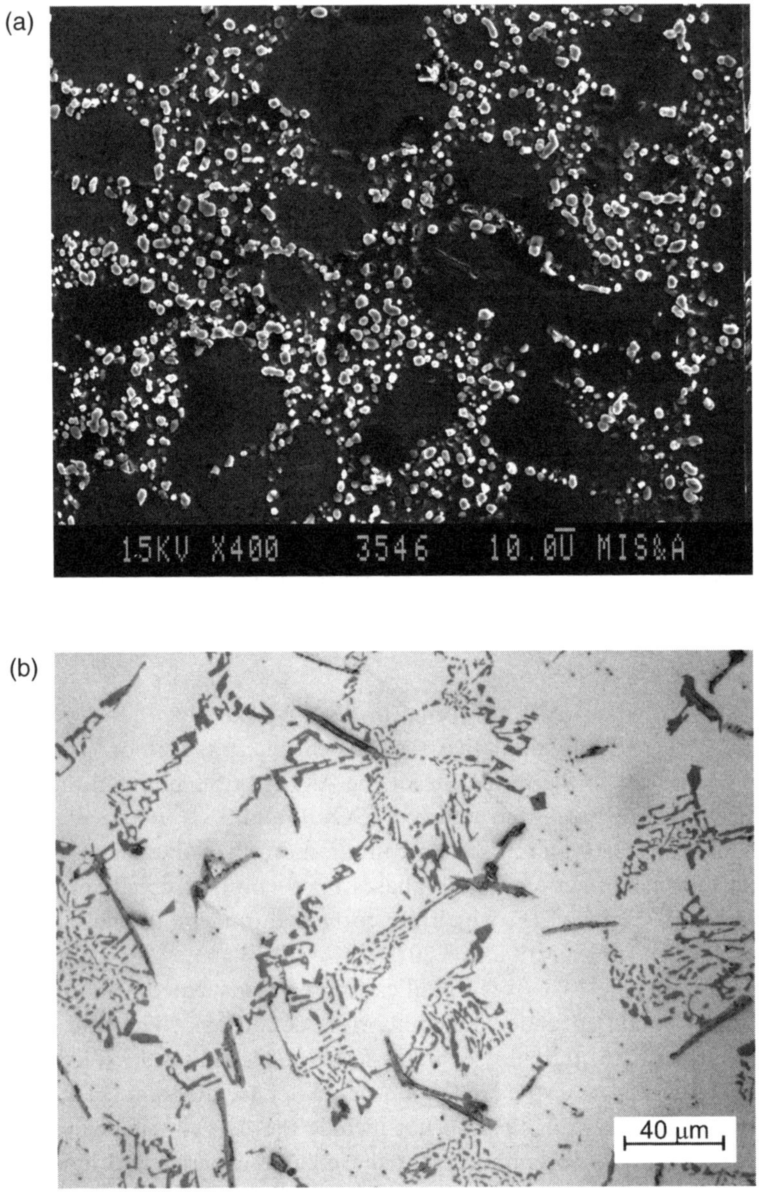

Figure 3.16. Microstructures of 3XX.0 and 2XX.0-type alloys (Al–Si–Cu–Mg–Fe): (a) 354.0 alloy (<0.1% Fe), T6, globular particles of (Si) phase, Cu, and Mg in (Al), SEM; (b) AK9M2 alloy (9% Si, 2% Cu, 1% Fe, 0.3% Mg), as-cast, optical microscope; (c) alloy containing 10% Si, 6% Cu, 0.5% Mg, and 0.15% Fe, $V_c \sim 10^{-2}$ K/s, SEM, Al_2Cu (large white particles), Al_5FeSi (needle), Q (gray); and (d) alloys containing 10% Cu, 2% Si, 0.5% Mg, and 0.5% Fe, $V_c \sim 10^{-2}$ K/s, SEM, Al_2Cu (large white large particles), Al_7Cu_2Fe (needle), Mg_2Si (small black particles) and S (compact gray particles) in a fine eutectic colony.

(c)

(d)

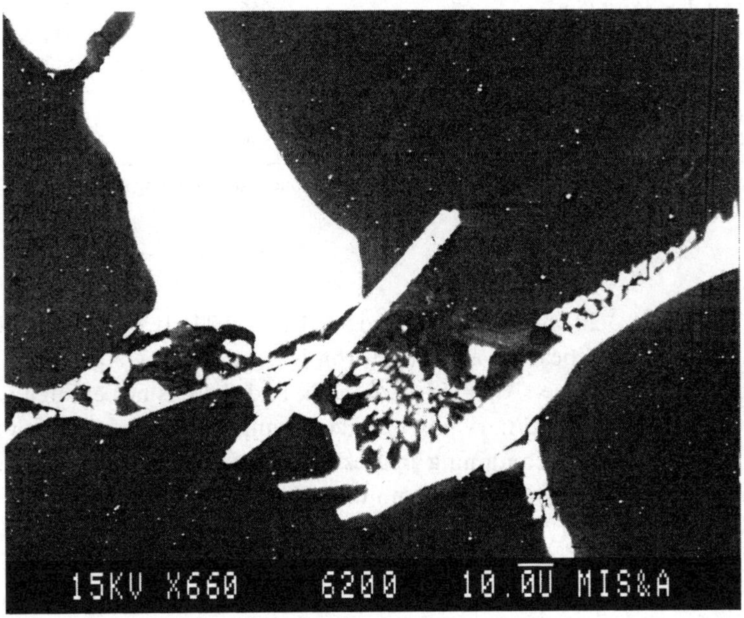

Figure 3.16 (*continued*)

Table 3.23. Solidification reactions under nonequilibrium conditions in a C355.2 alloy (5.05% Si, 1.04% Cu, 0.53% Mg, and 0.1%Fe) (Bäckerud et al., 1990)

Reaction	Temperatures (°C) at a cooling rate	
	0.3 K/s	5 K/s
$L \Rightarrow (Al)$	621–557	622–543
$L \Rightarrow (Al) + (Si)$	557–551	543–537
$L \Rightarrow (Al) + Al_5FeSi + (Si)$	551–535	537–532
$L + Al_5FeSi \Rightarrow (Al) + (Si) + Al_8FeMg_3Si_6$		
$L \Rightarrow (Al) + (Si) + Al_8FeMg_3Si_6 + Mg_2Si$	535–501	532–516
$L \Rightarrow (Al) + (Si) + Al_5FeSi + Al_2Cu$	501–489	516–477
$L \Rightarrow (Al) + (Si) + Al_5Cu_2Mg_8Si_6 + Al_2Cu$		
Solidus	489	477

Table 3.24. Solidification reactions under nonequilibrium conditions in a B390.1 alloy (17.45% Si, 4.81% Cu, 0.56% Mg, 0.74% Fe, and 0.27% Mn) (Bäckerud et al., 1990)

Reaction	Temperatures (°C) at a cooling rate	
	0.4 K/s	5 K/s
$L \Rightarrow (Si)$	634–617	622–562
$L \Rightarrow (Al)$, $L \Rightarrow (Al) + (Si)$	561	557
$L \Rightarrow (Al) + (Si) + Al_5FeSi$; $L \Rightarrow (Al) + (Si) + (AlMnFeSi)$	560–558	554–547
$L \Rightarrow (Al) + (Si) + Mg_2Si$	540–503	517–503
$L + Mg_2Si \Rightarrow (Al) + (Si) + Al_5Cu_2Mg_8Si_6 + Al_2Cu$*	503–493	503–483
$L \Rightarrow (Al) + (Si) + Al_5Cu_2Mg_8Si_6 + Al_2Cu$		
Solidus	493	483

* In our opinion this reaction should include only one Cu-containing phase, either Al_2Cu or $Al_5Cu_2Mg_8Si_6$

Tables 3.19 and 3.20) is not mentioned in Table 3.24. It is possible that during nonequilibrium solidification almost all iron is bound to the Al_5FeSi phase during the eutectic reaction $L \Rightarrow (Al) + (Si) + Al_5FeSi$, and there is not enough iron left for the formation of the quaternary phase. That assumption agrees well with the data in Table 3.24, where iron-containing phases do not form during lower temperature reactions. On the other hand, the amount of the π phase could be very small (by taking into account very low concentrations of iron in all multi-phase reactions, see Table 3.19), which makes the identification of this phase difficult.

In 2XX.0-series alloys, the coupled effect of Si, Mg, and Fe (neglecting Mn), can be traced by the phase distribution in solid state (Figure 3.9a). At a low concentration of silicon (and relatively high concentration of copper), the alloys fall into the

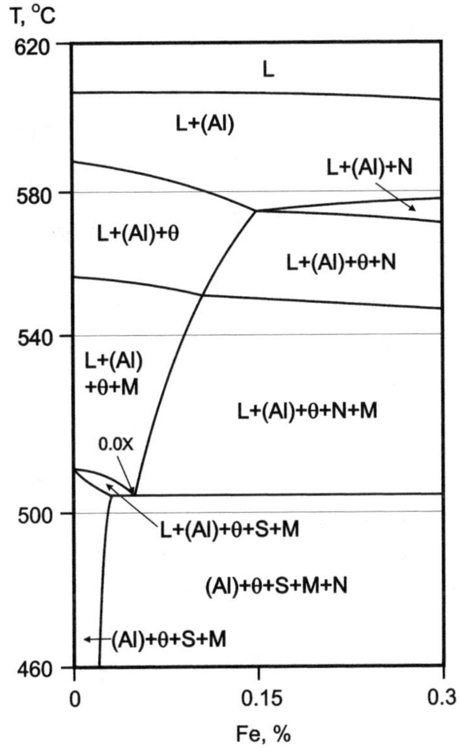

Figure 3.17. Polythermal section of Al–Cu–Fe–Mg–Si phase diagram at 10% Cu, 1.5% Mg, and 0.5% Si. M – Mg_2Si, θ – Al_2Cu, N – Al_7Cu_2Fe, S – Al_2CuMg.

five-phase region (Al) + Al_2Cu + Mg_2Si + Al_7Cu_2Fe + Al_2CuMg. The effect of iron impurity can be followed in the polythermal section in Figure 3.17 (Belov et al., 2002b). Even at small quantities of iron, the Al_7Cu_2Fe phase is formed earlier than Al_2Cu, through the reaction L $\Rightarrow$ (Al) + Al_7Cu_2Fe. Only after that the multi-phase eutectic reactions with participation of (Al) and Al_2Cu occur as shown in Figure 3.9b. At higher silicon concentrations, the phase composition can be different (see Figure 3.9) with all given phases remaining in the solid state. Different solidification paths are shown in Figure 3.18 as a flow chart (Belov and Koltsov, 2002). The best way to determine the phase composition of a specific alloy is to use calculations as shown in Appendix 4. As an example, Figure 3.19 gives the calculated dependence of the volume fractions of different phases on the concentration of silicon in a 222.0-type alloy (Table 3.10) (10% Cu, 0.3% Mg, 1% Fe) in the solid state at low temperatures. These dependences demonstrate that the variation of silicon concentration even within the allowed range can significantly change the phase

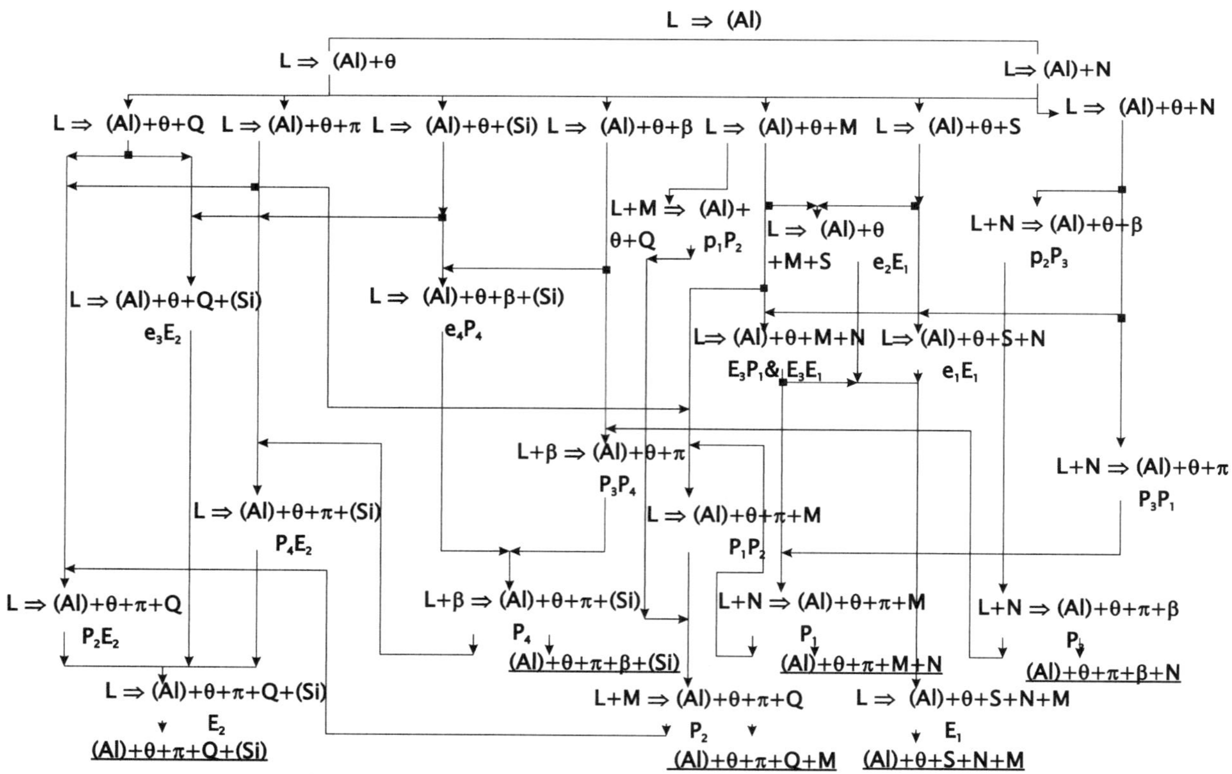

Figure 3.18. "Flow chart" for Al–Cu–Fe–Mg–Si system in the range of Al–Cu alloys. M – Mg_2Si, θ – Al_2Cu, β – Al_5FeSi, N – Al_7Cu_2Fe, S – Al_2CuMg, Q – $Al_5Cu_2Mg_8Si_6$, π – $Al_8FeMg_3Si_6$. Points correspond to invariant and monovariant reactions according to Table 3.22 and Figure 3.9b.

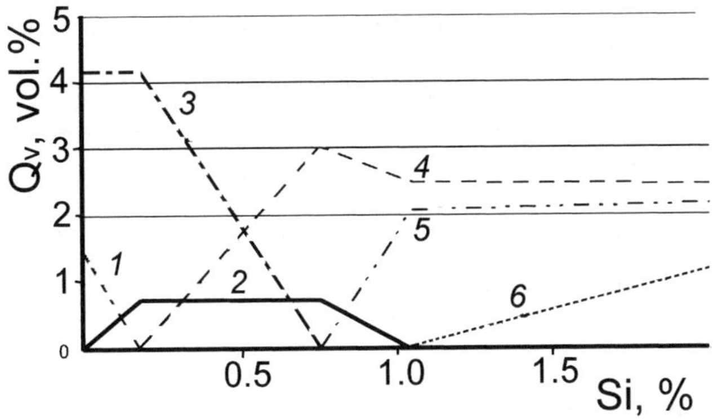

Figure 3.19. Calculated dependence of volume fractions of phases on the Si concentration in a 222.1 alloy at 10% Cu, 0.25% Mg, and 1% Fe (at temperatures below 100°C). 1 – Al_2CuMg, 2 – Mg_2Si, 3 – Al_7Cu_2Fe, 4 – Al_5FeSi, 5 – $Al_8FeMg_3Si_6$, and 6 – (Si). Q_V of the Al_2Cu phase is in the range 9–12 vol.% (not shown).

composition. Note that the Al_2CuMg and Al_7Cu_2Fe phases, that are present in a 222.1 alloy without silicon, disappear at 0.2% Si and 0.8% Si, respectively.

3.9. Al–Mg–Si–Cu WROUGHT ALLOYS OF 6XXX AND 2XXX SERIES

Several widely used commercial alloys of the 6XXX series contain copper which is added to improve strength. Examples of these alloys are given in Table 3.25. All these alloys are solution treated 5–10 K below the solidus, quenched, and then aged at 160–175°C for up to 18 h. Some alloys, e.g. 6009, 6010, and 6061, which are used for automotive body sheets are frequently aged during drying of paint, this process being called "bake" hardening. Al–Mg–Si alloys exhibit better mechanical properties when Si is in excess with respect to the stoichiometric composition of Mg_2Si (Mg:Si = 1.73 in wt%). Except for the balanced 6061 alloy, all other alloys presented in Table 3.25 contain Si in excess of Mg_2Si.

The solidification of 6XXX-series alloys containing copper can be described using the Al–Cu–Mg–Si phase diagram and is discussed in Section 3.4. Within the compositional range of commercial 6XXX alloys, the equilibrium solidification ends up with the aluminum solid solution with Mn, Ti, and Cr-containing particles formed during high-temperature eutectic (Mn) and peritectic (Ti, Cr) reactions. Only at the concentrations of alloying elements close to the upper limits, the alloy falls at different phase fields (containing Si and Mg_2Si) at the end of equilibrium solidification as shown in the isothermal section at 500°C (close to the solidus) for alloys containing 1% Si (Figure 3.11d). Impurity of Fe reacts with Si and Mn to form

Table 3.25. Chemical composition and liquidus and solidus temperatures of 6XXX-series alloys containing Cu (Davis, 1993)

Grade	Si, %	Mg, %	Cu, %	Mn, %	Other, %	T_{sol}, °C	T_{liq}, °C
6009	0.6–1.0	0.4–0.8	0.15–0.6	0.2–0.8	0.1Ti, 0.1Cr, <0.5Fe	560	650
6010	0.8–1.2	0.6–1.0	0.15–0.6	0.2–0.8	0.1Ti, 0.1Cr, <0.5Fe	585	650
6061	0.4–0.8	0.8–1.2	0.15–0.40	0.15	0.04–0.35 Cr, 0.15Ti, <0.7Fe	582	652
6066	0.9–1.8	0.8–1.4	0.7–1.2	0.6–1.1	0.2Ti, 0.4Cr, <0.5Fe	563	645
6070	1.0–1.7	0.5–1.2	0.15–0.40	0.4–1.0	0.15Ti, 0.1Cr, <0.5Fe	566	649

(AlFeMnSi) particles of eutectic origin. Under real casting conditions non-equilibrium eutectics containing Mg_2Si and Si are formed. Specific features of these transformations are discussed elsewhere in this book (Sections 1.4, 2.3, 3.4, 3.7). The main implication of copper addition to Al–Mg–Si alloys is the formation of the quaternary $Al_5Cu_2Mg_8Si_6$ (Q) phase through peritectic reactions with Mg_2Si and Si occurring at 529 and 512–514°C or by a multi-phase eutectic reaction at 505–507°C as shown in Tables 3.11 and 3.12. This phase is very stable and can be found in Al–Mg–Si alloys with an excess of Si even at as small additions of copper as 0.25% (Drits et al., 1977). According to Mondolfo (1976), this phase forms upon solidification and is in equilibrium with aluminum under the following conditions: Mg:Si < 1.73, (Mg) > 2(Cu), and (Cu) > 1%. Depending on the ratio between copper, magnesium, and silicon, this phase can coexist with Al_2Cu (θ) (that appears in Al–Cu–Mg–Si alloys with a sufficient amount of copper, >4%), Mg_2Si, and Si. Figure 3.20 shows isothermal sections of Al–Mg–Si–Cu phase diagram at 0.25% Cu and at two temperatures, 520°C and 175°C (though the low-temperature section is given with some mistakes, this is the only low-temperature section available in literature). The first temperature reflects the phase composition after solidification and solution heat treatment (in fact the homogenization of these alloys can be performed at temperatures as high as 565°C, hence even wider single-phase region of the aluminum solid solution in the high-temperature section). The maximum mutual solubility of Mg, Si, and Cu in solid aluminum allows all these elements to be transferred to the solid solution during high-temperature annealing. The second temperature corresponds to the equilibrium phase composition at a temperature of aging (precipitation hardening). Therefore, one may expect these sets of phases to be found among the products of decomposition of the supersaturated solid solution. It should be noted that most 6XXX-series alloys with copper (Table 3.25) also contain manganese that considerably affects the phase composition by forming (AlFeMnSi) phases during solidification and dispersoids during high-temperature annealing. However, the supersaturated solid solution after quenching contains mainly copper, magnesium, and silicon, which allows one to use the Al–Cu–Mg–Si phase diagram

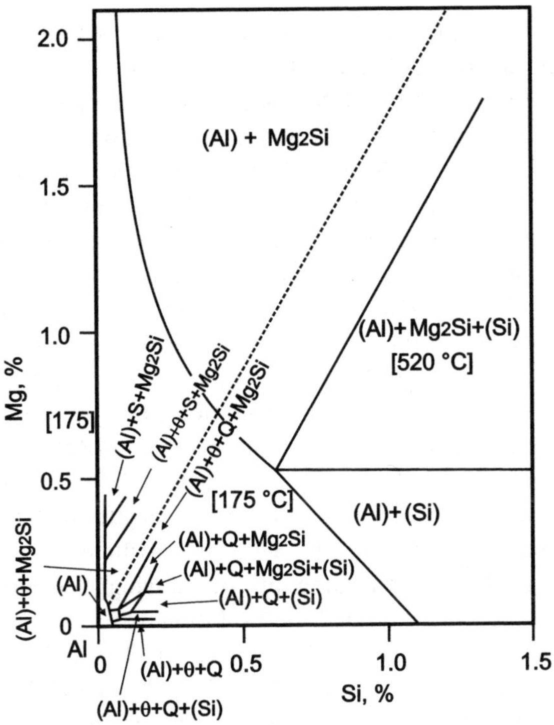

Figure 3.20. Isothermal sections of Al–Mg–Si–0.25% Cu phase diagram at 175 and 520°C (Drits et al., 1977).

for the analysis of (equilibrium) phase composition after aging. It is important to recall that the chemical composition of the solid solution, not of the alloy, should be applied to the phase diagram; and those compositions can be very different, especially by taking into account that silicon is bound with manganese and iron into insoluble particles of solidification origin.

In this section, we consider the phase composition of Al–Mg–Si alloys with copper in the temperature range of precipitation hardening. It is important to note that in this analysis the chemical composition of the supersaturated solid solution, rather than the alloy composition, is used.

Mondolfo, based on his studies performed in the 1950s, suggested that the Q phase is responsible for age hardening in alloys with excess Si and with the copper concentration lower than that of magnesium (Mondolfo, 1976). Since that time, numerous investigations have been conducted on the phase composition of Al–Mg–Si–(Cu) alloys after aging. It is clearly shown that several phases can simultaneously precipitate as a result of decomposition of the supersaturated solid solution, depending on the ratio and amount of alloying elements (Zolotorevskii et al., 1987, 1992, 2003; Chakrabarti and Laughlin, 2004). The situation is additionally complicated by the existence in Al–Mg–Si alloys of multiple metastable forms of Mg_2Si (β) with different composition, crystallography, and morphology (Maruyama et al., 1997; Matsuda et al., 2000, 2001; Gupta et al., 2001; Murayama et al., 2001). The occurrence of these phases is a function of alloy composition and aging regime.

It also has become obvious that the equilibrium phase diagram cannot reliably predict the phase composition of precipitation products. First of all, the decomposition of the supersaturated solid solution is a kinetic and non-equilibrium process. The precipitation occurs through several stages controlled by various factors such as diffusion and surface energy. In the case of multi-phase precipitation, as in Al–Mg–Si–Cu alloys, the sequence of phase precipitation and the gradual change of the solid solution composition are of paramount importance (Zolotorevskii et al., 1987; Dons, 2002). Secondly, and this is often forgotten, the chemical composition of the alloy determines the composition of the supersaturated solid solution but differs from it (Jena et al., 1993; Eskin, 1995, 2003; Hutchinson and Ringer, 2000). For many commercial alloys the amounts of alloying elements in the nominal composition and in the supersaturated solid solution differ by weight percents. And, finally, several phases and their modifications, which should not exist simultaneously, can coexist because of the kinetics of the process and different preferential sites of precipitation (Eskin, 1992; Eskin et al., 1999; Charai et al., 2000).

At the same time, the equilibrium phase diagram can give a clue as to what the phase composition should be with respect to the equilibrium phases. And this analysis can be performed using isothermal sections of the Al–Cu–Mg–Si phase diagram, e.g. those given in Figure 3.11b (200°C) and Figure 3.20 (175°C). The knowledge of phases that can precipitate in Al–Cu–Mg–Si alloys is necessary for further discussion. Metastable phases occurring in Al–Mg–Si alloys have been considered in Section 2.4 (Table 2.13).

The effect of copper addition on Al–Mg–Si alloys depends on the amount of copper and on the Mg:Si ratio. Let us consider here the characteristics of the quaternary Q phase and its possible precursors. The equilibrium Q phase has a hexagonal crystal structure with lattice parameters and composition given in Table 3.26 according to various sources.

It should be particularly noted that the β'_C phase in ternary Al–Mg–Si alloys with an excess of silicon and the Q phase observed in quaternary Al–Mg–Si–Cu alloys have the same crystal structure and differ only in the composition and distribution of atoms within the unit cell, the Q phase containing copper. Most of the authors report that the Q phase precipitates in Al–Cu–Mg–Si alloys in the form of laths whereas the β'_C phase forms rods. There is strong evidence that copper dissolves in the β'' phase which then evolves either as β (Mg_2Si) or as Q phase

Table 3.26. Composition and crystal structure of the Q phase and its precursors as compiled in Eskin (2003)

Phase	Composition			Lattice parameters, nm	
	Cu, %	Mg, %	Si, %	*a*	*c*
$Al_5Cu_2Mg_8Si_6$	20.3	31.1	27	1.032	0.405
$Al_xCu_2Mg_{12-x}Si_7$	20.3	31.1	31.4	1.0393	0.4017
$Al_4CuMg_6Si_6$ (Q′) precursor	–	–	–	1.04	0.406
$Al_3Cu_2Mg_9Si_7$	–	–	–	–	–
$Al_5Cu_2Mg_8Si_6$*	–	–	–	1.035	0.405
QC* (precursor)	–	–	–	0.67	0.405
QP* (precursor)	–	–	–	0.393	0.405
Q	20.6	32.6	30.2	–	–

* Precipitation in the matrix of an Al–4%Cu–1%Mg–0.5%Ag/SiC composite

Table 3.27. Effect of alloying elements' ratio on the equilibrium phase composition of Al–Mg–Si–Cu alloys and examples of commercial alloys with the specific phase composition (Chakrabarti and Laughlin, 2004)

Mg:Si	Cu	Phase region	Alloys
>1	High	(Al) + Al_2Cu + Mg_2Si + Q	6061, 6013, 2017, 2036
<1	High	(Al) + (Si) + Al_2Cu + Q	6009, 6010, 6111, 6066, 6016, 6351A, 6022, 2014, 2008
>1	Low (<0.5%)	(Al) + (Si) + Mg_2Si + Q	6061*, 6009*

* Lower Cu limit of the compositional range

depending on the alloy composition and conditions of precipitation (Murayama et al., 2001).

Chakrabarti and Laughlin (2004) report the equilibrium phase composition of quaternary Al–Mg–Si–Cu alloys at temperatures of artificial aging as shown in Table 3.27. Alloys can transit from one phase field to another within the grade (as shown for 6061 and 6009 in Table 3.27) or with changing amount of impurities (2024 without Si falls into the (Al) + Al_2Cu + Al_2CuMg phase field and 2024 with Si becomes 2017 and has an equilibrium phase composition of (Al) + Al_2Cu + Mg_2Si + Q (Eskin et al., 1989, 1992; Chakrabarti and Laughlin, 2004).

However, nonequilibrium phase compositions reflecting different stages of decomposition of supersaturated solid solution in Al–Mg–Si–Cu alloys are quite different from the equilibrium phase selection (Eskin, 1992, 2003; Chakrabarti and Laughlin, 2004). The equilibrium phase composition is the target towards which the metastable phases evolve. Table 3.28 gives the experimental results on precipitates found in different stages of aging and overaging as compared to the equilibrium phase composition (Eskin, 2003; Chakrabarti and Laughlin, 2004).

Table 3.28. Effect of alloying elements' ratio and aging regime on the nonequilibrium phase composition of Al–Mg–Si–Cu alloys (Eskin, 2003; Chakrabarti and Laughlin, 2004)

Mg:Si	Cu	Peak hardening	Overaging	Equilibrium
Balanced	Low	β''	$\beta' + B'(\lambda)$	Mg_2Si (β) + Al_2Cu (θ) + Q
Balanced	High	β'' (with Cu)	$\beta' + Q'$	Mg_2Si (β) + Al_2Cu (θ) + Q
Mg:Si < 1	Low	β''	β'_C or Q' + (Si)	Mg_2Si (β) + (Si) + Q
Mg:Si < 1	High	β'' (with Cu), θ'	Q' + (Si) + θ'	Q + (Si) + Al_2Cu (θ)

Figure 3.21a shows a diagram of phase distribution after aging for alloys containing up to 0.25% Cu, which reflects wrought alloys of 6XXX series like 6022, 6016, 6009, 6061, and foundry Al–Si–Mg alloys. Obviously, the main precipitating phase is β in its metastable modifications. Copper may dissolve in the β' phase. The hexagonal phase with lattice parameters $a = 1.04$ nm and $c = 0.405$ nm (β'_C or Q') is observed in alloys with Mg:Si < 1.2 (in wt%). In the alloys with an excess of silicon (Mg:Si < 1), (Si) forms its own particles. According to the equilibrium phase diagram (Figures 3.11b, 3.20, Table 3.27), the Q phase is present alongside β and (Si) in alloys with Mg:Si < 1.73 even at very small amounts of copper. However, there is no reliable evidence that it precipitates at the hardening stage of decomposition of the supersaturated solid solution.

With the increase in copper concentration in an alloy to 0.5–0.8% (alloys like 6013, 6111, 2008) the situation essentially remains the same, Figure 3.21b. The precipitating phase is mainly β in its modifications. However, Q' is reported to precipitate alongside β (Matsuda et al., 2001). This observed phase has the structure and morphology similar to the β'_C phase, but contains copper. The equilibrium phase composition in the considered compositional range changes on decreasing the Mg:Si ratio from $Mg_2Si + Al_2Cu + Q$ to $Q + Al_2Cu + (Si)$.

A further increase in the copper content in an alloy makes the precipitation pattern more complex, Figure 3.21c. Depending on the Mg:Si ratio the main hardening phase is S' or GPB (Mg:Si > 3), θ'(0.5 < Mg:Si < 2) or β''(Mg:Si < 0.5). The last phase may contain copper. In the alloys containing Mg:Si < 2 the equilibrium Q phase forms during solidification or sometimes is observed upon long annealing at high temperatures. Silicon forms its own particles at Mg:Si < 1.

Finally, Figure 3.21d shows the phase composition of alloys containing 2.5–4.5% Cu (alloys like 2036, 2017, 2024, 2014). The phase composition of precipitation products is in good agreement with the equilibrium phase composition for alloys with Mg:Si > 1, even at the hardening stage of precipitation (see Figure 3.11c). The θ' phase precipitates in all alloys containing Mg:Si < 8; S' phase is present at Mg:Si > 3; the $\beta''(\beta')$ is formed at 1 < Mg:Si < 8. However, in the alloys with an excess of silicon (Mg:Si < 1) where the Q phase is present according to the

(a)

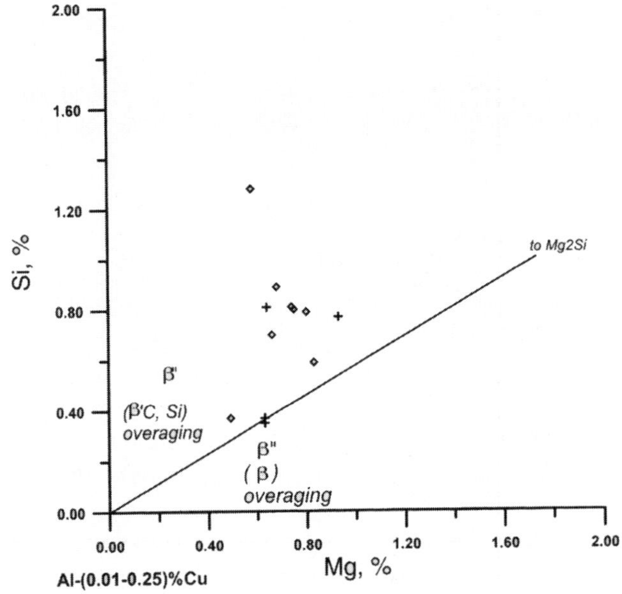

(b)

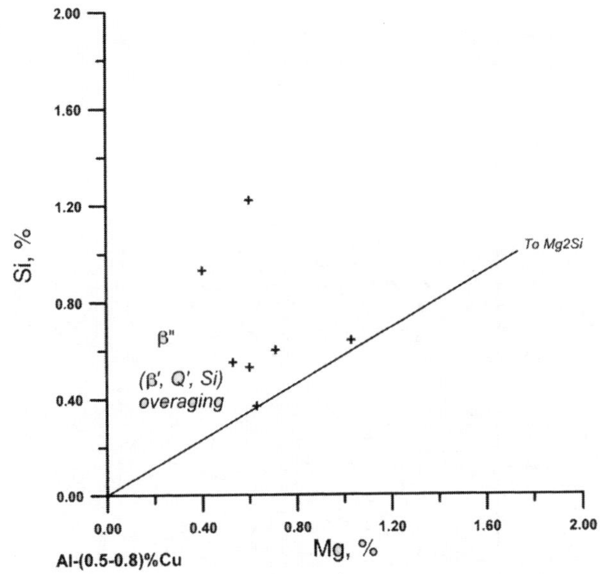

Figure 3.21. Distribution of phases after decomposition of supersaturated solid solution in Al–Mg–Si alloys containing different amounts of copper: (a) low-copper 6XXX alloys; (b) medium-copper 6XXX and 3XX.0 alloys; (c) high-copper 6XXX alloys, medium-copper 3XX.0 and low copper 2XXX alloys; and (d) high-copper 2XXX and 3XX.0 alloys. Dots reflect experimental compositions that have been processed (Eskin, 2003). Diagrams reflect peak hardening condition. Phases in parentheses correspond to overaging condition (if phase composition in different from that at peak hardening).

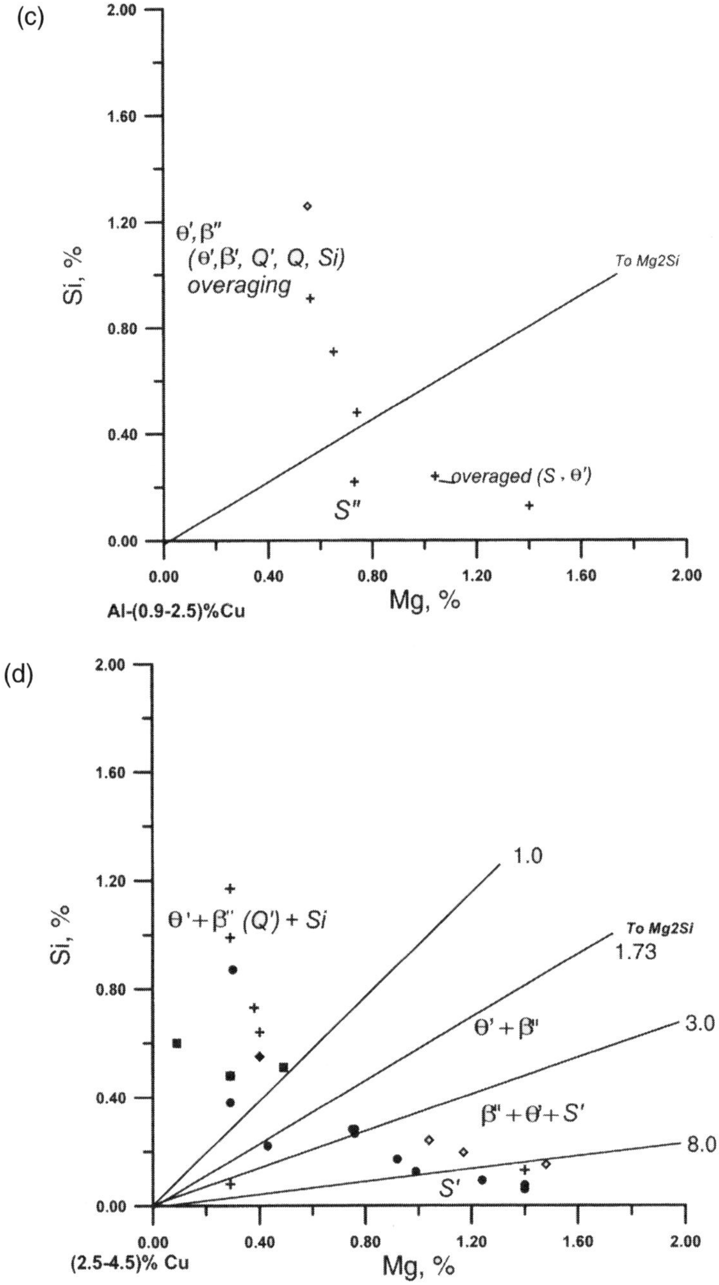

Figure 3.21 (*continued*)

equilibrium phase diagram, there is a controversy. At the hardening stage of precipitation, the main phases here are θ′ and β″ (Eskin et al., 1989, 1992, 2003). Traces of (Si) and, probably Q′ phase are also present, but do not contribute to the hardening. After long or high-temperature annealing the phase composition approaches the equilibrium – Al_2Cu, (Si), and Q.

Experimental results demonstrate that the decomposition of supersaturated solid solutions in Al–Mg–Si–Cu alloys may result in the precipitation of several phases, the particular selection of which is dependent on Mg:Si ratio, copper concentration, and annealing temperature. In the range of Mg:Si ratios above the Mg_2Si stoichiometry the phase composition agrees well with that predicted by the equilibrium phase diagram for the temperature of annealing. This suggests that the precipitation paths of the involved phases (Al_2Cu, Mg_2Si, Al_2CuMg) are straightforward from the zone stage to the equilibrium phase through the formation of one or two metastable modifications.

In the supersaturated solid solutions with an excess of silicon, the decomposition occurs with the formation of Mg_2Si, Al_2Cu, Q(AlMgSiCu), and (Si) phases. The coherent and semi-coherent modifications of the first two phases act as hardening agents upon aging at temperatures below 200°C. Silicon forms its own particles but does not contribute to hardening. In the hardening stage of precipitation, the coherent β″ and the semi-coherent θ′ phases are usually observed in Al–Cu–Mg_2Si–Si alloys, even if the equilibrium phase composition is different. Therefore, the equilibrium phase diagram cannot be directly applied to the interpretation of metastable phase composition. In the stage of softening and at temperatures above 200°C, the β″ phase gives place to semi-coherent β′ phases. The unique feature of Al–Mg_2Si–Si alloys is thc formation of scveral semi coherent modifications of the Mg_2Si (β) phase, different in structure and composition. One of these modifications has the structure similar to that of the equilibrium Q phase and differs from the latter only in composition. Copper, when added to Al–Mg_2Si–Si alloys, may dissolve in coherent and semi-coherent β-based phases. The incorporation of copper atoms in the metastable β-based phases and eventual formation of the equilibrium Q phase in some alloys give one grounds to consider these β-based phases as precursors of the Q phase.

Evidently, the final product of the precipitation path (equilibrium β (Mg_2Si) or Q(AlCuMgSi)) depends on the amount of copper and the Mg:Si ratio in the solid solution and on the temperature of annealing. The identity of metastable β'_C and equilibrium Q phases suggests that their crystallographic configuration is thermodynamically efficient. Moreover, the Q phase may be considered as the β′ (Mg_2Si-based) phase stabilized by copper. Of course this suggestion requires further exploration.

The sequence of precipitation in Al–Mg_2Si–Si–Cu alloys appears to be as follows:

1. Low-copper alloys: supersaturated solid solution (s.s.s.) – GPZ – β'' (with copper?) – θ' – (Si) – various modifications of β' (with copper) including β'_C – Mg_2Si, Al_2Cu, (Si).
2. High-copper alloys: s.s.s. – GPZ – β'' (with copper?) – θ' – (Si) – various modifications β' (with copper) including β'_C (Q′) – Q (AlCuMgSi), Al_2Cu, (Si).
3. At intermediate copper concentrations, in the case of natural aging prior to artificial aging, and at temperatures above 200°C, the combination of these two precipitation paths may occur.

Precipitation sequences in aging 3XX.0-series alloys containing magnesium and copper should be the same, though much less experimental data are available on that subject.

Chapter 4
Alloys of the Al–Mg–Mn–Si–Fe System

Commercial alloys of Al–Mg–Mn system constitute the 5XXX series of wrought alloys and 5XX.0 series of casting alloys. These alloys are characterized by high corrosion resistance, good technological plasticity and surface quality, excellent weldability, and moderate strength. Except for some special alloys, alloys of this group do not acquire additional strength due to precipitation hardening. Although the solubility of magnesium considerably decreases with temperature and a supersaturated solid solution can be easily obtained by quenching, the Al_8Mg_5 (β) phase precipitates upon annealing predominantly on dislocations and grain boundaries, rapidly grows and forms incoherent, relatively large particles which cannot contribute to hardening. The properties of 5XXX series alloys are achieved during casting, deformation, and annealing and are due to solid–solution hardening, work hardening, and controlled recrystallization. Casting 5XX.0 alloys are used in as-cast and annealed conditions.

The base phase composition of 5XXX and 5XX.0 series alloys can be analyzed using the ternary Al–Mg–Mn phase diagram and this diagram will be the first to be described in this chapter. Commercial wrought alloys of the 5XXX series contain 1 to 7% Mg, 0.1 to 1% Mn, less than 0.25% Cu and 0.25% Zn, and small additions of Ti, Cr, V, Be. Casting alloys of the 5XX.0 series may contain up to 12% Mg (typically 4–7%), up to 0.75% Mn and additions of Ti, Zr, Cr, V, and Be. Small additions of transition metals and beryllium do not affect the phase composition with regard to the main alloying elements. Transition metals form binary (or more complex) aluminides either during solidification (acting as grain refiners and forming phases more favorable in morphology with Fe) or during high-temperature annealing (acting as anti-recrystallizing agents); beryllium is added in order to reduce oxidation and loss of magnesium during melting.

However, as in the majority of aluminum alloys, impurities of iron and silicon can greatly contribute to the phase composition and properties of commercial alloys. Therefore, the corresponding phase diagrams will be considered in detail. After that, the phase composition of commercial wrought and casting alloys will be discussed, including some representatives of 3XXX- and 3XX.0-series alloys, the analysis of which can be performed using the Al–Fe–Mg–Mn–Si phase diagram.

4.1. Al–Mg–Mn PHASE DIAGRAM

The Al–Mg–Mn phase diagram has been thoroughly investigated on the Al–Mg side. In the Al corner of this system, the aluminum solid solution is in equilibrium with three phases: Al_6Mn (25.34% Mn), Al_8Mg_5 (Al_3Mg_2, β; 34.8–37.1% Mg), and $Al_{10}(MgMn)_3$ (a.k.a. $Al_{10}Mg_2Mn$, $Al_{18}Mg_3Mn_2$, T; 13.7% Mg, 13.5% Mn) (Mondolfo, 1976; Ran, 1993). Structure information on the first two phases can be found in Chapters 1 and 2. The $Al_{10}(MgMn)_3$ phase has a cubic structure, space group *Fd*3*m*, and lattice parameter $a = 1.4529$ nm.

Table 4.1 lists invariant equilibria in the aluminum corner of the Al–Mg–Mn system.

The ternary compound $Al_{10}(MgMn)_3$ is formed through peritectic reactions 1 and 3 in Table 4.1, eutectic transformation 4 is more likely to occur than reaction 5.

It should be, however, noted that the information on solidification reactions in this system is not certain and requires further investigation. For example, invariant eutectic reaction 4 occurring at 437°C and 0.1–0.2% Mn is sometimes considered to be eutectic reaction 5 at the same temperature but 1% Mn, whereas the geometry of the phase diagram suggests much lower concentration of Mn. There is also alternative peritectic reaction 2 that may occur instead of reaction 1.

The solubility of Mn in (Al) decreases when Mg is introduced in the system (Mondolfo, 1976). For example, an addition of 2% Mg to (Al) decreases the solubility of Mn at 597°C from 0.96 to 0.8%. The maximum solubility of Mg in (Al) is also affected by the presence of Mn, it measures 14% Mg in ternary alloys instead of 17.4% in the binary Al–Mg system. The solubilities of Mg in Al_6Mn and Mn in Al_8Mg_5 are negligibly small.

Table 4.1. Invariant equilibria in the Al corner of the Al–Mg–Mn system (Mondolfo, 1976; Ran, 1993)

No	Reaction	Point in Figure 4.1a	Concentrations in liquid phase		*T*, °C	Ref.
			Mg, %	Mn, %		
1	$L + Al_4Mn \Rightarrow Al_6Mn + Al_{10}(MgMn)_3$	P_2	18	2–3	–	Mondolfo, 1976
2	$L + Al_4Mn \Rightarrow Al_6Mn + Al_8Mg_5$*	–	29.5	1.2	–	Ran, 1993
3	$L + Al_6Mn \Rightarrow (Al) + Al_{10}(MgMn)_3$	P_1	22	<0.5	510**	Mondolfo, 1976
4	$L \Rightarrow (Al) + Al_8Mg_5 + Al_{10}(MgMn)_3$	E	33	0.1–0.2	437	Mondolfo, 1976
5	$L \Rightarrow (Al) + Al_8Mg_5 + Al_6Mn$*	–	28.3	1.0	437	Ran, 1993

* Reactions alternative to the preceding reactions
** Estimated from Figure 4.1a

Chemical compositions of some commercial alloys whose phase composition can be analyzed using the ternary phase diagram are listed in Table 4.2.

Figure 4.1 shows the solidification surface and isothermal section of the Al–Mg–Mn system. It can be easily seen that wrought and casting alloys listed in Table 4.2 fall into the equilibrium phase regions (Al) or (Al) + Al_6Mn. The solidification starts

Table 4.2. Chemical composition of some commercial alloys whose phase composition can be analyzed using Al–Mg–Mn phase diagram

Grade	Mg, %	Mn, %	Fe, %	Si, %	Other, %
AMg5(rus)	4.8–5.8	0.3–0.8	0.5	0.5	0.1Ti; 0.005Be
5456	4.7–5.5	0.5–1.0	0.4	0.25	0.12Cr; 0.2Ti
AMg5Mts(rus)	4.8–6.5	0.4–1.0	0.3	0.4	0.1Ti
520.0	9.5–10.6	0.15	0.3	0.25	0.005Be
3004	0.8–1.3	1.0–1.5	0.7	0.3	–
5454	2.4–3.0	0.5–1.0	0.4	0.25	0.2Ti
1541(rus)	3.8–4.8	0.2–0.5	0.1–0.3	<0.2	0.05Ti

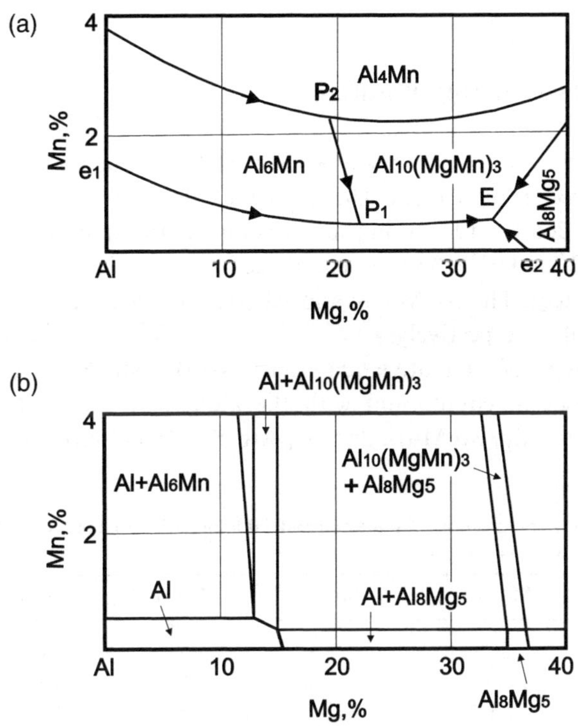

Figure 4.1. Aluminum corner of the Al–Mg–Mn phase diagram: (a) liquidus and (b) phase distribution at 450°C (Mondolfo, 1976).

with the formation of primary grains of aluminum solid solution, after that the binary eutectics (Al) + Al_6Mn is formed.

Under nonequilibrium solidification conditions, when cooling rates are sufficiently high, the solidification can continue towards formation of low-temperature eutectics 4 (in Table 4.1), peritectic reactions 1 and 3 are likely to be incomplete or even suppressed. There is also a possibility that the formation of the Al_6Mn phase is suppressed at earlier stages of solidification and the more Mn-rich Al_4Mn phase is formed instead. Then the solidification of alloys can be finished by the formation of the metastable eutectics (Al) + Al_4Mn + Al_8Mg_5 (Ran, 1993).

Figure 4.2 demonstrates two polythermal sections of the Al–Mg–Mn phase diagram in the range of magnesium and manganese concentrations typical of wrought and casting Al–Mg–Mn alloys. One can conclude that after the end of solidification, Al_8Mg_5 and $Al_{10}(MgMn)_3$ phases may precipitate in the solid state due to the decreasing solubility of magnesium and manganese in solid aluminum.

A typical heat treatment of Al–Mg–Mn alloys is annealing at 345 to 415°C. After such a processing, magnesium goes into the aluminum solid solution, nonequilibrium Mg-containing phases of eutectic origin being dissolved.

4.2. Al–Mg–Mn–Si PHASE DIAGRAM

Silicon is almost always present in Al–Mg–Mn alloys as an impurity. In this case, its concentration ranges from 0.05 to 0.5%. However, in some alloys silicon is deliberately added in order to improve casting characteristics, e.g. fluidity. Selected alloys of 6XXX and 3XX.0 series, e.g. those in Table 4.3, can be analyzed using this quaternary system. The Al–Mn–Si and Al–Mg–Si phase diagrams are considered in Chapters 1 and 2, respectively.

Mondolfo reports a tentative quaternary Al–Mg–Mn–Si phase diagram with the following phases in equilibrium with the aluminum solid solution: (Si), Mg_2Si, Al_6Mn, Al_8Mg_5, $Al_{10}(MgMn)_3$, and $Al_{15}Mn_3Si_2$ (Mondolfo, 1976).

Table 4.3. Chemical composition of some commercial alloys whose phase composition can be analyzed using Al–Mg–Mn–Si phase diagram

Grade	Mg, %	Mn, %	Fe, %	Si, %	Other, %
356.0	0.30–0.45	0.35	0.6	6.5–7.5	0.25Ti
1530(rus)	3.2–3.8	0.3–0.6	0.5	0.5–0.8	0.1T; 0.05Cr
512.0	3.5–4.5	0.35	0.5	1.4–2.2	0.25Ti
AMg5K(rus)	4.5–5.5	0.4–1.0	0.6	0.8–1.3	–

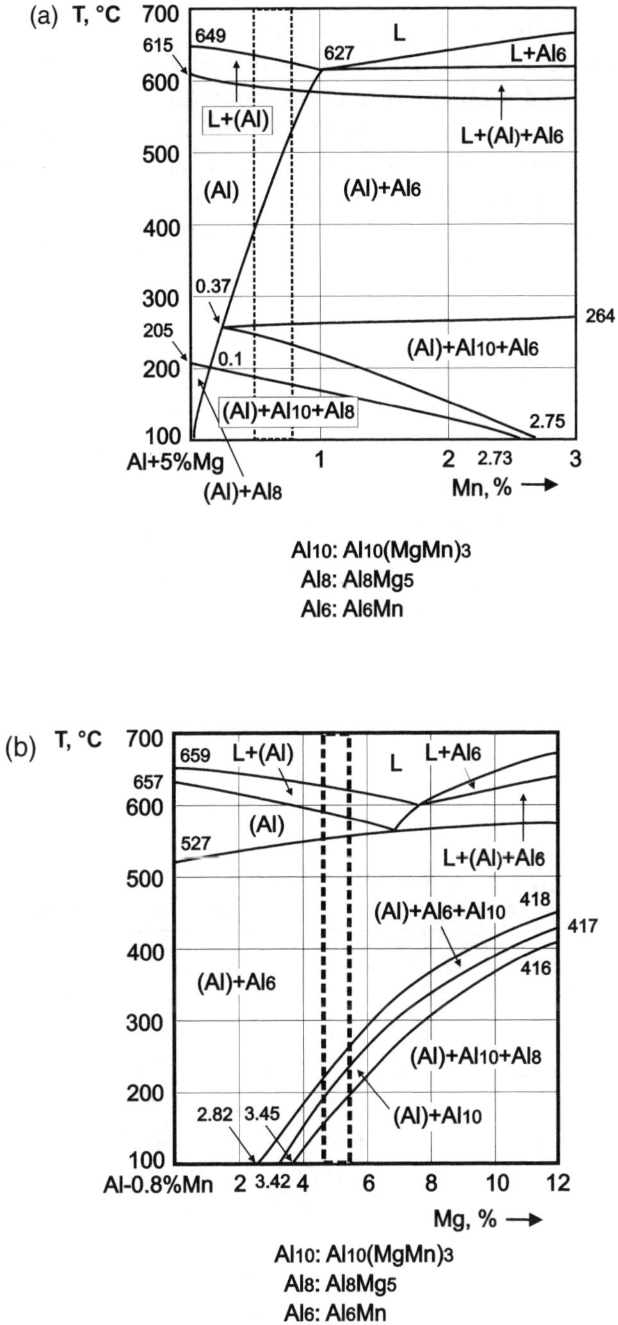

Figure 4.2. Polythermal sections of the Al–Mg–Mn phase diagram: (a) Al–5%Mg–Mn and (b) Al–0.8%Mn–Mg. Compositional ranges of 5456 and 535.0 alloys are marked.

Figure 4.3 shows the projection of solidification surface and the distribution of phases in the solid state. Invariant reactions that may occur in this system are given in Table 4.4 and monovariant reactions, in Table 4.5.

It is important to note that the Al–Mg–Mn–Si phase diagram is, in fact, poorly studied and the information given in Figure 4.3b and Table 4.5 can be considered only as a rough approximation. For example, the distribution of phase fields in the solid state (Figure 4.3a) suggests the possibility of a quasi-ternary section

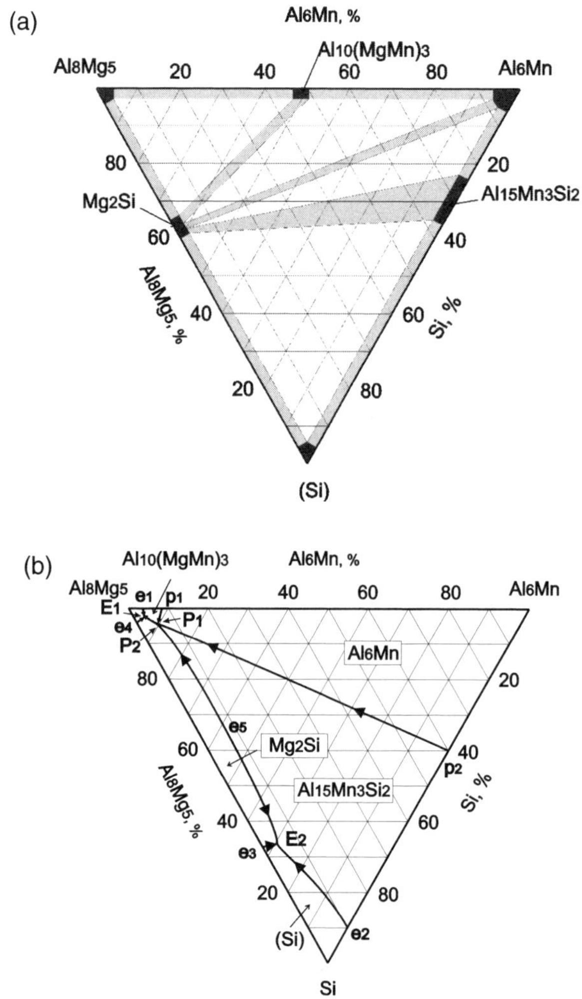

Figure 4.3. Phase distribution in the solid state (a), polythermal projection of solidification surface (b) Mondolfo (1976), and a suggested version of "triangulation" (c) in the Al–Mg–Mn–Si phase diagram.

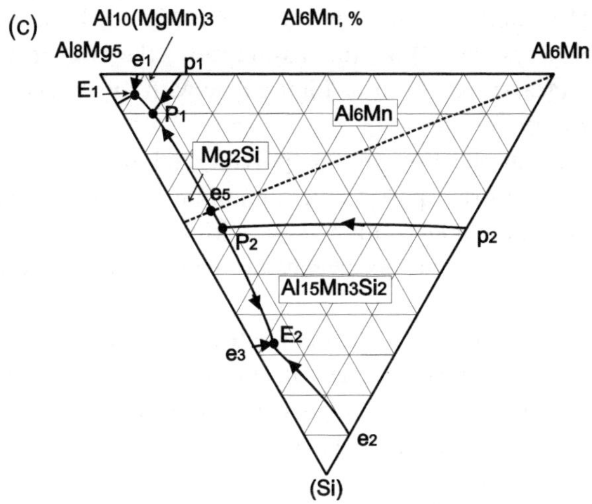

Figure 4.3 (*continued*)

Table 4.4. Invariant reactions in Al–Mg–Mn–Si system (after Mondolfo, 1976)*

Reaction	Point in Figure 4.3b	Concentrations in liquid phase			T,°C
		Mg, %	Mn, %	Si, %	
$L \Rightarrow (Al) + Al_8Mg_5 + Al_{10}(MgMn)_3 + Mg_2Si$	E_1	~30	~0.1	~0.2	~435
$L + Al_{15}Mn_3Si_2 \Rightarrow (Al) + Al_6Mn + Mg_2Si$	P_2	~30	~0.3	~0.5	~510
$L \Rightarrow (Al) + (Si) + Mg_2Si + Al_{15}Mn_3Si_2$	E_2	<5	<1	~12	~554
$L + Al_6Mn \Rightarrow (Al) + Al_{10}(MgMn)_3 + Al_{15}Mn_3Si_2$	P_1	<30	~0.1	<0.2	~500

* Modified by authors

Table 4.5. Monovariant reactions in Al–Mg–Mn–Si system (Mondolfo, 1976)

Reaction	Line in Figure 4.3b	T,°C
$L \Rightarrow (Al) + Al_8Mg_5 + Al_{10}(MgMn)_3$	e_1–E_1	437–435
$L \Rightarrow (Al) + Al_8Mg_5 + Mg_2Si$	e_4–E_1	449–435
$L \Rightarrow (Al) + Al_{10}(MgMn)_3 + Mg_2Si$	P_1–E_1	500–435
$L + Al_6Mn \Rightarrow (Al) + Al_{10}(MgMn)_3$	p_1–P_1	510–500
$L + Al_6Mn \Rightarrow (Al) + Al_{15}Mn_3Si_2$	p_2–P_2	648–510
$L \Rightarrow (Al) + (Si) + Al_{15}Mn_3Si_2$	e_2–E_2	575–554
$L \Rightarrow (Al) + (Si) + Mg_2Si$	e_3–E_2	555–554
$L \Rightarrow (Al) + Mg_2Si + Al_{15}Mn_3Si_2$	e_5–E_1	594–435
	e_5–E_2	594–554
$L \Rightarrow (Al) + Mg_2Si + Al_6Mn$	P_2–P_1	510–500

(Al)–Mg_2Si–Al_6Mn, similar to those in the Al–Fe–Mg–Si (Figure 2.4) and Al–Cu–Mg–Si (Figure 3.4) systems. With this assumption, the constitution of the phase diagram becomes clearer as shown in Figure 4.3c with two parts and two invariant reactions in each part.

Furthermore, the application of the polythermal projections of solidification surfaces available in Mg-rich part of the aluminum corner to the analysis of commercial alloys is questionable because these alloys, as a rule, contain less magnesium; than it is necessary for the occurrence of invariant reactions with participation of the Al_8Mg_5 phase.

When it comes to the structure, the main effect of adding Si to Al–Mg–Mn alloys is in the formation of Mg_2Si particles of eutectic origin. In addition, alloys rich in silicon contain (Si) and $Al_{15}Mn_3Si_2$ particles (iron being frequently dissolved in the latter phase); and alloys rich in magnesium – Al_8Mg_5 and $Al_{10}(MgMn)_3$ particles. During homogenizing annealing after casting, nonequilibrium Al_8Mg_5 and Mg_2Si (note that solid solubility of silicon in aluminum considerably decreases at magnesium concentrations over 3–4%) particles mostly dissolve in the solid solution, whereas Mn-containing crystals may only change the morphology by globularization. Eutectic silicon particles present in casting Al–Si-based alloys remain the main structure constituent, although their morphology may become more globular during homogenization as well.

Casting Al–Si alloys, e.g. A356.0-type, are subjected to hardening heat treatment by quenching and aging. As distinct from Al–Mg alloys, the concentration and ratio of magnesium and silicon in the supersaturated solid solution of Al–Si alloys is favorable for the formation of hardening metastable phases based on Mg_2Si. This is discussed in Chapters 2 and 3 in more detail.

It should be noted that most aluminum wrought alloys contain excess of iron over silicon. Such concentration ratio decreases the susceptibility of an alloy to hot cracking during casting. In addition, many casting alloys have relatively high allowable concentration of iron. Therefore, it is necessary to consider the Al–Fe–Mg–Mn system and then the Al–Fe–Mg–Mn–Si phase diagram.

4.3. Al–Fe–Mg–Mn PHASE DIAGRAM

This phase diagram is poorly studied and the evaluation based on the constituent ternary systems is shown in Figure 4.4. The phases Al_8Mg_5, Al_3Fe, $Al_6(FeMn)$, and $Al_{10}(MgMn)_3$ are in equilibrium with the aluminum solid solution (Mondolfo, 1976; Belov et al., 2002a).

Two invariant reactions are possible in the region close to the Al–Mg side of the concentration tetrahedron. Both occur at temperatures and concentrations close to

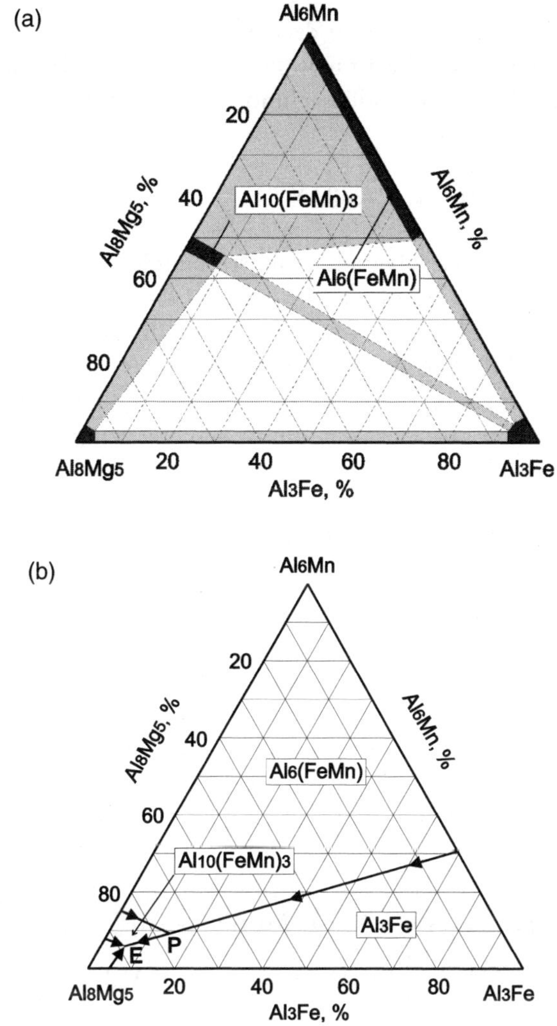

Figure 4.4. Phase distribution in the solid state (a) and polythermal projection of solidification surface (b) in the Al–Mg–Mn–Fe phase diagram (Belov et al., 2002a).

those of the Al–Mg binary eutectics, i.e. 450°C and 35% Mg:

$L + Al_6(FeMn) \Rightarrow (Al) + Al_3Fe + Al_{10}(MgMn)_3$ (point P in Figure 4.4b);

$L \Rightarrow (Al) + Al_3Fe + Al_8Mg_5 + Al_{10}(MgMn)_3$ (possibly at 435°C) (point E Figure 4.4b).

The solid solubility of Mn in (Al) decreases in the presence of iron (Mondolfo, 1976). Iron also suppresses the formation of the metastable $Al_{12}Mn$ phase upon decomposition of an aluminum solid solution supersaturated in manganese.

4.4. Al–Fe–Mg–Mn–Si PHASE DIAGRAM

This five-component phase diagram is most suitable for analyzing solidification paths and phase composition of Al–Mg–Mn and Al–Mg–Mn–Si alloys containing impurities of Si and Fe. It can be conditionally subdivided into two sections.

In the range of Mg-rich alloys, only three phases – Mg_2Si, Al_3Fe, and $Al_{10}(MgMn)_3$ – can be in equilibrium with (Al) and Al_3Mg_2. In these alloys the solidification ends (providing sufficient amount of alloying components) with the invariant eutectic reaction at $<435°C$:

$$L \Rightarrow (Al) + Al_3Fe + Mg_2Si + Al_8Mg_5 + Al_{10}(MgMn)_3.$$

As a result, the phase distribution in the solid state is very simple as depicted in Figure 4.5a.

Many commercial 5XXX-, 3XXX-, and 5XX.0-series alloys can be analyzed using this system as will be shown in Sections 4.5 and 4.6.

The situation in Al–Si-based alloys is complicated by the presence of a quaternary compound – $Al_8FeMg_3Si_6$ (π). Its presence results in the existence of two five-phase regions in the solid state (Figure 4.5b). Correspondingly, two invariant reactions are possible:

$$L \Rightarrow (Al) + (Si) + Mg_2Si + Al_{15}(FeMn)_3Si_2 + Al_8FeMg_3Si_6 \text{ at } <554°C;$$

$$L + Al_5FeSi \Rightarrow (Al) + (Si) + Al_{15}(FeMn)_3Si_2 + Al_8FeMg_3Si_6 \text{ at } <567°C.$$

The following phases can be in equilibrium with the aluminum solid solution in the solid state: (Si), Mg_2Si, Al_5FeSi, $Al_{15}(FeMn)_3Si_2$, and $Al_8FeMg_3Si_6$.

This portion of the phase diagram can be used for the prediction and analysis of solidification and phase composition of casting Al–Si alloys containing magnesium (3XX.0 series).

4.5. Al–Mg–Mn WROUGHT AND CASTING ALLOYS

Chemical compositions of commercial alloys of the Al–Mg–Mn system of 5XXX and 5XX.0 series and some representatives of wrought 6XXX-series and casting

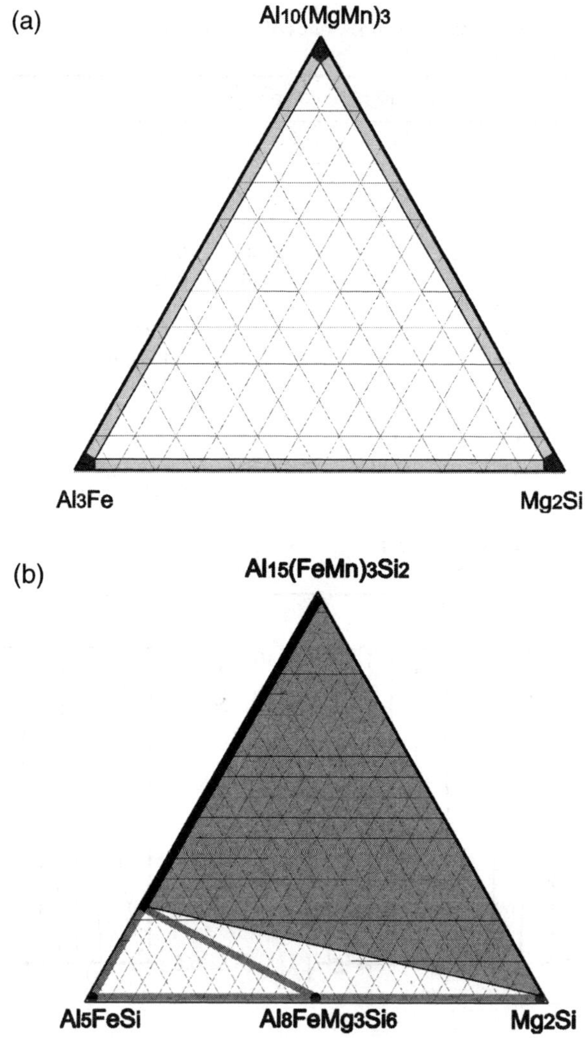

Figure 4.5. Phase distribution in the solid state in the Al–Mg–Mn–Fe–Si phase diagram in the range of Al–Mg (a) and Al–Si (b) alloys (Belov et al., 2002a).

3XX.0-series alloys with an increased amount of manganese and wrought 3XXX-series alloys with an increased amount of magnesium are given in Table 4.6. The equilibrium temperatures of solidus and liquidus (related to the average composition of these alloys) are shown according to ASM Specialty Book (Davis, 1993).

Let us first look at the general effect of magnesium and manganese on the solidus and liquidus of commercial alloys with minor silicon content (<0.5%). Plots in

Table 4.6. Chemical compositions and temperatures of solidus and liquidus for commercial alloys of the Al–Mg–Mn–Fe–Si system (Davis, 1993)

Grade	Average chemical composition					T_{sol},°C	T_{liq},°C
	Mg, %	Mn, %	Si, %	Fe, %	Other, %		
5457	1.0	0.3	<0.08	<0.10	–	629	654
5052	2.6	<0.1	<0.25	<0.40	0.25Cr	607	649
5454	2.7	0.8	<0.25	<0.40	0.12Cr	602	646
5083	4.4	0.7	<0.40	<0.40	0.15Cr	574	638
5182	4.5	0.35	<0.2	<0.35	–	577	638
5456	5.1	0.8	<0.25	<0.4	0.12Cr	570	638
535.0	6.8	0.8	<0.15	<0.15	0.15Ti; 0.005Be	550	630
520.0	10	<0.15	<0.25	<0.3	0.15Ti	450	605
3004	1.0	1.2	<0.30	<0.7	–	629	654
3105	0.5	0.55	<0.60	<0.70	<0.2Cr	638	657
6063	0.7	<0.1	0.40	<0.35	–	615	655
6205	0.5	0.1	0.75	<0.7	0.1Cr; 0.1Zr	613	645
6351	0.6	0.6	1.0	<0.5	–	555	650
356.0	0.35	<0.35	7.0	<0.6	–	555	615
359.0	0.60	<0.10	9.0	<0.2	–	555	615

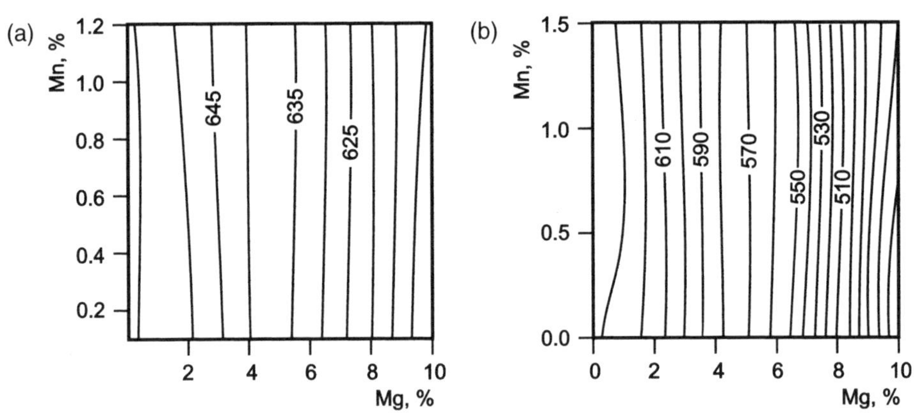

Figure 4.6. Isotherms of liquidus (a) and solidus (b) for commercial Al–Mg–Mn alloys containing iron and silicon on the impurity level (see Table 4.6).

Figure 4.6 are obtained by interpolation of data from Table 4.6. Obviously, magnesium has much stronger influence on both characteristic temperatures than manganese. These charts can serve as a directory on the correct choice of casting and annealing temperatures.

The phase composition of commercial alloys and the sequence of phase transformations during solidification can be considered starting from ternary and then going to more complex systems.

Polythermal sections of the ternary Al–Mg–Mn phase diagram shown in Figure 4.2 give us the first approximation of phase transformation history for alloys like 5083, 5182, 5456, and 535.0 containing 4–7% Mg and 0.3–0.8% Mn. Solidification starts at 638°C (630°C for 535.0 due to the higher concentration of Mg) with the formation of (Al) grains. After that, provided that the concentration of Mn is sufficient, the (Al) + Al_6Mn eutectics is formed in the range of temperatures from 627 to 617°C (Bäckerud et al., 1986). Other phases that may be present in the solid state are formed by precipitation from the aluminum solid solution. These reactions seldom occur during cooling after the end of solidification as, due to a relatively low diffusion coefficient, manganese usually remains in the supersaturated solid solution. During annealing of cast material, Mn-containing phases precipitate from the solid solution supersaturated during solidification and form dispersoids. The final equilibrium phase composition of alloys containing more than 4% Mg and 0.1% Mn is (Al) + Al_8Mg_5 + $Al_{10}(MgMn)_3$. If the concentration of Mg is less than 2–3%, the equilibrium phase composition at room temperature would be (Al) + Al_6Mn + (traces) $Al_{10}(MgMn)_3$.

The presence of silicon, as an impurity or an alloying element, in Al–Mg–Mn alloys results in the formation of Mg_2Si in addition to other phases that we have already considered. Figure 4.7a gives a polythermal section of the Al–Mg–Mn–Si system at a constant concentration of 5% Mg and 1% Mn that corresponds to the chemical compositions of such alloys as 5456, 535.0, AMg6(rus), AMg5Mts(rus) that allow up to 0.3% Si as an impurity (see Tables 4.2 and 4.6). After formation of primary aluminum grains, the binary eutectics (Al) + Al_6Mn is solidified. During further cooling, $Al_{15}Mn_3Si_2$ is formed. This phase then reacts with liquid to produce $Al_{10}(MgMn)_3$ and Mg_2Si phases (Table 4.4), and these phases remain in equilibrium with aluminum down to the room temperature. The Al_8Mg_5 in formed by precipitation from the aluminum solid solution upon cooling in the solid state.

Most of commercial alloys contain iron as an impurity. Figure 4.7b demonstrates the implications of iron addition on the phase composition of Al–Mg–Mn alloys of 5456, 535.0, AMg6(rus), AMg5Mts(rus) types that may contain up to 0.4% Fe. The solidification starts between 630 and 635°C with the formation of primary aluminum grains. The (Al) + $Al_6(FeMn)$ eutectics is formed in the temperature range from 600 to 570°C, and the solidification ends at approximately 570°C with the formation of the (Al) + $Al_6(FeMn)$ + Al_3Fe eutectics. Under equilibrium, other phases that are frequently observed in Al–Mg–Mn–Fe alloys, e.g. Al_8Mg_5 and $Al_{10}(MgMn)_3$, are formed only by precipitation from the aluminum solid solution upon cooling in the solid state. Under real, nonequilibrium conditions, these phases can form during solidification as a result of eutectic reactions.

The precipitation of Al_8Mg_5 and $Al_{10}(MgMn)_3$ phases due to the decreasing solubility of Mg and Mn in solid aluminum is illustrated in Figure 4.7c, d where

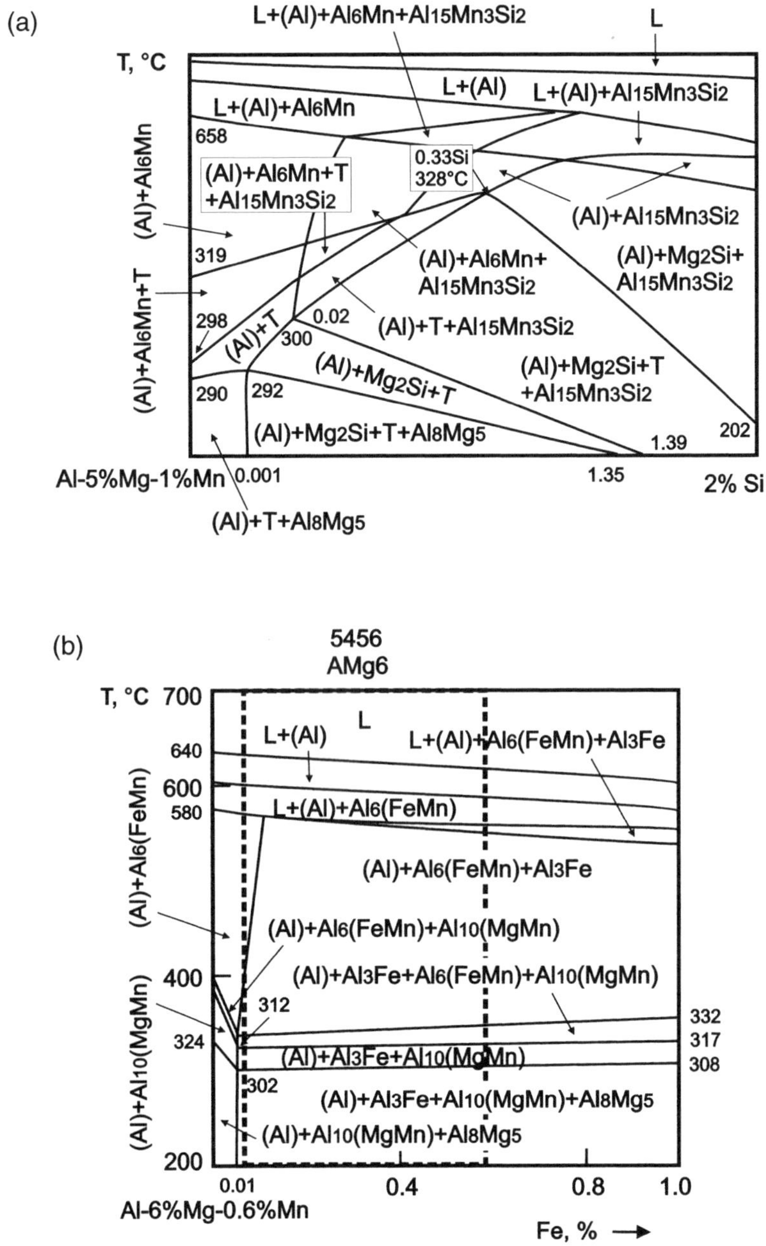

Figure 4.7. Polythermal sections of Al–Mg–Mn–Si (a) and Al–Mg–Mn–Fe (b) systems in the compositional range of commercial alloys of 5456 (AMg6(rus)) and 535.0 type. Isothermal sections of Al–Mg–Mn–1%Si system at 440°C (c) and 100°C (d), here T denotes $Al_{10}(MgMn)_3$.

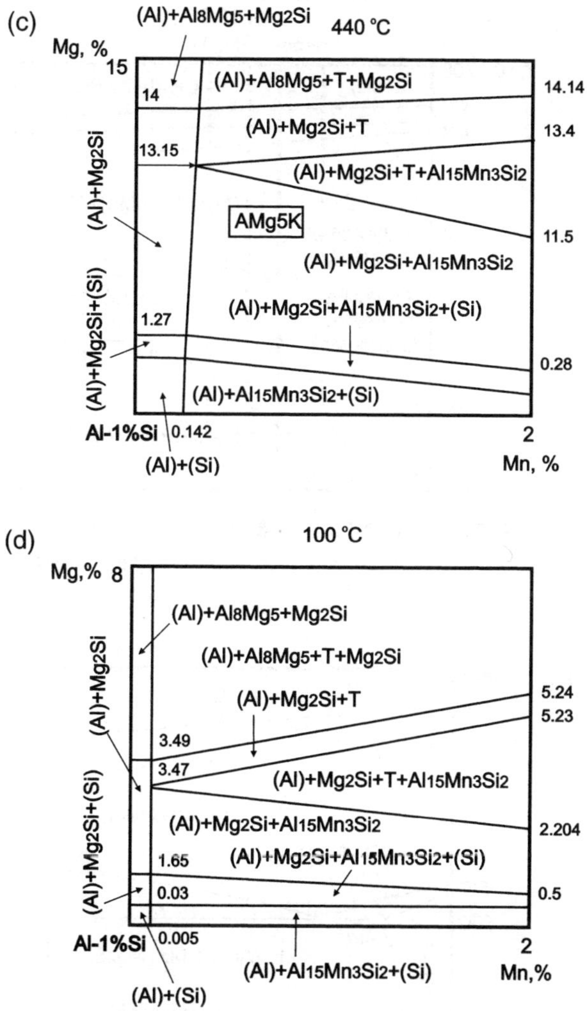

Figure 4.7 (*continued*)

two isothermal sections (at 440°C and 100°C) of the Al–Mg–Mn–Si system are given. The appearance of $Al_{10}(MgMn)_3$ phase and extension of phase fields with Al_8Mg_5 to lower concentrations of Mg are clearly seen.

Casting Al–Mg alloys may contain up to 10–12% Mg (520.0, AMg11rus) and allowed concentrations of Fe and Si can be as high as 1% (518.0, AMg11). The solidification and phase composition of such alloys can be analyzed using isothermal sections shown in Figure 4.8. In this group of high-magnesium alloys, the solidification starts, just as in wrought alloys of 5456 type, with the formation of

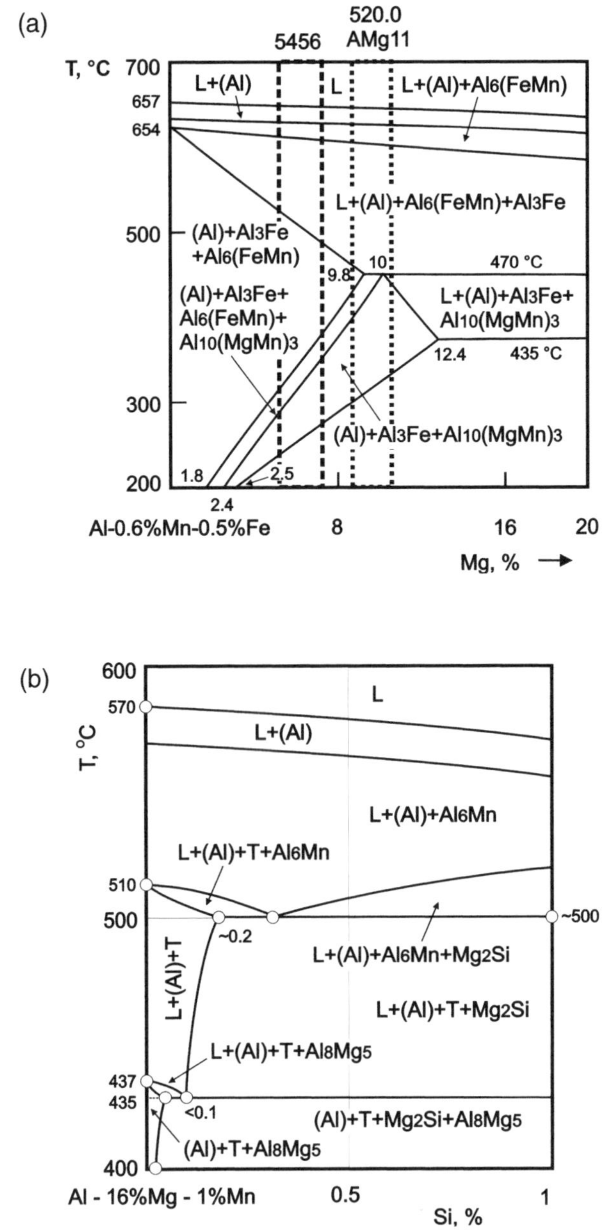

Figure 4.8. Polythermal sections of Al–0.6%Mn–0.5%Fe–Mg system with compositional ranges of 5456 wrought and 520.0 (AMg11(rus)) casting alloys (a) and Al–16% Mg–1% Mn–Si (b), here T denotes $Al_{10}(MgMn)_3$.

aluminum grains (Figure 4.8a). Then the (Al) + Al_6(FeMn) and (Al) + Al_6(FeMn) + Al_3Fe eutectics are formed. And the solidification ends at 470°C with the formation of the complex (Al) + Al_3Fe + $Al_{10}(MgMn)_3$ eutectics. In the case of higher than 10% concentration of magnesium, the solidification continues to lower temperatures and ceases only at 435°C with solidification of the (Al) + Al_3Fe + Al_8Mg_5 + $Al_{10}(MgMn)_3$ eutectics. Addition or impurity of silicon (Figure 4.8b) affects the solidification pattern by the formation of Mg_2Si-containing eutectics at 527 and 435°C, the latter reaction is more likely to occur in high-Mg alloys. Note that the isopleth in Figure 4.8b was constructed using our version of the Al–Mg–Mn–Si phase diagram as shown in Figure 4.3c.

Let us look at a more complicated case of five-component phase diagram Al–Mg–Mn–Fe–Si in the compositional range of a widely used 5182 alloy. Figure 4.9 shows a polythermal section corresponding to the compositional range of alloys 5457 (1% Mg) and 5182 (4.5–5% Mg) (Yan et al., 2001). The solidification path of these alloys include the following reactions: formation of primary (Al) grains (starts at 655 and 634°C for 5457 and 5182, respectively) and the binary eutectics (Al) + Al_3Fe at 643°C for 5457 and 622°C for 5182. The 5457 alloy is solid at 643°C (at these particular concentrations of Fe, Mn, and Si), but the 5182 alloy continues its solidification until 578–576°C when Mg_2Si is formed by a eutectic reaction.

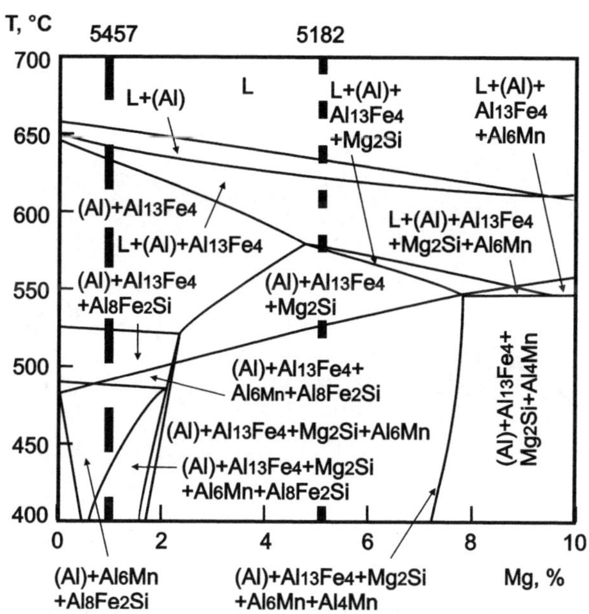

Figure 4.9. Isothermal section of Al–Mg–0.34%Mn–0.28%Fe–0.1%Si system calculated under equilibrium conditions (after Yan et al., 2001).

The solidification paths, that we considered here, describe the phase equilibrium and can hardly be accomplished under real casting conditions when cooling rates are high and the diffusion processes, especially in the solid phase, cannot be completed to such an extent that the compositions of the phases change with temperature in accordance with the equilibrium phase diagram. Local deviations from equilibrium result in microsegregation and eventually in the shift of local equilibrium to the concentrations where new phases are formed. In addition, some high-temperature peritectic reactions remain uncompleted, and high-temperature phases – that have to disappear as a result of these reactions – are retained at lower temperature and can be found in the solid sample.

These kinetic effects are frequently observed in practice. Phase particles that are formed upon nonequilibrium solidification can be dissolved during annealing that is an essential part of the heat treatment for Al–Mg alloys. As a result of annealing, nonequilibrium phases disappear and equilibrium phases may change the morphology of their particles. Figure 4.10 shows an example of Al–11%Mg–0.1%Si–0.2%Fe alloy in the as-cast condition and after annealing at 430°C for 16 h. The Al_8Mg_5 phase (light particles) that is formed under nonequilibrium solidification conditions during eutectic reaction at 435°C (Figure 4.8a) is dissolved, and the Mg_2Si phase (black hieroglyphic particles) that is equilibrium in this compositional range changes the morphology. Gray particles of Al_3Fe and $Al_{10}(FeMn)_3$ are not affected by the annealing.

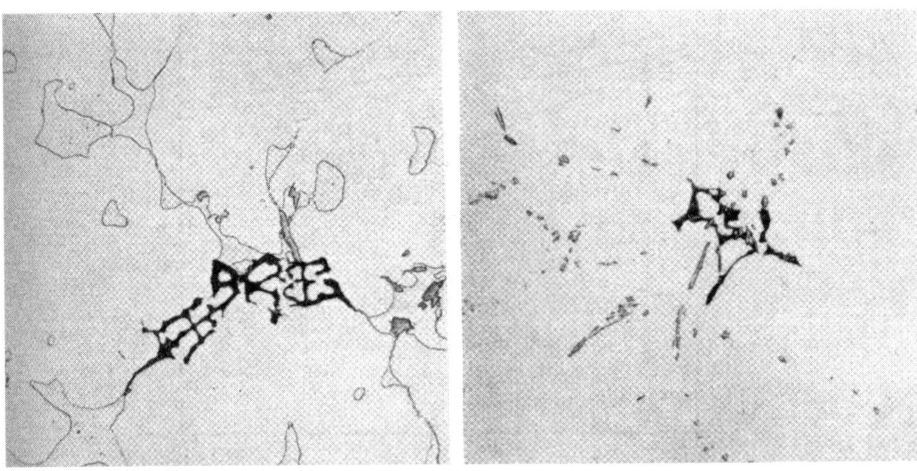

(a) (b)

Figure 4.10. Effect of annealing at 430°C on the phase composition and morphology of an Al–11%Mg alloy: (a) as-cast [Al_8Mg_5; Mg_2Si; Al_3Fe; $Al_{10}(MgMn)_3$] and (b) annealed [rounded Mg_2Si; unchanged Al_3Fe and $Al_{10}(MgMn)$].

The consequences of nonequilibrium solidification for the phase transformations and phase selection are demonstrated below for three alloys.

A 5182 alloy has been tested upon solidification at different cooling rates, from 0.3 to 11 K/s, these cooling rates being characteristic of direct-chill and die casting (Bäckerud et al., 1986). The equilibrium phase diagram gives a solidus temperature of 576–578°C. However, thermal and phase analyses performed during and after solidification clearly demonstrate that the solidification continues at lower temperatures. Table 4.7 gives solidification reactions and corresponding temperature ranges at different cooling rates for this alloy (Bäckerud et al., 1986). The solidification starts with the formation of aluminum grains. The liquidus temperature slightly decreases with an increasing cooling rate due to increased undercooling. Then two concurrent reactions may occur at 620°C (only one of these reactions is shown in Figure 4.8a). After that the eutectics containing Mg_2Si and $Al_3(FeMn)$ is formed at about 580°C. The equilibrium solidification ends at this point. However, experimental results show that, after reaching the equilibrium solidus, the alloy still contains liquid phase which undergoes a complex eutectic reaction and completely vanishes only at 470°C, extending thereby the solidification range by almost 100°C. Thompson et al. (2004) studied the nonequilibrium solidification of a 5182 alloy cast at different cooling rates. They observed that the temperature of $(Al) + Al_6(FeMn)$ eutectics decreases from 588 to 575°C and the solidus decreases from 510 to 461°C on increasing the cooling rate from 0.5 to 2 K/s. The resultant phase composition of a 5182 alloy is (Al), Al_3Fe, Mg_2Si, Al_8Mg_5, and $Al_6(FeMn)$, the latter phase is frequently replaced by $Al_4(FeMn)$ due to the incomplete high-temperature peritectic reaction (reaction 2 in Table 4.1) and enrichment of liquid in Mg and Mn (Figure 4.9) (Yan et al., 2001). All these data testify that the $Al_{15}(FeMn)_3Si_2$ or $Al_{15}Mn_3Si_2$ phases do not form during nonequilibrium solidification, and are substituted for another Mn-containing phase $Al_6(FeMn)$ (Table 4.7). This confirms, though indirectly, our version of the

Table 4.7. Solidification reactions under nonequilibrium conditions in a 5182 alloy (4.74% Mg, 0.34% Mn, 0.28% Fe, and 0.1% Si) (Bäckerud et al., 1986)

Reaction	Temperatures (°C) at a cooling rate	
	0.3 K/s	11 K/s
$L \Rightarrow (Al)$	632	632–623
$L \Rightarrow (Al) + Al_6(FeMn)$ and/or $L \Rightarrow (Al) + Al_3(FeMn)$	621–617	620
$L \Rightarrow (Al) + Al_3Fe + Mg_2Si$	586	583
$L \Rightarrow (Al) + Al_3Fe + Mg_2Si + Al_8Mg_5$	557–470	543–470
Solidus	470	470

Al–Mg–Mn–Si phase diagram that shows that the $Al_{15}Mn_3Si_2$ phase cannot be formed in Mg-rich alloys, even under nonequilibrium solidification conditions (Figure 4.3c).

An essentially binary 518.2 alloy contains about 8% Mg and impurities of Si (0.1%), Fe (0.15%), and Mn (0.01%). The solidification starts at about 610°C with the formation of (Al) and is expected to end at 450°C with the eutectic reaction $L \Rightarrow (Al) + Al_8Mg_5$ at 450°C (Bäckerud et al., 1990). However, even relatively small concentrations of Fe and Si result in the enrichment of liquid with these elements up to the concentrations required for the lower temperature eutectics to be formed. Therefore, the solidification continues with the formation of Al_3Fe and Mg_2Si phases and ends only in the temperature range of 430–440°C with the eutectic reaction $L \Rightarrow (Al) + Al_3Fe + Mg_2Si + Al_8Mg_5$ (Bäckerud et al., 1990). Table 4.8 gives solidification reactions and corresponding temperatures for a 518.2 alloy. The increasing cooling rate during solidification decreases the solidus of the alloy and increases the amount of the last, nonequilibrium eutectics.

Another casting alloy that is examined by Bäckerud et al. (1990) is a 512.0 alloy containing 4% Mg, 0.35% Mn, 1.8% Si, and 0.55% Fe. According to the equilibrium phase diagram, this alloy ends its solidification with the eutectic reaction $L \Rightarrow (Al) + Al_{15}Mn_3Si_2$ (Figure 4.7a) at approximately 617°C (Bäckerud et al., 1990). However, the solidification sequence is influenced by the presence of Mn and Fe, these elements starting to form their phases with Al and Si immediately after the beginning of primary (Al) solidification. The resultant $Al_{15}(FeMn)_3Si_2$ phase is formed during eutectic reaction that starts at 617°C and continues down to 587–589°C, transforming then to the ternary eutectics with the precipitation of Mg_2Si in a temperature range of 590–580°C. Table 4.9 shows the solidification reactions for the 512.0 alloy (Bäckerud et al., 1990). Solidification reactions in the 512.0 alloy agree well with our version of the Al–Mg–Mn–Si phase diagram shown in Figure 4.3c. At the Mg:Si ratio as in the 512.0 alloy, this alloys falls into the field

Table 4.8. Solidification reactions under nonequilibrium conditions in a 518.2 alloy (7.6% Mg, 0.01% Mn, 0.16% Fe, and 0.1% Si) (Bäckerud et al., 1990)

Reaction	Temperatures (°C) at a cooling rate	
	0.2 K/s	6 K/s
$L \Rightarrow (Al)$	612	610
$L \Rightarrow (Al) + Al_3Fe$	611–532	609–548
$L \Rightarrow (Al) + Al_3Fe + Mg_2Si$	532–518	548–511
$L \Rightarrow (Al) + Al_3Fe + Mg_2Si + Al_8Mg_5$	437–435	439–428
Solidus	435	428

Table 4.9. Solidification reactions under nonequilibrium conditions in a 512.0 alloy (4.0% Mg, 0.35% Mn, 0.55% Fe, and 1.8% Si) (Bäckerud et al., 1990).

Reaction	Temperatures (°C) at a cooling rate	
	0.3 K/s	6 K/s
$L \Rightarrow (Al)$	627–624	626–623
$L \Rightarrow (Al) + Al_{15}(FeMn)_3Si_2$	617–589	609–587
$L \Rightarrow (Al) + Al_{15}(FeMn)_3Si_2 + Mg_2Si$	589–580	587–579
Solidus	580	579

e_2–E_2–P_2–p_2* (Figure 4.3c). Therefore, the first eutectic reaction in this alloy is $L \Rightarrow (Al) + Al_{15}(FeMn)_3Si_2$ just as it is shown in Table 4.9. After that, the ternary eutectics with participation of Mg_2Si has to form along the P_2–E_2 line in Figure 4.3c and finish the solidification sequence.

4.6. ALLOY 3004

A 3004 alloy is a representative of 3XXX-series (Al–Mn) alloys that additionally contain magnesium, which improves mechanical properties while preserving good corrosion resistance. The ternary Al–Mg–Mn phase diagram (Figures 4.1, 4.2b) suggests that the solidification starts at 658°C with the formation of aluminum grains, then Al_6Mn is formed through a eutectic reaction.

The presence of unavoidable impurities of Fe and Si makes some alterations to the solidification reactions and temperatures. First of all, the Al_6Mn phase is replaced by $Al_6(FeMn)$, then the $Al_{15}(MnFe)_3Si_2$ and Mg_2Si phases appear. The characteristic temperatures of solidification reactions are decreased somewhat by the presence of Fe and Si. The entire sequence of solidification is summarized in Table 4.10 based on experiments reported by Bäckerud et al. (1986). The solidification starts with the formation of primary aluminum grains at 652°C. The $Al_6(FeMn)$ phase forms at about 644°C by a eutectic reaction, the temperature of which is about 10°C lower than in a Mg-free alloy (3003). $Al_6(FeMn)$ reacts with liquid through a peritectic reaction at 630°C to form $Al_{15}(FeMn)_3Si_2$. An interesting observation that is made by Bäckerud et al. (1986) is that, contrary to usual trend,

* In this analysis, one should take into account that (1) all iron in bound to the $Al_{15}(FeMn)_3Si_2$ phase and the effective concentration of Mn in the Al–Mg–Mn–Si system would be [Fe] + [Mn] = 0.9% and (2) due to the considerable solubility of Mg in (Al) the effective concentration of Mg is less than the nominal one and could be taken as 2% Mg (the rest of Mg remains in the solid solution). Thus the alloy composition for the analysis is Al–2% Mg–0.9% Mn–1.8% Si.

Table 4.10. Solidification reactions under nonequilibrium conditions in a 3004 alloy (1.0% Mg, 0.99% Mn, 0.43% Fe, and 0.14% Si) (Bäckerud et al., 1986)

Reaction	Temperatures (°C) at a cooling rate	
	0.4 K/s	9.0 K/s
$L \Rightarrow (Al)$	652–648	652–649
$L \Rightarrow (Al) + Al_6(FeMn)$	644–643	639
$L + Al_6(FeMn) \Rightarrow (Al) + Al_{15}(FeMn)_3Si_2$	630	630
$L \Rightarrow (Al) + Al_{15}(FeMn)_3Si_2 + Mg_2Si$	587	586
Solidus	583	580

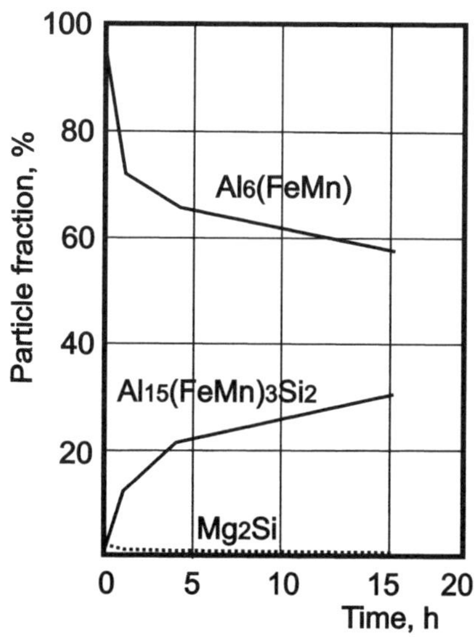

Figure 4.11. Variation in phase fractions during annealing of a cast 3004 alloy at 585°C (after Trømborg et al., 1993).

the peritectic reaction fully completes at higher cooling rates and only partially – upon slower cooling. This is explained from the size of $Al_6(FeMn)$ particles that become coarser at low cooling rates. The solidification ends at about 580°C with the formation of Mg_2Si by a multi-phase eutectic reaction.

Particles of $Al_6(FeMn)$ remaining in as-cast material due to the incomplete peritectic reaction should dissolve during annealing giving place to the more stable $Al_{15}(FeMn)_3Si_2$ phase. The occurrence of this process was confirmed by Trømborg et al. (1993) by measuring the amount of phase particles after different annealing times at 585°C. The results are given in Figure 4.11.

4.7. CASTING 3XX.0 ALLOYS CONTAINING MAGNESIUM AND MANGANESE

Casting alloys of the Al–Mg–Si system are discussed in Chapter 2, here we only consider the effect of Mn on solidification paths in A356.0-type alloys. The main implication of Mn on the sequence of solidification in this type of alloys is the occurrence of two eutectic reactions with the formation of the $Al_{15}(FeMn)_3Si_2$ phase. The amount of this phase is larger at higher cooling rates.

Bäckerud et al. (1990) suggested schematic solidification paths that are relevant to A356-type alloys with and without manganese. These diagrams are given in Figure 4.12.

Table 4.11 summarizes the effects of composition and cooling rate on solidification reactions in two A356.0-type alloys.

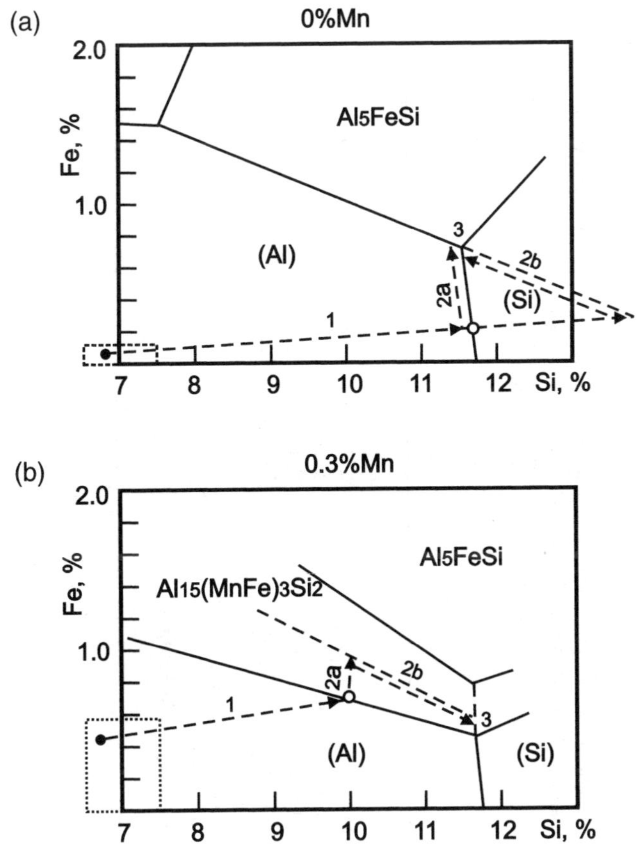

Figure 4.12. Diagrams of solidification paths for two A356-type alloys containing 6.7–6.8% Si, 0.30–0.35% Mg and (a) 0.08% Fe, 0.0% Mn and (b) 0.44% Fe, 0.30% Mn (after Bäckerud et al., 1990). Numbers on lines correspond to solidification reactions in Table 4.11.

Table 4.11. Solidification reactions under nonequilibrium conditions in 356.0-type alloys containing 6.7–6.8% Si and 0.30–0.35% Mg ("pure": 0.08% Fe, 0.0% Mn; "with Mn": 0.44% Fe, 0.30% Mn) (Bäckerud et al., 1990)

Reaction	Notation in Figure 4.12	Temperatures (°C) at a cooling rate			
		0.2–0.3 K/s		5.0 K/s	
		Pure	With Mn	Pure	With Mn
$L \Rightarrow (Al)$	1	615–609	612–610	614–608	614–611
$L \Rightarrow (Al) + Si$	2a (Figure 4.12a)	609–563	–	608–555	–
$L \Rightarrow (Al) + Al_{15}(FeMn)_3Si_2$	2a (Figure 4.12b)	–	610–570	–	611–560
$L \Rightarrow (Al) + Al_5FeSi$	2b (Figure 4.12a)	n/a	–	n/a	–
$L \Rightarrow (Al) + Al_5FeSi + Al_{15}(FeMn)_3Si_2$	2b (Figure 4.12b)	–	594–569	–	590–560
$L \Rightarrow (Al) + Si + Al_5FeSi$	3 (Figure 4.12a,b)	575	575–555	575	575–555
$L + Al_5FeSi \Rightarrow (Al) + Si + Al_8FeMg_3Si_6$	–	564–561	–	554–555	–
$L \Rightarrow (Al) + Si + Mg_2Si$	–	561–553	555–539	554–534	555–541
$L \Rightarrow (Al) + Si + Mg_2Si + Al_8FeMg_3Si_6$	–	553–544	–	534–519	541–505
Solidus	–	544	539	519	505

The "pure" alloy starts solidification with the formation of the aluminum solid solution, then the binary (Al) + (Si) and ternary (Al) + (Si) + Al_5FeSi eutectics are formed. Due to the presence of Mg in the composition of the alloy, the solidification continues to lower temperatures and ceases at 554°C (equilibrium temperature) with the formation of a complex eutectics as shown in Table 4.11. Kinetic effects that are observed in this alloy include the formation of coarse Al_5FeSi particles before the onset of the ternary eutectic reaction (extension 2b in Figure 4.12a), decreasing temperatures of solidification reactions and, eventually, a lower solidus with increasing cooling rates. The peritectic transformation of Al_5FeSi to $Al_8FeMg_3Si_6$ is observed only in the Mn-free alloy.

According to Bäckerud et al. (1990) the addition of Mn (and Fe) to the alloys results in the alteration of phase transformations during solidification. After the formation of primary aluminum grains, the $Al_{15}(FeMn)_3Si_2$ and Al_5FeSi phases are formed through binary and ternary eutectic reactions. Only after that the main eutectic containing (Si) and Al_5FeSi phases is solidified. The rest of solidification occurs as in the alloy without Mn. The amount of the $Al_8FeMg_3Si_6$ phase increases with increasing cooling rate in both alloys; and in the alloy with Mn this phase is observed only at high cooling rates.

It is worth to note that, according to the experimental data of Bäckerud et al. (1990) the addition of Mn considerably decreases the solidus of A356.0-type alloys, especially at high cooling rates occurring during chill and die casting, which contradicts the phase diagram, as manganese does not form low-temperature eutectics, and the heat treating practice when 356.0-type alloys are solution treated in the range of 530–540°C without any problems.

Chapter 5
Alloys of the Al–Cu–Mn–(Mg, Fe, Si) System

This chapter considers the phase composition of alloys with copper and manganese as the main components. Many casting and wrought alloys of the 2XX.0 and 2XXX series belong to this system. As these alloys often contain magnesium, silicon and iron (as alloying elements or impurities), in most cases the analysis of at least quaternary diagrams is required. First and foremost, this is the Al–Cu–Mg–Mn diagram that is essential for the correct analysis of the phase composition of important commercial alloys of the 2024 type. In these alloys, manganese has a significant effect on the phase composition, which makes insufficient the use of the ternary Al–Cu–Mg phase diagram only.

5.1. Al–Cu–Mn PHASE DIAGRAM

This phase diagram can be used to correctly analyze the phase composition of 224.0-type casting alloys and 2219-type wrought alloys at low concentrations of magnesium, iron, and silicon impurities in them (Table 5.1). The use of only the binary Al–Cu diagram is inadequate due to pronounced effects of Mn on the phase composition and solidification reactions.

The aluminum corner of the Al–Cu–Mn diagram contains the Al_2Cu and Al_6Mn phases and a ternary compound usually designated as T. The ternary T phase has a homogeneity range of 12.8–19% Cu and 19.8–24% Mn. Two formulae of the compound – $Al_{20}Cu_2Mn_3$ (15.3% Cu, 19.8% Mn) and $Al_{12}CuMn_2$ (12.8% Cu, 22.1% Mn) – are possible within the limits of this concentration range. Up to 0.1% Mn dissolves in the Al_2Cu phase, and about 0.2% Cu, in Al_6Mn.

The invariant reactions occurring in Al-rich ternary alloys are listed in Table 5.2, and the respective monovariant reactions – in Table 5.3. The Al–Cu–Mn diagram in the Al-rich region is shown in Figure 5.1. Due to the close temperatures of the ternary and binary (Al–Cu) eutectics, an addition of manganese does not noticeably decrease the solidus of commercial alloys.

The mutual solubility of copper and manganese in solid aluminum is given in Table 5.4. These data suggest that in the range of 2XX.0 and 2XXX-series alloys the equilibrium solubility of Mn in (Al) is significantly lower than that in 3XXX-series

Table 5.1. Chemical composition of some commercial alloys whose phase composition can be analyzed using Al–Cu–Mn phase diagram

Grade	Cu, %	Mn, %	Other		
			Mg, %	Si, %	Fe, %
224.0	4.5–5.5	0.2–0.5	–	0.06	0.1
AM5(rus)	4.5–5.3	0.6–1.0	0.05	0.3	0.3
2219	5.8–6.8	0.2–0.4	0.02	0.2	0.3

Table 5.2. Invariant reactions in ternary alloys of Al–Cu–Mn system (Mondolfo, 1976; Drits et al., 1977)

Reaction	Point in Figure 5.1a	*T*, °C	Concentrations in liquid phase	
			Cu, %	Mn, %
$L + Al_4Mn \Rightarrow Al_6Mn + Al_{20}Cu_2Mn_3$	P_1	625	15.6	2.1
$L + Al_6Mn \Rightarrow (Al) + Al_{20}Cu_2Mn_3$	P_2	616	14.8	0.9
$L \Rightarrow (Al) + Al_2Cu + Al_{20}Cu_2Mn_3$	E	547.5	32.5	0.6

Table 5.3. Monovariant reactions in ternary alloys of Al–Cu–Mn system

Reaction	Line in Figure 5.1a	*T*, °C
$L \Rightarrow (Al) + Al_2Cu$	e_2–E	548–547
$L \Rightarrow (Al) + Al_{20}Cu_2Mn_3$	P_2–E	616–547
$L \Rightarrow (Al) + Al_6Mn$	e_1–P_2	658–616

Table 5.4. Limit solid solubility of Cu and Mn in (Al) in Al–Cu–Mn alloys (Drits et al., 1977)

T, °C	$(Al) + Al_6Mn + Al_{20}Cu_2Mn_3$		$(Al) + Al_2Cu + Al_{20}Cu_2Mn_3$	
	Cu, %	Mn, %	Cu, %	Mn, %
623.5	1.4	1.17	–	–
616	1.3	1.0	–	–
610	1.3	1.0	–	–
600	1.1	0.9	–	–
550	0.85	0.6	–	–
547.5	–	–	5.5	0.2
525	0.95	0.44	4.95	0.2
500	0.65	0.4	4.05	0.2
450	0.5	0.2	2.55	0.15
400	0.4	0.1	1.5	0.1

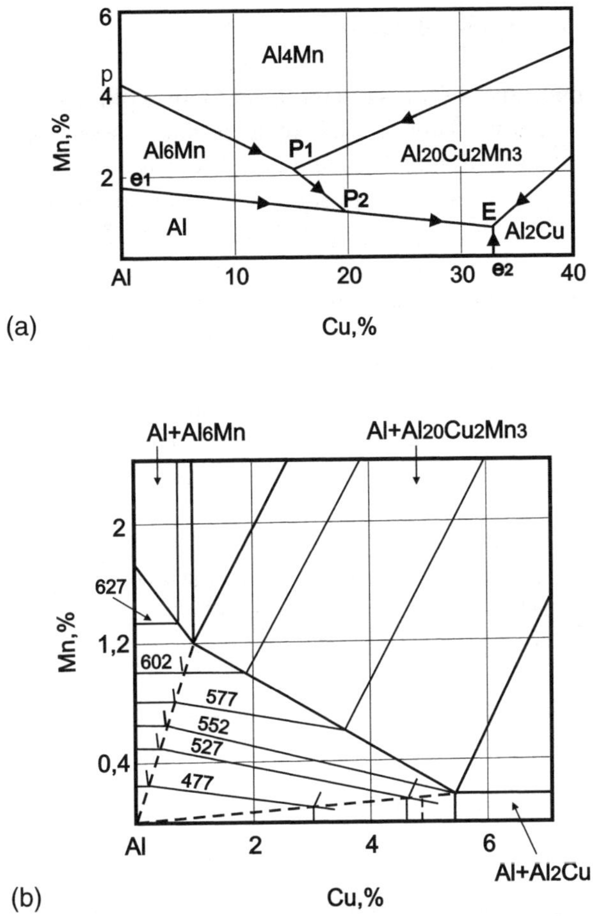

Figure 5.1. Phase diagram of Al–Cu–Mn system: (a) liquidus; (b) solidus.

alloys. However, this does not affect much the supersaturation of (Al) in Mn during nonequilibrium solidification. In particular, according to our data on casting alloys containing 5% Cu, the concentration of manganese in the solid solution supersaturated during solidification can reach 2%. Major deviations from the equilibrium during solidification are due to the formation of the nonequilibrium (Al) + Al_2Cu eutectics and a supersaturated solid solution of Mn in (Al). The decomposition of the latter during homogenization or another heat treatment associated with the heating to over 300–350°C leads to the formation of Mn-containing dispersoids, mainly represented by $Al_{20}Cu_2Mn_3$.

5.2. Al–Cu–Mg–Mn PHASE DIAGRAM

In spite of the importance of this system for the analysis of many 2XX.0- and 2XXX-series commercial alloys (Table 5.5), it remains poorly examined. Distribution of phases in the solid state (Figure 5.2a) and liquidus projection (Figure 5.2b) given by Mondolfo (1976) as well as invariant solidification reactions listed in Table 5.6 are largely hypothetical. According to this version of the phase diagram, only the phases from the constituent binary and ternary systems can be in equilibrium with (Al) in the quaternary system (see Sections 3.2, 4.1, 5.1). All monovariant lines of the quaternary phase diagram lie close to the Al–Cu–Mg face of the concentration tetrahedron, and the corresponding invariant points of the quaternary system are close to those of the Al–Cu–Mg ternary system. As the effect of manganese on the liquidus and solidus can be considered negligible, the Al–Cu–Mg diagram can serve as a reference in determining these temperatures for quaternary alloys (see Section 3.2).

Table 5.5. Chemical composition of some commercial alloys whose phase composition can be analyzed using Al–Cu–Mg–Mn diagram

Grade	Cu, %	Mn, %	Mg, %	Other	
				Si, %	Fe, %
201.0	4.0–5.0	0.2–0.4	0.15–0.35	0.05	0.1
206.0	4.2–5.0	0.2–0.5	0.15–0.35	0.1	0.15
2037	1.4–2.2	0.1–0.4	0.3–0.8	0.5	0.5
2048	2.8–3.8	0.2–0.6	1.2–1.8	0.15	0.2
2224	3.8–4.4	0.3–0.9	1.2–1.8	0.12	0.15
2024	3.8–4.9	0.3–0.9	1.2–1.8	0.5	0.5
2001	5.2–6.0	0.15–0.50	0.20–0.45	0.20	0.20

Table 5.6. Invariant reactions in quaternary alloys of Al–Cu–Mg–Mn system (Mondolfo, 1976)

Reaction	Point in Figure 3.4b	Concentrations in liquid phase			*T*, °C
		Cu, %	Mg, %	Si, %	
$L \Rightarrow (Al) + Al_2Cu + Al_2CuMg + Al_{20}Cu_2Mn_3$	E_1	~32	~6	~0.5	~503
$L + MnAl_6 \Rightarrow (Al) + Al_2CuMg + Al_{20}Cu_2Mn_3$ or $L + Al_{20}Cu_2Mn_3 \Rightarrow (Al) + Al_2CuMg + MnAl_6$	P_1*	–	–	–	–
$L + AlCuMg_2 \Rightarrow (Al) + Al_6CuMg_4 + Al_6Mn$	P_2	~10	~25	~0.3	~467
$L \Rightarrow (Al) + Al_6CuMg_4 + Al_{10}(MgMn)_3 + Al_8Mg_5$	E_2	~2.5	~30	~0.2	~447
$L + Al_6Mn \Rightarrow (Al) + Al_8Mg_5 + Al_{10}(MgMn)_3$	P_3	<2.5	<32	<0,2	–

* Mondolfo gives the second reaction (Mondolfo, 1976)

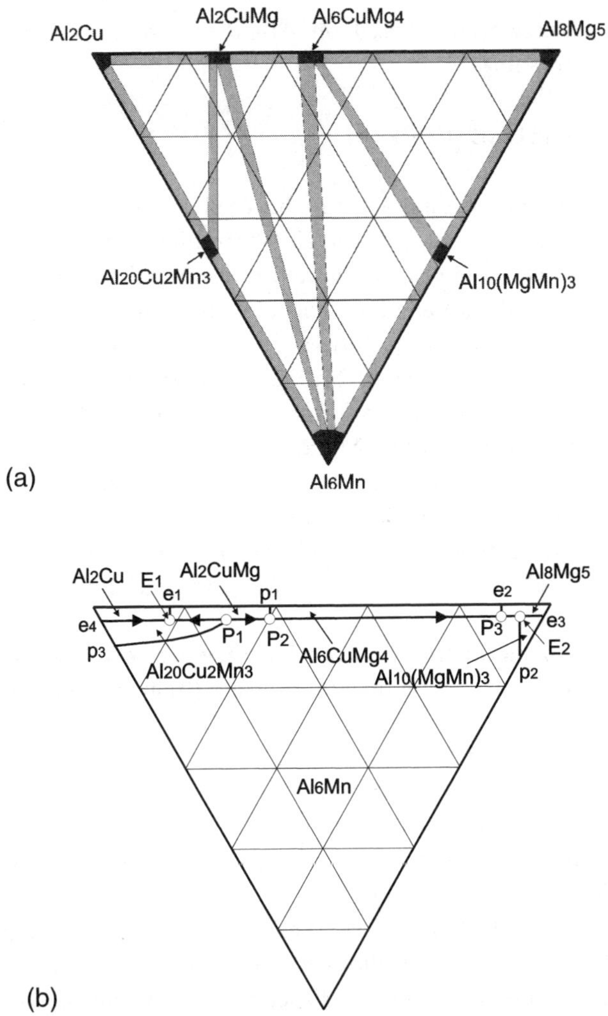

Figure 5.2. Phase diagram of Al–Cu–Mg–Mn system: (a) distribution of phase fields in the solid state and (b) polythermal projection of liquidus.

Within the concentration limits typical of commercial alloys (<1% Mn), manganese completely goes into the solid solution during nonequilibrium solidification. In following heating, e.g. homogenization, Mn-containing dispersoids are formed and the phase composition approaches equilibrium as shown in Figure 5.2a. As copper and magnesium participate in the formation of two Mn-containing phases ($Al_{20}Cu_2Mn_3$ and $Al_{10}(MgMn)_3$), the precipitation of these dispersoids can decrease the amount of free copper and magnesium in the solid solution available for the formation of precipitates based on Al_2Cu and Al_2CuMg, which may affect strengthening upon aging.

5.3. Al–Cu–Fe–Mn PHASE DIAGRAM

The Al–Cu–Fe–Mn phase diagram can be used in analyzing the effect of iron impurity on the phase composition of 2219-type alloys at a low concentration of silicon (Table 5.1). This quaternary diagram is also required for the construction of five-component phase diagrams of systems containing manganese.

No true quaternary phases are formed in the aluminum corner of the Al–Cu–Fe–Mn system. However, as the Al_6Mn and $Al_6(FeCu)$ phases are isomorphic, a continuous series of solid solutions is formed between them. The resultant phase field is designated as $(AlCu)_6(FeCuMn)$ (Mondolfo, 1976). Mondolfo (1976) reports that $(AlCu)_6(FeCuMn)$ crystals extracted from an alloy containing 7.76% Cu, 0.75% Mn, and 1.5% Fe have an orthorhombic crystal structure with lattice parameters $a = 0.7473$ nm, $b = 0.6452$ nm, $c = 0.8794$ nm. These values are in good agreement with the lattice parameters of the Al_6Mn and $Al_6(FeCu)$ phases (Sections 1.2 and 3.3).

Figure 5.3 shows the distribution of phase regions in the solid state (a) and a polythermal projection of the solidification surface (b) in the aluminum corner of the Al–Cu–Fe–Mn system.

This quaternary system is characterized by two invariant five-phase reactions involving (Al) as shown in Table 5.7. The monovariant reactions are given in Table 5.8. It should be noted that the monovariant line e_3–p_3 (Figure 5.3b) changes its character from eutectic in point e_3 ($L \Rightarrow (Al) + Al_3Fe + Al_6Mn$) to peritectic in point p_3 ($L + Al_3Fe \Rightarrow (Al) + Al_6(FeCu)$).

The presence of the $(AlCu)_6(FeCuMn)$ phase within a broad compositional range strongly impedes the analysis of alloys belonging to this system, because without direct experimental determination it is difficult to assess how much of copper, iron, and manganese is bound in this phase. Additional challenges are presented by non-equilibrium solidification of quaternary alloys in this system. In particular, the following deviations from the equilibrium phase composition can occur: (a) formation of a supersaturated solid solution of Mn in (Al); (b) incomplete peritectic reaction (Table 5.7); and (c) formation of nonequilibrium eutectics involving the Al_2Cu phase at a comparatively low concentration of copper. The ternary Al–Cu–Mn phase diagram cannot give correct answers because it does not take into account reactions between Cu, Mn, and Fe. For example, the concentration of Mn in a supersaturated solid solution of a quaternary alloy can be much lower than in the Al–Cu–Mn system (Section 5.2), because some manganese binds to Fe-containing phases during solidification.

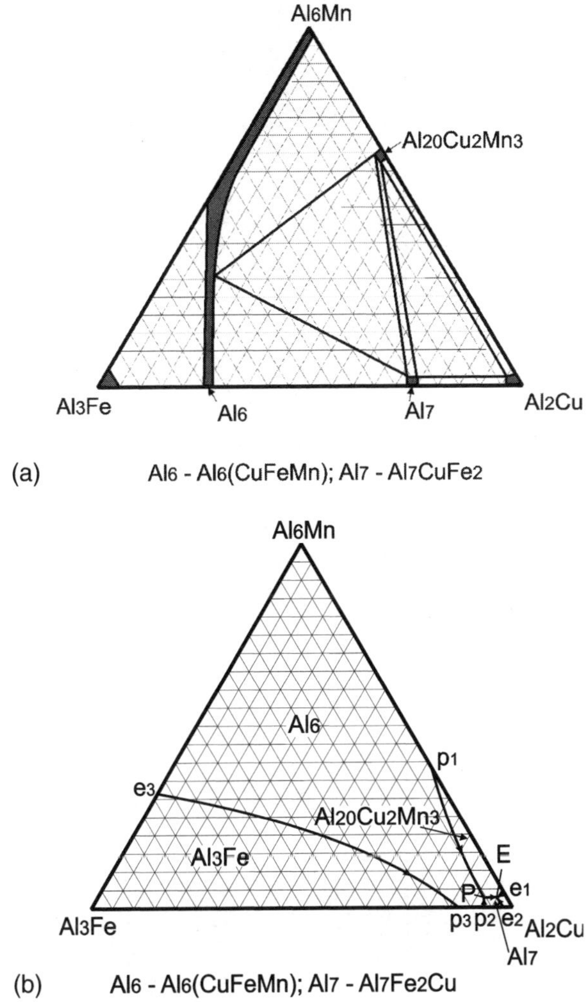

Figure 5.3. Phase diagram of Al–Cu–Fe–Mn system: (a) distribution of phase fields in the solid state and (b) polythermal projection of liquidus.

Table 5.7. Invariant reactions in quaternary alloys of Al–Cu–Fe–Mn system

Reaction	Point in Figure 5.3b	Concentrations in liquid phase			*T*, °C
		Cu, %	Mn, %	Fe, %	
$L \Rightarrow (Al) + Al_2Cu + Al_7Cu_2Fe + Al_{20}Cu_2Mn_3$	E	31–33	<0.5	<0.5	>537
$L + Al_6Mn^* \Rightarrow (Al) + Al_7Cu_2Fe + Al_{20}Cu_2Mn_3$	P	15–20	<1	<1	<587

* $(AlCu)_6(CuFeMn)$

Table 5.8. Bivariant and monovariant reactions in quaternary alloys of Al–Cu–Fe–Mn system

Bivariant reactions	Field in Figure 5.3b	T, °C	Monovariant reactions	Line in Figure 5.3b	T, °C
$L \Rightarrow (Al) + Al_2Cu$	Al_2Cu–e_2–E–e_1	548–537	$L \Rightarrow (Al) + Al_2Cu + Al_7Cu_2Fe$	e_2–E	545–537
$L \Rightarrow (Al) + Al_{20}Cu_2Mn_3$	e_1–E–P–p_1	616–537	$L \Rightarrow (Al) + Al_2Cu + Al_{20}Cu_2Mn_3$	e_1–E	547–537
$L \Rightarrow (Al) + Al_7Cu_2Fe$	e_2–p_2–P–E	590–537	$L \Rightarrow (Al) + Al_7Cu_2Fe + Al_{20}Cu_2Mn_3$	P–E	587–537
$L \Rightarrow (Al) + Al_6Mn$	Al_6Mn –p_1–P–p_2–p_3–e_3	658–587	$L + Al_6Mn \Rightarrow (Al) + Al_{20}Cu_2Mn_3$	p_1–P	616–587
$L \Rightarrow (Al) + Al_3Fe$	Al_3Fe– e_3–p_3	655–620	$L + Al_6Mn \Rightarrow (Al) + Al_7Cu_2Fe$	p_2–P	590–587
			$L \Rightarrow (Al) + Al_3Fe + Al_6Mn$*	e_3–p_3	654–620

* Reaction transforms to the peritectic reaction $L + Al_3Fe \Rightarrow (Al) + Al_6Mn$

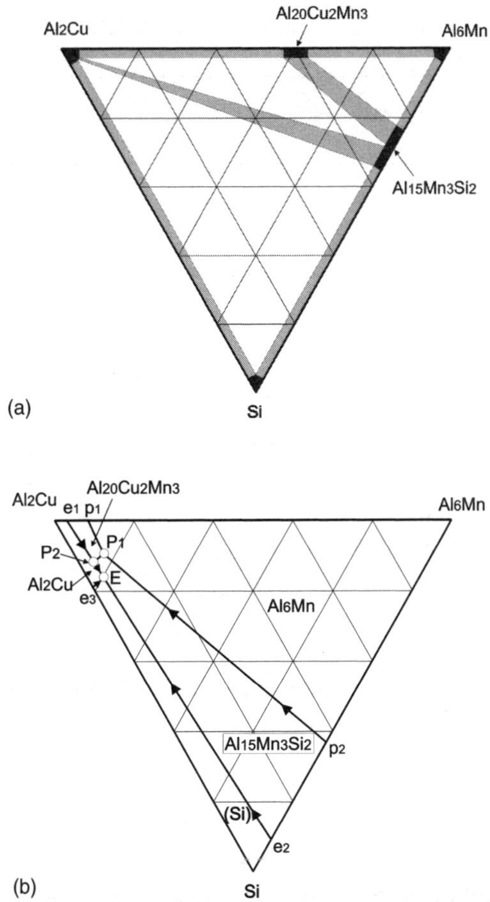

Figure 5.4. Phase diagram of Al–Cu–Mn–Si system: (a) distribution of phase fields in the solid state and (b) polythermal projection of liquidus.

5.4. Al–Cu–Mn–Si PHASE DIAGRAM

The phase diagram of this system can be used for the analysis of phase composition of alloys containing copper, manganese, and silicon at a low content of iron impurity, examples of such alloys are given in Table 5.9. This quaternary diagram is also required for the evaluation of quinary alloys, i.e. the Al–Cu–Fe–Mn–Si system.

As in the case of the previous system, this phase diagram is also largely hypothetical. The distribution of phases in the solid state (Figure 5.4a) and liquidus projection (Figure 5.4b), as well as solidification reactions (Tables 5.10 and 5.11), are given according to the assessment by Mondolfo (1976). According to this variant, only the phases from the constituent ternary systems can be in equilibrium with (Al)

Table 5.9. Chemical composition of some commercial alloys whose phase composition can be analyzed using Al–Cu–Mn–Si diagram

Grade	Cu, %	Mn, %	Si, %	Other	
				Mg, %	Fe, %
2003	4.0–5.0	0.3–0.8	<0.3	0.02	0.30
2025	3.5–5.0	0.4–1.2	0.5–1.2	0.05	1.0
2021	5.6–6.8	0.2–0.4	0.2	0.02	0.30

Table 5.10. Invariant reactions in quaternary alloys of Al–Cu–Mn–Si system (Mondolfo, 1976)

Reaction	Point in Figure 5.4b	Concentrations in liquid phase			*T*, °C
		Cu, %	Mn, %	Si, %	
$L \Rightarrow (Al) + Al_2Cu + (Si) + Al_{15}Mn_3Si_2$	E	~25	~1	~5	~517
$L + Al_{20}Cu_2Mn_3 \Rightarrow (Al) + Al_2Cu + Al_{15}Mn_3Si_2$	P_2	~20	~1	~3	~547
$L + Al_6Mn \Rightarrow (Al) + Al_{20}Cu_2Mn_3 + Al_{15}Mn_3Si_2$	P_1	~15	~1.5	~4	~597

in the Al–Cu–Mn–Si system. As manganese only slightly affects the liquidus and solidus temperatures, the Al–Cu–Si diagram can be the reference in determining these temperatures (Section 3.1).

In the range of Al–Cu alloys, depending on the ratio between silicon and manganese, no more than two of the following three phases – $Al_{20}Cu_2Mn_3$, $Al_{15}Mn_3Si_2$, and (Si) – can be in equilibrium with (Al) and Al_2Cu. All these phases can form during the solidification (mainly by eutectic reactions) and also precipitate from (Al). In Al–Si alloys, only Al_2Cu and $Al_{15}Mn_3Si_2$ can be in equilibrium with (Al) and (Si). The Al_2Cu phase can be both of solidification and secondary origin, and the ternary compound mainly forms upon solidification as eutectic or primary structure constituent.

Under nonequilibrium conditions, as in other systems with manganese, a supersaturated solid solution of Mn in (Al) can be formed during solidification and cooling in the solid state. As the concentration of silicon increases, the amount of the $Al_{15}Mn_3Si_2$ phase formed during solidification goes up.

5.5. Al–Cu–Fe–Mn–Si PHASE DIAGRAM (FOR Al–Cu AND Al–Si ALLOYS)

The phase diagram of this quinary system provides sufficient information for the correct analysis of the phase composition of many commercial alloys of

Table 5.11. Bivariant and monovariant reactions in quaternary alloys of the Al–Cu–Mn–Si system

Bivariant reactions	Field in Figure 5.4b	T, °C	Monovariant reactions	Line in Figure 5.4b	T, °C
$L \Rightarrow (Al) + Al_2Cu$	$Al_2Cu–e_1–E–e_2$	548–517	$L \Rightarrow (Al) + Al_2Cu + (Si)$	$e_3–E$	525–517
$L \Rightarrow (Al) + (Si)$	$(Si)–e_3–E–e_2$	577–517	$L \Rightarrow (Al) + Al_2Cu + Al_{15}Mn_3Si_2$	$P_2–E$	547–517
$L \Rightarrow (Al) + Al_{15}Mn_3Si_2$	$p_2–P_1–P_2–E–e_2$	649–517	$L \Rightarrow (Al) + (Si) + Al_{15}Mn_3Si_2$	$e_2–E$	574–517
$L \Rightarrow (Al) + Al_6Mn$	$Al_6Mn–p_2–P_1–p_1$	658–597	$L \Rightarrow (Al) + Al_{20}Cu_2Mn_3 + Al_{15}Mn_3Si_2$	$P_1–P_2$	597–547
$L \Rightarrow (Al) + Al_{20}Cu_2Mn_3$	$p_1–P_1–P_2–e_1$	616–547	$L \Rightarrow (Al) + Al_2Cu + Al_{20}Cu_2Mn_3$	$e_1–P_2$	447.5–547
			$L + Al_6Mn \Rightarrow (Al) + Al_{20}Cu_2Mn_3$	$p_1–P_1$	616–597
			$L + Al_6Mn \Rightarrow (Al) + Al_{15}Mn_3Si_2$	$p_2–P_1$	648–497

Table 5.12. Chemical composition of some commercial alloys whose phase composition can be analyzed using Al–Cu–Fe–Mn–Si phase diagram

Grade	Cu, %	Mn, %	Si, %	Fe, %	Other		
					Mg, %	Ni, %	Zn, %
2003	4.0–5.0	0.3–0.8	0.3–0.1	0.3	0.02	–	0.10
2021	5.6–6.8	0.2–0.4	0.2	0.30	0.02	–	0.10
2219	5.8–6.8	0.2–0.4	0.2	0.3	0.02	–	0.10
308.0	4.0–5.0	0.5	5.0–6.0	1.0	0.1	–	1.0
383.0	2.0–3.0	0.5	9.5–11.5	1.3	0.1	0.3	–
208.2	3.5–4.5	0.3	2.5–3.5	0.8	0.03	–	0.2

2XXX-, 2XX.0-, and 3XX.0-series that do not contain magnesium and nickel (Tables 5.1, 5.5, and 5.12).

Assuming that only phases from the constituent quaternary systems can be in equilibrium with (Al) in the Al–Cu–Fe–Mn–Si system, we suggest the distribution of phase regions for Al–Si-rich and Al–Cu-rich alloys belonging to this system according to the method described in Appendix 3.

Silicon-rich alloys. According to the quaternary diagrams Al–Fe–Mn–Si, Al–Cu–Fe–Si, and Al–Cu–Mn–Si, only three phases – Al_2Cu, Al_5FeSi, $Al_{15}Mn_3Si_2$ – can be in equilibrium with (Al) and (Si). This conforms to one of the two following invariant reactions:

$$L + Al_5FeSi \Rightarrow (Al) + (Si) + Al_2Cu + Al_{15}(FeMn)_3Si_2 \text{ or}$$

$$L \Rightarrow (Al) + (Si) + Al_2Cu + Al_5FeSi + Al_{15}(FeMn)_3Si_2.$$

Figure 5.5 presents the second variant. The assumed composition of the eutectic point is as follows: ~25% Cu, ~5% Si, ~1% Mn, and ~0.4% Fe, and the reaction occurs at ~516°C. The Al–Si-rich portion of the phase diagram is characterized by a wide homogeneity range of the $Al_{15}(FeMn)_3Si_2$ phase (Figure 5.5a). If the concentrations of iron and manganese are small, i.e. the silicon phase forms earlier, and if the concentration of copper exceeds 4%, then only eutectic reactions occur during equilibrium solidification of quinary Si-rich alloys, as illustrated in Figure 5.5b and Tables 5.13 and 5.14.

It should be noted that the suggested version differs from that given by Phragmen (1950) according to whom the Al_7Cu_2Fe phase, not Al_5FeSi, is in equilibrium with (Al) and (Si), which leads to the hypothetical eutectic equilibrium: $L \Rightarrow (Al) + (Si) + Al_2Cu + Al_7Cu_2Fe + Al_{15}(FeMn)_3Si_2$.

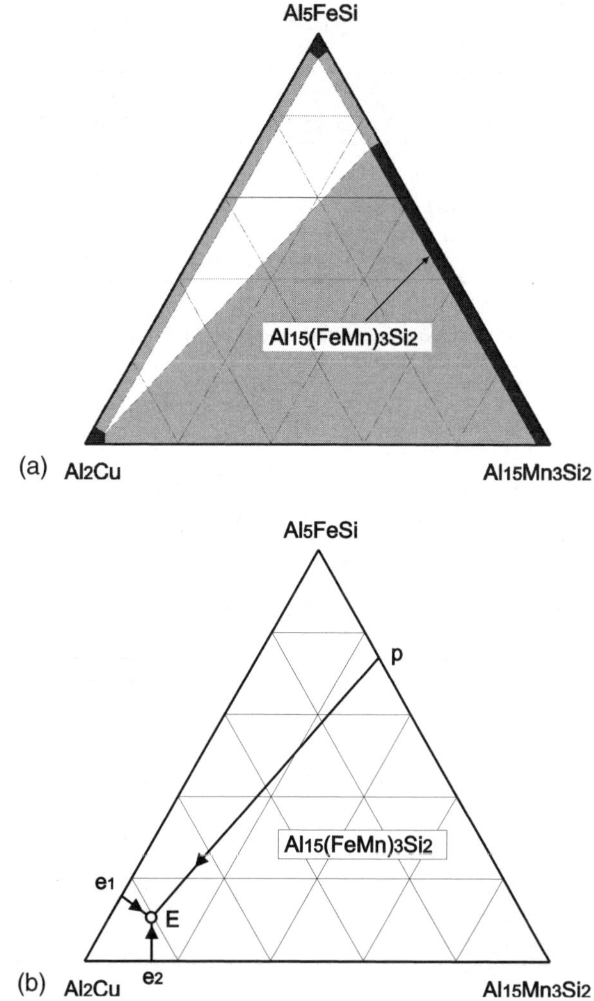

Figure 5.5. Phase diagram of Al–Cu–Fe–Mn–Si system in the range of Al–Si alloys: (a) distribution of phase fields in the solid state and (b) polythermal projection of liquidus.

Copper-rich alloys. In the quaternary systems Al–Cu–Fe–Mn, Al–Cu–Fe–Si, and Al–Cu–Mn–Si the following phases – (Si), Al_7Cu_2Fe, $Al_{20}Cu_3Mn_2$, Al_5FeSi, and $Al_{15}Mn_3Si_2$ – can be in equilibrium with (Al) and Al_2Cu. This suggests the occurrence of three invariant reactions in the Al–Cu-rich region of the quinary system (Table 5.15).

These invariant reactions in Table 5.15 differ from those given by Mondolfo (1976), according to whom the eutectic transformation $L \Rightarrow (Al) + (Si) + Al_2Cu +$

Table 5.13. Monovariant reactions in quinary alloys of Al–Cu–Fe–Mn–Si system with participation of (Al) and (Si) phases

Reaction	Line in Figure 5.5b	*T*, °C
$L \Rightarrow (Al) + (Si) + Al_2Cu + Al_5FeSi$	e_1–E	525–516
$L \Rightarrow (Al) + (Si) + Al_2Cu + Al_{15}(FeMn)_3Si_2$	e_2–E	517–516
$L + Al_5FeSi \Rightarrow (Al) + (Si) + Al_{15}(FeMn)_3Si_2$ transforms to $L \Rightarrow (Al) + (Si) + Al_{15}(FeMn)_3Si_2 + Al_5FeSi$	p–E	575–516

Table 5.14. Bivariant reactions in quinary alloys of Al–Cu–Fe–Mn–Si system with participation of (Al) and (Si) phases

Reaction	Field in Figure 5.5b	*T*, °C
$L \Rightarrow (Al) + (Si) + Al_2Cu$	Al_2Cu–e_1–E–e_2	515–516
$L \Rightarrow (Al) + (Si) + Al_5FeSi$	Al_5FeSi–p–E–e_1	576–516
$L \Rightarrow (Al) + (Si) + Al_{15}(FeMn)_3Si_2$	$Al_{15}Mn_3Si_2$–e_2–E–p	575–516

Table 5.15. Invariant reactions in quinary alloys of Al–Cu–Fe–Mn–Si system with participation of (Al) and Al_2Cu phases

Reaction	Point in Figure 5.6b	Concentrations in liquid phase				*T*, °C
		Cu, %	Fe, %	Mn, %	Si, %	
$L \Rightarrow (Al) + Al_2Cu + (Si) + Al_5FeSi + Al_{15}(FeMn)_3Si_2$	E	~25	~0.4	~1	~5	~516
$L + Al_7Cu_2Fe \Rightarrow (Al) + Al_2Cu + Al_5FeSi + Al_{15}(FeMn)_3Si_2$;	P_2	~25	~0.4	~1	~4	~533
$L + Al_{20}Cu_3Mn_2 \Rightarrow (Al) + Al_2Cu + Al_{15}(FeMn)_3Si_2 + Al_7Cu_2Fe$	P_1	~20	~0.3	~1	~3	~546

$Al_7Cu_2Fe + Al_{15}(FeMn)_3Si_2$ takes place. Mono- and bivariant reactions in this system are listed in Tables 5.16 and 5.17, respectively.

Figure 5.6 presents a version of this system based on the constitution of the constituent quaternary diagrams, by taking into account a wide homogeneity range of the $Al_{15}(FeMn)_3Si_2$ phase.

5.6. Al–Cu–Mg–Mn–Si PHASE DIAGRAM (FOR Al–Cu AND Al–Si ALLOYS)

This quinary system makes it possible to analyze the phase composition of many commercial 2XXX-, 2XX.0-, 6XXX-, and 3XX.0-series alloys with minor content of

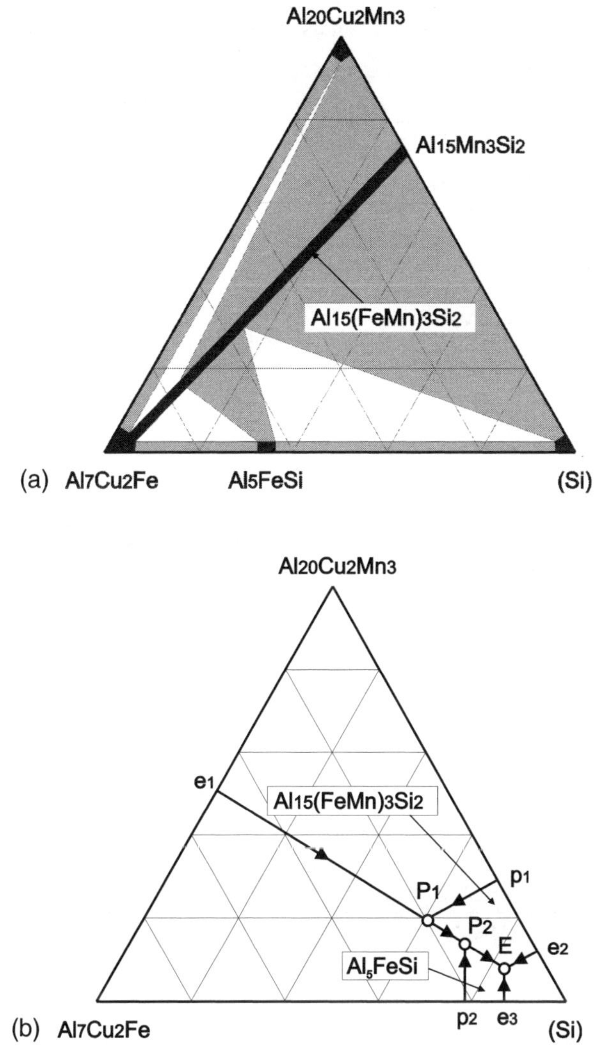

Figure 5.6. Phase diagram of Al–Cu–Fe–Mn–Si system in the range of Al–Cu alloys: (a) distribution of phase fields in the solid state and (b) polythermal projection of liquidus.

iron and nickel (Table 5.18). It is especially important for the analysis of 2214-type alloys, in which Cu, Mg, Mn, and Si are intentional additions and have a significant effect on the phase composition.

Assuming that only those phases that are available in the constitutive quaternary systems can be in equilibrium with (Al) in the Al–Cu–Mg–Mn–Si system, we suggest the distribution of phase regions for Al–Si and Al–Cu alloys according to the method described in Appendix 3.

Table 5.16. Monovariant reactions in quinary alloys of the Al–Cu–Fe–Mn–Si system with participation of (Al) and Al_2Cu phases

Reaction	Line in Figure 5.6b	T, °C
$L \Rightarrow (Al) + Al_2Cu + (Si) + Al_{15}(FeMn)_3Si_2$	e_2–E	517–516
$L \Rightarrow (Al) + Al_2Cu + (Si) + Al_5FeSi$	e_3–E	525–516
$L \Rightarrow (Al) + Al_2Cu + Al_5FeSi + Al_{15}(FeMn)_3Si_2$	P_2–E	533–516
$L + Al_{20}Cu_3Mn_2 \Rightarrow (Al) + Al_2Cu + Al_{15}(FeMn)_3Si_2$	p_1–P_1	547–546
$L \Rightarrow (Al) + Al_2Cu + Al_{20}Cu_3Mn_2 + Al_7Cu_2Fe$	e_1–P_1	537–546
$L + Al_7Cu_2Fe \Rightarrow (Al) + Al_2Cu + Al_5FeSi$	p_2–P_2	534–533
$L \Rightarrow (Al) + Al_2Cu + Al_{15}(FeMn)_3Si_2 + Al_7Cu_2Fe$	P_1–P_2	546–533

Table 5.17. Bivariant reactions in quinary alloys of the Al–Cu–Fe–Mn–Si system with participation of (Al) and Al_2Cu phases

Reaction	Field in Figure 5.6b	T, °C
$L \Rightarrow (Al) + Al_2Cu + Al_7Cu_2Fe$	Al_7Cu_2Fe–e_1–P_1–P_2 –p_2	545–533
$L \Rightarrow (Al) + Al_2Cu + Al_{20}Cu_2Mn_3$	$Al_{20}Cu_2Mn_3$– p_1–P_1 –e_1	547–546
$L \Rightarrow (Al) + Al_2Cu + Al_{15}(FeMn)_3Si_2$	p_1 –e_2–E–P_2–P_1	547–533
$L \Rightarrow (Al) + Al_2Cu + Al_5FeSi$	p_2 –P_2–E–e_3	525–516
$L \Rightarrow (Al) + Al_2Cu + (Si)$	(Si)–e_3–E–e_2	525–516

Table 5.18. Chemical composition of some commercial alloys whose phase composition can be analyzed using Al–Cu–Mg–Mn–Si phase diagram

Grade	Cu, %	Mn, %	Si, %	Mg, %	Other		
					Fe, %	Ni, %	Zn, %
2038	0.8–1.2	0.1–0.4	0.5–1.5	0.4–1	0.6	–	0.25
2017	2.5–4.5	0.4–1.0	0.2–0.8	0.4–0.8	0.7	–	0.25
2214	3.9–5.0	0.4–1.2	0.5–1.2	0.2–0.8	0.3	–	0.25
222.1	9.2–10.7	0.5	2	0.2–0.35	1.2	0.5	0.8
6066	0.7–1.2	0.6–1	0.9–1.8	0.8–1.4	0.5	–	0.25
6111	0.5–0.9	0.15–0.45	0.7–1.1	0.5–1.0	0.4	–	0.25
6013	0.6–1.1	0.2–0.8	0.6–1.0	0.8–1.2	0.5	–	0.25
B319.1	3.0–4.0	0.8	5.5–6.5	0.15–0.5	0.9	0.5	1.0
328.1	1.0–2.0	0.2–0.6	7.5–8.5	0.25–0.6	0.8	0.25	1.5
392.1	0.4–0.8	0.2–0.6	18–20	0.9–1.2	1.1	0.5	0.4

Silicon-rich alloys. According to the Al–Mg–Mn–Si, Al–Cu–Mg–Si, and Al–Cu–Mn–Si quaternary diagrams, only four phases – Mg_2Si, Al_2Cu, $Al_5Cu_2Mg_8Si_6$, and $Al_{15}Mn_3Si_2$ – can be in equilibrium with (Al) and (Si). This suggests the possibility of two invariant reactions shown in Table 5.19 (Mondolfo, 1976). Bi- and monovariant reactions which can proceed in Al–Si-rich alloys are given in Figure 5.7b and Tables 5.20 and 5.21. The distribution of the phases in the solid state is characterized by the presence of only two five-phase regions as shown in Figure 5.7a.

Copper-rich alloys. In the quaternary systems Al–Cu–Mn–Mg, Al–Cu–Mn–Si, and Al–Cu–Mg–Si, the following phases – (Si), $Al_{20}Cu_3Mn_2$, $Al_{15}Mn_3Si_2$, Mg_2Si, Al_2CuMg, and $Al_5Cu_2Mg_8Si_6$ – can be in equilibrium with (Al) and Al_2Cu. We

Table 5.19. Invariant reactions in quinary alloys of Al–Cu–Mg–Mn–Si system with (Al) and (Si) phases

Reaction	Point in Figure 5.7b	Concentrations in liquid phase				*T*, °C
		Si, %	Cu, %	Mg, %	Mn, %	
$L \Rightarrow (Al) + (Si) + Al_2Cu + Al_5Cu_2Mg_8Si_6 + Al_{15}Mn_3Si_2$	E	~6	~28	~2	~1	~505
$L + Mg_2Si + (Si) \Rightarrow (Al) + Al_5Cu_2Mg_8Si_6 + Al_{15}Mn_3Si_2$	P	~10	~14	~3	~1	~528

Table 5.20. Monovariant reactions in quinary alloys of Al–Cu–Mg–Mn–Si system with (Al) and (Si) phases

Reaction	Line in Figure 5.7b	*T*, °C
$L \Rightarrow (Al) + (Si) + Al_2Cu + Al_5Cu_2Mg_8Si_6$	e_1–E	507–505
$L \Rightarrow (Al) + (Si) + Al_2Cu + Al_{15}Mn_3Si_2$	e_3–E	517–505
$L \Rightarrow (Al) + (Si) + Al_5Cu_2Mg_8Si_6 + Al_{15}Mn_3Si_2$	P–E	528–517
$L \Rightarrow (Al) + (Si) + Mg_2Si + Al_{15}Mn_3Si_2$	e_2–P	567–528
$L + (Si) + Mg_2Si \Rightarrow (Al) + Al_5Cu_2Mg_8Si_6$	p–P	529–528

Table 5.21. Bivariant reactions in quinary alloys of Al–Cu–Mn–Mg–Si system with (Al) and (Si) phases

Reaction	Field in Figure 5.7b	*T*, °C
$L \Rightarrow (Al) + (Si) + Al_2Cu$	Al_2Cu–e_1–E–e_3	525–516
$L \Rightarrow (Al) + (Si) + Al_5Cu_2Mg_8Si_6$	p–P–E–e_1	529–516
$L \Rightarrow (Al) + (Si) + Al_{15}Mn_3Si_2$	$Al_{15}Mn_3Si_2$–e_3–E–P–e_2	573–517
$L \Rightarrow (Al) + (Si) + Mg_2Si$	Mg_2Si–e_2–P–p_1	555–528

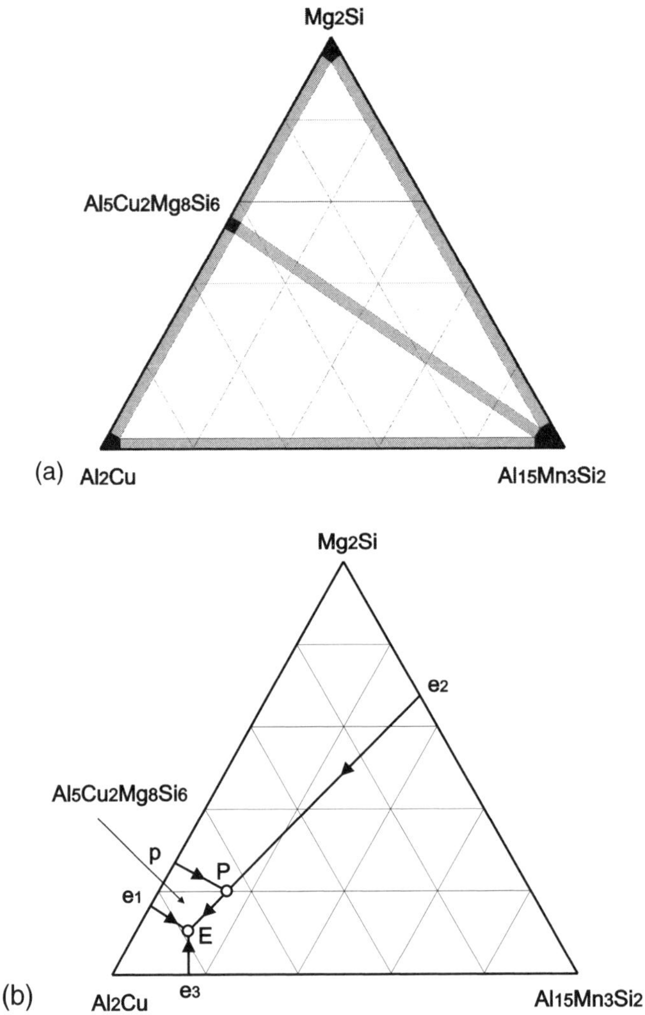

Figure 5.7. Phase diagram of Al–Cu–Mg–Mn–Si system in the range of Al–Si alloys: (a) distribution of phase fields in the solid state and (b) polythermal projection of liquidus.

suggest a variant of polyhedration of this system shown in Figure 5.8. Table 5.22 lists invariant reactions that occur in Al–Cu-rich alloys of this system. Note that the eutectic reaction E_2 differs from that given by Mondolfo, with the $Al_{15}Mn_3Si_2$ phase participating instead of $Al_{20}Cu_3Mn_2$. In our version of the phase diagram, we take into account that the $Al_{20}Cu_3Mn_2$ phase disappears through a peritectic reaction from the Al–Cu–Mn–Si system (P_2 in Table 5.10). Mono- and bivariant reactions which can proceed in Al–Cu-rich alloys are given in Tables 5.23 and 5.24, respectively.

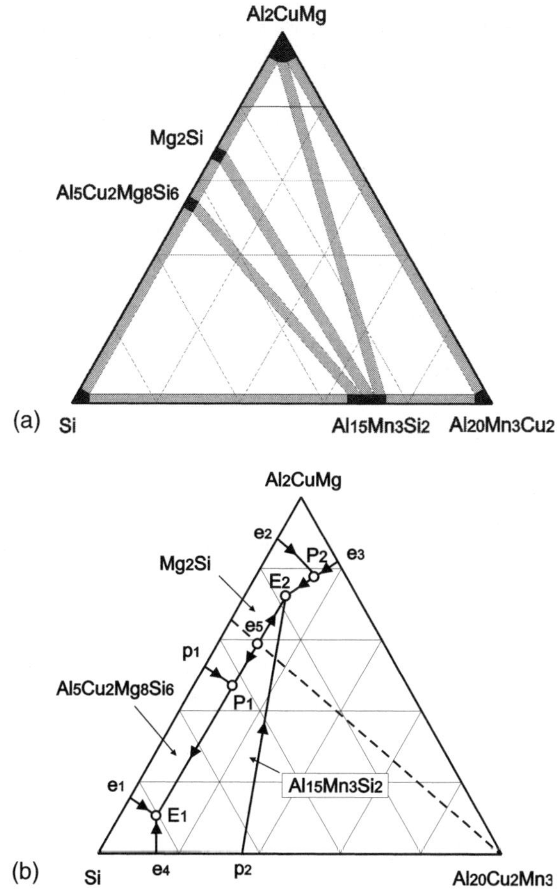

Figure 5.8. Phase diagram of Al–Cu–Mg–Mn–Si system in the range of Al–Cu alloys: (a) distribution of phase fields in the solid state and (b) polythermal projection of liquidus.

Table 5.22. Invariant reactions in quinary alloys of Al–Cu–Mg–Mn–Si system with (Al) and Al_2Cu phases

Reaction	Point in Figure 5.8b	Concentrations in liquid phase				*T*, °C
		Si, %	Cu, %	Mg, %	Mn, %	
$L \Rightarrow (Al) + Al_2Cu + (Si) + Al_5Cu_2Mg_8Si_6 + Al_{15}Mn_3Si_2$	E_1	~6	~28	~2	~1	~505
$L + Mg_2Si \Rightarrow (Al) + Al_2Cu + Al_5Cu_2Mg_8Si_6 + Al_{15}Mn_3Si_2$	P_1	~3	~31	~3	~1	~511
$L \Rightarrow (Al) + Al_2Cu + Al_2CuMg + Mg_2Si + Al_{15}Mn_3Si_2$	E_2	~0.3	~30	~7	~1	~500
$L + Al_{20}Cu_3Mn_2 \Rightarrow (Al) + Al_2Cu + Mg_2Si + Al_{15}Mn_3Si_2$	P_2	~0.4	~30	~6	~1	~502

Table 5.23. Monovariant reactions in quinary alloys of Al–Cu–Mg–Mn–Si system with (Al) and Al_2Cu phases

Reaction	Line in Figure 5.8b	*T*, °C
L ⇒ (Al) + Al_2Cu + (Si) + $Al_5Cu_2Mg_8Si_6$	e_1–E_1	507–505
L ⇒ (Al) + Al_2Cu + (Si) + $Al_{15}Mn_3Si_2$	e_4–E_1	517–505
L ⇒ (Al) + Al_2Cu + $Al_5Cu_2Mg_8Si_6$ + $Al_{15}Mn_3Si_2$	P_1–E_1	511–505
L ⇒ (Al) + Al_2Cu + Al_2CuMg + Mg_2Si	e_2–E_2	505–500
L ⇒ (Al) + Al_2Cu + Al_2CuMg + $Al_{20}Cu_3Mn_2$	e_3–P_2	503–502
L ⇒ (Al) + Al_2Cu + $Al_{15}Mn_3Si_2$ + Al_2CuMg	P_2–E_2	502–500
L + Mg_2Si ⇒ (Al) + Al_2Cu + $Al_5Cu_2Mg_8Si_6$	p_1–P_1	512–511
L + $Al_{20}Cu_3Mn_2$ ⇒ (Al) + Al_2Cu + $Al_{15}Mn_3Si_2$*	p_2–P_2	547–502
L ⇒ (Al) + Al_2Cu + Mg_2Si + $Al_{15}Mn_3Si_2$	e_5–E_2	514–500
	and e_5–P_1	514–511

* May transform to a eutectic reaction

Table 5.24. Bivariant reactions in quinary alloys of the Al–Cu–Mn–Mg–Si system with (Al) and Al_2Cu phases

Reaction	Field in Figure 5.8b	*T*, °C
L ⇒ (Al) + Al_2Cu + (Si)	(Si)–e_1–E_1–e_4	525–505
L ⇒ (Al) + Al_2Cu + $Al_5Cu_2Mg_8Si_6$	p_1–P_1–E_1–e_1	512–505
L ⇒ (Al) + Al_2Cu + $Al_{15}Mn_3Si_2$	p_2–e_4–E_1–P_1–e_5–E_2–P_2	547–505
L ⇒ (Al) + Al_2Cu + Al_2CuMg	Al_2CuMg–e_3–E_2–e_2	505–499
L ⇒ (Al) + Al_2Cu + Mg_2Si	p_1–e_2–E_2–e_5–P_1	515–499
L ⇒ (Al) + Al_2Cu + $Al_{20}Cu_3Mn_2$	$Al_{20}Cu_3Mn_2$–p_2–P_2–e_3	547–514

5.7. Al–Cu–Mn–(Mg, Si) WROUGHT AND CASTING ALLOYS (2XXX, 2XX, AND 3XX SERIES)

The easiest alloys for the analysis are those containing only copper and manganese. However, they are not too many (Table 5.1), as commercial alloys usually have impurities of Fe and Si. The isothermal sections at 540 and 200°C appear to be the most characteristic sections and are shown in Figure 5.9a, b. The section at 540°C shows that 224.0-type alloys in T4 state contain only $Al_{20}Cu_2Mn_3$ as an excess phase, whereas 2219-type alloys have also the Al_2Cu phase. The section at 200°C demonstrates that in T7 state all alloys of this group are three-phase alloys.

During solidification, copper participates in eutectic reactions. The resultant eutectics is usually divorced and appears as Al_2Cu veins at dendritic cell boundaries. Polythermal section at 0.6% Mn in Figure 5.9c demonstrates that this eutectics is nonequilibrium in 224.0-type alloys. Manganese, on the contrary, can be completely

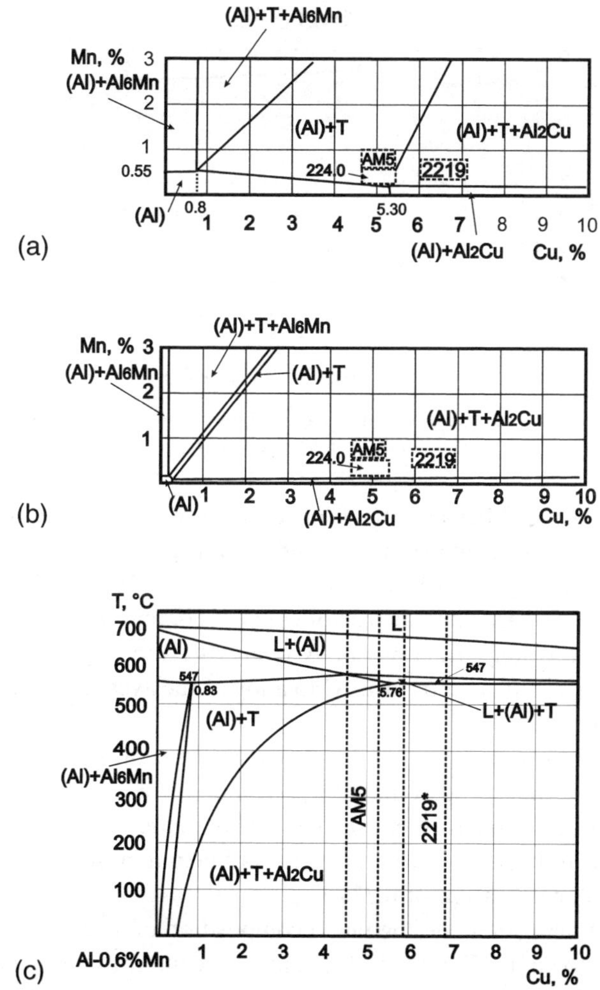

Figure 5.9. Isothermal (a, b) and polythermal (c) sections of Al–Cu–Mn phase diagram: (a) 540°C; (b) 200°C; and (c) 0.6%Mn with compositional ranges of AM5rus, 224.0, and 2219 alloys (note that the Mn content in a 2219 alloy in (c) is above the grade limit). T – $Al_{20}Cu_2Mn_3$.

dissolved in (Al) during nonequilibrium solidification, even though its maximum equilibrium solubility in (Al) at room temperature does not exceed 0.05%. During high-temperature anneals, nonequilibrium Al_2Cu particles dissolve in (Al), while $Al_{20}Cu_2Mn_3$ dispersoids precipitate as a result of decomposition of the aluminum solid solution supersaturated in Mn (in accordance with Figure 5.9a). These dispersoids remain virtually unchanged during downstream processing and use. So, the as-quenched structure consists of the aluminum solid solution supersaturated

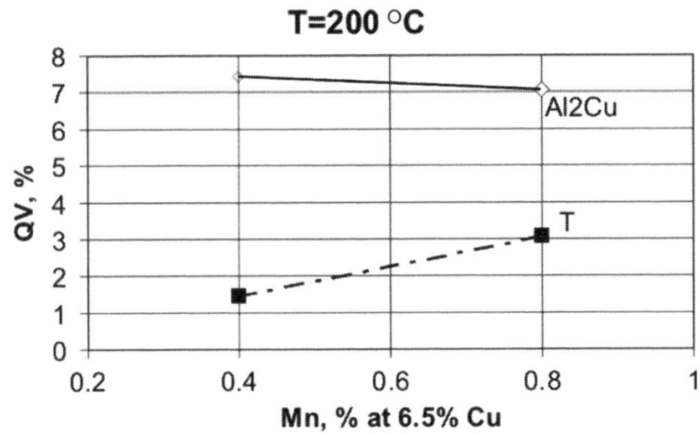

Figure 5.10. Calculated dependence of volume fractions of phases on Mn content in AM5 (rus) (Table 5.1) alloy at 6.5% Cu (200°C)

with copper and (AlCuMn) dispersoids of sub-micron size. During subsequent aging, copper precipitates from the solid solution, forming hardening, metastable phases θ″ and θ′ (Al_2Cu). Figure 5.10 shows that additions of Mn decrease the amount of copper available for hardening as part of copper is bound in the $Al_{20}Cu_2Mn_3$ phase. For example, the volume fractions of $Al_{20}Cu_2Mn_3$ and Al_2Cu particles are 3 and 5 vol.%, respectively, in an annealed AM5 alloy of the average composition (5% Cu, 0.8% Mn, Table 5.1).

An impurity of iron (>0.1%) in 224.0-type alloys results in the formation of Fe-containing phases (Bäckerud et al., 1990). At low silicon concentration, the appearance of the Al_7Cu_2Fe phase is most likely. This phase is formed through eutectic reactions listed in Tables 5.7 and 5.8, and its maximum volume fraction is about 1.2 vol.% at a concentration of 0.3% of Fe.

The effect of manganese on the phase composition of a 224.0-type alloy (5% Cu, 0.2% Fe) is shown in Figure 5.11a, and the combined influence of iron and manganese can be traced in an isothermal section at 5% Cu in Figure 5.11b. At a temperature of homogenization (540°C), 224.0- and AM5-type alloys (Table 5.1), irrespective of the Fe:Mn ratio, fall into the phase region (Al) + Al_7FeCu_2 + $Al_{20}Cu_2Mn_3$ in Figure 5.11b.

The analysis of alloys containing silicon starts with the Al–Cu–Mn–Si phase diagram, Figure 5.12 giving some of the relevant sections. The isothermal sections at 0.5% Mn (Figure 5.12a, b) show that silicon should be completely dissolved in the solid solution during homogenization of 2003-type alloys containing relatively small amounts of silicon (see compositions in Table 5.9). At higher Si concentrations, e.g. in 208.2-type alloys (Table 5.12), silicon participates in the formation of $Al_{15}Mn_3Si_2$ and (Si) phases (Figure 5.12b). Note that silicon considerably decreases the solidus of Al–Cu alloys. Therefore, the maximum homogenization temperature

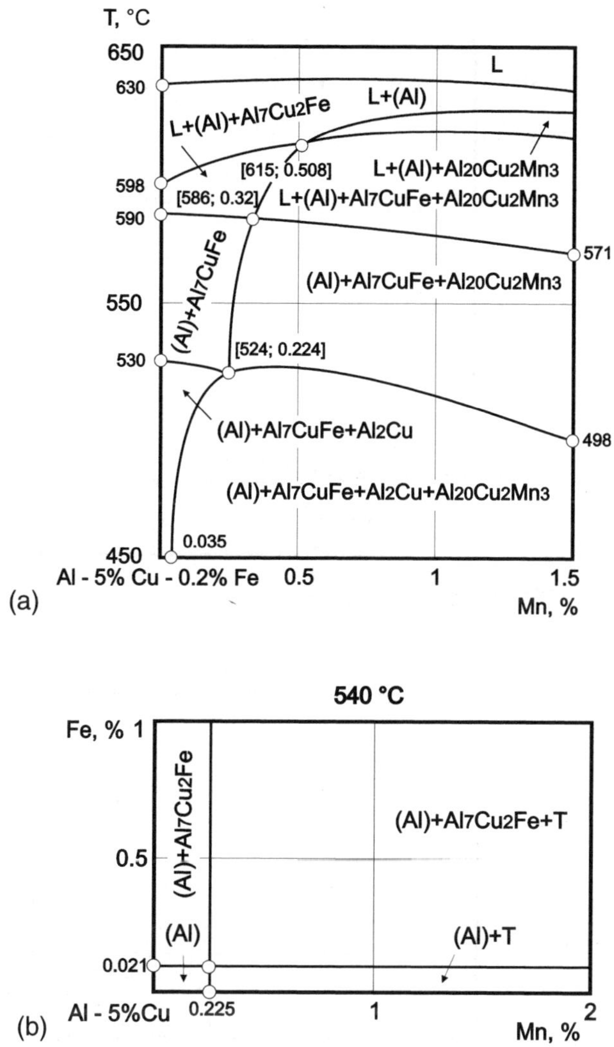

Figure 5.11. Polythermal (a) and isothermal (b) sections of Al–Cu–Mn–Fe phase diagram at 5% Cu: (a) 0.2% Fe and (b) 540°C.

should be strictly controlled and lowered in the case of alloys containing more than 1% Si (see Figure 5.12b).

The effect of silicon on the solidification sequence in 2XX.0-series alloys is illustrated in Figure 5.12c with an isopleth at 5% Cu and 1.5% Mn. The higher than usual concentration of Mn is necessary to show the peritectic reaction $L + Al_6Mn \Rightarrow (Al) + Al_{20}Cu_2Mn_3 + Al_{15}Mn_3Si_2$ (P_1 in Figure 5.4b and Table 5.10) and to assure

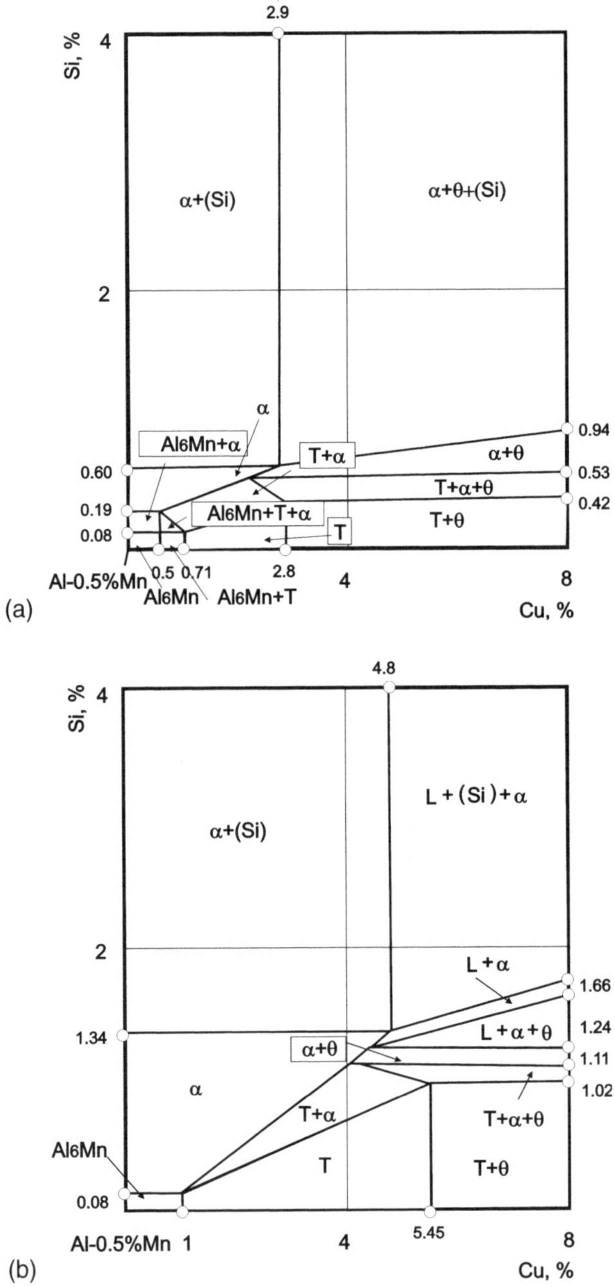

Figure 5.12. Isothermal (a, b) and polythermal (c, d) sections of Al–Cu–Mn–Si phase diagram: (a) 0.5% Mn, 450°C; (b) 0.5% Mn, 540°C; (c) 5% Cu and 1.5% Mn; and (d) 3% Si and 0.5% Mn. α – $Al_{15}Mn_3Si_2$, T – $Al_{20}Cu_2Mn_3$, θ – Al_2Cu. All phase fields in (a) and (b) contain also (Al).

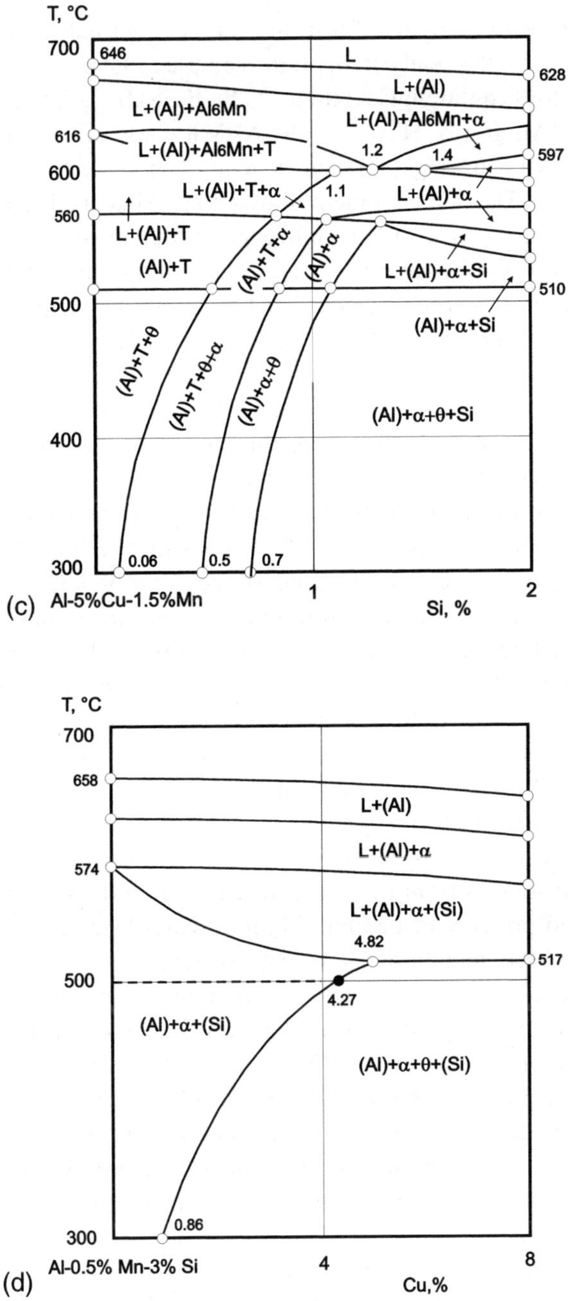

Figure 5.12 (*continued*)

the primary solidification of (Al). With this, the polythermal section in Figure 5.12c can also be used for the analysis of more complex alloys containing copper and manganese. In alloys containing less than 2% Si, the (Al) + Al_6Mn eutectics solidifies next to primary (Al) grains. However, the Al_6Mn phase is not retained in the solid state as it disappears during peritectic reactions shown in Tables 5.10 and 5.11 (p_1–P_1, p_2–P_1, P_1). The same isopleth shows that $Al_{20}Cu_2Mn_3$ is present only in alloys with less than 1% Si. By taking into account that most of 1.5% Mn (almost all of it in low-silicon alloys) remains in aluminum solid solution during nonequilibrium solidification, this section can be used to determine the phase composition of dispersoids, e.g. an alloy with 5% Cu, 1.5% Mn, and 0.7% Si does not contain $Al_{20}Cu_2Mn_3$ precipitates.

Manganese present in Cu-containing, low-iron 308.0-type alloys (Table 5.12) can form only one phase – $Al_{15}Mn_3Si_2$. According to the Al–Cu–Mn–Si phase diagram, this phase is formed either by binary or ternary eutectic reaction (p_2–P_1–P_2–E–e_2 region or e_2–E line in Figure 5.4b and Table 5.11). The equilibrium solidification (at a copper concentration less than 4.5%) ends with the formation of the ternary eutectics. At a higher copper concentration (upper limit of a 308.0 alloy), and at any copper concentration under nonequilibrium conditions, the solidification ceases with the invariant eutectic reaction $L \Rightarrow (Al) + Al_2Cu + (Si) + Al_{15}Mn_3Si_2$ at 517°C (point E in Figure 5.4b and Table 5.10). Figure 5.12d shows relevant polythermal sections at 3% Si and 0.5% Mn. By further increasing the concentration of Si (>4%), the (Al) + $Al_{15}Mn_3Si_2$ eutectics is substituted for the (Al) + (Si) eutectics, otherwise the boundaries in Figure 5.12d remain unchanged.

To analyze the effects of iron and silicon impurities on the phase composition of 2219-type alloys (compositions in Table 5.12), one should use the Al–Cu–Fe–Mn–Si phase diagram. It follows from the phase distribution in the solid state (Figure 5.6a) that the combined presence of Fe and Si in most cases leads to the formation of the $Al_{15}(FeMn)_3Si_2$ phase, mainly of the eutectic origin.

The combined effect of copper and magnesium can be followed by isothermal sections of the Al–Cu–Mg–Mn phase diagram at 0.5% Mn (i.e. at the average manganese concentration in most 2XX.0- and 2XXX-type alloys). The section at 500°C in Figure 5.13a shows that in a 206.0-type alloy (Table 5.5), containing copper and magnesium at the upper limit, a minor amount of eutectic-origin Al_2Cu and Al_2CuMg particles can be preserved after quenching, in addition to $Al_{20}Cu_2Mn_3$ dispersoids. After aging, this alloy contains Al_2Cu and Al_2CuMg phases in the form of metastable precipitates (Figure 5.13b).

At a higher magnesium content, e.g. in a 2224-type alloy (Table 5.5), the phases Al_2Cu, Al_2CuMg, $Al_{20}Cu_2Mn_3$, and Al_6Mn can be found after solution treatment (Figure 5.13a). At least one of the Mn-containing phases is also present as dispersoids. The same selection of phases forms the phase composition after aging as

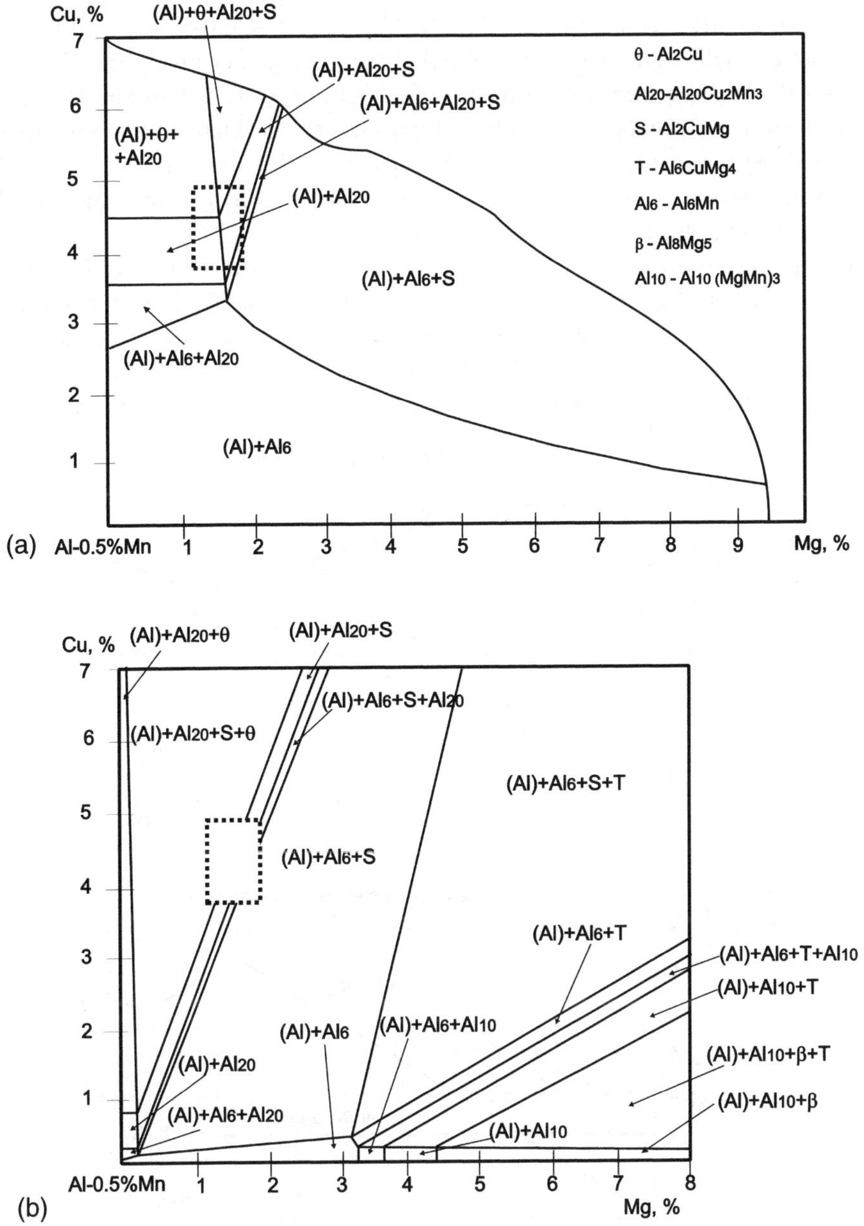

Figure 5.13. Isothermal sections of Al–Cu–Mg–Mn phase diagram at 0.5% Mn: (a) 500°C and (b) 200°C. Compositional range of a 206.0 alloy is shown.

shown in Figure 5.13b, though the Al_2Cu and Al_2CuMg phases (mostly in metastable modifications) are now represented by secondary precipitates.

The calculated dependences of the volume fractions (Q_v) of phases on Mn and Mg concentration in 2024 alloy are given in Figure 5.14. It is interesting to note that the increased Mn concentration not only increases the amount of $Al_{20}Cu_2Mn_3$ dispersoids (which is expected) but also considerably decreases the amount of the main hardening phase Al_2CuMg.

The phase composition of 2014-type alloys at a low concentration of iron impurity can be analyzed using the distribution of phase fields in the solid state shown in Figure 5.8a. As this alloy has a broad compositional range (see Table 5.18), its phase composition in the solid state can vary. In equilibrium with (Al), besides the Al_2Cu phase, can be all six phases – Mg_2Si, (Si), $Al_5Cu_2Mg_8Si_6$, $Al_{15}Mn_3Si_2$,

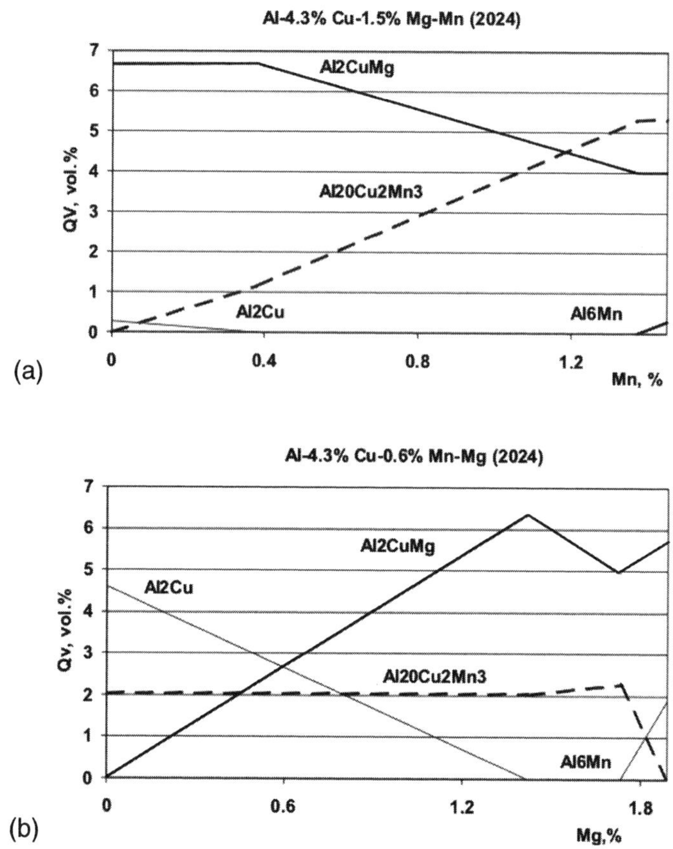

Figure 5.14. Calculated dependences of volume fractions of phases on Mn (a) and Mg (b) content in a 2024 alloy at 4.3% Cu (<100°C): (a) 1.5%Mg; (b) 0.6%Mn.

Al_2CuMg, and $Al_{20}Cu_3Mn_2$ – occurring in the corresponding region of the Al–Cu–Mn–Mg–Si system.

Alloys of the 6XXX series containing copper and manganese and an excess of Si over Mg_2Si, e.g. a 6066 alloy in Table 5.18, can be in first approximation analyzed using the Si-rich part of the Al–Cu–Mg–Mn–Si phase diagram (Figure 5.7a). As the Al_2Cu phase is not likely to be formed in these alloys, only three phases can be in equilibrium with (Al) and (Si), i.e. Mg_2Si, $Al_5Cu_2Mg_8Si_6$, and $Al_{15}Mn_3Si_2$. An impurity of iron can completely dissolve in the last phase. The same applies to casting 3XX.0-series alloys with copper and manganese (Table 5.18) with only one difference – the Al_2Cu phase does form in these alloys.

A number of phases in the as-cast (nonequilibrium) state can be larger than that under equilibrium conditions, but the sequence of solidification reactions is in general agreement with the corresponding phase diagrams. Tables 5.25 and 5.26 give as an example the solidification reactions identified during nonequilibrium solidification of 206.2 and 2024 alloys (Bäckerud et al., 1986, 1990). These alloys belong to the Al–Cu–Mg–Mn system but the presence of Fe and Si impurities requires the analysis of the six-component system. Even small amounts of Fe (0.03%) and Si (0.05%) in a 206.2 alloy cause the formation of phases containing these elements (Table 5.25). The first reactions are in agreement with the Al–Cu–Fe–Mn phase diagram (Figure 5.3, Tables 5.7 and 5.8) with the sequential formation of (Al), $Al_6(MnCuFe)$, $Al_{20}Cu_2Mn_3$, and Al_7Cu_2Fe phases. The solidification ends with the formation of Al_2Cu, Mg_2Si, and Al_2CuMg phases, though the analysis of quinary Al–Cu–Fe–Mg–Mn (Figure 5.6) and Al–Cu–Mg–Mn–Si (Figure 5.8) phase diagrams suggests that there should be more than one reaction with participation of these phases. The version suggested by Bäckerud et al. (1986) may be a consequence of very small amounts of Mg_2Si and Al_2CuMg phases in the as-cast structure.

The as-cast structure of a 2024 alloy that contains more magnesium (1.56%), iron (0.23%), and silicon (0.21%) exhibits particles of the $Al_{15}(MnCuFe)_3Si_2$ phase that is

Table 5.25. Solidification reactions under nonequilibrium conditions in a 206.2 alloy (4.36% Cu, 0.30% Mg, 0.26% Mn, 0.05% Si, and 0.03% Fe) (Bäckerud et al., 1990)

Reaction	Temperatures (°C) at a cooling rate	
	0.3 K/s	4.5 K/s
$L \Rightarrow (Al)$	651–649	651–641
$L \Rightarrow (Al) + Al_6(MnCuFe)$	649–625	641–618
$L + Al_6(MnCuFe) \Rightarrow (Al) + Al_{20}Cu_2Mn_3$	625–529	618–527
$L \Rightarrow (Al) + Al_{20}Cu_2Mn_3 + Al_2Cu + Al_7Cu_2Fe$	529–521	527–505
$L \Rightarrow (Al) + Al_2Cu + Al_2CuMg + Mg_2Si$	–	505–495
Solidus	521	495

Table 5.26. Solidification reactions under nonequilibrium conditions in a 2024 alloy (4.44% Cu, 1.56% Mg, 0.55% Mn, 0.21% Si, and 0.23% Fe) (Bäckerud et al., 1986)

Reaction	Temperatures (°C) at a cooling rate	
	0.8 K/s	13 K/s
$L \Rightarrow (Al)$	637–633	637–627
$L \Rightarrow (Al) + Al_{15}(MnCuFe)_3Si_2$	633–613	613
$L \Rightarrow (Al) + Al_{15}(MnCuFe)_3Si_2 + Al_{20}Cu_2Mn_3$	551–538	544
$L + Al_{20}Cu_2Mn_3 \Rightarrow (Al) + Al_{15}(MnCuFe)_3Si_2 + Al_2Cu$		
$L \Rightarrow (Al) + Al_2Cu + Mg_2Si$	486	480
$L \Rightarrow (Al) + Al_2Cu + Al_2CuMg + Mg_2Si$		
Solidus	486	480

formed after primary (Al) grains (Table 5.26). As iron is completely bound in this phase, the early solidification reactions can be analyzed using the Al–Cu–Mn–Si phase diagram (Figure 5.4, Tables 5.10 and 5.11). Therefore, the binary eutectic reaction $L \Rightarrow (Al) + Al_{15}Mn_3Si_2$ shall transform to the ternary one $L \Rightarrow (Al) + Al_{20}Cu_2Mn_3 + Al_{15}Mn_3Si_2$ following the line P_1–P_2 in Figure 5.4b. Then the $Al_{20}Cu_2Mn_3$ must disappear through the peritectic reaction $L + Al_{20}Cu_2Mn_3 \Rightarrow (Al) + Al_{15}Mn_3Si_2 + Al_2Cu$ (P_2 in Figure 5.4b). These reactions are in good agreement with those listed in Table 5.26. The next reactions can be analyzed using the Al–Cu–Mg–Si phase diagram (Figure 3.4) as manganese and iron are already consumed by the earlier formed phases.

Figure 5.15 demonstrates some typical microstructures of cast Al–Cu–Mn alloys showing Cu- and Mn-containing phases.

In 3XX.0-series alloys the main Mn-containing phase is $Al_{15}(MnFe)_3Si_2$. This phase can be formed during a binary eutectic reaction after the formation of primary (Al) grains in the compositional range of a 319.1 alloy (Bäckerud et al., 1990). Table 5.27 shows solidification reactions that are observed during non-equilibrium solidification of a 319.1 alloy (Bäckerud et al., 1990). The first reactions are in good agreement with the Al–Fe–Mn–Si phase diagram (Section 1.4, Figure 1.5). Table 5.27 shows that the Al_5FeSi phase is formed during a ternary eutectic reaction. It means that the liquid composition falls onto the P_2–P_1 line in Figure 1.5 and Table 1.16. One can then expect the peritectic reaction $L + Al_5FeSi \Rightarrow (Al) + (Si) + Al_{15}(FeMn)_3Si_2$ corresponding to point P_1 (Table 1.15) and after that a reaction with participation of the $Al_{15}(FeMn)_3Si_2$ phase rather than Al_5FeSi as shown in Table 5.27 after Bäckerud et al. (1990) This discrepancy might be the effect of nonequilibrium solidification, i.e. incomplete peritectic reaction.

At lower temperatures (when manganese and iron are almost completely bound to the relevant phases) the rest of the solidification sequence can be analyzed using the Al–Cu–Mg–Si phase diagram. The solidification ends with the eutectic reaction

(a)

(b)

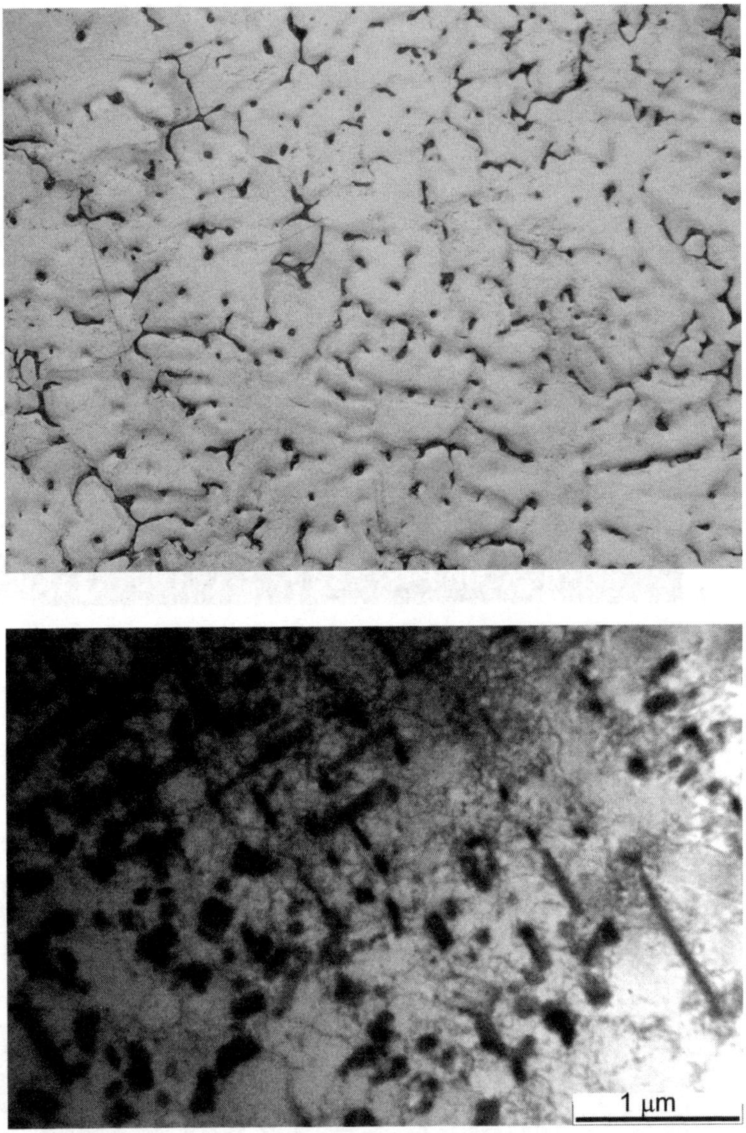

Figure 5.15. Typical microstructures of Al–Cu–Mn alloys: (a) as-cast AM5 alloy (Al–5%Cu–1%Mn) alloy, optical microscope, ×200, veins of (Al) + Al_2Cu nonequilibrium eutectics, Mn in (Al); (b) ingot of an Al–2%Cu–2% Mn alloy annealed at 550°C for 3 h, TEM, dispersoids of the $Al_{20}Cu_2Mn_3$ phase; (c) sheet of a 2219 alloy, T7, SEM, particles of Al_2Cu phase (eutectic origin), not dissolved in (Al) during annealing at 540°C; and (d) ingot of an Al–5%Cu–1%Mn–0.6%Fe alloy, T4, SEM, particles of the $Al_6(MnCuFe)$ and Al_7Cu_2Fe phases (eutectic origin), not dissolved in (Al) during annealing at 540°C.

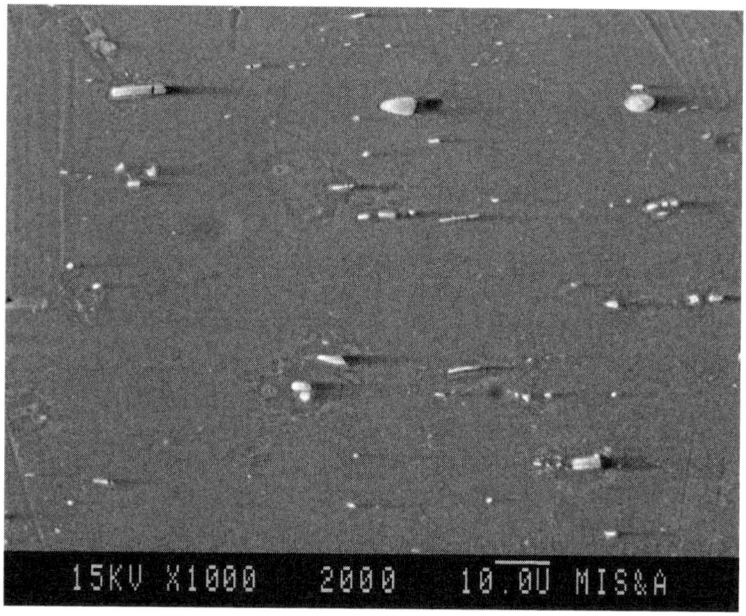

(c)

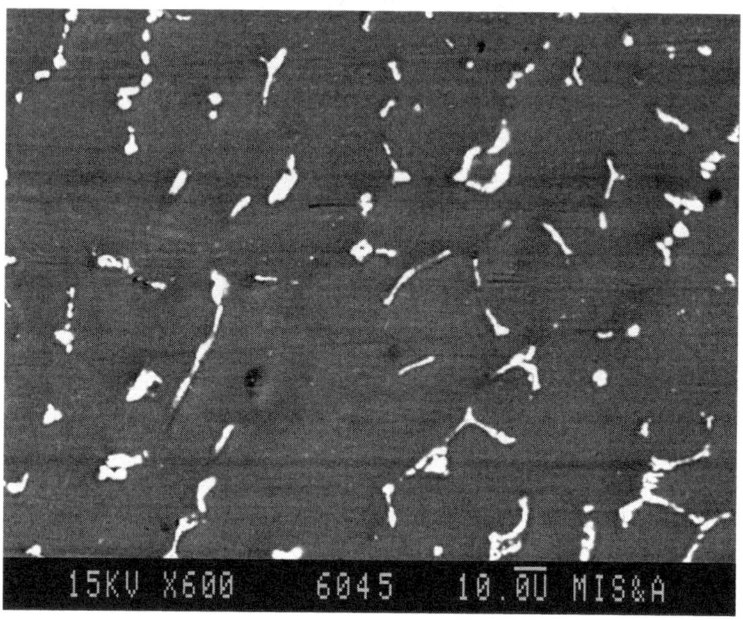

(d)

Figure 5.15 (*continued*)

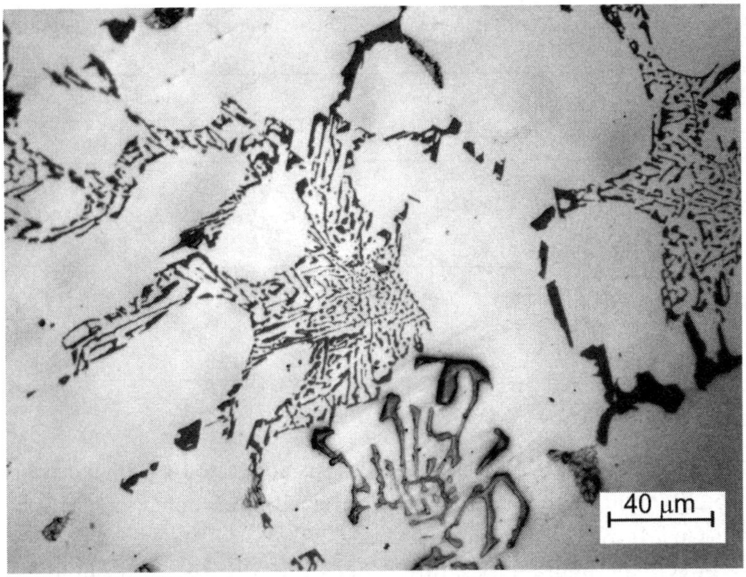

(a)

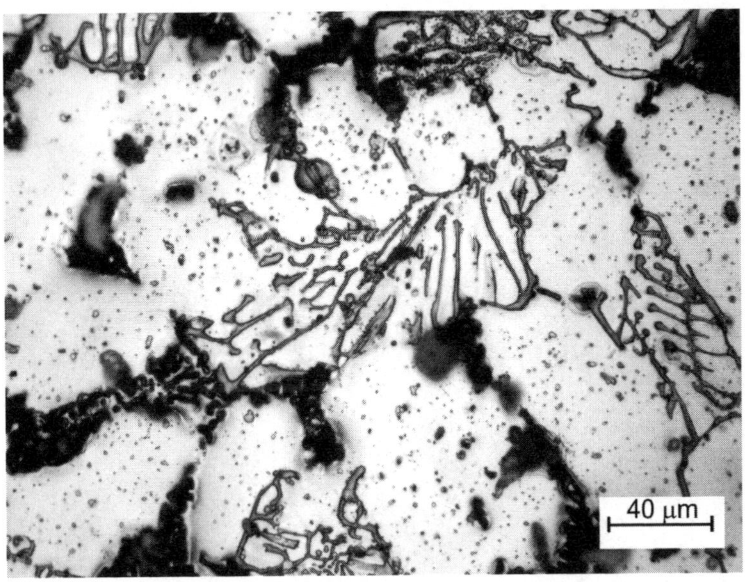

(b)

Figure 5.16. Microstructure of an AK5M alloy (Al–5%Si–1.3%Cu–0.5%Mg–0.4%Mn–0.5%Fe) alloy: (a) as-cast, $Al_{15}(MnFe)_3Si_2$ skeleton, (Si) particles (gray), and multiphase colony (Al) + (Si) + $Al_{15}(MnFe)_3Si_2$ [+ Al_2Cu + Q], optical microscope, mechanical polishing and (b) T4 (annealed 500°C, 10 h), unchanged $Al_{15}(MnFe)_3Si_2$ skeleton, globular (Si) particles, Cu- and Mg-containing phases are dissolved in (Al), optical microscope, electrolytic polishing.

Table 5.27. Solidification reactions under nonequilibrium conditions in a 319.1 alloy (5.7% Si, 3.4% Cu, 0.36% Mn, 0.1% Mg, and 0.62% Fe) (Bäckerud et al., 1990)

Reaction	Temperatures (°C) at a cooling rate	
	0.25 K/s	5 K/s
L ⇒ (Al)	609–583	610–585
L ⇒ (Al) + $Al_{15}(MnCuFe)_3Si_2$	583–554	585–548
L ⇒ (Al) + $Al_{15}(MnCuFe)_3Si_2$ + Al_5FeSi		
L ⇒ (Al) + Al_5FeSi + (Si)	554–516	554–542
L ⇒ (Al) + Al_5FeSi + (Si) + Al_2Cu	516–505	542–504
L ⇒ (Al) + (Si) + $Al_5Cu_2Mg_8Si_6$ + Al_2Cu	505–492	504–468
Solidus	492	468

L ⇒ (Al) + (Si) + $Al_5Cu_2Mg_8Si_6$ + Al_2Cu as is suggested by Bäckerud et al., (1990) (Table 5.27) or at even lower temperature with the eutectic reaction L ⇒ (Al) + (Si) + $Al_5Cu_2Mg_8Si_6$ + Al_2Cu + $Al_{15}Mn_3Si_2$ (Table 5.19).

The effect of high-temperature annealing on the morphology and phase composition of excess phases is demonstrated in Figure 5.16 for an AK5M2 casting alloy (similar to 319.0 alloy).

Chapter 6

Alloys with a High Content of Zinc

This chapter considers the phase composition of alloys that contain zinc and magnesium as obligatory components. Many of these alloys also contain copper; therefore, the Al–Mg–Zn and Al–Cu–Mg–Zn phase diagrams are basic for this group of alloys, which includes mainly wrought alloys of the 7XXX series, e.g. high-strength 7075 and 7055 alloys widely used in aircraft structures. Casting alloys of the 7XX.0 series with an increased zinc concentration have limited application. The high-strength alloys of this group contain usually only small amounts of iron and silicon impurities, so that their analysis can be restricted to the basic diagrams. When the amount of these impurities is significant, i.e. the phase composition is affected, respective phase diagrams with Fe and Si should be considered. In addition, alloys of the 7XXX series usually contain transition metals (Mn, Zr, Cr, Ti), which are mainly present as dispersoids.

6.1. Al–Mg–Zn PHASE DIAGRAM

The Al–Mg–Zn phase diagram is the basic diagram for such alloys as 7104, 7005, 7008, etc. (Table 6.1), and can be also, albeit with some restrictions, applied to high-strength Al–Zn–Mg–Cu alloys containing less than ~1% Cu, e.g. 7076 and 7016 alloys.

The Al–Mg–Zn phase diagram has been studied in sufficient detail (Phillips, 1959; Mondolfo, 1976; Drits et al., 1977) and the published versions can be used for commercial alloys.

If we accept the most probable (in our view) version of the Al–Zn phase diagram (i.e. without the AlZn phase that Mondolfo (1976) has included into this binary system) then in the ternary system (Al) can be in equilibrium with the following phases: Al_8Mg_5, $Al_2Mg_3Zn_3$, $MgZn_2$, Mg_2Zn_{11}, and (Zn).

The Al_8Mg_5 phase (discussed in detail in Section 2.1) dissolves up to 10% Zn. The compound $MgZn_2$ (84.32% Zn) is a prototype of the hexagonal Laves phase. It belongs to the space group $P6_3/mmc$ (12 atoms per unit cell) with parameters $a = 0.516$–0.522 nm and $c = 0.849$–0.856 nm. Up to 3% Al can be dissolved in it. The Mg_2Zn_{11} phase (6.33% Mg) has a cubic structure (space group $Im3$, 39 atoms per unit cell) with lattice parameter $a = 0.855$ nm. This phase dissolves less than 1% Al. The composition of the ternary phase $Al_2Mg_3Zn_3$ changes within the range of 20–35% Mg and 22–65% Zn, and can be also described by the formula

Table 6.1. Chemical composition of some commercial alloys whose phase composition can be analyzed using the Al–Mg–Zn phase diagram

Grade	Zn, %	Mg, %	Other*			
			Cu, %	Mn, %	Fe, %	Si, %
7104	3.6–4.4	0.5–0.9	0.03	–	0.4	0.25
7019	3.5–4.5	1.5–2.5	0.2	0.15–0.5	0.45	0.35
7039	3.5–4.5	2.3–3.3	0.1	0.1–0.4	0.4	0.3
7004	3.8–4.6	1.0–2.0	–	0.2–0.7	0.35	0.25
7024	3.0–5.0	0.5–1.0	0.1	0.1–0.6	0.4	0.3
7025	3.0–5.0	0.8–1.5	0.1	0.1–0.6	0.4	0.3
7005	4.0–5.0	1.0–1.8	0.1	0.2–0.7	0.4	0.35
7017	4.0–5.2	2.0–3.0	0.2	0.05–0.2	0.45	0.35
7008	4.5–5.5	0.7–1.4	0.05	0.05	0.1	0.1
7003	5.0–6.5	0.5–1.0	0.2	0.3	0.35	0.3
7046	6.6–7.6	1.0–1.6	0.25	0.3	0.4	0.2
V92ts(rus)	2.9–3.6	3.9–4.6	0.05	0.6–0.1	0.3	0.2
VAL11(rus)	2–2.5	6–7	–	0.1–0.2	0.3	0.2

* Some grades contain Cr, Zr, and Ti

$(AlZn)_{49}Mg_{32}$. It has a cubic structure (space group *Im*3, 162 atoms per unit cell). The lattice parameter can change from 1.429 to 1.471 nm with the Zn content increasing. This phase is usually designated as T and is isomorphic to the similar phase from the Al–Cu–Mg system. The (Zn) phase is a solid solution of Al and Mg in Zn; the maximal solubility of magnesium does not exceed 0.1%, and that of aluminum is about 0.5%.

The general appearance of the Al–Mg–Zn phase diagram, and also the liquidus, solidus, and solvus isotherms (for the aluminum corner of the diagram) are given in Figure 6.1. The invariant reactions involving (Al) are given in Table 6.2, and the respective monovariant reactions, in Table 6.3. Two quasi-binary sections, Al–$Al_2Mg_3Zn_3$ (489°C) and Al–$MgZn_2$ (475°C), can be singled out in the Al–Mg–Zn system. In the case of Al–$MgZn_2$, the three-phase invariant transformation coincides with the four-phase transformation. The solubilities of Mg and Zn in (Al) decrease significantly as the temperature lowers (Table 6.4). This determines the considerable effect of precipitation hardening due to the formation of GP zones and metastable modifications of the phases $Al_2Mg_3Zn_3$ (T′) and $MgZn_2$ (η′).

By the time the solidification is completed, almost all commercial alloys of the 7XXX series get into the single-phase region, i.e. all reactions represented in Tables 6.2 and 6.3 should not occur under equilibrium conditions. However, in real solidification the nonequilibrium eutectics are formed, usually involving the phases $Al_2Mg_3Zn_3$ and $MgZn_2$. As the temperatures of these eutectics are rather low and the liquidus of most alloys exceeds 600°C, the casting properties of the alloys of

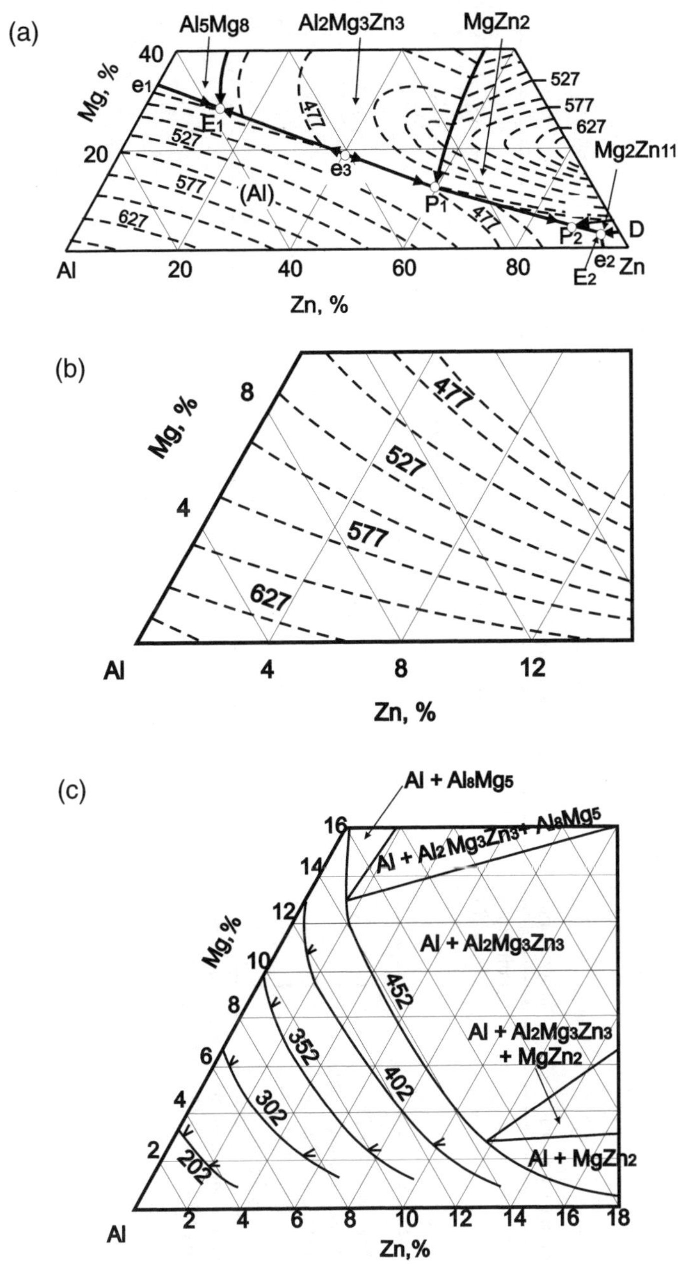

Figure 6.1. Phase diagram of Al–Mg–Zn system (Mondolfo, 1976; Drits et al., 1977; Phillips, 1959): projection of (a) liquidus surface and liquidus isotherms; (b) solidus isotherms; and (c) solvus isotherms.

Table 6.2. Invariant reactions in ternary alloys of the Al–Mg–Zn system (Mondolfo, 1976; Drits et al., 1977)

Reaction	Point in Figure 6.1a	T, °C	Concentration in phases, %*							
			1		2		3		4	
			Mg	Zn	Mg	Zn	Mg	Zn	Mg	Zn
$L \Rightarrow (Al) + Al_8Mg_5 + Al_2Mg_3Zn_3$	E_1	447	30	12	13	2	34	10	30	26
$L \Rightarrow (Al) + Al_2Mg_3Zn_3$ (quasi-binary)	e_3	489	18	45	5	12	21	54	–	–
$L + Al_2Mg_3Zn_3 \Rightarrow (Al) + MgZn_2$**	P_1	475	11.3	60.4	3	14	16	83	20	64
$L + MgZn_2 \Rightarrow Mg_2Zn_{11} + (Al)$	P_2	368	3.5	92	15	85	7	92	1	78
$L \Rightarrow (Al) + (Zn) + Mg_2Zn_{11}$	E_2	343	3	93	1	80	7	92	0.5	99

* 1–4 – sequential numbers of the phases in the reactions from left to right

** Alternative reactions $L \Rightarrow (Al) + MgZn_2$ or $L \Rightarrow (Al) + MgZn_2 + Al_2Mg_3Zn_3$

Table 6.3. Monovariant reactions in ternary alloys of the Al–Mg–Zn system

Reaction	Line in Figure 6.1a	*T*, °C
$L \Rightarrow (Al) + Al_2Mg_3Zn_3$	e_3–E_1 and e_3–P_1	489–447 and 489–475
$L \Rightarrow (Al) + Al_8Mg_5$	e_1–E_1	450–447
$L \Rightarrow (Al) + MgZn_2$	P_1–P_2	475–368
$L \Rightarrow (Al) + Mg_2Zn_{11}$	P_2–E_2	468–343
$L \Rightarrow (Al) + (Zn)$	e_2–E_2	382–343

Table 6.4. Limit solubility of Mg and Zn in solid aluminum in the Al–Mg–Zn system (Drits et al., 1977)

Phase region	*T*, °C	475	460	447	440	400	350	300	200
$(Al) + Al_8Mg_5 + Al_2Mg_3Zn_3$	Mg, %	–	–	12.5	12.3	10.5	8.4	6.0	2.8
	Zn, %	–	–	1.8	1.6	1.1	0.6	0.4	0.2
$(Al) + MgZn_2 + Al_2Mg_3Zn_3$	Mg, %	2.8	2.6	–	2.3	1.7	1.1	0.7	0.2
	Zn, %	14.3	12.5	–	11.4	8.6	6.0	3.7	1.0

this group are quite poor due to the wide solidification range. During annealing in the single-phase region, the phases containing magnesium and zinc readily dissolve in (Al), which is due to the fast diffusion of these elements in solid aluminum.

6.2. Al–Cu–Zn PHASE DIAGRAM

The Al–Cu–Zn phase diagram is of no great importance from the practical viewpoint, because no commercially significant alloys are based on this system. For us, this diagram is required for understanding the processes occurring in the quaternary Al–Cu–Mg–Zn system.

Adopting the most probable version of the Al–Zn phase diagram (Mondolfo, 1976; Drits et al., 1977), we can conclude that the phases Al_2Cu, $Al_3Cu_5Zn_2$, $CuZn_5$, and (Zn) are in equilibrium with (Al). The last three phases are formed only in alloys with high concentrations of copper and zinc.

The $Al_3Cu_5Zn_2$ (T) phase has a homogeneity range of 56–58% Cu and 10–30% Zn, but only a composition of 60.1% Cu and 24.7% Zn corresponds to the exact stoichiometric ratio. The phase has a cubic structure (space group *Pm*3*m*, 2 atoms per unit cell) with the parameters that vary from $a = 0.291$ nm at 57% Cu and 10% Zn up to $a = 0.294$ nm at 57% Cu and 25% Zn (Mondolfo, 1976). The $CuZn_5$ phase (78–87% Zn) has a hexagonal structure (space group *P*6/*mmc*) with the parameters varying around $a = 0.274$ nm and $c = 0.492$ nm. This phase can dissolve up to 5% Al.

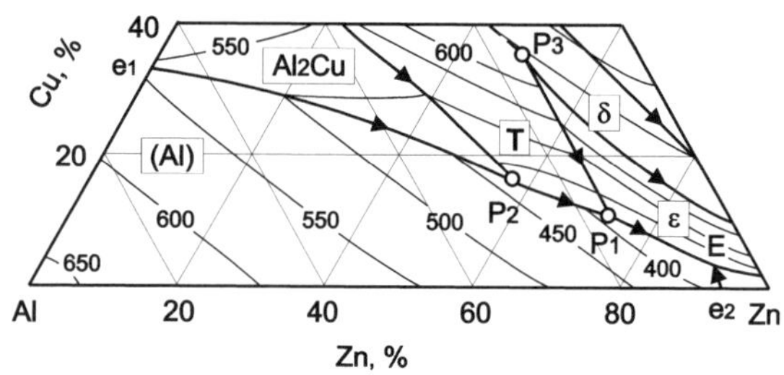

Figure 6.2. Projection of liquidus surfaces in the aluminum corner of the Al–Cu–Zn system (Drits et al., 1977). T – $Al_3Cu_5Zn_2$; ε – $CuZn_5$; and δ – solid solution based on AlCu.

The (Zn) phase is a solid solution of Al and Cu in Zn. At the concentrations of commercial importance (<15% Cu, <15% Zn), no other excess phases besides Al_2Cu (see Section 3.1) can be formed. According to the data by Mondolfo (1976), 2–3% Zn is dissolved in this phase.

By taking into account that we adopted the version of the Al–Zn phase diagram without the AlZn compound, Figure 6.2 shows the projection of liquidus surfaces in the Al–Cu–Zn system. The invariant reactions involving (Al) are given in Table 6.5, and the limit solubilities of copper and zinc in (Al) are in Table 6.6. It follows from Table 6.6 that the mutual solubility of Cu and Zn in solid (Al) is approximately the same as in the respective binary systems.

Table 6.5. Invariant reactions in Al-rich ternary alloys of the Al–Cu–Zn system (Drits et al., 1977)

Reaction	Point in Figure 6.2	T, °C	Concentration in phases, %*							
			1		2		3		4	
			Cu	Zn	Cu	Zn	Cu	Zn	Cu	Zn
$L+CuAl_2 \Rightarrow (Al)+Al_3Cu_5Zn_2$	P_2	420	15	60	52	2	1.5	65	55	13
$L+Al_3Cu_5Zn_2 \Rightarrow (Al)+CuZn_5$	P_1	396	10.7	74	55.5	14	1.8	72	23	72
$L \Rightarrow (Al)+CuZn_5+(Zn)$	E	379.5	3.7	89.3	1.5	78.1	15.5	83.32	2.75	96

* 1–4 are the sequential numbers of the phases in the reactions

Table 6.6. Limit solubility of Cu and Zn in solid (Al) of the Al–Cu–Zn system (Mondolfo, 1976)

T, °C	427	377	352	327	277	227
Cu, %	2.7	1.8	1.5	1.3	0.7	0.45
Zn, %	70	47	43	29	14	6

6.3. Al–Cu–Mg–Zn PHASE DIAGRAM

Although this system is the basis of the strongest aluminum alloys, it is insufficiently studied even with respect to the compositional range of commercial alloys. Though numerous experimental data are compiled by Mondolfo (1976), the essential information on the constitution of the aluminum corner of the system is still absent.

A specific feature of this quaternary diagram is the existence of three domains of continuous solid solutions, which are formed by the phases from the Al–Mg–Zn and Al–Cu–Mg ternary systems, i.e. between Al_6CuMg_4 and $Al_2Mg_3Zn_3$, between $MgZn_2$ and AlCuMg, and between $Al_5Cu_6Mg_2$ and Mg_2Zn_{11} (Figure 6.3a). Note that in the Al–Cu–Mg system, the CuMgAl and $Al_5Cu_6Mg_2$ phases are not in equilibrium with (Al), and an addition of Zn is required for the equilibrium to be established.

The Al_6CuMg_4 and $Al_2Mg_3Zn_3$ phases exist in a wide homogeneity range even in the respective ternary systems, and in the quaternary system the homogeneity region of the mutual solid solution (phase T) is rather vast as well. The T phase has a cubic structure (space group *Im*3, 162 atoms per unit cell) with the lattice parameter *a* varying from 1.415 up to 1.471 nm. The quaternary solution between compounds AlCuMg and $MgZn_2$ (designated as the M phase) has a hexagonal structure (space group $P6_3/mmc$, 12 atoms per unit cell) with approximate lattice parameters $a = 0.518$ nm and $c = 0.852$ nm. The solid solution formed by compounds $Al_5Cu_6Mg_2$ and Mg_2Zn_{11} (the Z phase) has a cubic structure (space group *Im*3, 39 atoms per unit cell) with parameter $a = 0.831$–0.855 nm. The phase $CuZn_5$ from the Al–Cu–Zn system has been considered in Section 6.2; here we would like to note that this phase can dissolve up to 5% Al.

The characteristics of the Al_2CuMg (S) and Al_2Cu (θ) phases from the Al–Cu–Mg system are also given in Sections 3.2 and 3.1, respectively. The θ phase virtually does not dissolve magnesium, and the solubility of zinc in the S phase does not exceed 1%.

In the alloys containing 4–8% Zn and 0.5–1.0% Cu, the lattice parameter of (Al) increases with the Mg content in the solid solution and reaches 0.407–0.408 nm at 6–7% Mg (Mondolfo, 1976).

The distribution of the phase regions in the solid state is given in Figure 6.3b following the version suggested by Mondolfo with the exception of the AlZn phase (1976). Numerous experimental data on commercial alloys of the 7XXX series show that they contain at least one of the two phases – M or T. Considering this fact, commercial alloys can get only into the following two four-phase regions: (Al) + T + S + M and (Al) + Z + M + S. Table 6.7 gives chemical compositions of some commercial alloys of the 7XXX series.

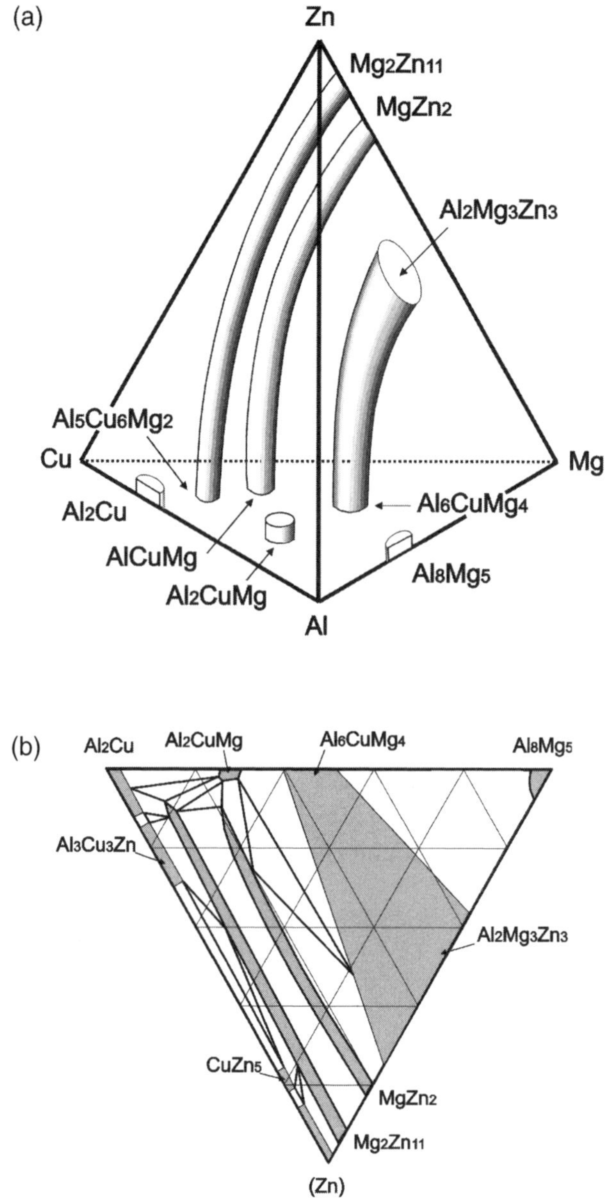

Figure 6.3. Phase diagram of Al–Cu–Mg–Zn system: (a) compositional ranges of single phases in a 3D diagram (Mondolfo, 1976); (b) distribution of phase fields in the solid state (Mondolfo, 1976); (c) polythermal projection of solidification surfaces. T (Al_6CuMg_4 – $Al_2Mg_3Zn_3$), M ($MgZn_2$ – AlCuMg), Z ($Al_5Cu_6Mg_2$ and Mg_2Zn_{11}), S – Al_2CuMg, and θ – Al_2Cu.

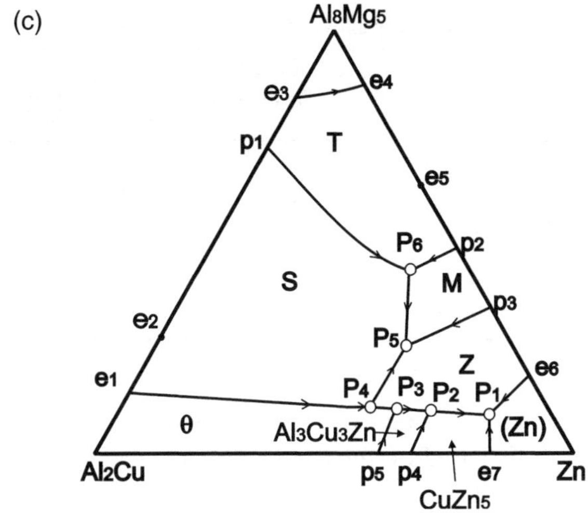

Figure 6.3 *(continued)*

Table 6.7. Chemical composition of some commercial alloys whose phase composition can be analyzed using the Al–Cu–Mg–Zn phase diagram

Grade	Zn, %	Mg, %	Cu, %	Other*		
				Mn, %	Fe, %	Si, %
7179	3.8–4.8	2.9–3.7	0.4–0.8	0.1–0.3	0.2	0.15
7016	4.0–5.0	0.8–1.4	0.45–1.0	0.03	0.12	0.1
7229	4.2–5.2	1.3–2.0	0.5–0.9	0.03	0.08	0.06
7075	5.1–6.1	2.1–2.9	1.2–2.0	0.3	0.5	0.4
7475	5.2–6.2	1.9–2.6	1.2–1.9	0.06	0.12	0.1
7012	5.8–6.5	1.8–2.2	0.8–1.2	0.08–0.15	0.25	0.15
7109	5.8–6.5	2.2–2.7	0.8–1.3	0.1	0.15	0.1
7010	5.7–6.7	2.1–2.6	1.5–2.0	0.1	0.15	0.12
7050	5.7–6.7	1.9–2.6	2.0–2.6	0.1	0.15	0.12
7278	6.6–7.4	2.5–3.2	1.6–2.2	0.02	0.2	0.15
7060	6.1–7.5	1.3–2.1	1.8–2.6	0.2	0.2	0.15
7064	6.8–8.0	1.9–2.9	1.8–2.4	–	0.15	0.15
7001	6.8–8.0	2.6–3.4	1.6–2.6	0.2	0.4	0.35
7076	7.0–8.0	1.2–2.0	0.3–1.0	0.3–0.8	0.6	0.4
7149	7.2–8.2	2.0–2.9	1.2–1.9	0.2	0.2	0.15
7055	7.7–8.4	1.8–2.3	2.0–2.6	0.05	0.15	0.1
V95och(rus)	5.0–6.5	1.8–2.8	1.4–2.0	0.2–0.6	0.15	0.1
1933(rus)	6.5–7.3	1.6–2.8	0.8–1.2	–	0.06–0.15	0.1
V96ts-3(rus)	7.6–8.6	1.7–2.3	1.4–2.0	–	0.2	0.1

* Some grades also contain Cr, Zr, and Ti

Table 6.8. Invariant reactions in quaternary alloys of the Al–Cu–Mg–Zn system

Reaction	Point on Figure 6.3a	Composition of liquid			*T*, °C
		Zn, %	Mg, %	Cu, %	
$L + CuZn_5 \Rightarrow (Al) + (Zn) + Z$	P_1	91.1	2.2	3.4	350
$L + Al_3Cu_3Zn \Rightarrow (Al) + CuZn_5 + Z$	P_2	82.6	2.4	10.1	363
$L + Al_2Cu \Rightarrow (Al) + AlCu_3Zn_3 + Z$	P_3	77.2	3.0	9.8	377
$L + S + Al_2Cu \Rightarrow (Al) + Z$	P_4	6.5	6.5	38.9	482
$L + S \Rightarrow (Al) + Z + M$	P_5	–	–	–	<467
$L + T \Rightarrow (Al) + S + M$ or $L \Rightarrow (Al) + T + S + M$	P_6	–	–	–	<467

The polythermal diagram of the Al–Cu–Mg–Zn system, shown in Figure 6.3c, is suggested by the authors based on the distribution of the phase regions in the solid state (Figure 6.3b) and our own experimental data for the concentration range of up to 90% Zn, 40% Cu, and 30% Mg. The invariant reactions, possible in this quaternary system, are given in Table 6.8. We should note that these reactions occur at concentrations quite different from the compositions of commercial alloys (Table 6.7). By taking into account the existence of quasi-binary sections Al–S and Al–T in the Al–Cu–Mg and Al–Mg–Zn systems, respectively (Sections 3.2 and 6.1), one can assume the presence of a quasi-ternary section (Al)–T–S in the Al–Cu–Mg–Zn system.

Under real conditions, the solidification in 7XXX-series alloys completes with the formation of nonequilibrium eutectics at a temperature as low as 465–469°C (Bäckerud et al., 1986; own data). Probably, these are binary or ternary eutectics, i.e. bi- or monovariant eutectics but forming in a narrow temperature range. The formation of the $L \Rightarrow (Al) + T + M + Al_2Cu$ eutectics at 466–470°C in a 7075 alloy suggested by Bäckerud et al. (1986) is unlikely.

During aging after quenching the metastable phases T′, M′ (η′), and S′ can be formed in commercial alloys of the Al–Cu–Mg–Zn system. The crystallographic characteristics of these phases are given in Table 6.9.

6.4. Al–Fe–Mg–Zn PHASE DIAGRAM

Iron is a major impurity (along with silicon) in alloys of 7XXX series (Tables 6.1 and 6.7). The assessment of its influence on the phase composition and solidification reactions requires an analysis of the respective phase diagrams. In some Russian grades, e.g. V95pch, iron though in a small amount (0.1–0.25%) is an additive. However, information on the constitution of multicomponent phase diagrams for aluminum with magnesium, zinc, and iron is very scarce. For example,

Table 6.9. Crystal structure of metastable phases formed in commercial alloys of the Al–Cu–Mg–Zn system (Graf, 1957; Mondolfo et al., 1959; Mondolfo, 1976; Katgerman and Eskin, 2003)

Phase	Crystal structure	Lattice parameters		
		a, nm	*c*, nm	β
M′ (η′),	Hexagonal	0.496	0.868	–
	Hexagonal	0.496	$6d(111)_{(Al)}$ (1.403)	–
	Hexagonal	0.515–0.523	0.848–0.862	–
	Monoclinic	0.497	0.554	120°
	Hexagonal	0.496	0.702	–
T′	Cubic	1.42–1.44	–	–
	Hexagonal	1.39	2.75	–
S′	Orthorhombic	0.405	0.720	*b* = 0.906 nm

Mondolfo (1976) mentions the AlFeZn compound. On the other hand, in ternary systems Al–Fe–Mg and Al–Fe–Zn the only Fe-containing phase in equilibrium with (Al) is Al_3Fe, which seems to be the most trustworthy one (Mondolfo, 1976; Drits et al., 1977). This phase composition is also supported by experimental data on a 7005 alloy (Bäckerud et al., 1986).

As the solubilities of magnesium and zinc in Al_3Fe are very low (as is the solubility of iron in Mg- and Zn-containing phases), predicting the constitution of the aluminum corner of the Al–Fe–Mg–Zn phase diagram presents no problem. Figure 6.4a shows the distribution of phase regions in solid state, and Figure 6.4b demonstrates the polythermal projection of solidification surfaces. Several invariant reactions can occur in quaternary alloys as listed in Table 6.10. These reactions by the temperature and, possibly, by the composition are close to the corresponding reactions of the Al–Mg–Zn system (Table 6.2). The concentration of iron in the liquid, as estimated from the respective ternary diagrams, should be low. The effect of this element on the liquidus and solidus temperatures of quaternary alloys is rather small. The Al_3Fe phase is formed already at a small iron concentration by the binary eutectic reaction within a broad temperature range (>150°C). This phase appears in the structure as rather coarse inclusions.

6.5. Al–Mg–Si–Zn PHASE DIAGRAM

Silicon (together with iron) is the major impurity in 7XXX-series alloys (Tables 6.1 and 6.7), and the analysis of its effect on the phase equilibria requires the knowledge of the respective phase diagrams. According to the available data, addition of silicon to Al–Mg–Zn alloys does not form any other phases than Mg_2Si and (Si), which suggests the constitution of the aluminum corner of the quarternary diagram as

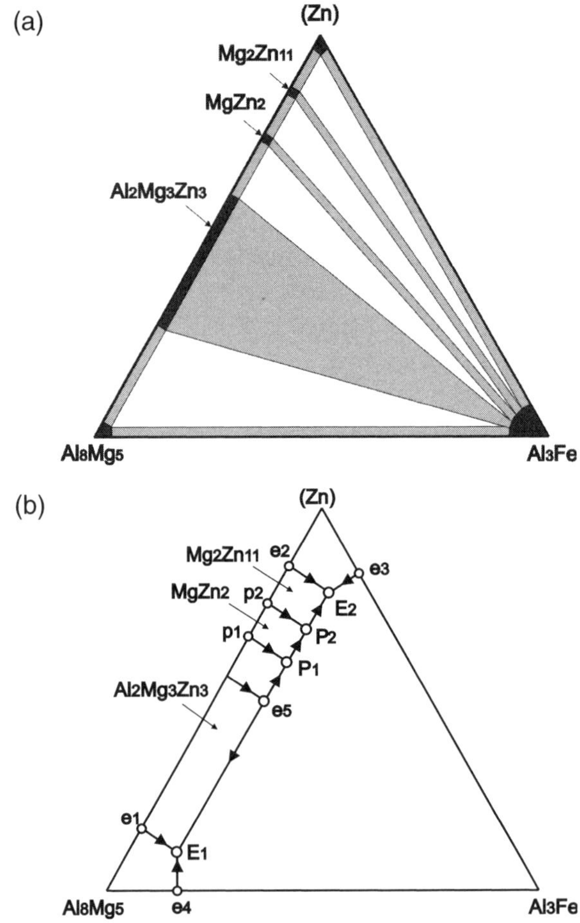

Figure 6.4. Phase diagram of Al–Fe–Mg–Zn system: (a) distribution of phase fields in the solid state and (b) polythermal projection of solidification surfaces.

Table 6.10. Invariant reactions in quaternary alloys of the Al–Fe–Mg–Zn system

Reaction	Point in Figure 6.4b	T, °C	Concentrations in liquid phase		
			Zn, %	Mg, %	Fe, %
$L \Rightarrow (Al) + Al_8Mg_5 + Al_2Mg_3Zn_3 + Al_3Fe$	E_1	~446	~12	~30	<1
$L \Rightarrow (Al) + Al_2Mg_3Zn_3 + Al_3Fe$ (quasi-binary)	e_5	~488	~45	~18	<1
$L + Al_2Mg_3Zn_3 \Rightarrow (Al) + MgZn_2 + Al_3Fe$*	P_1	~474	~60	~11	<1
$L + MgZn_2 \Rightarrow (Al) + Mg_2Zn_{11} + Al_3Fe$	P_2	~367	~92	~3.5	<1
$L \Rightarrow (Al) + (Zn) + Mg_2Zn_{11} + Al_3Fe$	E_2	~342	~93	~3	<1

* or $L \Rightarrow (Al) + MgZn_2 + Al_3Fe$, or $L \Rightarrow (Al) + MgZn_2 + Al_2Mg_3Zn_3 + Al_3Fe$

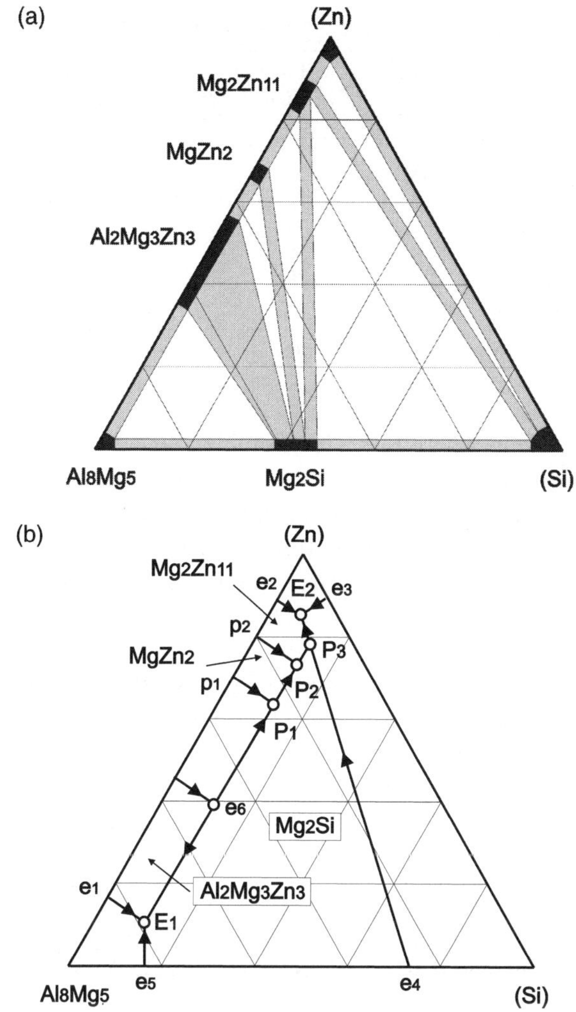

Figure 6.5. Phase diagram of Al–Mg–Si–Zn system: (a) distribution of phase fields in the solid state and (b) polythermal projection of solidification surfaces.

shown in Figure 6.5 (Mondolfo, 1976; Drits et al., 1977). This phase diagram includes the quasi-ternary sections involving magnesium silicide. According to the data compiled by Drits et al. (1977) the ternary eutectic point at the Al–$MgZn_2$–Mg_2Si section almost coincides with the binary eutectics Al–$MgZn_2$ (Table 6.2). Probably, the same is valid for the other quasi-ternary sections, i.e. Al–$Al_2Mg_3Zn_3$–Mg_2Si and Al–Mg_2Zn_{11}–Mg_2Si. The invariant solidification reactions in quaternary alloys of the Al–Mg–Si–Zn system are given in Table 6.11. Note that these reactions are very close to the similar reactions in Table 6.10. This is a typical feature of

Table 6.11. Invariant reactions in quaternary alloys of the Al–Mg–Si–Zn system

Reaction	Point in Figure 6.5b	T, °C	Concentrations in liquid phase		
			Zn, %	Mg, %	Si, %
$L \Rightarrow (Al) + Al_8Mg_5 + Al_2Mg_3Zn_3 + Mg_2Si$	E_1	~446	~12	~30	<1
$L \Rightarrow (Al) + Al_2Mg_3Zn_3 + Mg_2Si$ (quasi-binary)	e_6	~488	~45	~18	<1
$L + Al_2Mg_3Zn_3 \Rightarrow (Al) + MgZn_2 + Mg_2Si$*	P_1	~474	~60	~11	<1
$L + MgZn_2 \Rightarrow (Al) + Mg_2Zn_{11} + Mg_2Si$	P_2	~367	~92	~3.5	<1
$L + Mg_2Si \Rightarrow (Al) + Mg_2Zn_{11} + (Si)$	P_3	~367	~92	~3.5	<1
$L \Rightarrow (Al) + (Zn) + Mg_2Zn_{11} + (Si)$	E_2	~342	~93	~3	<1

* or $L \Rightarrow (Al) + MgZn_2 + Mg_2Si$ or $L \Rightarrow (Al) + MgZn_2 + Al_2Mg_3Zn_3 + Mg_2Si$

Al–Mg–Zn–H systems where H is a low-soluble element. It is also interesting that Si forms two phases – (Si) and Mg_2Si – in this system.

The low concentration of silicon in the liquid phase suggests that the Mg_2Si phase in 7XXX-series alloys mainly forms through the binary eutectic reaction within a wide temperature range (>100 K). The solubility of silicon in (Al) can be estimated in the first approximation from the data on the Al–Mg–Si system (Section 2.1), i.e. the solubility of Si in (Al) at 500°C and 2% Mg does not exceed 0.1–0.15% (this corresponds to the admissible level in most 7XXX alloys). Hence, a larger amount of silicon impurity in an alloy will decrease the concentration of magnesium in (Al) due to the formation of insoluble inclusions of the Mg_2Si phase. These inclusions have a tendency of spheroidization during annealing at temperatures above 500°C, which has a favorable effect on the mechanical properties (Belov et al., 1992).

6.6. Al–Mg–Zn–(Cu) WROUGHT AND CASTING ALLOYS (7XXX AND 7XX.0 SERIES)

Copper-less 7XXX-series alloys, e.g. 7004 and 7005 contain, as a rule, less than 6–7% (Zn + Mg), as higher concentrations facilitate stress corrosion (Table 6.1). The phase diagram shown in Figure 6.1 and the isopleth at 5% Zn in Figure 6.6a suggest that the alloys of this group can be easily transformed into the single-phase state upon homogenization. The phase composition after aging can be (in first approximation) estimated using the isothermal section at 202°C (Figure 6.1c). At a high Zn:Mg ratio, the $MgZn_2$ phase and its metastable modifications η' and η'' can be expected as aging products. At a ratio Zn:Mg <2, the T ($Al_2Mg_3Zn_3$) phase is formed.

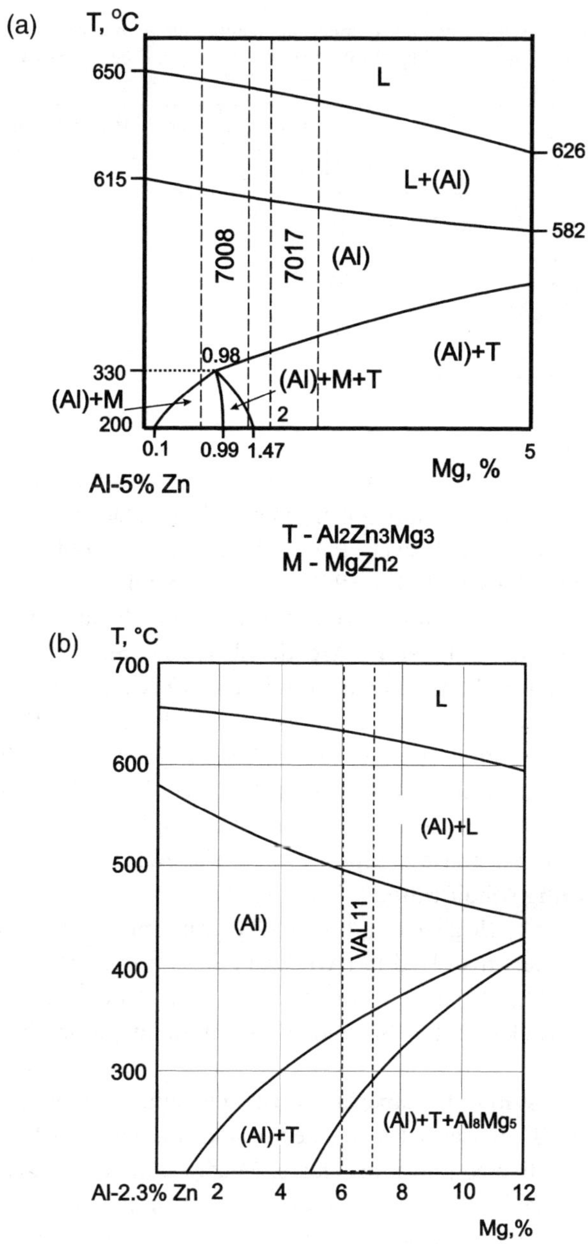

Figure 6.6. Polythermal sections of Al–Mg–Zn phase diagram at 5% Zn (a) and 2.3% Zn (b).

Table 6.12. Solidification reactions under nonequilibrium conditions in a 7005 alloy containing 5.14% Zn, 1.12% Mg, 0.18% Mn, 0.19% Fe, and 0.08% Si (Bäckerud et al., 1986)

Reaction	Temperatures (°C)* at a cooling rate	
	0.3 K/s	15 K/s
$L \Rightarrow (Al)$	641	638
$L \Rightarrow (Al) + Al_3Fe$	632–596	610
$L \Rightarrow (Al) + Mg_2Si$	596	560
$L \Rightarrow (Al) + Al_2Mg_2Zn_3$	470	470
Solidus	470	470

* Start of reaction (author' remark)

The presence of iron and silicon impurities causes the formation of Al_3Fe and Mg_2Si phases as it has been shown experimentally by Bäckerud et al. (1986) for a 7005 alloy (Table 6.12). However, the Al_8Fe_2Si phase should form in this type of alloys at Si:Fe > 3 (Mondolfo, 1976). The probability of Al_8Fe_2Si formation increases at high cooling rates as one can conclude from the analysis of non-equilibrium solidification of Al–Fe–Mg–Si alloys (see Section 2.3). Slow cooling favors the formation of Al_3Fe, as magnesium in 7XXX alloys is in excess to Mg_2Si and, therefore the solidification reactions follow the line e_6–E_1 in Figure 2.4b. Additions of Mn in some commercial 7XXX alloys promote the formation of $Al_{15}(FeMn)_3Si_2$ particles that can hardly be dissolved during heat treatment and are retained in the structure of the final product. Small additions of Cr and Zr do not affect much the as-cast phase composition, as these elements enter the aluminum solid solution during solidification.

Inclusions of the η ($MgZn_2$) and T ($Al_2Mg_3Zn_3$) phases formed during solidification are completely dissolved in (Al) during homogenization in the temperature range 435–445°C. The structure of the semifinished, worked product exhibits, as a rule, insoluble particles of Fe, Si, and Mn-containing phases broken during the deformation.

During decomposition of a supersaturated solid solution (aging after quenching), metastable η' and T' phases act as hardening phases, which at late stages of aging turn into the respective equilibrium phases, $MgZn_2$ and $Al_3Mg_3Zn_3$.

A Russian grade V92Ts contains more magnesium than 7XXX alloys (see Table 6.1). As a result, its structure in the as-cast state resembles that of the 7005 alloy, but contains more T ($Mg_3Zn_3Al_2$) phase formed during the nonequilibrium solidification by the eutectic reaction $L \Rightarrow (Al) + Al_2Mg_3Zn_3$. The metastable T' phase is the main hardening agent in the V92Ts alloy.

Casting alloys of the Al–Mg–Zn system are far less in use than the wrought alloys, mainly because of their high susceptibility to hot cracking during solidification. Table 6.1 shows the chemical composition of a Russian casting alloy VAL11 that is characterized by a high magnesium concentration. During nonequilibrium solidification, in addition to the T phase (Figure 6.6b) the Al_8Mg_5 phase is formed by the invariant eutectic reaction (point E_1 in Table 6.2 and Figure 6.1a). Upon solution treatment, magnesium and zinc completely enter the solid solution and form, upon subsequent aging, hardening particles of the T′ phase.

High-strength wrought aluminum alloys have a complex chemical composition, i.e. contain at least five or six components (see Tables 6.1 and 6.7). With certain assumptions, these alloys can be assigned to the Al–Cu–Mg–Zn system.

According to Bäckerud et al. (1986) the nonequilibrium solidification of a 7075 alloy ends in the range of 466–469°C with the invariant eutectic reaction $L \Rightarrow (Al) + Al_2Cu + MgZn_2 + Al_2Mg_2Zn_3$ (Table 6.13). In our opinion, the reaction with the participation of the Al_2CuMg (S) phase is more likely (see Table 6.8).

Typical microstructures of as-cast and homogenized commercial V95och(rus) and 1933rus alloys (with low concentrations of Fe and Si) are shown in Figure 6.7.

From the isothermal sections of the quaternary diagram shown in Figure 6.8 it follows that copper in 7075-type high-strength alloys is partially dissolved in (Al), and partially participates in the formation of the S (Al_2CuMg) phase from the Al–Cu–Mg system. Note that no quaternary phases are formed in the aluminum corner of the Al–Cu–Mg–Zn system.

At a temperature of 460°C, close to the most often used solution treatment temperature, the composition of a 7074 alloy (Figure 6.8d) falls on the border between the single phase domain (Al) and the two-phase region $(Al) + Al_2CuMg$; whereas a 7012-type alloy remains completely in the solid solution region (Al).

Table 6.13. Solidification reactions under nonequilibrium conditions in a 7075 alloy (5.72% Zn, 2.49% Mg, 1.36% Cu, 0.19% Cr, 0.28% Fe, and 0.11% Si) (Bäckerud et al., 1986)

Reaction	Temperatures (°C)* at a cooling rate	
	0.3 K/s	2.3 K/s
$L \Rightarrow (Al)$	630–623	628
$L \Rightarrow (Al) + Al_3Fe$	618–615	–
$L \Rightarrow (Al) + Mg_2Si$	568–563	558–550
$L \Rightarrow (Al) + Al_2Cu^{**} + MgZn_2 + Al_2Mg_2Zn_3$	469	466
Solidus	469	466

* Start of reaction

** Al_2CuMg according to Table 6.8

(a)

(b)

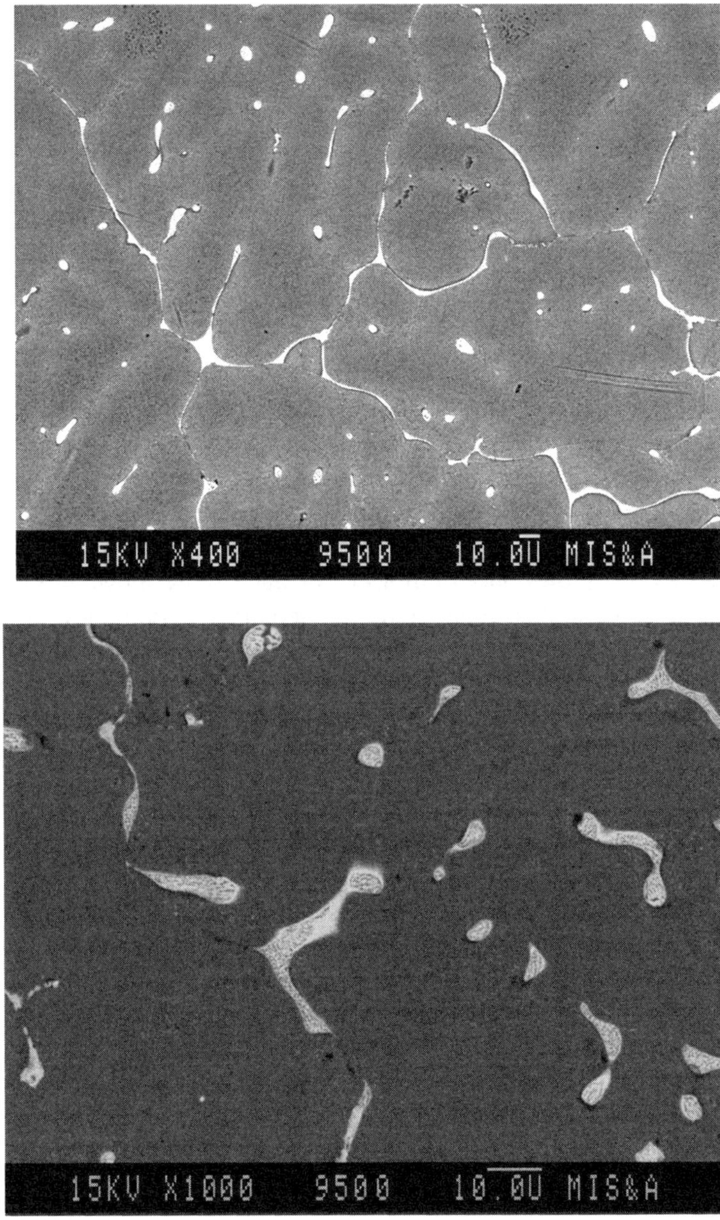

Figure 6.7. SEM structure of ingots of 7075 type-alloys: (a) as-cast 1933(rus) alloy (6.9% Zn; 2.0% Mg; 1.1% Cu; 0.08% Fe; 0.1% Si; 0.12% Zr), veins of the M phase; (b) as-cast V95och(rus) alloy (5.8% Zn; 2.3% Mg; 1.7% Cu; 0.15% Fe; 0.35% Mn; 0.12% Cr; <0.05% Si), eutectic colony with M and T phases; (c) a 1933(rus) alloy (6.6% Zn;1.8% Mg; 0.95% Cu; 0.13% Fe; 0.08% Si; 0.03% Ti; 0.13%Zr) annealed at 465°C for 24 h, undissolved particles of the Al_8Fe_2Si phase; and (d) the same alloy as in (c) annealed at 400°C for 8 h, secondary precipitates of the M phase.

(c)

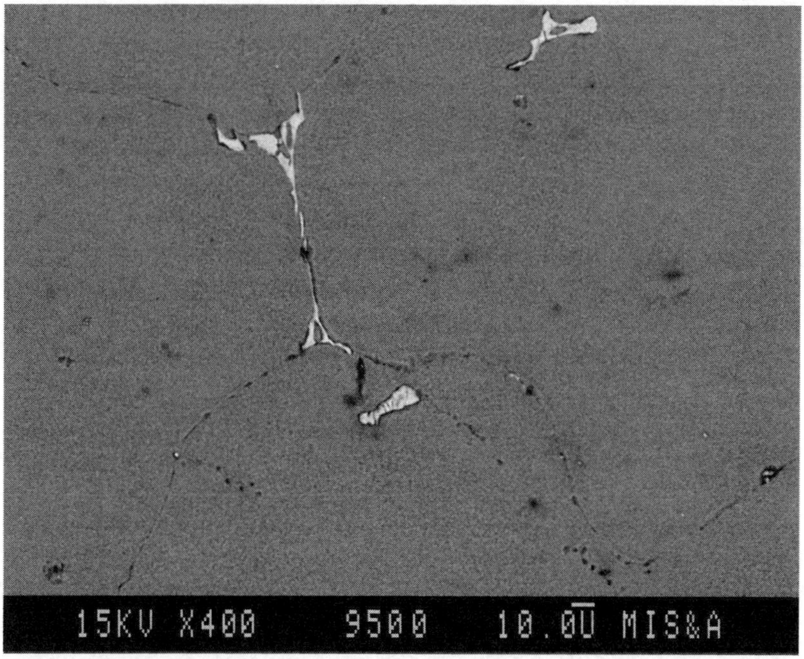

(d)

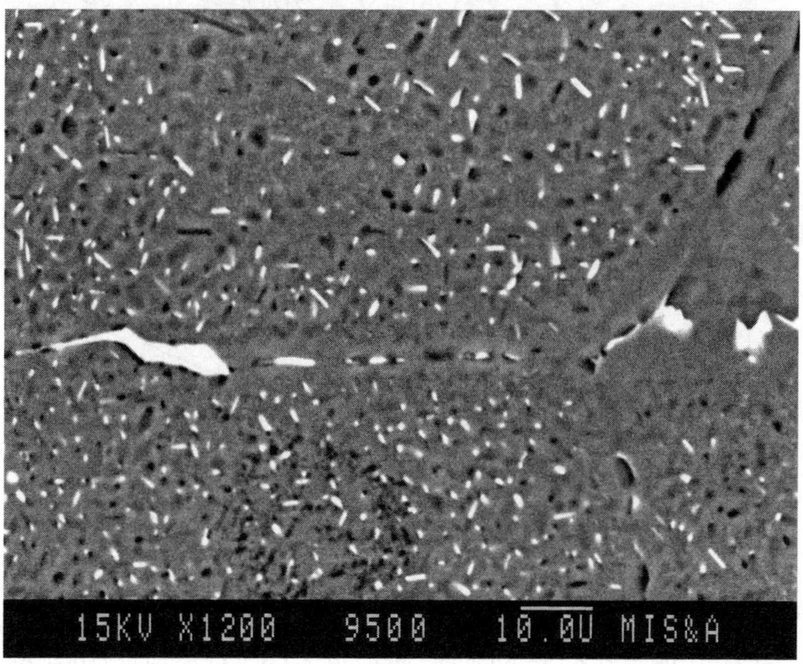

Figure 6.7 (*continued*)

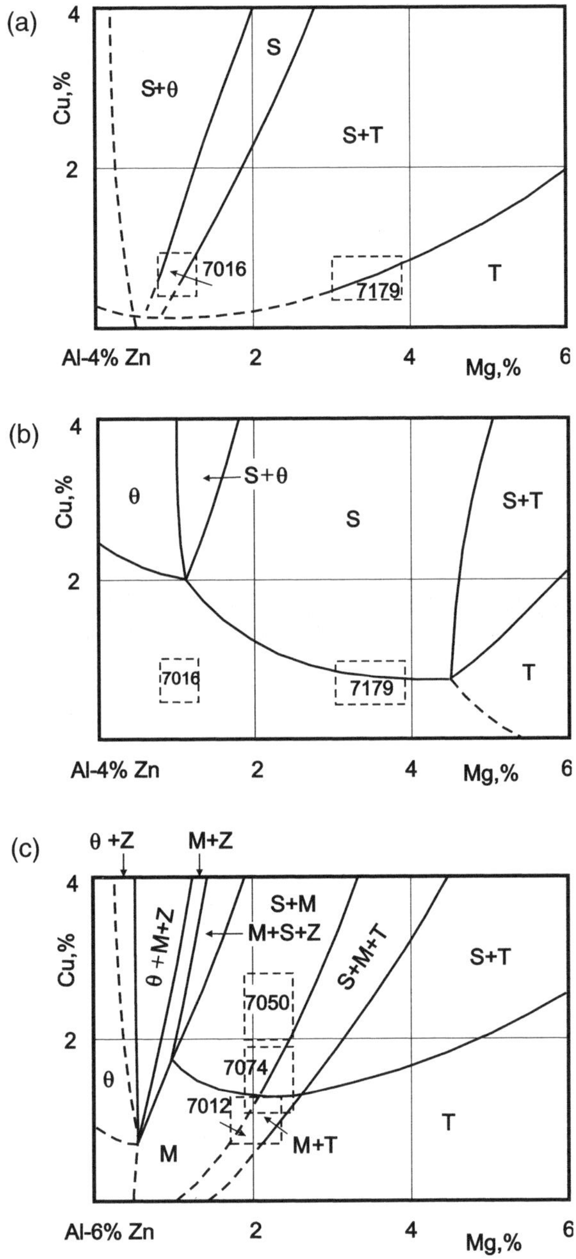

Figure 6.8. Isothermal sections of Al–Cu–Mg–Zn phase diagram at 4% Zn (a, b), 6% Zn (c, d), and 8% Zn (e, f) at 200°C (a, c, e), and 460°C (b, d, f) (Zakharov et al., 1961). T – (Al_6CuMg_4 – $Al_2Mg_3Zn_3$), M – ($MgZn_2$ – AlCuMg), Z – ($Al_5Cu_6Mg_2$ – Mg_2Zn_{11}), S – Al_2CuMg, and θ – Al_2Cu. All phase fields contain (Al).

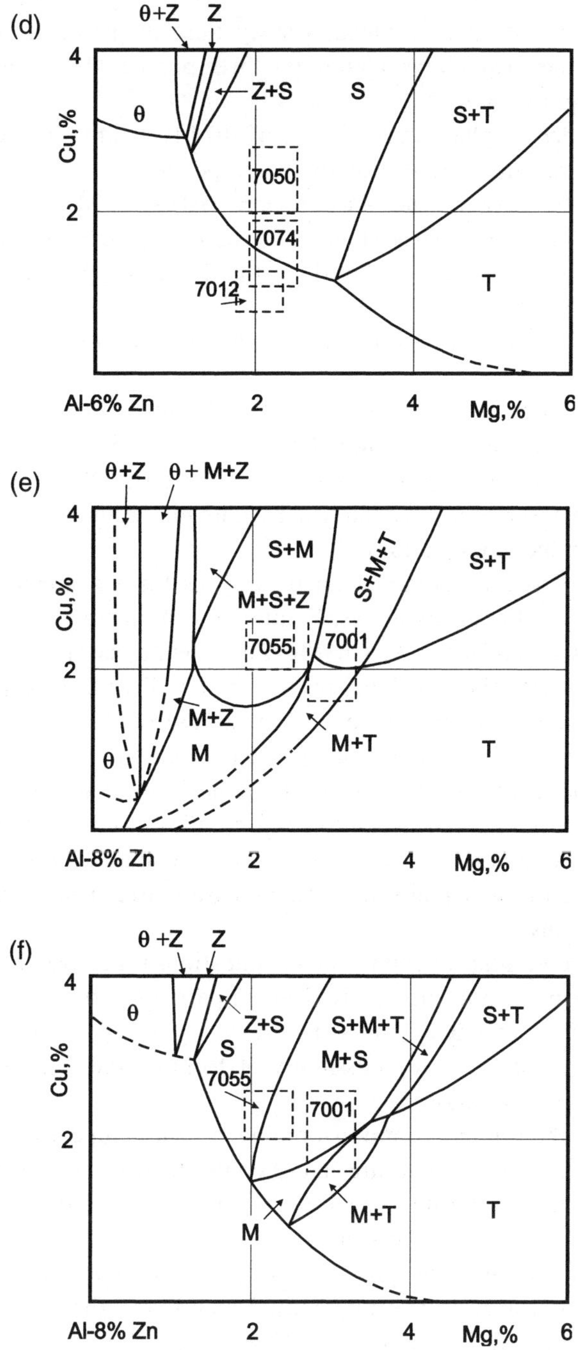

Figure 6.8 (*continued*)

At a temperature of 200°C, which is fairly close to the aging temperature, the compositions of these two alloys fall into the three-phase and two-phase regions with the M phase (Figure 6.8c).

The main hardening phase during aging of 7074-type alloys is the M phase (or η ($MgZn_2$)). However, the composition of 7074-type alloys at 200°C (Figure 6.8c) are close to the border between the phase regions (Al) + S + M (η) and (Al) + S + M (η) + T, (Al) + M (η), and (Al) + M (η) + T. Therefore, a certain role in the hardening of these alloys can also be played by the T ($Al_2Mg_3Zn_3$) and S (Al_2CuMg) phases.

Similar analysis can be performed for the phase composition of the other high-strength alloys.

For example, 7055 and 7001 alloys (Table 6.7) fall into the phase region (Al) + S + M (η) at a quenching temperature of 460°C (Figure 6.8f), and appear to be at the border of the phase regions (Al) + S + M (η) and (Al) + S + M (η) + T at 200°C (Figure 6.8e). Therefore, the most likely hardening phases in this alloy are M (η) and S in their metastable modifications.

Calculations of phase diagrams can be very useful for practical assessment of phase composition of complex alloys. Figure 6.9 demonstrates several polythermal and isothermal sections of the Al–Cu–Mg–Zn system constructed by authors using Thermocalc software. The calculated results agree well with the experimental data.

Strictly speaking, it should be noted that the use of the equilibrium phase diagram for the forecast of the phase composition of aging products is not correct (see Section 3.9). One has to use the chemical composition of the supersaturated solid solution (not the nominal composition of the alloy) and the nonequilibrium or metastable phase diagram that, typically, is not yet available. Otherwise, the conclusions made based on the equilibrium phase diagram are, at best, valid only for the overaged conditions.

Despite the broad range of commercial applications for Al–Zn–Mg–Cu alloys, the nature of their aging is not sufficiently clear. In particular, there are many questions regarding the effect of Cu on the structure of aged alloys. Many issues are yet unanswered with respect to the possibility of forming the S (Al_2CuMg) phase in commercially important alloys.

Besides that, there is no common opinion on the temperature range of precipitation of the η′ phase. There are many contradictory data on the sequence of precipitation of the equilibrium η-phase with various orientation relationships with the matrix in the Al–Cu–Mg–Zn system.

We thoroughly studied the precipitation phenomena in Al–Zn–Mg–Cu alloys with the aim to construct some polythermal sections of the phase diagram under conditions of metastable equilibrium (Aksenov et al., 1992; Kuznetsov et al., 1992). Three alloys, i.e. Al–6%Zn–1.6%Mg–1%Cu; Al–6%Zn–1.6%Mg; and

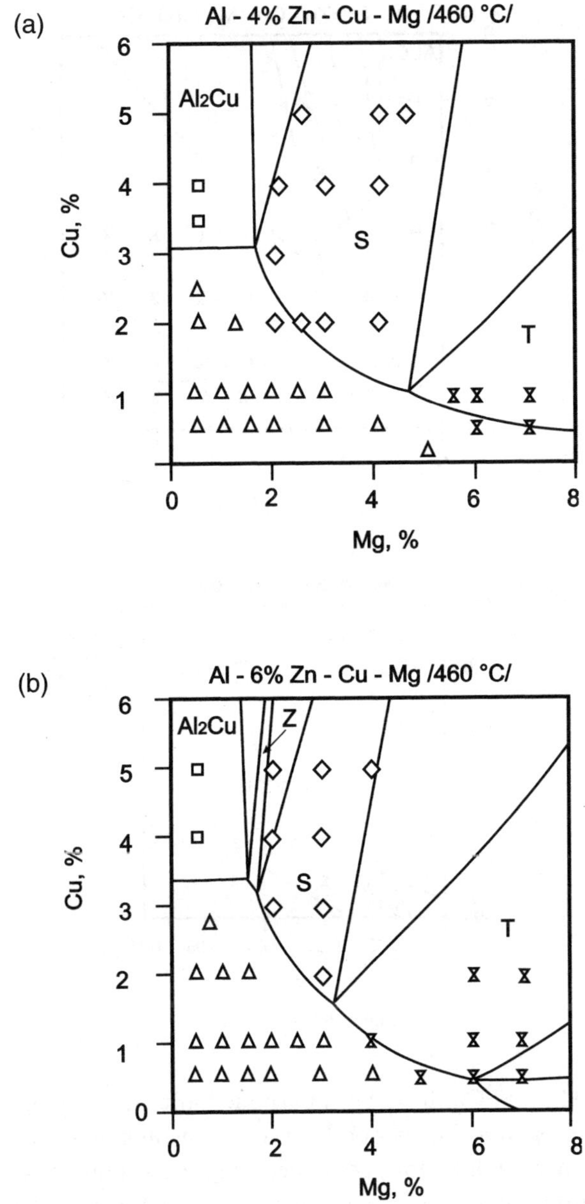

Figure 6.9. Isothermal (a–c) and polythermal (d) sections of Al–Cu–Mg–Zn phase diagram calculated by Thermocalc: (a) 4% Zn, 460°C; (b) 6% Zn, 460°C (c) 8% Zn, 460°C; and (d) 8% Zn and 2% Cu. Experimental values from Zakharov et al. (1961) are marked as points. T – ($Al_6CuMg_4 - Al_2Mg_3Zn_3$), M – ($MgZn_2$ – AlCuMg), Z ($Al_5Cu_6Mg_2 - Mg_2Zn_{11}$), and S – Al_2CuMg. All phase fields in isothermal sections contain (Al).

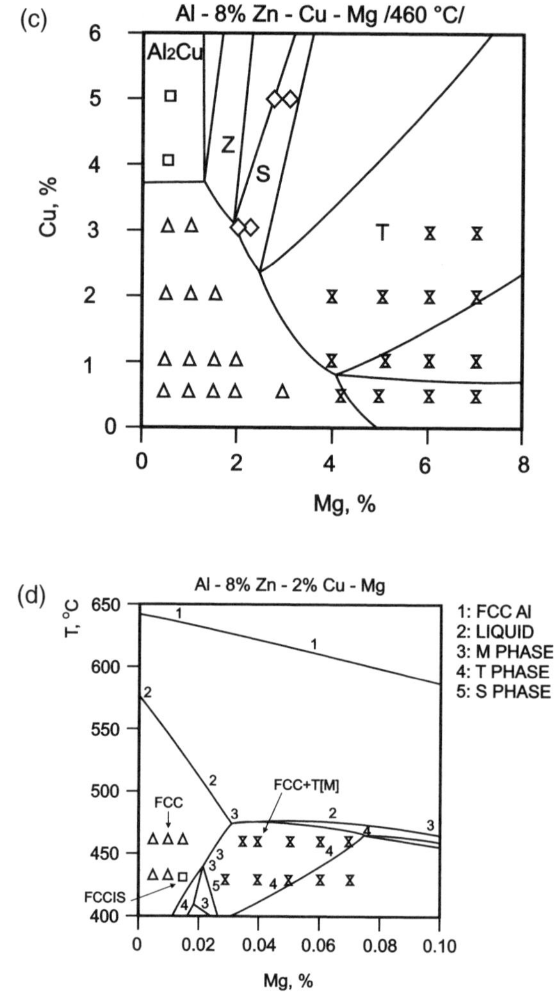

Figure 6.9 (*continued*)

Al–6%Zn–1.6%Mg–2% Cu, have been examined after quenching from 460°C and aging by different regimes. The period between quenching and aging did not exceed 10 min. After quenching, all alloys contained only the aluminum solid solution.

The compositions of the supersaturated solid solutions as determined by the electron microprobe analysis were close to the average chemical compositions of the alloys (Table 6.14).

The products of decomposition of the supersaturated solid solution were identified using selected-area electron diffraction patterns and electron micrographs. The temperature, above which the metastable phase is not observed at any holding time,

Table 6.14. Composition of the supersaturated solid solution of the tested alloys

Alloy	Content of components, wt.%					
	Zn	Mg	Cu	Fe	Si	Al
Al–6% Zn–1.6% Mg	6.2	1.4	–	<0.1	<0.08	balance
Al–6% Zn–1.6% Mg–1% Cu	6	1.5	1.1	0.05	0.08	balance
Al–6% Zn–1.6% Mg–2% Cu	6.2	1.5	1.98	<0.1	0.08	balance

is usually adopted as the solvus temperature of this phase (Zolotorevskii et al., 1984). However, it is known that at temperatures close to the solvus of a metastable phase, the nucleation period of its formation can be larger than that of a stable phase. Then the metastable phase will form later than the stable phase (Novikov, 1982). This may occur within a narrow temperature range close to the solvus of the metastable phase, where the supersaturation of the solid solution with respect to the metastable phase is low. As a result, due to the preceding formation of the stable phase and consuming of the alloying components from the solid solution, the metastable phase may not form at all; and the given annealing temperature may mistakenly be taken as its solvus. For a more accurate assessment of the solvus for the η′ and η phases, we used isothermal transformation curves, i.e. C-curves. To avoid mistakes, we first made a forecast based on calculations and then validated the results with experiments.

The position (temperature) of a minimal nucleation period for the nucleation of a phase can be determined by the phase solvus as (Davydov et al., 1973):

$$(T_{\mathrm{min}})/(T_{\mathrm{S}}) = \mathrm{K},$$

where T_{min} and T_{S} are the temperatures (in Kelvin) of the minimal nucleation period and the solvus, respectively, for a given phase; K = const.

Analysis of numerous literature data on the C-curves for various aluminum alloys shows that this ratio is correct (Davydov et al., 1973; own data). For Al–Zn–Mg–Cu alloys, K = 0.87 in the range of T-phase precipitation and K = 0.93–0.95 for alloys in the (Al) + η phase region. These data were obtained from the isothermal transformation diagrams for different Al–Zn–Mg alloys (Davydov et al., 1973; Davydov, 1984) and from the solvuses published elsewhere (Phillips, 1961). The ratio remains the same for the C-curves constructed by mechanical properties, electric conductivity, and lattice parameter. The ratio failed only for the C-curves constructed using susceptibility to intercrystalline corrosion, probably, due to the fact that intercrystalline corrosion is affected by grain-boundary precipitates rather than the precipitates in the bulk of the grain, and as a result the minimal nucleation period of the C-curve is shifted to lower temperatures.

Thus, the search for the solvus temperature of the metastable phase is limited to the determination of the temperature of the minimal nucleation period for this phase. Then, the considered ratio can be used. Note that this ratio is valid both for the equilibrium phase and for all its metastable modifications formed during the decomposition of the solid solution. This assumption is based on the fact that all phase modifications contain the same alloying elements, and only the surface energy of a precipitate changes.*

When identifying the reflections in electron diffraction patterns, we considered all possible variants of phases (equilibrium and metastable) that may occur in the Al–Cu–Mg–Zn system (see Section 6.1 and Table 6.9).

Transmission electron microscopy shows that the following phases are precipitated in all given alloys during aging:

1. The hexagonal η'_2 phase with parameters $a = 0.496$, $c = 1.403$ nm, with the orientation relationship: (00.1) $\eta'_2//(111)_{(Al)}$; (10.0) $\eta'_2//(110)_{(Al)}$ and
2. The hexagonal phase η with parameters: $a = 0.52$ nm, $c = 0.857$ nm in two orientation relationships: $(00.1)_{\eta 1}//(110)_{(Al)}$; $(10.0)_{\eta 1}//(001)_{(Al)}$ and $(00.1)_{\eta 2}//(111)_{(Al)}$; $(10.0)_{\eta 2}//(110)_{(Al)}$.

The major distinction of the alloys containing copper is the precipitation of a phase identified as S′ (Al_2CuMg) (Figure 6.10). In addition, copper has a significant effect on the formation kinetics of the η'- and η-phases.

The entire sequence of precipitation and the effect of copper on the kinetics can be traced on the isothermal transformation curves given in Figure 6.11.

Using the method of thermodynamical calculation of phase equilibria (Kuznetsov et al., 1992) and the analysis of equilibrium solidification in the Al–Mg–Zn and Al–Cu–Mg–Zn systems, we determined the critical points in the studied alloys. These critical points were used to construct the polythermal section that was also confirmed experimentally as shown in Figure 6.12a.

It is important to note that the calculated equilibrium polythermal section is consistent with the data reported elsewhere (Zakharov et al., 1961; Zakharov, 1980). The main difference is the presence of the S phase in the quaternary alloys at 200°C, i.e. as an equilibrium phase after aging.

To confirm the calculated section experimentally, we determined the solvus temperatures of the alloys with 1 and 2% Cu. The alloys were annealed at 340, 360,

---

* This assumption may not be valid for the metastable phases that significantly differ from the stable phase in composition, e.g. β″, β′, Mg_2Si.

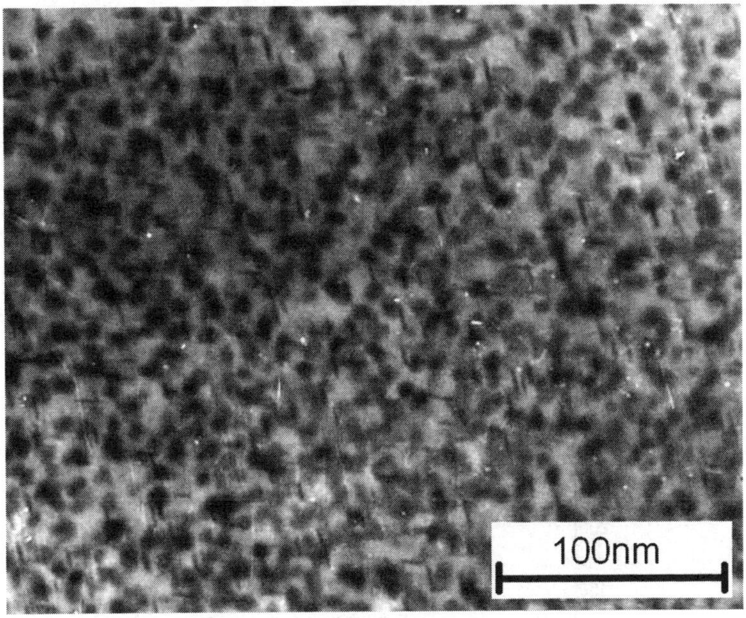

Figure 6.10. Fine structure of a 7075-type alloy after aging.

380, 400, 430°C for 100 h. In the alloy with 1% Cu at temperatures ≤400°C the T ($Al_2Mg_3Zn_3$) phase was found in the structure. The S (Al_2CuMg) phase was not observed, due to its small volume fraction. In the alloy with 2% Cu, this phase was present after annealing at temperatures up to 430°C.

Therefore, the polythermal section constructed using the thermodynamical calculation adequately describes the solid-state phase transformations in these alloys under equilibrium conditions.

However, this section fails to reflect the processes that occur during real aging. The experimental data on the phase composition of the aging products shows that the T phase does not form during decomposition of the supersaturated solid solution (contrary to the conclusion that may be made based on the low-temperature sections of the equilibrium phase diagram). Therefore, the polythermal section was re-calculated without T phase (Kuznetsov et al., 1992). The so-called metastable section is given in Figure 6.12b. One can see that the solvus of the η phase increases in the range from 0 to 2% Cu (360°C at 1% Cu). At higher copper concentrations, the equilibrium and metastable solvuses of the η phase merge.

The calculation results were verified experimentally by a reverse treatment. In these experiments, the maximum temperature of reverse stage of aging (at a minimal holding time) up to which the η phase is stable shall correspond to its solvus.

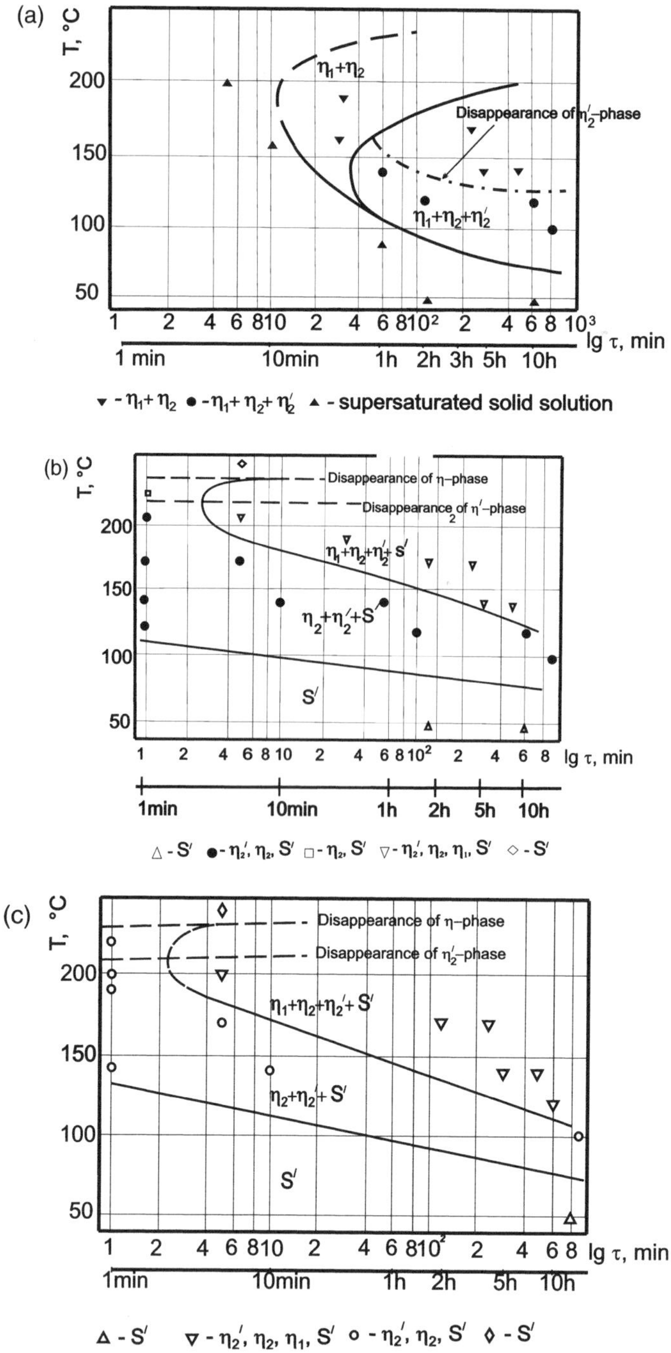

Figure 6.11. C-curves of isothermal transformations upon aging in (a) Al–6%Zn–1.6%Mg; (b) Al–6%Zn–1.6%Mg–1%Cu; and (c) Al–6%Zn–1.6%Mg–2% Cu alloys.

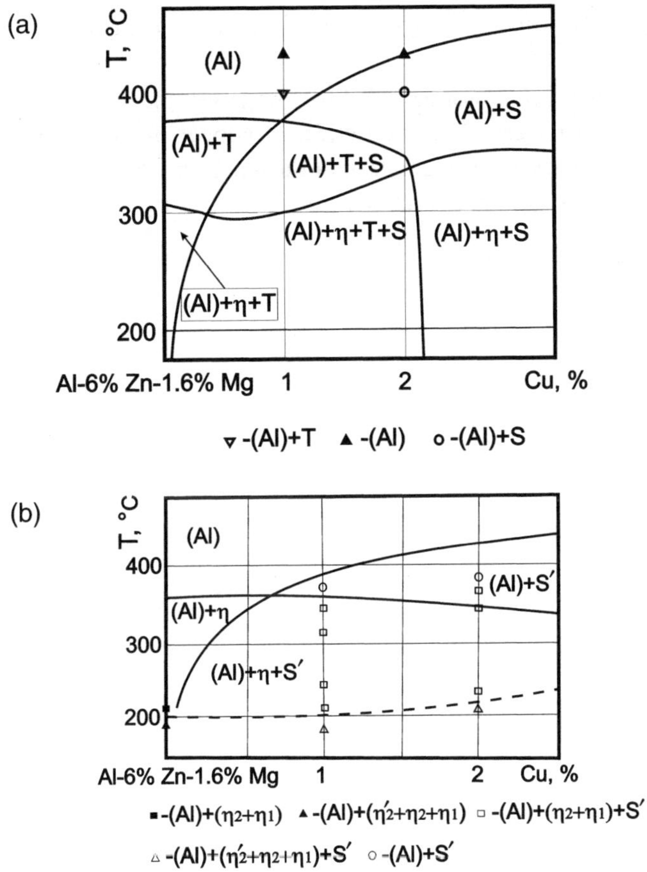

Figure 6.12. Polythermal section of Al–Cu–Mg–Zn phase diagram at 6% Zn and 1.6% Mg: (a) equilibrium and (b) metastable state.

The experimental solvus of the η phase for the alloy with 1% Cu is 360 ± 10°C that agrees well with the calculation. For the alloy with 2% Cu, the experimental solvus temperature of the η phase was determined as 380 ± 10°C. A slight discrepancy with the calculation in this case can be explained by the fact that, as the concentration of copper increases, its solubility in the $MgZn_2$ phase goes up, which was not taken into account in the calculations.

Thus, the calculated metastable polythermal section adequately reflects the real processes occurring upon aging.

The solvus for the η'_2 phase can also be calculated and validated experimentally. As the activation energies for this phase and for η and η′ phases are different, coefficient K = 0.86–0.88. The experimental solvuses for the η'_2 phase were determined as 210 ± 10°C at 1% Cu and 230 ± 10°C at 2% Cu (in Al–6%Zn–1.6% Mg–Cu alloys).

We can conclude that copper slightly increases the solvuses of stable and metastable η-based phases. The main effect of copper on the structure of Al–Zn–Mg–Cu alloys is, however, the formation of the hardening S′ phase and the change of precipitation kinetics that can be traced in C-curves in Figure 6.11.

Chapter 7
Alloys with Nickel

Some aluminum, mainly special-purpose alloys, both wrought (Table 7.1) and casting (Table 7.2), contain nickel as an additive. Nickel has low solubility in solid (Al), and during solidification enters various phases, in particular, those containing iron and copper. Analysis of the phase composition of such alloys requires the use of multicomponent phase diagrams involving nickel. This chapter is an attempt to collect all available information on such diagrams. As only a small number of Al-based alloying systems with nickel are experimentally studied, the variants of phase diagrams suggested in this chapter are largely assessments. We should note that most commercial alloys with the nickel addition belong to systems with five or more components; therefore, the phase diagrams considered here make it possible to only partially analyze the phase composition of such alloys.

7.1. Al–Fe–Ni PHASE DIAGRAM

This phase diagram can be used for the analysis of the phase composition of an 8001 alloy (Table 7.1) that contains only nickel and iron as the alloying elements. This phase diagram is also necessary for the analysis of more complex systems.

In the Al–Fe–Ni ternary system, the Al_3Fe, Al_3Ni, and Al_9FeNi phases are in equilibrium with the aluminum solid solution. The Al_3Fe phase dissolves up to 3–4% Ni, and no more than 1% Fe can be dissolved in the Al_3Ni phase.

The Al_3Ni phase (42% Ni) has an orthorhombic structure (space group *Pnma*, 16 atoms per unit cell) with parameters $a = 0.6611$ nm, $b = 0.7366$ nm, $c = 0.4812$ nm (Mondolfo, 1976). The density of this phase in the binary Al–Ni system is 3.95–3.96 g/cm^3. The Al_3Ni phase can be considered as heat resistant with microhardness at room temperature of 5.95 GPa and 1-h microhardness at 300°C, 3.54 GPa (Kolobnev, 1973).

The Al_9FeNi (T) compound has a monoclinic structure (space group $P2_1/c$, 22 atoms per unit cell) with parameters $a = 0.6207$ nm, $b = 0.6271$ nm, $c = 0.8598$ nm, $\beta = 94°66'$ (Budberg and Price, 1992), its density is 3.4 g/cm^3, and Vickers hardness is 6.9–7.4 GPa (Mondolfo, 1976). The homogeneity range of the Al_9FeNi phase is from 4.5% Fe, 28% Ni up to 14% Fe, 18% Ni (Mondolfo, 1976).

The (Al) + Al_3Ni eutectics is formed in Al-rich alloys of the Al–Ni system at 640°C and 6% Ni (Mondolfo, 1976). This eutectic has a fine structure with certain

Table 7.1. Chemical composition of some commercial wrought alloys with nickel

Grade	Ni, %	Fe, %	Si, %	Cu, %	Mg, %	Mn, %
8001	0.4–0.9	0.45–0.70	0.17	0.15	–	–
2618	0.9–1.2	0.9–1.3	0.10–0.25	1.9–2.7	1.3–1.8	–
2031	0.6–1.4	0.6–1.2	0.5–1.3	1.8–2.8	0.6–1.2	0.5
2018	1.7–2.3	0.1	0.9	3.5–4.5	0.45–0.9	0.2
2218	1.7–2.3	0.1	0.9	3.5–4.5	1.2–1.8	0.2

Table 7.2. Chemical composition of some commercial casting alloys with nickel

Grade*	Ni, %	Fe, %	Si, %	Cu, %	Mg, %	Mn, %
361.0	0.2–0.3	1.1	9.5–10.5	0.5	0.4–0.6	0.25
393.0	2.0–2.5	1.3	21–23	0.7–1.1	0.7–1.3	0.1
336.0	2.0–3.0	1.2	11–13	0.5–1.5	0.7–1.3	0.35
339.1	0.5–1.5	0.9	11–13	1.5–3.0	0.6–1.5	0.5
242.0	1.7–2.3	1.0	0.7	3.5–4.5	1.2–1.8	0.35
516.0	0.25–0.4	0.35–1.0	0.3–1.5	0.3	2.5–4.5	0.15–0.4
FM 109	0.8–1.1	0.5	11.5–12.5	0.9–1.3	1.1–1.3	0.05–0.2
FM 113	0.8–1.2	0.35	11.5–12.5	3.0–3.3	0.9–1.2	0.15
FM 135	0.8–1.2	0.35	12.7–13.7	4.8–5.3	0.9–1.2	0.1
FM 120	0.7–1.3	0.65	12.0–13.5	0.8–1.5	0.9–1.3	0.05–0.3
FM 180	0.8–1.3	0.57	17.0–19.0	0.8–1.5	0.8–1.3	0.05–0.2
FM S2N	2.1–2.5	0.4	11.4–12.4	3.1–3.5	0.6–1.0	0.15
FM S2	2.3–2.8	0.5	11.0–12.0	3.3–3.8	0.6–0.9	0.15–0.25
FM B1	2.3–2.8	0.5	12.5–13.5	4.9–5.4	0.6–0.9	0.15–0.25
FM B2	2.7–3.0	0.5	12.2–12.6	3.9–4.3	0.6–0.9	0.15
FM 2500	4.2–5.0	0.5	22.8–23.8	6.3–7.2	1.7–2.0	0.35
FM 2393	4.8–6.0	0.7	22.8–23.8	6.6–7.5	1.7–2.0	0.35–0.5

* FM- specifications of Federal-Mogul Corporation Powertrain Systems (Russia)

orientation relations between the phases (Belov and Zolotorevskii, 2001; Belov et al., 2004). A natural composite material with special properties can be formed during directional crystallization (Martin et al., 1997). However, even a small amount of iron impurity (~0.2%) has a significant effect on the structure and phase composition of Al–Ni alloys (Belov et al., 2002a). This follows from the Al–Fe–Ni phase diagram. Two invariant transformations – eutectic and peritectic – occur in the aluminum corner of this ternary system (Table 7.3). The temperature ranges for the formation of the Al_3Fe, Al_3Ni, and Al_9FeNi phases during monovariant solidification reactions are given in Table 7.4. Figure 7.1 shows the projection of the solidification surfaces and the distribution of phase regions in the solid state in the aluminum corner of the Al–Fe–Ni phase diagram. According to the equilibrium phase diagram, the solidus temperature of ternary alloys cannot be below 637°C,

Table 7.3. Invariant reactions in ternary alloys of Al–Fe–Ni system (Mondolfo, 1976; Drits et al., 1977)

Reaction	Points in Figure 7.1a	*T*, °C	Concentrations in liquid phase	
			Ni, %	Fe, %
$L + Al_3Fe \Rightarrow (Al) + Al_9FeNi$	P	649	1.7	1.7
$L \Rightarrow (Al) + Al_9FeNi + Al_3Ni$	E	637	6.0	0.1–0.3

Table 7.4. Monovariant reactions in ternary alloys of Al–Fe–Ni system

Reaction	Lines in Figure 7.1a	*T*, °C
$L \Rightarrow (Al) + Al_9FeNi$	P–E	649–637
$L \Rightarrow (Al) + Al_3Fe$	e_1–P	655–649
$L \Rightarrow (Al) + Al_3Ni$	e_2–E	640–637

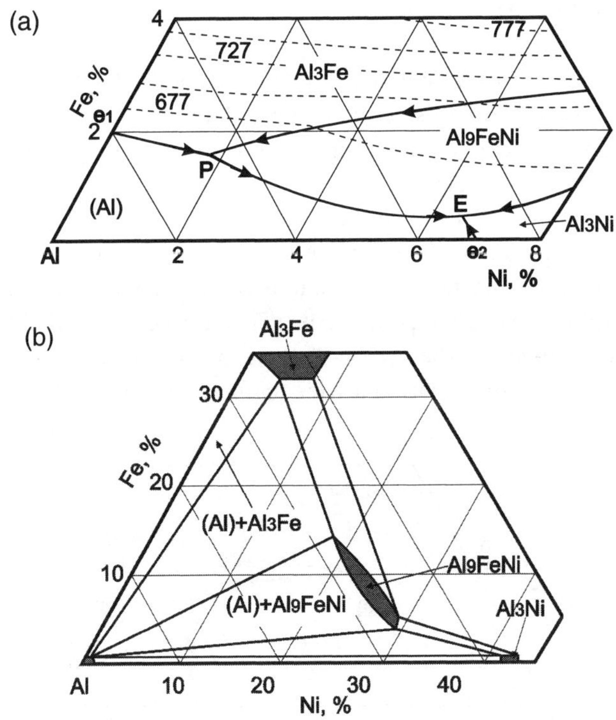

Figure 7.1. Phase diagram of Al–Fe–Ni system: (a) projection of the solidification surface and (b) distribution of phase regions in the solid state.

so all three binary eutectics in Table 7.4 solidify within a relatively narrow temperature range. According to Mondolfo (1976) the concentration of iron in the ternary eutectics is 0.3%, but according to our data it should be slightly less, as we observe an appreciable amount of primary Al_9FeNi crystals in an Al–6% Ni–0.3% Fe alloy.

The solubility of nickel in solid aluminum is 0.05% at the eutectic temperature and decreases down to 0.03% at 627°C and to 0.006% at 527°C. However, even this low solubility can lead to a noticeable strengthening due to the precipitation of a metastable modification of Al_3Ni (Tsubukino et al., 1996).

On increasing the cooling rate during solidification, the region of (Al) primary solidification in ternary Al–Fe–Ni alloys widens, which is especially pronounced upon rapid solidification processing (RS/PM). The metastable phases Al_6Fe and Al_mFe can appear instead of the stable Al_3Fe phase. However, the phases Al_3Ni and Al_9FeNi remain in their equilibrium forms even after rapid solidification. The solid solubilities of iron and nickel in aluminum upon rapid solidification may increase up to 0.3% Fe and 0.4% Ni (Belov et al., 2002a).

7.2. Al–Ni–Si PHASE DIAGRAM

Although commercial alloys belonging solely to the Al–Ni–Si system are virtually nonexistent, knowledge of this ternary phase diagram is required for the analysis of multicomponent alloys involving nickel and silicon, particularly piston alloys of the 3XX.0 series (Table 7.2).

No ternary compounds form in the aluminum corner of the Al–Ni–Si system, so only phases from the binary systems – Al_3Ni and (Si) – can be in equilibrium with (Al). The only invariant transformation involving (Al) is of eutectic character (Figure 7.2, Table 7.5). The ternary eutectic temperature (557°C) determines the solidus of most alloys of the Al–Ni–Si system. If silicon is present, the binary (Al) + Al_3Ni eutectics forms within a wide temperature range (Table 7.5), which has

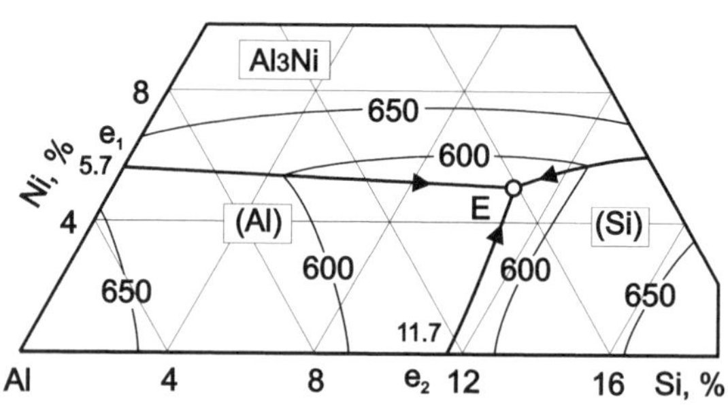

Figure 7.2. Phase diagram of Al–Ni–Si system: projection of the solidification surface.

Table 7.5. Invariant and monovariant reactions in ternary alloys of Al–Ni–Si system (Mondolfo, 1976; Drits et al., 1977)

Reaction	Point or line in Figure 7.2	T, °C	Concentrations in liquid phase	
			Ni, %	Si, %
L⇒(Al) + Al_3Ni + (Si)	E	567	5	11–12
L⇒(Al) + Si	e_1–E	577–557	–	–
L⇒(Al) + Al_3Ni	e_2–E	640–557	–	–

a negative effect on its fineness: its structure becomes coarser as compared to binary alloys (Belov and Zolotorevskii, 2003). The solubility of silicon in the Al_3Ni phase is about 0.4–0.5% (Mondolfo, 1976). Nickel decreases the solubility of silicon in (Al) (Drits et al., 1977).

7.3. Al–Cu–Ni PHASE DIAGRAM

This phase diagram is helpful in the analysis of 2618-type heat-resistant alloys and 339.0-type piston alloys that contain nickel, copper, and other alloying components (Tables 7.1 and 7.2).

The ternary Al_7Cu_4Ni phase forms in the aluminum corner of the Al–Cu–Ni system. Besides it, the phases Al_2Cu, Al_3Ni, and Al_3Ni_2 from the corresponding binary systems are in equilibrium with the aluminum solid solution (Mondolfo, 1976; Drits et al., 1977).

The ternary phase Al_7Cu_4Ni exists in the homogeneity range of 38.7–50.7% Cu, 11.8–22.2% Ni. Because of such a wide homogeneity range, different stoichiometric compositions are assigned to this phase besides Al_7Cu_4Ni, e.g. Al_6Cu_3Ni, Al_7CuNi, Al_3Cu_2Ni, and Al_9Cu_3Ni. Accordingly, various crystal structures are reported for Al_7Cu_4Ni, rhombohedral, hexagonal, or cubic (Mondolfo, 1976; Hatch, 1984). This phase has a density of 5.48 g/cm^3 and is heat resistant with a 1-h microhardness at 300°C of 5.8 GPa (Kolobnev, 1973).

The binary Al_3Ni_2 phase also has a broad range of homogeneity, which spreads in the ternary system up to 31.2% Cu, 29.9% Ni or up to the composition of Al_3CuNi. This compound has a hexagonal crystal structure (space group $P\bar{3}m1$) with lattice parameters $a = 0.4036$ nm and $c = 0.490$ nm. The density is 4.76 g/cm^3 (Hatch, 1984). It should be noted that Al_3Ni_2 is not in equilibrium with (Al) in the binary Al–Ni system.

Tables 7.6 and 7.7 present invariant and monovariant reactions in ternary alloys of the Al–Cu–Ni system, and Figure 7.3 shows the liquidus and solidus projections.

Table 7.6. Invariant reactions in ternary alloys of Al–Cu–Ni system (Mondolfo, 1976; Drits et al., 1977)

Reaction	Point in Figure 7.3a	T, °C	Concentrations in liquid phase	
			Cu, %	Ni, %
$L + Al_3Ni \Rightarrow (Al) + Al_3Ni_2$ (Al_3CuNi)	P_1	630	16	4
$L + Al_3Ni_2 \Rightarrow (Al) + Al_7Cu_4Ni$	P_2	590	22	2
$L \Rightarrow (Al) + Al_2Cu + Al_7Cu_4Ni$	E	546	32.5	0.9

Table 7.7. Monovariant reactions in ternary alloys of Al–Cu–Ni system

Reaction	Lines in Figure 7.3a	T, °C
$L \Rightarrow (Al) + Al_3Ni$	e_1–P_1	640–630
$L \Rightarrow (Al) + Al_3Ni_2$	P_1–P_2	630–590
$L \Rightarrow (Al) + Al_7Cu_4Ni$	P_2–E	590–546
$L \Rightarrow (Al) + Al_2Cu$	e_2–E	547–546

The solubility of nickel in (Al) is very small, and that of copper in ternary alloys depends on the phase region into which an alloy falls (Table 7.8).

Under real solidification conditions, due to the incomplete peritectic reactions (Table 7.6), solidification of most alloys ends at the ternary eutectic temperature. The ternary eutectics is constituted mainly by Al_2Cu crystals.

7.4. Al–Mg–Ni PHASE DIAGRAM

This phase diagram is required for the analysis of commercial alloys containing nickel and magnesium (Tables 7.1 and 7.2).

Though the experimental data available on the Al–Mg–Ni system are scarce, the absence of ternary compounds makes it possible to relatively easily assess its constitution. The only invariant transformation in this system is close, by its temperature and composition, to the binary eutectics from the Al–Mg system (Mondolfo, 1976; Drits et al., 1977)

$L \Rightarrow (Al) + Al_3Ni + Al_8Mg_5$ at 449°C, 1.7% Ni and 32% Mg (point E in Figure 7.4).

Table 7.9 shows monovariant reactions proceeding in the Al–Mg–Ni system. The liquidus and solidus projections are shown in Figure 7.4. In the presence of

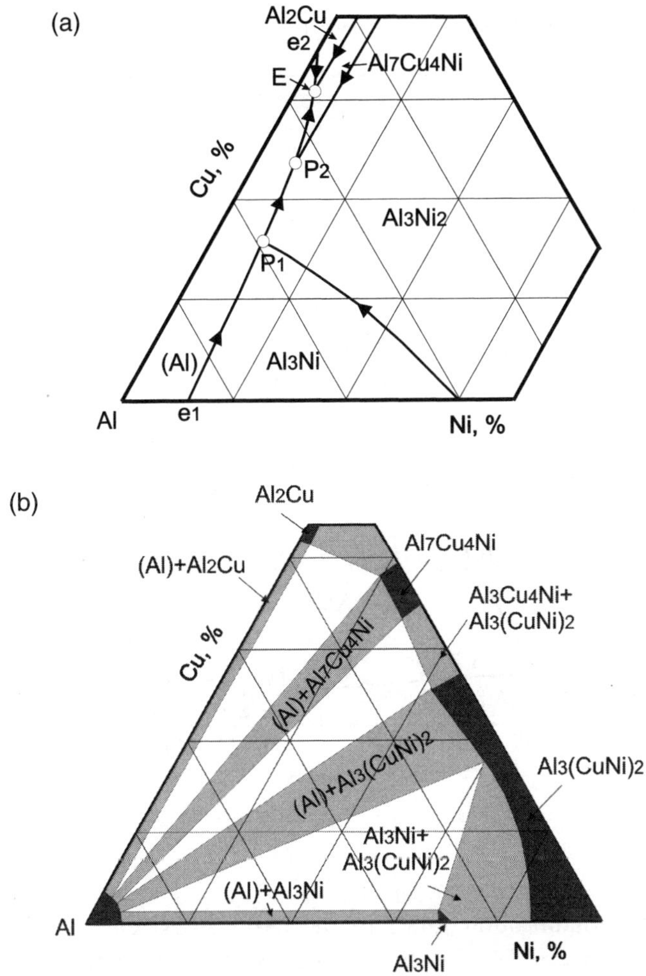

Figure 7.3. Phase diagram of Al–Cu–Ni system: (a) projection of the solidification surface and (b) distribution of phase regions in the solid state.

Table 7.8. Limit solid solubility of Cu in (Al) in the ternary phase fields of Al–Cu–Ni system (Figure 7.3b) (Drits et al., 1977)

T, °C	(Al) + Al_3Ni + Al_3Ni_2	(Al) + Al_3Ni_2 + Al_7Cu_4Ni	(Al) + Al_2Cu + Al_7Cu_4Ni
561	–	4.35	–
554	1.7	–	–
547	–	–	5.3
527	1.5	3.3	3.8
427	1.2	1.5	1.9

Table 7.9. Monovariant reactions in ternary alloys of Al–Mg–Ni system

Reaction	Lines in Figure 7.4	T, °C
L ⇒(Al) + Al_8Mg_5	e_2–E	450–449
L ⇒(Al) + Al_3Ni	e_1–E	640–449

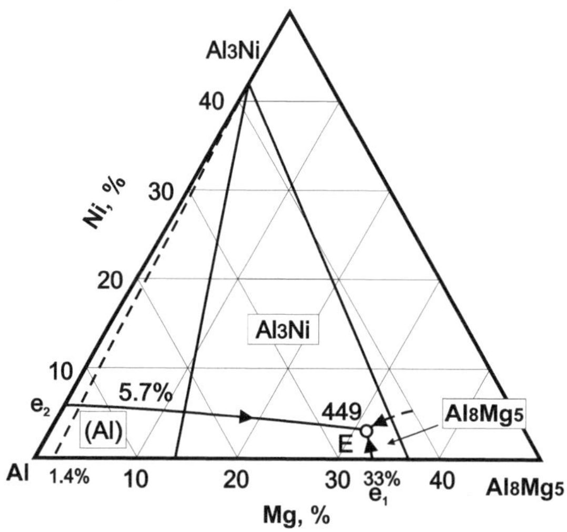

Figure 7.4. Phase diagram of Al–Mg–Ni system.

magnesium the solidification range of the (Al) + Al_3Ni eutectics is considerably broadened, therefore Al_3Ni eutectic crystals become coarser than in binary Al–Ni alloys.

The solubilities of magnesium and nickel in solid (Al) in ternary alloys are probably close to those in the binary systems.

7.5. Al–Mn–Ni PHASE DIAGRAM

Consideration of this ternary phase diagram is necessary because of the presence of manganese in some Ni-containing commercial alloys (usually as an impurity) (Tables 7.1 and 7.2). Moreover, this phase diagram is the basis for promising casting heat-resistant alloys considered elsewhere (Belov et al., 1993b, c; Belov, 1994, 1996; Belov et al., 1994; Belov and Zolotorevskii, 2003; Lin et al., 2004).

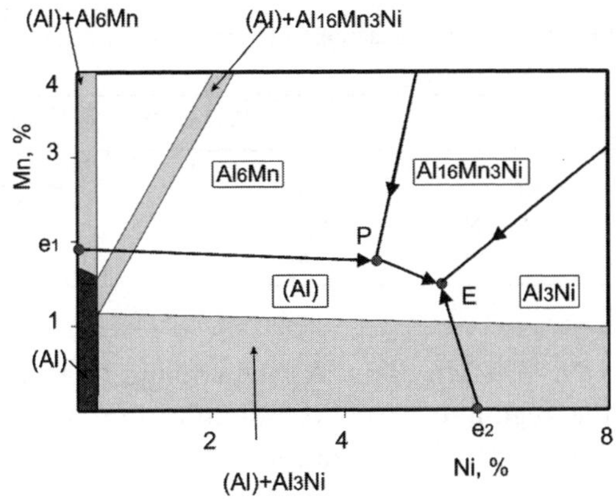

Figure 7.5. Phase diagram of Al–Mn–Ni system: projection of the solidification surface and distribution of phase regions in the solid state at 627°C.

Table 7.10. Invariant reactions in ternary alloys of Al–Mn–Ni system (Mondolfo, 1976; Drits et al., 1977)

Reaction	Point in Figure 7.5	*T*, °C	Concentrations in liquid phase	
			Mn, %	Ni, %
$L + Al_6Mn \Rightarrow (Al) + Al_{16}Mn_3Ni$	P	645	1.7	4.5
$L \Rightarrow (Al) + Al_3Ni + Al_{16}Mn_3Ni$	E	637	1.3	5.3

Scarce data on the Al Mn Ni system suggest that a ternary compound with the formula $Al_{16}Mn_3Ni$ can be in equilibrium with (Al) in addition to the binary aluminides Al_3Ni and Al_6Mn (Mondolfo, 1976). This compound contains 23–26% Mn and 5.6–9.5% Ni and has an orthorhombic structure (space group *Bbmm*, *Bbm*2, or *Bb*2*m*, ~160 atoms per unit cell) with lattice parameters $a = 2.38$ nm, $b = 1.25$ nm, $c = 0.755$ nm; and density, 3.62 g/cm^3.

The projection of liquidus surface and the distribution of phase fields at 627°C are shown in Figure 7.5. Two invariant reactions (eutectic and peritectic) may proceed in Al-rich alloys (Table 7.10). Table 7.11 shows monovariant reaction occurring in the Al corner of the system. The solubility of nickel in solid aluminum is very small. The solubility of manganese in (Al) decreases in the presence of nickel from 1% at 627°C in a binary alloy to about 0.8% in a ternary alloy with nickel. Less than 0.05% Ni dissolves in the Al_6Mn phase. The Al_3Ni compound dissolves a maximum of 0.26% Mn (Mondolfo, 1976).

In the as-cast state, the solubility of manganese in (Al) can be significantly higher than in the equilibrium one: according to our data up to 1.5% Mn can dissolve

Table 7.11. Monovariant reactions in ternary alloys of Al–Mn–Ni system

Reaction	Line in Figure 7.5	*T*, °C
$L \Rightarrow (Al) + Al_6Mn$	e_1–P	658–645
$L \Rightarrow (Al) + Al_{16}Mn_3Ni$	P–E	645–637
$L \Rightarrow (Al) + Al_3Ni$	e_2–E	640–637

in (Al) in an Al–4% Ni–2% Mn alloy. As the cooling rate increases, so does the solubility; besides, the region of primary solidification of (Al) extends, mainly towards the increase in the Mn concentration.

7.6. Al–Fe–Ni–Si PHASE DIAGRAM

This quaternary phase diagram makes it possible to completely analyze the effect of silicon impurity on the phase composition of 8001-type alloys, and partially analyze the combined effect of Ni, Fe, and Si on the phase composition of Ni-containing multicomponent alloys (Tables 7.1 and 7.2).

No quaternary compounds have been found in the aluminum corner of the Al–Fe–Ni–Si system. This suggests that only the phases from the binary and ternary systems – Al_3Fe, Al_3Ni, Al_9FeNi, Al_8Fe_2Si, Al_5FeSi, and (Si) – can be in equilibrium with (Al).

The solubility of nickel in the Al_5FeSi phase is insignificant – less than 1%, the solubility of silicon in the Al_9FeNi phase can be up to 4%. As the silicon content in an alloy increases, the Ni:Fe ratio in the Al_9FeNi phase goes up (Zolotorevskii et al., 1989). According to the results of electron microprobe analysis of casting and heat-treated quaternary alloys containing 1% Fe and 1% Ni, we found a very large scatter of concentrations of these elements in the Al_9FeNi (T) phase, which approximately corresponds to the homogeneity range of phase T in the Al–Fe–Ni system (Figure 7.1b). Due to the slow diffusion of Fe and Ni in aluminum, this scatter is preserved during heat treatment (550°C) for a relatively long time (24 h).

Using the data from ternary systems, and from our own research (Belov et al., 2002a), we suggest the constitution of the aluminum corner of this quaternary system as shown in Figure 7.6. Due to a broad region of homogeneity of the Al_9FeNi phase, a considerable part of the Al–Fe–Ni–Si phase diagram in solid state is occupied by the region $(Al) + (Si) + Al_9FeNi$. The other phase regions in the solid state are unambiguously determined by the rules of polyhedration, and correspond to the experimental data. In particular, all four-phase regions contain the phase Al_9FeNi (Figure 7.6a).

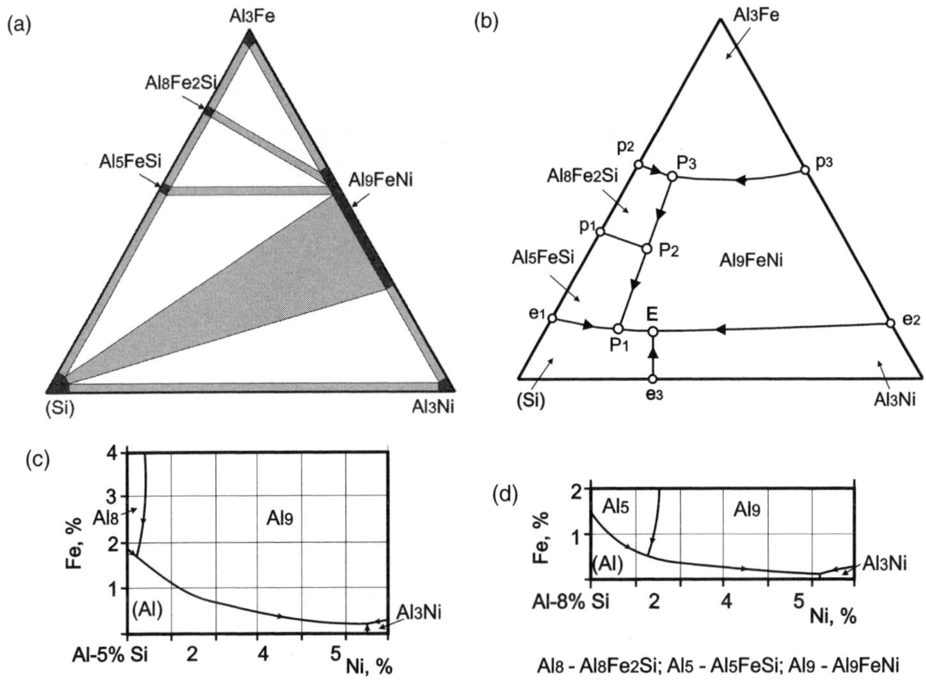

Figure 7.6. Phase diagram of Al–Fe–Ni–Si system: (a) distribution of phase fields in the solid state; (b) polythermal projection of the solidification surfaces; (c) projection of the liquidus surface at 5% Si; and (d) projection of the liquidus surface at 8% Si.

Table 7.12. Invariant reactions in quaternary alloys of Al–Fe–Ni–Si system (Belov et al., 2002a)

Reaction	Point in Figure 7.6b	Concentrations in liquid phase			*T*, °C
		Fe, %	Ni, %	Si, %	
$L \Rightarrow (Al) + (Si) + Al_3Ni + Al_9FeNi$	E	0.2–0.4	4–5	12–14	~556
$L + Al_5FeSi \Rightarrow (Al) + (Si) + Al_9FeNi$	P_1	0.6–1	2.5–3	13–14	573–576
$L + Al_8Fe_2Si \Rightarrow (Al) + Al_5FeSi + Al_9FeNi$	P_2	3–4	1–2	6–8	600–610
$L + Al_3Fe \Rightarrow (Al) + Al_9FeNi + Al_8Fe_2Si$	P_3	3–5	1–1.5	4–6	620–628

A version of the liquidus projection in the Al–Fe–Ni–Si system shown in Figure 7.6b is based on the constitution of the constitutive ternary phase diagrams. In the aluminum corner of the Al–Fe–Ni–Si system, we can assume the occurrence of four invariant five-phase reactions: three peritectic (P_1, P_2, P_3) and one eutectic (E). All reactions involving (Al) that may occur during solidification of quaternary alloys are summarized in Tables 7.12 and 7.13.

Table 7.13. Bivariant and monovariant reactions in quaternary alloys of Al–Fe–Ni–Si system (Belov et al., 2002a)

Bivariant reactions	Field in Figure 7.6b	*T*, °C	Monovariant reactions	Line in Figure 7.6b	*T*, °C
L ⇒ (Al) + (Si)	(Si)–e_1–P_1–E–e_3	577–556	L ⇒ (Al) + (Si) + Al_3Ni	e_3–E	557–556
L ⇒ (Al) + Al_3Ni	Al_3Ni–e_3–E–e_2	640–556	L ⇒ (Al) + (Si) + Al_5FeSi	e_1–P_1	577–573
L ⇒ (Al) + Al_9FeNi	p_3–P_3–P_2–P_1–E–e_2	649–556	L ⇒ (Al) + (Si) + Al_9FeNi	P_1–E	573–556
L ⇒ (Al) + Al_5FeSi	p_1–P_2–P_1–e_1	611–573	L ⇒ (Al) + Al_3Ni + Al_9FeNi	e_2–E	637–556
L ⇒ (Al) + Al_8Fe_2Si	p_2–P_3–P_2–p_1	629–600	L ⇒ (Al) + Al_5FeSi + Al_9FeNi	P_2–P_1	600–573
L ⇒ (Al) + Al_3Fe	Al_3Fe–p_3–P_3–p_2	655–620	L + Al_8Fe_2Si ⇒ (Al) + Al_5FeSi	p_1–P_2	611–600
			L ⇒ (Al) + Al_8Fe_2Si + Al_9FeNi	P_3–P_2	620–600
			L + Al_3Fe ⇒ (Al) + Al_8Fe_2Si	p_2–P_3	629–620
			L + Al_3Fe ⇒ (Al) + Al_9FeNi	p_3–P_3	649–620

Information on the range of (Al) primary solidification is also useful. It is reflected in Figures 7.6c and 7.6d, which show the boundaries of the liquidus surfaces at 5 and 8% Si. Evidently, the concentration of silicon has a significant effect on the position of primary solidification regions. As the concentration of silicon decreases from 8 to 5%, the region of (Al) primary solidification widens and the region of β(AlFeSi)-phase primary solidification vanishes giving place to the α(AlFeSi) primary phase. At a concentration of 8% Si (and, apparently, at a higher content of silicon), primary crystals of the Al_5FeSi or Al_9FeNi phases can be found at small concentrations of Fe and Ni (0.6–0.8%), which in some cases can make inefficient or even harmful the use of nickel addition as a modifier of an Fe-containing phase (Zolotorevskii et al., 1989).

It should be noted that incomplete peritectic reactions can result in the appearance of more phases in the as-cast structure of Al–Fe–Ni–Si alloys than it follows from Figure 7.6a. Due to slow diffusion of iron and nickel in aluminum, such "excess" phases can be retained even after high-temperature annealing. In particular, needle-like Al_5FeSi crystals are found in as-cast alloys with 8% Si and 0.8% Fe at nickel concentrations over 1% (Belov et al., 2002a), though alloys of this composition should contain under equilibrium conditions only (Al), (Si), and Al_9FeNi.

7.7. Al–Cu–Fe–Ni PHASE DIAGRAM

In spite of the importance of this quaternary phase diagram for the analysis of 2618-type alloys (Table 7.1), the information on the phase equilibria is too scarce to yield a substantiated prediction. According to Mondolfo, only the phases from the constituent binary and ternary systems – Al_3Ni, Al_3Fe, Al_2Cu, Al_7Cu_2Fe, Al_7Cu_4Ni, Al_9FeNi, $Al_6(FeCu)$, and $Al_3(CuNi)_2$ – can be in equilibrium with (Al) (Mondolfo, 1976). However, the solubility of the fourth component in some ternary phases can be quite significant. In particular, nickel can replace up to 6.5–6.8% Fe the Al_7Cu_2Fe phase. Some data indicate a substantial solubility (4–5%) of iron in the $Al_3(CuNi)_2$ compound and that of nickel in the $Al_6(FeCu)$ phase. By taking into account such wide ranges of homogeneity for at least some phases; obviously large number of phase regions; and possible numerous invariant reactions, the constitution of the quaternary phase diagram is presumably rather complex.

The distribution of phase regions in the solid state, as suggested by Mondolfo (1976) and plotted in Figure 7.7, shows that in Al–Cu-rich alloys (in the presence of the Al_2Cu phase) iron and nickel should be bound in the phases Al_7Cu_2Fe and Al_7Cu_4Ni. This rules out the presence of Al_9FeNi and contradicts experimental data on the phase composition of 2618-type alloys that usually contain Al_9FeNi

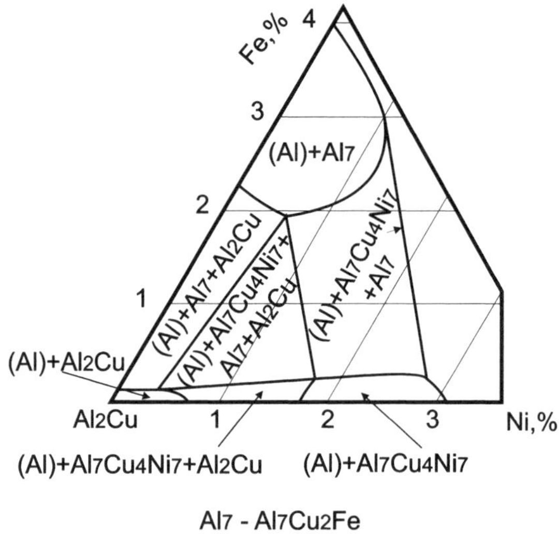

Figure 7.7. Phase diagram of Al–Cu–Fe–Ni system: distribution of phase fields in the solid state (Mondolfo, 1976).

particles. On the other hand, according to Drits et al. (1977), addition of up to 15% copper to Al–Fe–Ni alloys does not change the phase composition that remains $(Al) + Al_3Fe + Al_3Ni + Al_9FeNi$ with the only effect that the range of Al_9FeNi primary solidification narrows on increasing the amount of copper. The liquidus surface of Al–Cu–Fe–Ni alloys at Cu concentrations up to 3% virtually coincides with the liquidus surface in the Al–Fe–Ni system (Drits et al., 1977).

7.8. Al–Mg–Ni–Si PHASE DIAGRAM

This system can be used for the analysis of 3XX.0-series casting alloys, containing silicon, nickel, and magnesium additions (Table 7.2).

The Al–Mg–Ni–Si phase diagram was experimentally studied in the region of the Al–Mg_2Si–Si–Al_3Ni tetrahedron, inside which no new phases were found, Figure 7.8a (Belov, 1993a). The Mg_2Si, (Si), and Al_3Ni phases have almost the same compositions as in the corresponding ternary systems. The Mg-rich portion of the phase diagram can be predicted with a certain confidence, even in the absence of experimental data. This is due to the fact that in the ternary Al–Mg–Ni and Al–Mg–Si systems the concentrations of the third component, i.e. nickel and silicon, respectively, in the ternary eutectics involving (Al) and the phase Al_8Mg_5 are relatively low (see Sections 7.4 and 2.1). Three invariant eutectic reactions (one of

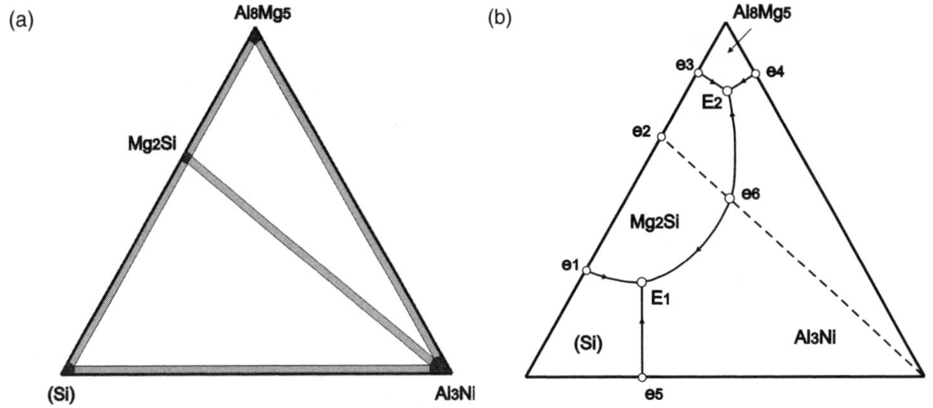

Figure 7.8. Phase diagram of Al–Mg–Ni–Si system: (a) distribution of phase fields in the solid state and (b) polythermal projection of the solidification surfaces.

Table 7.14. Invariant reactions in quaternary alloys of Al–Mg–Ni–Si system (Belov, 1993a)

Reaction	Point in Figure 7.8b	Concentrations in liquid phase			*T*, °C
		Mg, %	Ni, %	Si, %	
$L \Rightarrow (Al) + Mg_2Si + (Si) + Al_3Ni$	E_1	3.5	2	13	550
$L \Rightarrow (Al) + Mg_2Si + Al_3Ni$ (quasi-ternary)	e_6	7.4	3	4.8	590
$L \Rightarrow (Al) + Mg_2Si + Al_3Ni + Al_8Mg_5$	E_1	~32	<1.7	<0.4	~447

Table 7.15. Bivariant and monovariant reactions in quaternary alloys of Al–Mg–Ni–Si system (Belov, 1993a)

Bivariant reactions	Field in Figure 7.8b	*T*, °C	Monovariant reactions	Lines in Figure 7.8b	*T*, °C
$L \Rightarrow (Al) + Mg_2Si$	e_1–E_1–e_6–E_2–e_3	595–447	$L \Rightarrow (Al) + Mg_2Si + (Si)$	e_1–E_1	555–550
$L \Rightarrow (Al) + (Si)$	(Si)–e_1–E_1–e_5	577–550	$L \Rightarrow (Al) + Mg_2Si + Al_3Ni$	e_6–E_1	590–550
				e_6–E_2	590–447
$L \Rightarrow (Al) + Al_3Ni$	Al_3Ni–e_5–E_1–e_6–E_2–e_4	640–447	$L \Rightarrow (Al) + Al_3Ni + (Si)$	e_5–E_1	557–550
$L \Rightarrow (Al) + Al_8Mg_5$	Al_8Mg_5–e_4–E_2–e_3	450–447	$L \Rightarrow (Al) + Mg_2Si + Al_8Mg_5$	e_3–E_2	449–447
			$L \Rightarrow (Al) + Al_3Ni + Al_8Mg_5$	e_4–E_2	449–447

them quasi-ternary) occur in quaternary alloys as shown in Figure 7.8b and in Table 7.14; corresponding bi- and monovariant reactions are listed in Table 7.15.

As peritectic reactions do not occur in this system, the phase composition of as-cast and, especially heat-treated quaternary alloys is close to the equilibrium phase composition as shown in Figure 7.8a.

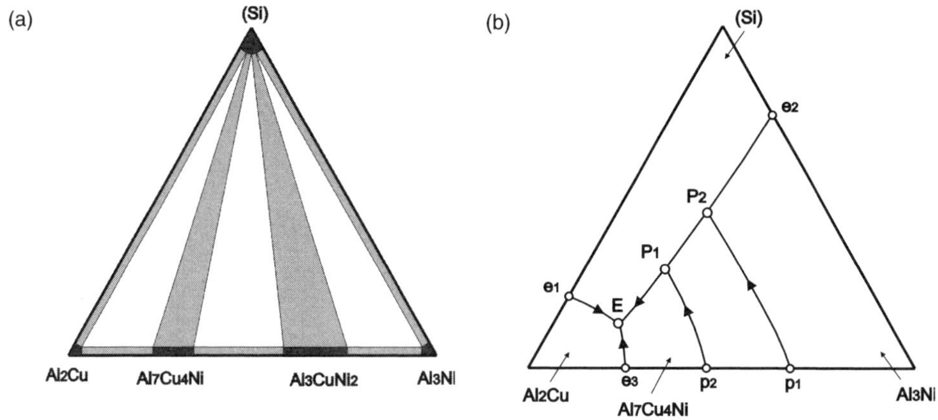

Figure 7.9. Phase diagram of Al–Cu–Ni–Si system: (a) distribution of phase fields in the solid state and (b) polythermal projection of solidification surfaces.

Table 7.16. Invariant reactions in quaternary alloys of Al–Cu–Ni–Si system

Reaction	Point in Figure 7.9b	Concentrations in liquid phase			*T*, °C
		Cu, %	Ni, %	Si, %	
$L + Al_3Ni \Rightarrow (Al) + Al_2Cu + Al_3(CuNi)_2 + (Si)$	P_1	~16	~4	1–2	~540
$L + Al_3(CuNi)_2 \Rightarrow (Al) + Al_7Cu_4Ni + (Si)$	P_2	~22	~2	3–4	~530
$L \Rightarrow (Al) + Al_2Cu + Al_7Cu_4Ni + (Si)$	E	~30	~1	~5	~520

7.9. Al–Cu–Ni–Si PHASE DIAGRAM

This quaternary phase diagram is required for the analysis of piston alloys of the 3XX.0 series and some casting 2XX.0 series alloys (Table 7.2). There are no experimental data available on the phase equilibria in quaternary alloys of this system, so an evaluation of the phase diagram is given here.

Assuming that only the phases from the constituent binary and ternary systems are in equilibrium with (Al) in the Al–Cu–Ni–Si system, the most probable distribution of phase regions in the solid state is shown in Figure 7.9a. It suggests that addition of silicon to Al–Cu–Ni alloys in amounts exceeding the Si solubility in (Al) leads to the formation of only one phase – (Si). The solubility of silicon in the ternary compounds Al_7Cu_4Ni and $Al_3(CuNi)_2$ is, probably as low as in the binary compounds Al_2Cu and Al_3Ni.

In the aluminum corner of the Al–Cu–Ni–Si system, we assume the presence of three invariant reactions: two peritectic and one eutectic listed in Table 7.16 and shown in Figure 7.9b. The respective bi- and monovariant reactions are given in Table 7.17.

7.10. Al–Mg–Ni–Zn PHASE DIAGRAM

The phase diagram of this quaternary system is considered in this chapter mainly because it can help to analyze the phase composition of promising high-strength alloys that can be used for both cast shapes and deformed semifinished items (Kubicek et al., 1993; Tagiev et al., 1996; Belov and Zolotorevskii, 2002, 2003; Aksenov et al., 2003). The Al–Mg–Ni–Zn phase diagram can be used to estimate the effect of nickel on the phase composition of 7XXX-series alloys containing a relatively small amount of copper (up to 1%). This diagram can also be used for the analysis of some promising rapidly solidified alloys (Dobatkin et al., 1995).

Though only few reference data are available on the Al–Mg–Ni–Zn phase diagram, its constitution, at least in the aluminum corner can be predicted with sufficient accuracy. This is facilitated by two factors: (i) nickel has a low solubility in solid (Al) and (ii) nickel does not form phases with zinc and magnesium that could be in equilibrium with (Al). In addition, the authors accumulated a considerably large experimental data on the effect of nickel (up to 6%) on the structure and phase composition of materials close to 7075 and 7005 alloys (Belov and Zolotorevskii, 2002, 2003).

Based on the data available on the constitutive ternary systems (Mondolofo, 1976), we suggest the distribution of phase regions in the solid state as follows (Figure 7.10a). According to this distribution, only the Al_3Ni compound from the

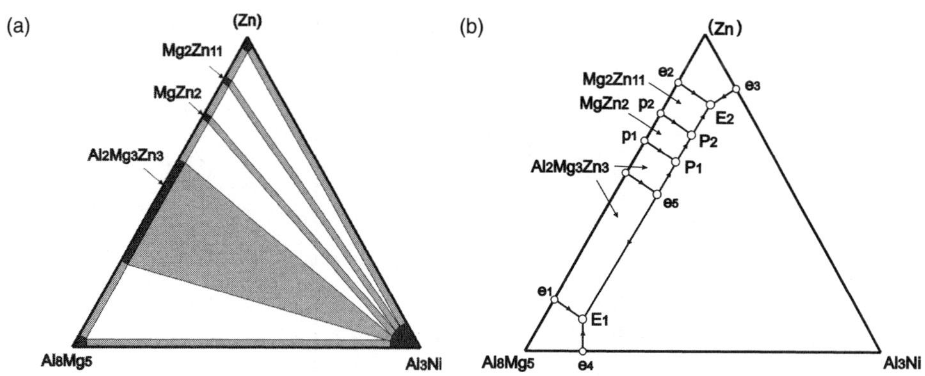

Figure 7.10. Phase diagram of Al–Mg–Ni–Zn system: (a) distribution of phase fields in the solid state and (b) polythermal projection of solidification surfaces.

Table 7.17. Bivariant and monovariant reactions in quaternary alloys of Al–Cu–Ni–Si system

Bivariant reactions	Field in Figure 7.9b	*T*, °C	Monovariant reactions	Line in Figure 7.9b	*T*, °C
$L \Rightarrow (Al) + Al_3Ni$	$Al_3Ni–e_2–P_1–p_1$	640–540	$L \Rightarrow (Al) + Al_3Ni + (Si)$	$e_2–P_1$	557–540
$L \Rightarrow (Al) + Al_3(CuNi)_2$	$p_2–P_2–P_1–p_1$	630–530	$L \Rightarrow (Al) + Al_3(CuNi)_2 + (Si)$	$P_1–P_2$	540–530
$L \Rightarrow (Al) + Al_7Cu_4Ni$	$e_3–E–P_2–p_2$	590–520	$L \Rightarrow (Al) + Al_7Cu_4Ni + (Si)$	$P_2–E$	530–520
$L \Rightarrow (Al) + Al_2Cu$	$Al_2Cu–e_1–E–e_3$	548–520	$L \Rightarrow (Al) + Al_2Cu + (Si)$	$e_1–E$	525–520
$L \Rightarrow (Al) + (Si)$	$(Si)–e_2–P_1–P_2–E–e_1$	577–520	$L \Rightarrow (Al) + Al_2Cu + Al_7Cu_4Ni$	$e_3–E$	546–520
			$L + Al_3(CuNi)_2 \Rightarrow (Al) + Al_7Cu_4Ni$	$p_2–P_2$	590–530
			$L + Al_3Ni \Rightarrow (Al) + Al_3(CuNi)_2$	$p_1–P_1$	630–540

Table 7.18. Invariant reactions in quaternary alloys of Al–Mg–Ni–Zn system (prediction)

Reaction	Point in Figure 7.10b	T, °C	Concentrations in liquid phase		
			Mg, %	Zn, %	Ni, %
$L \Rightarrow (Al) + Al_8Mg_5 + Al_2Mg_3Zn_3 + Al_3Ni$	E_1	~446	~30	~12	<2%
$L \Rightarrow (Al) + Al_2Mg_3Zn_3 + Al_3Ni$ (quasi-ternary)	e_5	~487	~18	~45	<2%
$L + Al_2Mg_3Zn_3 \Rightarrow (Al) + MgZn_2 + Al_3Ni$	P_1	~473	~11.3	~60.4	<2%
$L + MgZn_2 \Rightarrow (Al) + Mg_2Zn_{11} + Al_3Ni$	P_2	~366	~3.5	~92	<2%
$L \Rightarrow (Al) + Mg_2Zn_{11} + (Zn) + Al_3Ni$	E_2	~341	~3	~93	<2%

Al–Ni system is in equilibrium with (Al) and with phases from the Al–Mg–Zn system, i.e. Al_8Mg_5, (Zn), $MgZn_2$, Mg_2Zn_{11}, and $Al_2Mg_3Zn_3$. Hence, the addition of Ni to Al–Mg–Zn alloys may result only in the formation of Al_3Ni.

Prediction of the invariant reactions, as listed in Table 7.18 and shown in Figure 7.10b, is based on the fact that the concentration of nickel in the invariant point is relatively small and, as a result, the temperatures are close to those of invariant reactions in the Al–Mg–Zn system. The same feature is observed in the Al–Mg–Ni system (Section 7.4) when the $L \Rightarrow (Al) + Al_8Mg_5 + Al_3Ni$ eutectics contains 32% Mg and 1.7% Ni and forms at 449°C (close to the $(Al) + Al_8Mg_5$ eutectics). Table 7.19 suggests that the binary $(Al) + Al_3Ni$ eutectics, which largely determines the structure of quaternary alloys, can form within a very wide temperature range. In particular, in alloys with high zinc concentration that complete solidification by the reaction $L \Rightarrow (Al) + MgZn_2 + Al_3Ni$, the solidification range of thc binary $(Al) + Al_3Ni$ cutcctics can be as large as 150°C.

7.11. WROUGHT ALLOYS OF 8001 TYPE

Isothermal and polythermal sections of the Al–Fe–Ni phase diagram will suffice to analyze the phase composition of 8001-type alloys (Table 7.1). The isopleth at 1% Ni shown in Figure 7.11a demonstrates that within the entire range of iron concentrations and at any temperature only one phase – Al_9FeNi – can be in equilibrium with (Al) in an 8001 alloy. Another feature of this type of alloys is a narrow solidification range of 640–660°C. As a result, 8001-type alloys have good casting properties, with a potential application in production of complex cast shapes.

Belov et al. (1994) showed that alloys based on the $(Al) + Al_9FeNi$ eutectics with additions of chromium and zirconium exhibited good combination of casting and high-temperature properties alongside sufficient tensile properties at room temperature. Eutectic crystals of Al_9FeNi can change the morphology (spheroidize,

Table 7.19. Bivariant and monovariant reactions in quaternary alloys of Al–Mg–Ni–Zn system

Bivariant reactions	Field in Figure 7.10b	*T*, °C	Monovariant reactions	Line in Figure 7.10b	*T*, °C
$L \Rightarrow (Al) + Al_8Mg_5$	Al_8Mg_5–e_1–E_1–e_4	450–446	$L \Rightarrow (Al) + Al_8Mg_5 + Al_2Mg_3Zn_3$	e_1–E_1	447–446
$L \Rightarrow (Al) + Al_2Mg_3Zn_3$	p_1–P_1–e_5–E_1–e_1	489–446	$L \Rightarrow (Al) + Al_8Mg_5 + Al_3Ni$	e_4–E_1	449–446
$L \Rightarrow (Al) + MgZn_2$	p_2–P_2–P_1–p_1	475–366	$L \Rightarrow (Al) + Al_3Ni + Al_2Mg_3Zn_3$	e_5–E_1;	487–446
				e_5–P_1	487–473
$L \Rightarrow (Al) + Mg_2Zn_{11}$	e_2–E–P_2–p_2	368–341	$L + Al_2Mg_3Zn_3 \Rightarrow (Al) + MgZn_2$	p_1–P_1	489–487
$L \Rightarrow (Al) + (Zn)$	(Zn)–e_3–E_2–e_2	382–341	$L \Rightarrow (Al) + Al_3Ni + MgZn_2$	P_1–P_2	473–366
$L \Rightarrow (Al) + Al_3Ni$	Al_3Ni–e_4–E_1–e_5–P_1–P_2–E_2–e_3	640–341	$L + MgZn_2 \Rightarrow (Al) + Mg_2Zn_{11}$	p_2–P_2	368–366
			$L \Rightarrow (Al) + Al_3Ni + Mg_2Zn_{11}$	P_2–E_2	366–341
			$L \Rightarrow (Al) + Al_3Ni + (Zn)$	e_3–E_2	380–341
			$L \Rightarrow (Al) + (Zn) + Mg_2Zn_{11}$	e_2–E_2	343–341

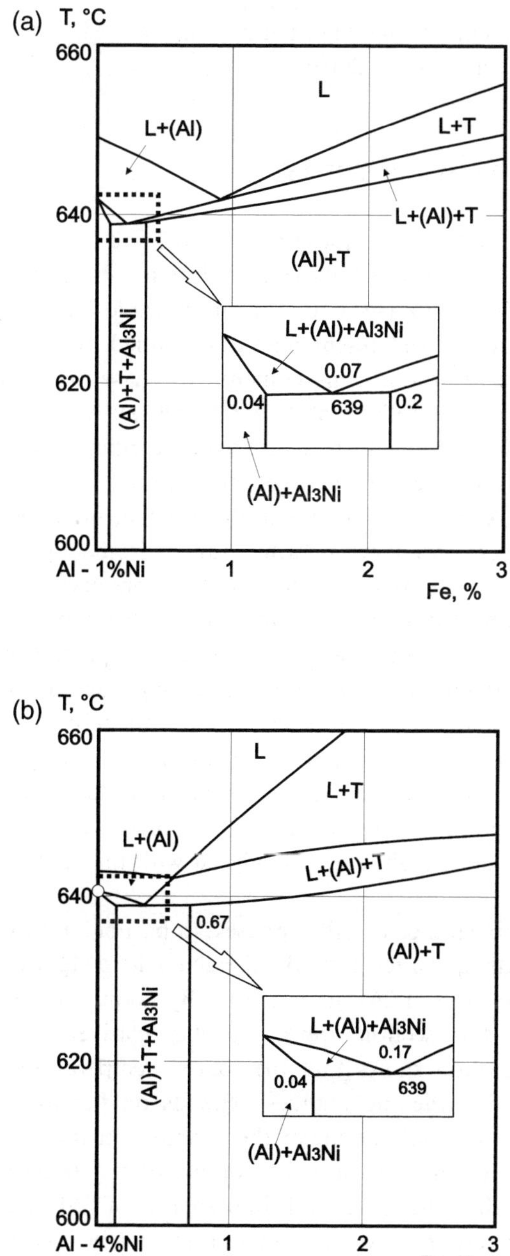

Figure 7.11. Polythermal sections of Al–Fe–Ni phase diagram at 1% Ni (a) and 4% Ni (b). T denotes Al_9FeNi.

fragment, and coarsen) during high-temperature anneals at temperatures above 450–500°C (Belov et al., 1996b, 2002a).

7.12. WROUGHT ALLOYS OF 2618 TYPE

A stringent analysis of 2618-type alloys requires the five-component Al–Cu–Fe–Mg–Ni phase diagram, because these alloys contain enough copper and magnesium to produce the Al_2CuMg phase (see composition in Table 7.1). As this phase diagram is yet to be constructed, we present here a simplified analysis of the phase composition of this type of alloys, using constitutive phase diagrams and considerable experimental data on the structure of 2618-type alloys. According to these data, the microstructure of these alloys in the as-cast state contains particles of the Al_9FeNi and Al_2CuMg phases. The former, probably, forms through the binary eutectic reaction L ⇒(Al) + Al_9FeNi which, due to the presence of copper and magnesium in an alloy, occurs over a wide temperature range, from approximately 640–645°C down to 505–515°C. The Al_9FeNi phase is then preserved in the final structure of deformed semifinished items. A subsequent treatment can only change the morphology of particles that, as a rule, is rather compact. The appearance of the Al_2CuMg phase in the as-cast structure is a consequence of nonequilibrium solidification. During homogenizing annealing, it completely dissolves in solid (Al). Metastable modifications of this phase (usually S″ and S′) precipitate during age hardening as a result of decomposition of a supersaturated solid solution. Therefore, the effect of copper and magnesium on the phase composition of aluminum matrix after heat treatment should be analyzed with the Al–Cu–Mg phase diagram (Section 3.2).

If the possible appearance of other phases, in particular Al_2Cu, is excluded, it is convenient to use the quasi-ternary Al–Al_9FeNi–Al_2CuMg section for the analysis of the phase composition of 2618-type alloys. By taking into account the available experimental data, this section appears to be relatively simple as depicted in Figure 7.12. This section shows that the solidus temperature that determines the homogenization and hardening regimes depends on the total amount of copper and magnesium, i.e. on the amount of the S phase. In the first approximation, at a low silicon content, one can use the solidus of the ternary Al–Cu–Mg phase diagram (Figure 3.2b). The nonequilibrium solidus of 2618-type alloys, according to Figure 7.12, corresponds to the temperature of the quasi-ternary L ⇒(Al) + Al_9FeNi + Al_2CuMg eutectics and is about 515°C. At a maximum copper concentration (within the alloy nominal composition), the Al_2Cu phase can form, in this case the solidification completes at a lower temperature, ~505°C as results from the Al–Cu–Mg phase diagram (Section 3.2). Figure 7.12 also demonstrates that the

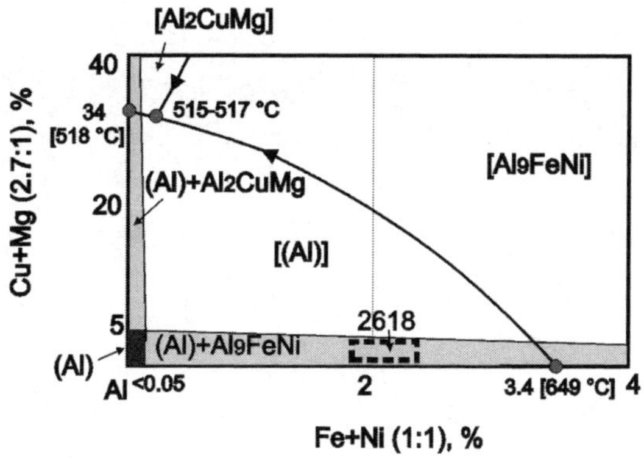

Figure 7.12. Quasi-ternary section Al–Al_9FeNi–Al_2CuMg of the Al–Cu–Fe–Mg–Ni phase diagram (assessment).

formation of Al_9FeNi primary particles is unlikely in the entire compositional range of a 2618 alloy.

A typical microstructure of a 2618 type alloys exhibit Al_9FeNi phase as shown in Figure 7.13.

7.13. PISTON CASTING ALLOYS OF 339.0 TYPE

A strict analysis of casting piston alloys of the 3XX.0 series requires the six-component Al–Cu–Fe–Mg–Ni–Si phase diagram, as all elements of this system are present in most commercial alloys with compositions given in Table 7.2 and, more importantly, they all have a strong effect on the phase composition. Analysis of piston alloys is complicated by the formation of primary crystals of the silicon phase and often occurrence of "primary" Ni-containing phases. A simplified analysis of the phase composition of piston alloys can be performed using quinary phase diagrams in the range of Al–Si alloys, using some assumptions.

Evaluation of the equilibrium phase distribution in the solid state of quinary alloys with nickel (Figure 7.14) can be made based on the knowledge of all quaternary diagrams with silicon, i.e. Al–Fe–Ni–Si, Al–Cu–Ni–Si, Al–Mg–Ni–Si, Al–Cu–Mg–Si, Al–Cu–Fe–Si, and Al–Fe–Mg–Si. All these systems are considered in this book.

Analysis of the phase composition of 393-type and FM piston alloys (Table 7.2) at a low concentration of iron impurity can be performed with the Al–Cu–Mg–Ni–Si diagram in the Si-rich region (Figure 7.14a). According to the constituent quaternary diagrams, the following phases can be in equilibrium with (Al) and (Si): Al_3Ni, $Al_3(CuNi)_2$, Al_7Cu_4Ni, Al_2Cu, Mg_2Si, and $Al_5Cu_2Mg_8Si_6$. According to the assumed

(a)

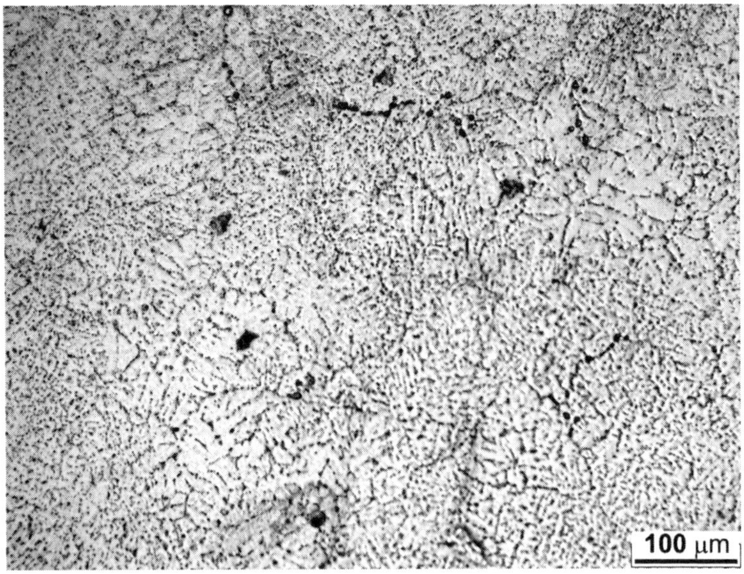

(b)

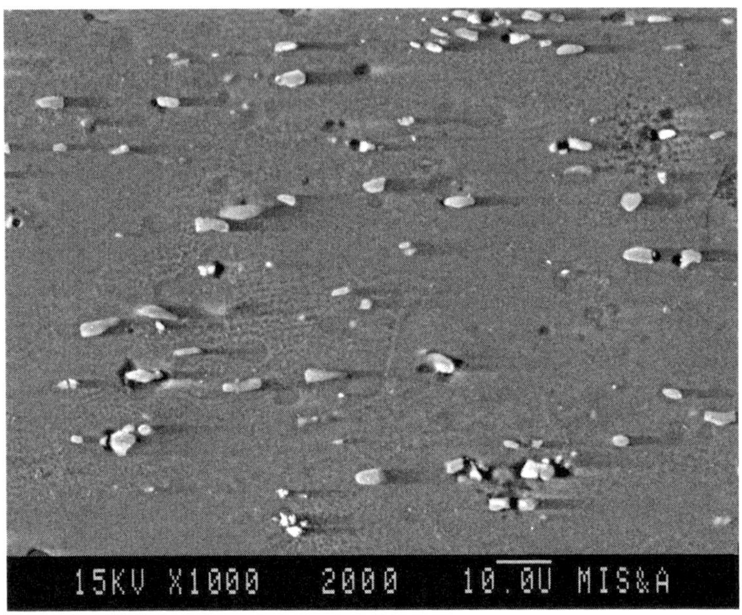

Figure 7.13. Microstructure of AK4-1ch (a) and AK4-2ch (b) (Russian grades of the 2618 type): (a) ingot annealed at 490°C, 10 h, eutectic particles of Al_9FeNi phase and precipitates of S phase, optical microscope and (b) sheet (T7), particles of Al_9FeNi phase, SEM.

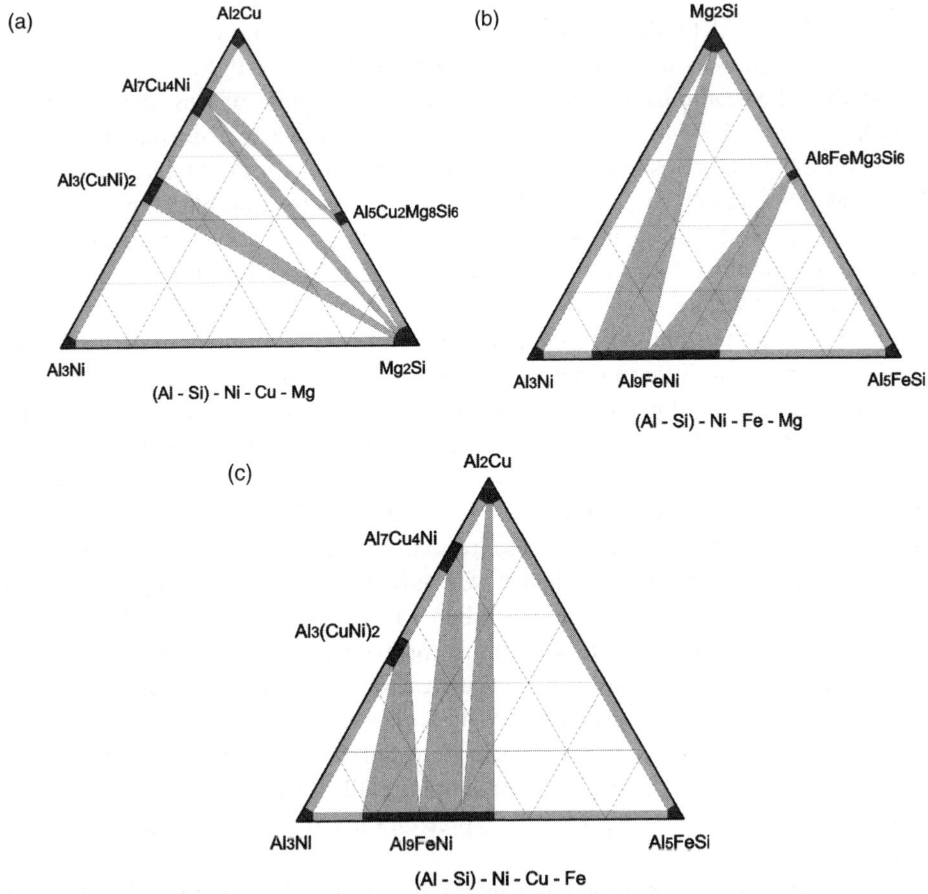

Figure 7.14. Distribution of phase fields in the solid state in quinary systems with Ni in Al–Si alloys: (a) Al–Cu–Mg–Ni–Si; (b) Al–Fe–Mg–Ni–Si; and (c) Al–Cu–Fe–Ni–Si. All phase fields contain (Al) and (Si).

distribution of the phase regions, given in Figure 7.14a, all these phases can, in various combinations, be present in commercial piston alloys. Under conditions of nonequilibrium solidification, the total number of phases can be more than five, because the constitution of the polythermal projection suggests the presence of several peritectic reactions (apparently there should be more peritectic reactions than in the constituent quaternary systems).

If iron is present in an alloy to such an extent that it influences the phase composition (and that happens at a relatively low iron concentration, $\leq 0.5\%$), then the Al_9FeNi and $Al_8FeMg_3Si_6$ phases can appear in addition to the already considered phases. The more the magnesium in the alloy, the more probable is the presence of the $Al_8FeMg_3Si_6$ compound. A solution treatment should lead to complete or partial dissolution of Al_2Cu, Mg_2Si, and $Al_5Cu_2Mg_8Si_6$ in (Al), depending on the alloy composition. Remaining particles of these phases may acquire a globular shape. The Ni-containing phases do not change the morphology. After aging, the phase composition of the aluminum matrix can be analyzed using the Al–Cu–Mg–Si diagram by taking into account the composition of a supersaturated solid solution as discussed in Sections 3.4 and 3.9. If the alloys are heat treated without quenching, for the proper analysis one should know the composition of (Al) in the as-cast state.

The distribution of phase fields in alloys without copper is shown in Figure 7.14b.

Alloys containing little magnesium can be analyzed using the Al–Cu–Fe–Ni–Si phase diagram. According to the constitutive quaternary diagrams, the following phases – Al_3Ni, $Al_3(CuNi)_2$, Al_7Cu_4Ni, Al_2Cu, Al_9FeNi, and Al_5FeSi – can be in equilibrium with (Al) and (Si). The most probable distribution of phase regions in the solid state is shown in Figure 7.14c. It suggests that the phase composition of such alloys strongly depends on the iron concentration. If the concentration of iron is low, then, most probably, nickel will be bound to the phases $Al_3(CuNi)_2$ and Al_7Cu_4Ni. At a higher iron content, when the ratio Fe:Ni$\approx$1 is achieved, the formation of the Al_9FeNi phase is most probable. Under conditions of non-equilibrium solidification and the nickel concentration at a lower nominal level, one can also expect the appearance of the Al_5FeSi phase, as it follows from the Al–Fe–Ni–Si phase diagram (Section 7.6). During the solution treatment, which is rarely done with piston alloys, the Al_2Cu phase should totally dissolve in (Al). Other phases remain largely intact. During aging, the precipitation of Al_2Cu and its metastable modifications occur.

For the analysis of piston alloys containing considerable amount of nickel (for example FM2500 and FM2393 in Table 7.2), besides multicomponent phase diagrams that are available only as assessments, the polythermal section of the Al–Fe–Ni phase diagram at 4% Ni (Figure 7.11b) can be used. This section shows that the increasing amount of iron in an alloy causes the formation of primary Al_9FeNi crystals.

Bäckerud et al. (1990) examined the as-cast structure of a piston 339.1 alloy containing 11.9% Si, 0.99% Ni, 0.75% Fe, 0.95% Cu, 1.16% Mg, 0.2% Mg, and 0.33% Zn. The presence of Mn complicates the picture due to the formation of phases containing Mn and Fe. Table 7.20 shows that the Mn-containing phase forms just after the formation of (Al) and (Si). During subsequent solidification, only reactions with participation of already considered in this section phases occur. The identification of all the phases reported by Bäckerud et al. (1990) seems substantiated, except for Al_3Ni the occurrence of which at the Fe:Ni ratio close to unity is little probable. One should expect rather the formation of Al_9FeNi. The

Table 7.20. Solidification reactions under nonequilibrium conditions in a 339.1 alloy containing 11.9 % Si, 0.99% Ni, 0.75% Fe, 0.95% Cu*, 1.16% Mg, 0.2% Mn, and 0.33% Zn (Bäckerud et al., 1990)

Reaction	Temperatures (°C) at a cooling rate	
	0.3 K/s	4 K/s
$L \Rightarrow (Si)$, $L \Rightarrow (Al)$, $L \Rightarrow (Al) + (Si)$	563–560	561–559
$L \Rightarrow (Al) + (Si) + Al_{15}(MnFe)_3Si_2$	560–544	559–544
$L \Rightarrow (Al) + (Si) + Al_5FeSi$		
$L \Rightarrow (Al) + (Si) + Mg_2Si + Al_8FeMg_3Si_6$	544–538	544–534
$L \Rightarrow (Al) + Al_3Ni$	538–530	534–583
$L + Al_3Ni \Rightarrow (Al) + Al_3(CuNi)_2$		
Complex reaction with Al_2Cu and other phases	530–499	
Solidus	499	483

* Lower than the nominal lower limit (see Table 7.2)

structure of an AL30rus alloy (which is an analog of 339.1) contains considerable amount of Al_9FeNi crystals (Prigunova et al., 1996). In this alloy, the eutectic colonies $(Al) + (Si) + Al_9FeNi$ are the main structure constituents as shown in Figure 7.15a. It should be noted that commercial piston alloys containing 11–13% Si and modified with phosphorus frequently contain considerable amount of primary silicon as a result of nonequilibrium solidification (Figure 7.15b). On the other hand, the presence of Al_5FeSi needles in a 339.1 alloy seems logical from the analysis of the Al–Fe–Ni–Si phase diagram, as a result of suppressed peritectic reaction (P_1 in Table 7.12 and Figure 7.6b).

A peculiar feature of many piston alloys is the presence of numerous branched crystals of $Al_8FeMg_3Si_6$. This is a result of high magnesium concentration (Table 7.2). Table 2.10 (Chapter 2) shows that the volume fraction of this quaternary phase is three times the volume fraction of Al_5FeSi at the same iron concentration. As an example, Figures 7.15c, d give microstructures of an FMS2N alloy (Table 7.2) showing the particles of the $Al_8FeMg_3Si_6$ phase. This alloy has a high Ni:Fe ratio (>6), therefore nickel is mostly bound in Cu-containing phases ($Al_3(CuNi)_2$ and/or Al_7Cu_4Ni) (Figure 7.15d) rather than in Al_9FeNi.

7.14. HIGH-STRENGTH CASTING ALLOY AZ6N4

Experimental alloys based on the Al–Mg–Ni–Zn and Al–Cu–Mg–Ni–Zn systems can be used to produce both cast shapes and deformed semifinished items (Belov et al., 1992; Kubicek et al., 1993; Tagiev et al., 1996; Belov and Zolotorevskii, 2002; Aksenov et al., 2003). The obtained combination of mechanical and technological properties makes these alloys promising materials that can compete with existing

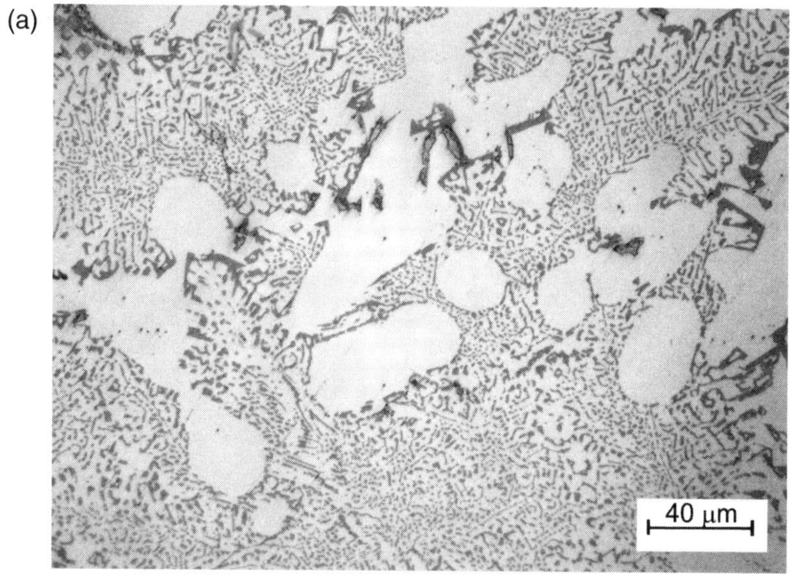

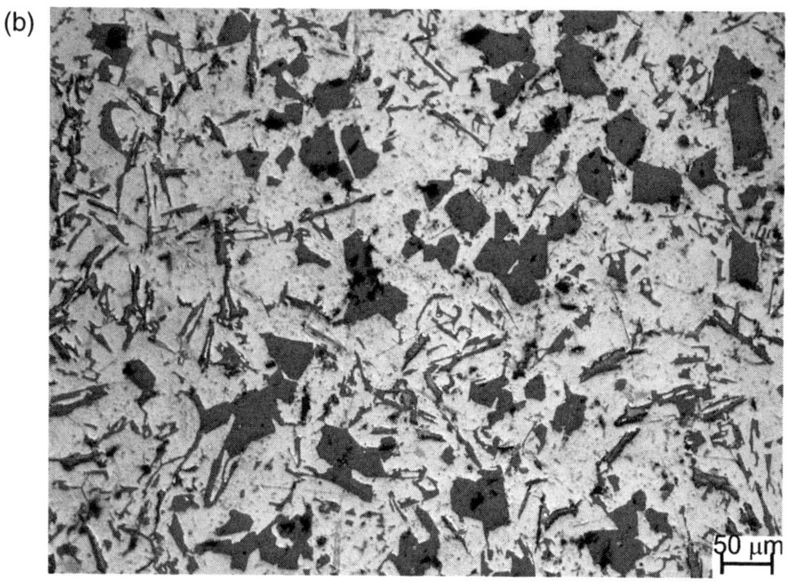

Figure 7.15. Microstructure of piston alloys, LM (a, b) SEM (c, d): (a) AL30rus (339.1) – fine eutectics (Al) + (Si) + Al_9FeNi + [Al_2Cu, Q, π, Mg_2Si]; (b) FMS2N – primary crystals of (Si), (AlCuNi) phases were found but not Al_9FeNi; (c) FM135 – π particle (gray) and eutectic colony containing presumably (Al) + (Si) + π + Al_2Cu + Q + Al_7Cu_4Ni, and (d) FM135, multiphase structure, the same phases as in (c) + Mg_2Si (black), big white skeleton of Al_7Cu_4Ni. Note in (c–d) (Si) particles have very little contrast with aluminum matrix (almost invisible).

(c)

(d)

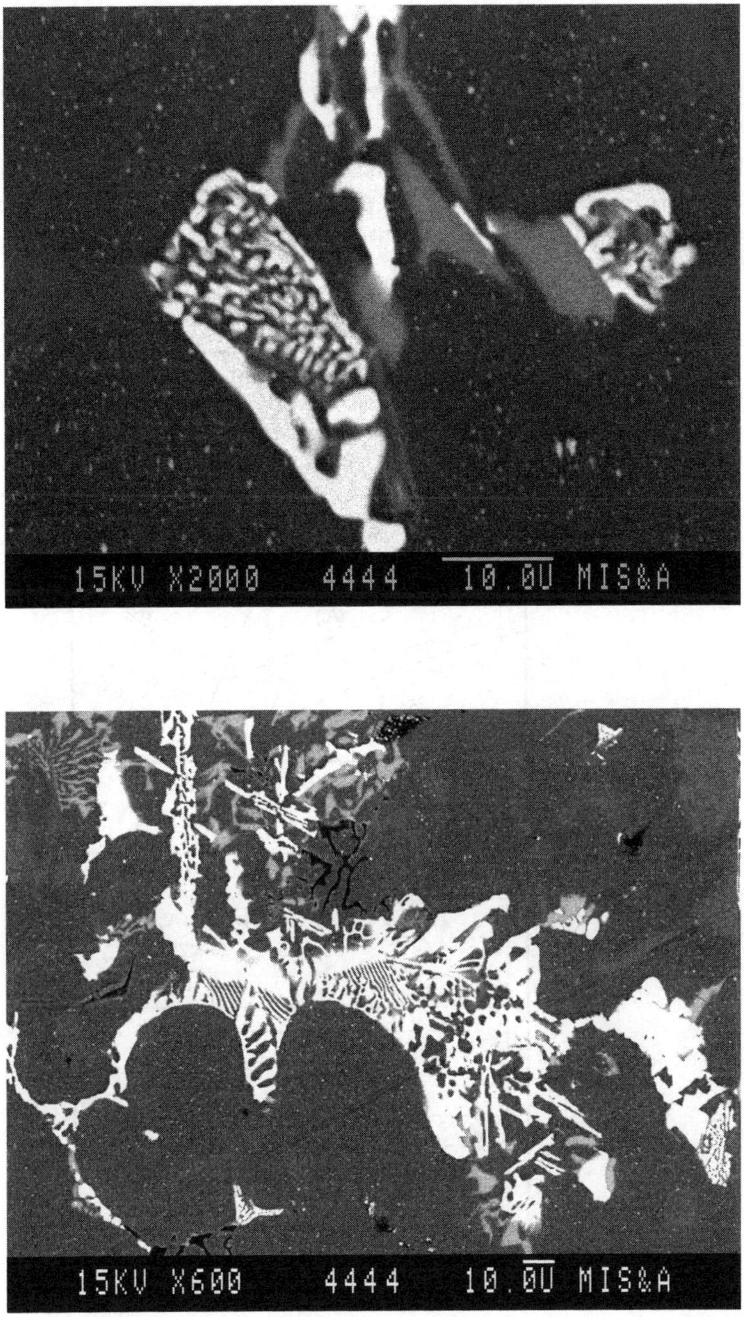

Figure 7.15 (*continued*)

alloys of the 7XXX series. A characteristic feature of the structure of new alloys (so-called nikalins) is the presence of a large volume fraction of the Al_3Ni phase.

Projections of liquidus and solidus surfaces constructed as Al–Σ(Zn + Mg)–Ni and Al–Σ(Zn + Mg + Cu)–Ni diagrams are most suitable for characterization of the phase composition of such alloys. As an example, Figure 7.16 shows liquidus and solidus projections of Al–Σ(Zn + Mg + Cu)–Ni alloys containing alloying elements at the following ratio Zn:Mg:Cu = 6:2:1.

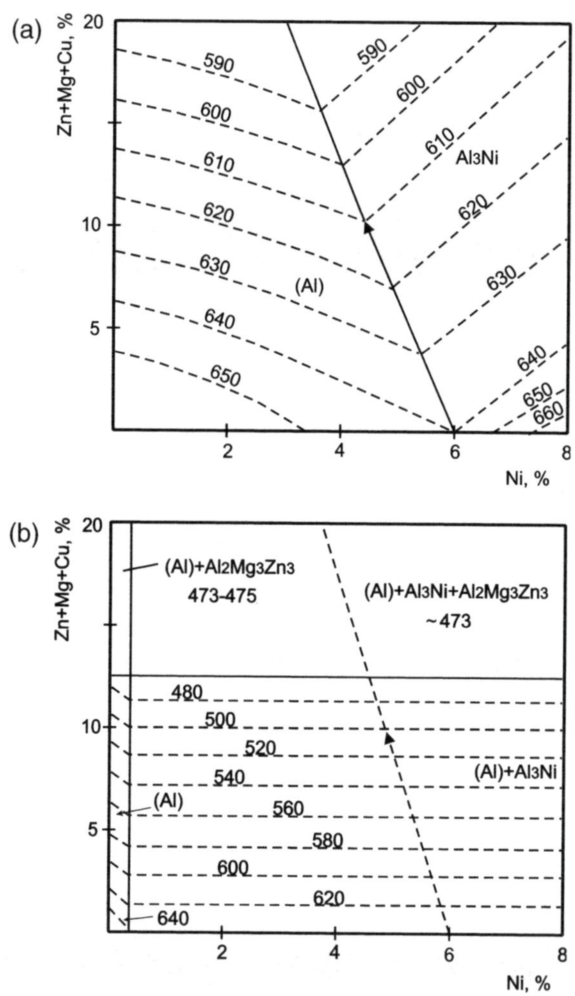

Figure 7.16. Section Al–Σ(Zn + Mg + Cu)–Ni of Al–Cu–Mg–Ni–Zn phase diagram (at the ratio Zn:Mg:Cu = 6:2:1): (a) projection of liquidus and (b) projection of solidus.

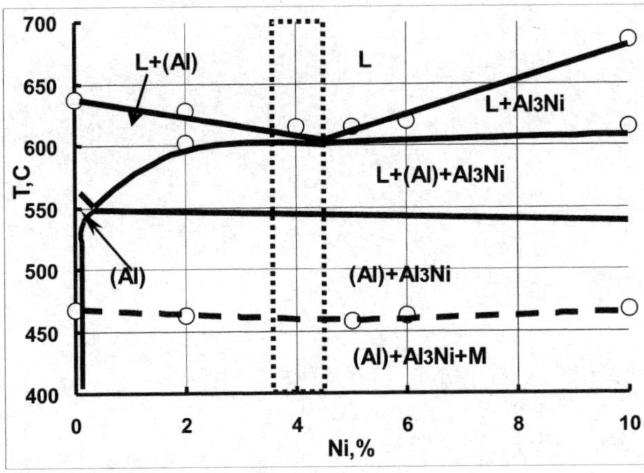

Figure 7.17. Polythermal section of Al–Cu–Mg–Ni–Zn phase diagram at 6% Zn, 2% Mg, and 1% Cu. Dashed line shows non-equilibrium solidus.

The liquidus projection in Figure 7.16a clearly demonstrates that the only primary phases that can solidify in the given composition range are (Al) and Al_3Ni. On further solidification, the eutectic reaction L ⇒(Al) + Al_3Ni occurs in a temperature range as shown in the polythermal section (Al–6%Zn–2%Mg–1%Cu)–Ni in Figure 7.17. Under nonequilibrium conditions, solidification ceases at 473–475°C with the eutectic reaction L ⇒(Al) + Al_3Ni + M(AlZnMgCu) the temperature and composition of which are close to those of the L ⇒(Al) + M ($MgZn_2$) reaction in the ternary Al–Mg–Zn system (Section 6.1). Hence, the nonequilibrium solidus (T_{NES}) of alloys in the given compositional range is virtually unaffected by the concentration of alloying elements, and the solidification range ($\Delta T = T_L - T_{NES}$) becomes solely the function of the liquidus temperature T_L. The polythermal section in Figure 7.17 shows that the least solidification range (and, therefore the best casting properties) corresponds to a Ni concentration of 4.5%, i.e. the eutectic concentration. This concentration is also the limit beyond which the formation of coarse, primary Al_3Ni crystals becomes possible. Eutectic (Al) + Al_3Ni colonies with M(AlZnMgCu) precipitates (from divorced nonequilibrium eutectics) are main structure constituents of an Al–6% Zn–4.5% Ni–2% Mg–1% Cu alloy in the as-cast state (Figure 7.18a).

Figure 7.17 suggests also the temperature of the first stage of solution treatment (T_1), which has to be above the solvus (T_{ss}) but below T_{NES}. In this case, the nonequilibrium M(AlZnMgCu) phase is dissolved without melting. Due to a slow diffusion of Ni in (Al), the fragmentation and spheroidization of eutectic Al_3Ni particles do not occur upon anneals below 500°C, i.e. below T_{NES}. Therefore, the solution treatment of high-strength nikalins is performed in two stages: at T_1 for

(a)

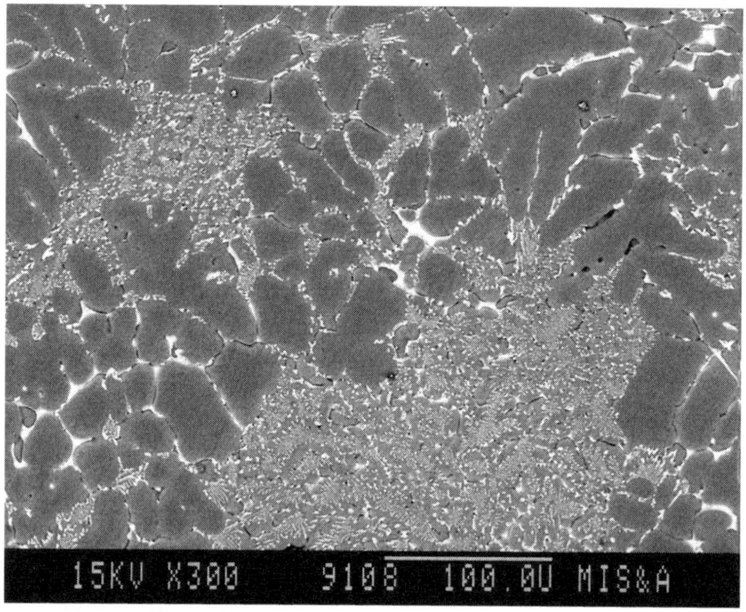

(b)

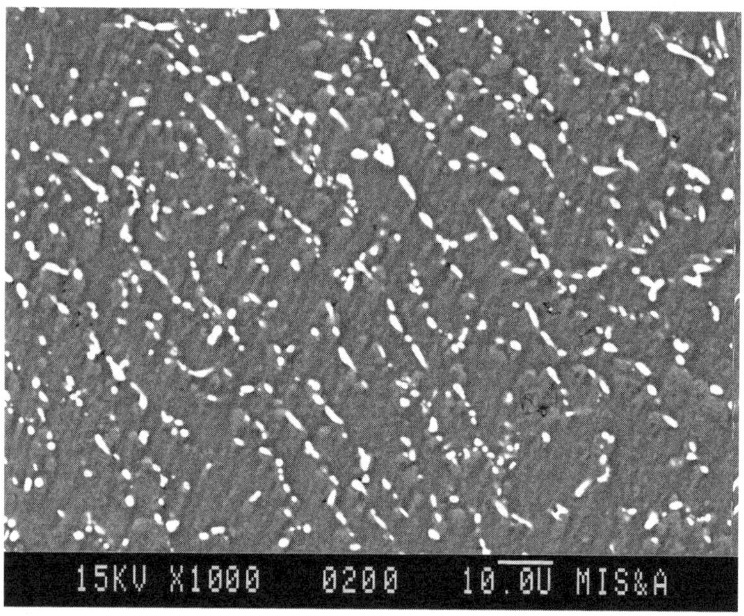

Figure 7.18. Microstructure of a AZ6N4 casting alloy (nikalin): (a) as-cast state, SEM; eutectic colonies (Al) + Al_3Ni and M(AlZnMgCu) phase from divorced eutectics; (b) T4 (450°C, 3 h + 500°C, 3 h), SEM; globular particles of Al_3Ni phase; (c) T6 (130°C, 10 h), TEM; globular particles of Al_3Ni phase; and (d) T7 (120°C, 6 h + 160°C, 3 h), TEM; large globular particle of Al_3Ni and fine precipitates of M′.

(c)

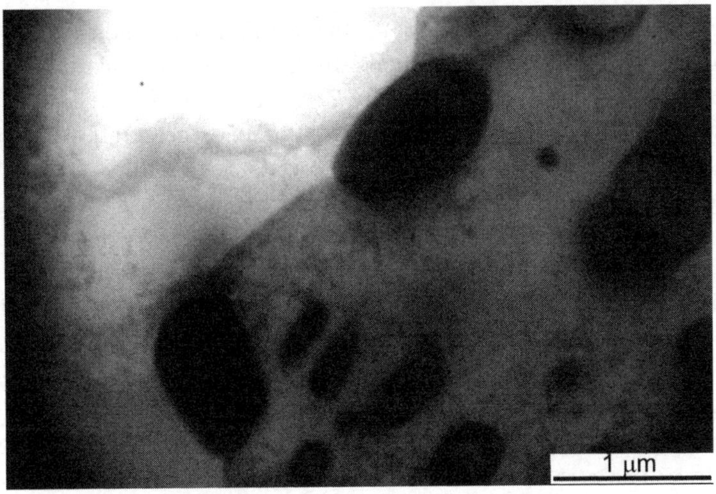

(d)

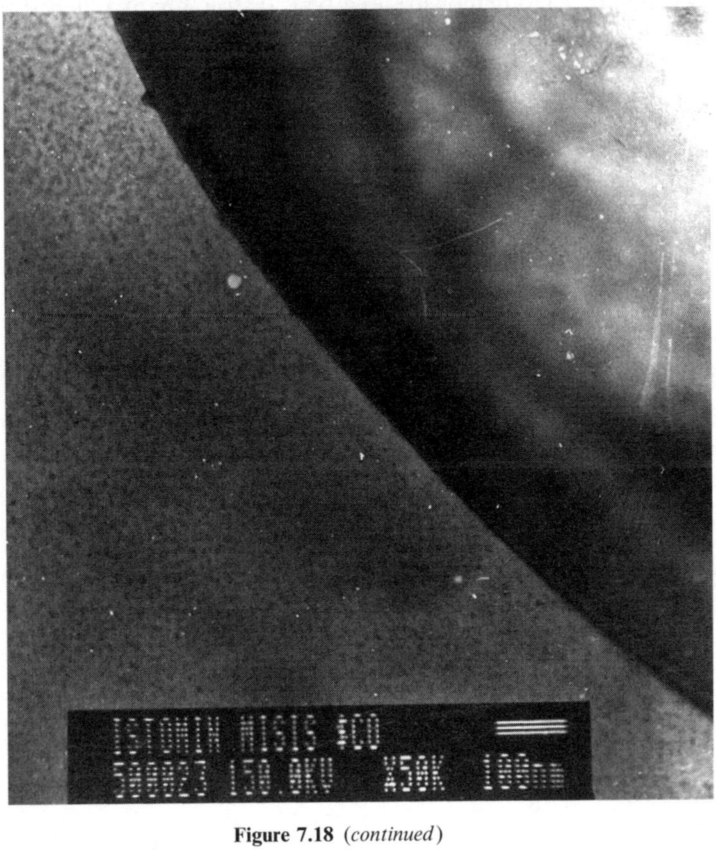

Figure 7.18 (*continued*)

homogenization and at T_2 (close to the equilibrium solidus, Figure 7.17) for spheroidization of intermetallic particles. In the case of fine as-cast structure, eutectic Al_3Ni particles become globular, Figure 7.18b, c.

On increasing $\Sigma(Zn + Mg + Cu)$ the optimum concentration of Ni decreases along the line of the binary (Al) + Al_3Ni eutectics (Figure 7.16a). The solidus projection (Figure 7.16b) shows that with increasing the total amount of alloying elements $\Sigma(Zn+Mg+Cu)$ the equilibrium solidus decreases that makes it difficult to obtain globular Al_3Ni particles in conventional chill castings as the temperature is too low and the eutectic fineness is insufficient for the morphology change within reasonable time frame. Less alloyed materials give more opportunities for structure modification during high-temperature anneals but they are less attractive with respect to the strength properties.

Quenching and subsequent aging of nikalins are similar to those for heat-treatable Al–Si alloys. The best combination of strength, ductility and corrosion resistance is obtained through two-stage annealing facilitating uniform distribution of hardening precipitates. Figure 7.18d demonstrates a typical structure of a high-strength nikalin quenched and aged to the best combination of mechanical properties.

Chapter 8
Alloys with Lithium

Aerospace, aircraft, and automotive industries demand light, stiff, high-strength materials. Aluminum alloys containing lithium as a main alloying element are the response to these demands. Starting from the 1960s this group of alloys is under development. Each per cent of lithium added to aluminum decreases the density of the alloy by 3% and increases the elastic modulus by 5–6%. Tensile strength increases almost linearly with Li additions as well. Al–Li alloys also exhibit excellent fatigue endurance and cryogenic toughness. The anomalous increase of elastic modulus in solid-solution type aluminum alloys with additions of Li and Mg (elements having elastic moduli lower than that of aluminum) is believed to be caused by atomic ordering in solid solutions. A favorable combination of mechanical properties of commercial Al–Li alloys is usually achieved after heat treatment and is a result of corresponding phase composition and structure formation.

Binary Al–Li alloys have not found commercial application. However, alloys containing additionally Cu, Mg, or a combination of these two elements proved to be suitable for special applications (such as aerospace structures, military aircraft, racing cars) despite manufacturing difficulties. Further improvement of physical and service characteristics of these alloys can be achieved by small additions of transition elements.

In this chapter we consider equilibrium phase diagrams of Al–Cu–Li, Al–Li–Mg, Al–Li–Mn, Al–Li–Si, Al–Li–Zr, and Al–Cu–Li—Mg, and some other quaternary systems, application of these diagrams to commercial Al–Li alloys including effects of some additions, and the phase composition and structure of Al–Li alloys after precipitation hardening.

8.1. Al–Cu–Li PHASE DIAGRAM

Commercial alloys that belong to this system contain 1–2.5% Li, 2.5–5.5% Cu, and small additions of Zr (AA2090) or Mn and Cd (VAD23rus).

Two binary and three ternary phases are in equilibrium with the aluminum solid solution in the aluminum corner of the Al–Cu–Li system. The structure parameters, composition, and density of these phases are listed in Table 8.1. The ternary compounds have narrow homogeneity ranges.

The T_2 and R phases are frequently confused with respect to their stability and equilibrium with (Al). The present understanding is that these phases are indeed

Table 8.1. Crystal structure and chemical composition of phases that occur in the aluminum corner of the Al–Cu–Li phase diagram (Mondolfo, 1976; Drits et al., 1977; Harmelin and Legendre, 1991a)

Phase	Crystal structure, space group	Lattice parameters, nm		Composition, %	Formation temperature, °C	Density, g/cm^3
		a	*c*			
Al_2Cu (θ)	Tetragonal, *I5/mcm*	0.6063	0.4872	52.5–53.9Cu	591	4.34
AlLi (δ)	Cubic, *Fd3m*	0.637	–	17.8–25Li	717	1.725
$Al_{7.5}Cu_4Li$ (T_B)	Cubic, *Fm3m*	0.583	–	54.9Cu, 1.5Li	542–550	3.64–3.76
Al_2CuLi (T_1)	Hexagonal, $P\bar{6}m2$ or *P6/mmm*	0.497	0.935	52.8Cu, 5.4Li	>635	–
Al_6CuLi_3 (T_2)	Icosahedral	–	–	26.9Cu; 8.8Li	610–620	–
Al_5CuLi_3 (R)	Cubic, *Im3*	1.3914	–	28.8Cu, 9.6Li	>635	–

different stable phases, though the T_2 phase is formed as an icosahedral quasi-crystalline compound (Harmelin and Legendre, 1991a). In the adopted version of the Al–Cu–Li phase diagram the R phase is not in equilibrium with (Al). The T_B phase can be considered as the metastable θ′ (Al_2Cu) phase stabilized by replacement of Al atoms by Li atoms (Mondolfo,1976).

The solidification reactions in Al-rich alloys are shown in Figure 8.1 and invariant reactions are listed in Table 8.2. Temperatures of the solidification reactions are not well established and the temperature ranges are given to accommodate different reference data. Figure 8.2 shows the distribution of phase fields at 500 and 350°C.

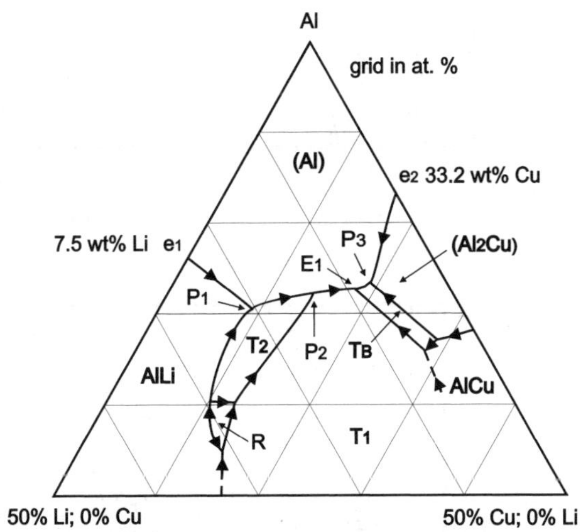

Figure 8.1. Projection of liquidus surface in the Al-rich portion of the Al–Cu–Li system (after Harmelin and Legendre, 1991a). Note that grid and axis are in at.%, concentrations of binary eutectic reactions are given in wt% for scale comparison. T_B – $Al_{7.5}Cu_4Li$, T_1 – Al_2CuLi, and T_2 – Al_6CuLi_3.

Table 8.2. Invariant equilibria in the Al corner of Al–Cu–Li phase diagram (Harmelin and Legendre, 1991a)

Reaction	Point in Figure 8.1	Concentrations in liquid phase, at.%		Temperature, °C
		Cu	Li	
$L+AlLi \Rightarrow (Al)+T_2$	P_1	9	20.5	564 or 572
$L+T_2 \Rightarrow (Al)+T_1$	P_2	14.5	13.5	540 or 542
$L+Al_2Cu \Rightarrow (Al)+T_B$	P_3	19.6	7.3	522–535
$L \Rightarrow (Al)+T_1+T_B$	E_1	18.8	8.7	518–528

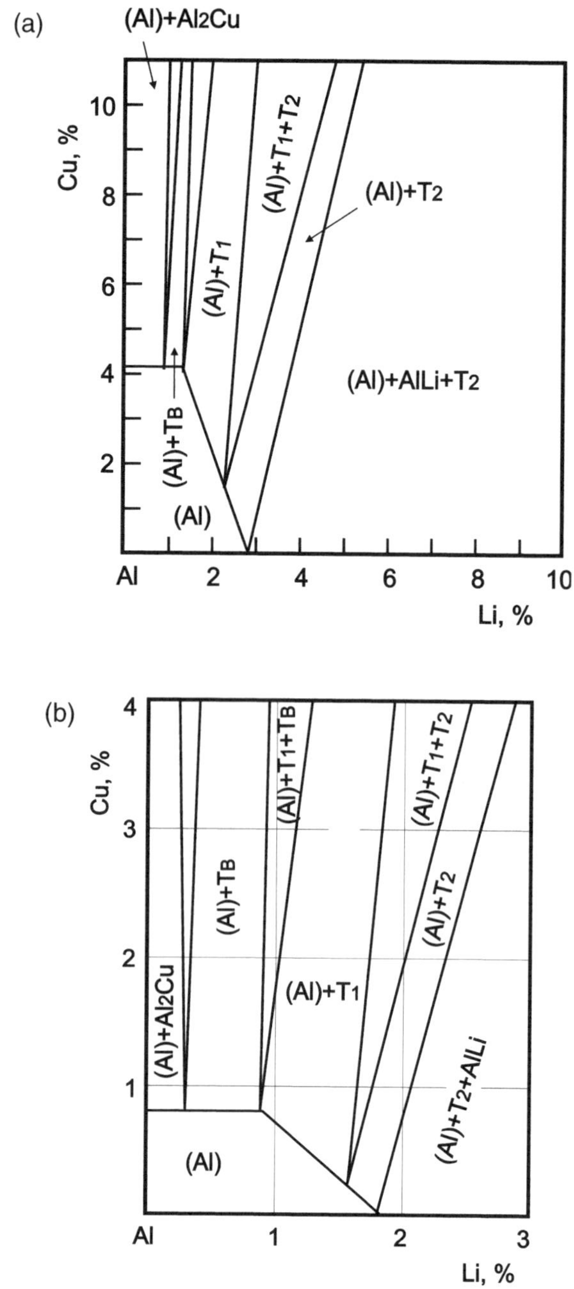

Figure 8.2. Isothermal sections of the Al–Cu–Li system at 500°C (a) and 350°C (b) (Drits et al., 1977).

Figure 8.2 shows that the addition of copper decreases the solid solubility of Li in aluminum, which in binary Al–Li alloys equals to 4.3% at 602°C, 3.1% at 527°C, 2.2 at 427°C, 1.6% at 327°C, and 1.1% at 227°C (Mondolfo, 1976).

The solidification of a VAD23-type alloy (1.15% Li, 5.15% Cu) starts with the formation of primary (Al) grains, then a small amount of binary (Al) + Al_2Cu eutectics precipitates. The remaining liquid reacts with Al_2Cu according to the transition reaction P_3 (Figure 8.1, Table 8.2). During that reaction (under equilibrium conditions) Al_2Cu disappears and T_1 and T_B phases are formed. The solidification is likely to end with the formation of (Al) + T_B binary eutectics. During cooling in the solid state, T_B and T_1 phases precipitate from the aluminum solid solution as a result of the decreasing solubility of Cu and Li in (Al). The resultant phase composition of a VAD23-type alloy is (Al) + T_1 + T_B, the ternary phases forming different structure constituents.

In the case of a 2090-type alloy with a higher concentration of Li (2.25%) and lower concentration of copper (2.7%), after primary formation of (Al) grains, the (Al) + AlLi binary eutectics solidifies. The alloy then undergoes transition reactions P_1 and P_2 (Figure 8.1, Table 8.2) and solid-state precipitation of T_2 and T_1 phases. Hence, the final equilibrium phase composition at room temperature of a 2090-type alloy is (Al) + T_1 + T_2.

Under nonequilibrium solidification conditions, the amount of the binary eutectics in both the alloys becomes larger. Peritectic reactions (P_3 in VAD23 and P_1 in 2090) may not complete, hence some particles of Al_2Cu and AlLi, respectively, remain in the fully solidified structure. The formation of the ternary eutectics E_1 (Figure 8.1, Table 8.2) becomes possible. The structure of all eutectics in wrought Al–Cu–Li alloys is divorced and appears as individual particles at grain and dendritic cell boundaries. Nonequilibrium phases Al_2Cu and AlLi will dissolve during homogenization annealing of commercial alloys.

Some of the Al–Cu–Li alloys contain small amounts of magnesium (less than 1%). The implications of this addition on the phase composition and solidification is discussed in Section 8.6.

8.2. Al–Li–Mg PHASE DIAGRAM

The Al–Li–Mg phase diagram is very important for the analysis of commercial Al–Li alloys containing magnesium such as Russian Grade 1420 (5.5% Mg and 2% Li). This system has been thoroughly studied, and it is found that the following phases can be in equilibrium with the aluminum solid solution: Al_8Mg_5, $Al_{12}Mg_{17}$, Al_2LiMg, and AlLi. Therefore, only one ternary compound is formed in this system. The Al_2LiMg phase has a cubic structure with a lattice parameter of 2.031 nm

(Ghosh, 1993c) and exists in the compositional range from 10.3 to 11.3% Li and from 27.1 to 24% Mg (Mondolfo, 1976) or 32–34.2 at.% Li and 13.5–14 at.% Mg (Ghosh, 1993c). This compound is formed during a peritectic reaction at 536°C. The addition of Li to Al–Mg alloys narrows the compositional range of Al_8Mg_5 and broadens the homogeneity range of $Al_{12}Mg_{17}$ (cubic, $a = 1.05547$ nm). As a result, the latter phase comes in equilibrium with (Al) in ternary Al–Li–Mg alloys. This phase can dissolve Li at the expense of Mg, changing the composition to $Al_4Li_2Mg_3$ (7.1% Li, 36.6% Mg) (Mondolfo, 1976) or to about 20 at.% Li (Ghosh, 1993c). The Al_8Mg_5 phase also dissolves some Li but to a considerably lower level, about 7 at.% (Ghosh, 1993c).

The solidification surface in the aluminum corner of the Al–Li–Mg system and the corresponding invariant reactions are given in Figure 8.3 and Table 8.3 according

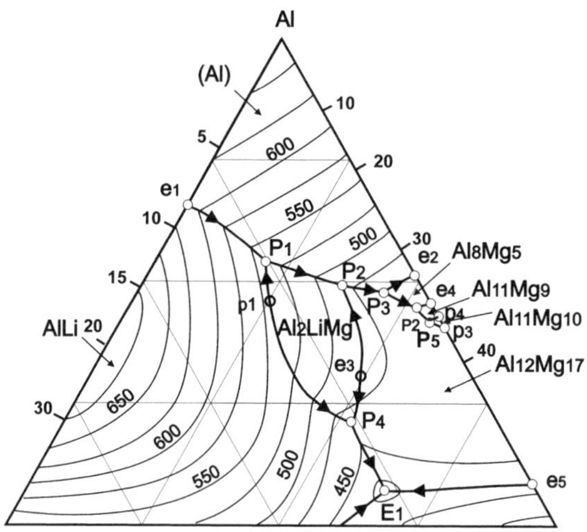

Figure 8.3. Projection of liquidus surface in the Al corner of the Al–Li–Mg system (after Ghosh, 1993c). Note that grid is at.% and axes are in wt%.

Table 8.3. Invariant equilibria in the Al corner of Al–Li–Mg phase diagram (Ghosh, 1993c)

Point in Figure 8.3	Reaction	Concentrations in liquid phase, at.%		Temperature, °C
		Li	Mg	
e_1	$L \Rightarrow (Al) + AlLi$	7.5	–	602
e_2	$L \Rightarrow (Al) + Al_8Mg_5$	–	34	450
P_1	$L + AlLi \Rightarrow (Al) + Al_2LiMg$	19.4	14.6	536
P_2	$L + Al_2LiMg \Rightarrow (Al) + Al_{12}Mg_{17}$	10.8	27.7	483
P_3	$L + (Al) + Al_{12}Mg_{17} \Rightarrow Al_8Mg_5$	6.0	33.5	458
P_4	$L + Al_2LiMg \Rightarrow AlLi + Al_{12}Mg_{17}$	20.1	40.1	464

to Ghosh (1993c). This version differs significantly from previously reported solidification reactions as compiled by Drits et al. (1977) and is based on a careful assessment of recently reported data including thermodynamic calculations of binary phase diagrams constituting the ternary system.

Figure 8.4 demonstrates isothermal sections of Al–Li–Mg system at two temperatures. Limit solubilities of Mg and Li in solid aluminum at different temperatures are given in Table 8.4 and Figure 8.5. Additions of Mg to Al–Li alloys

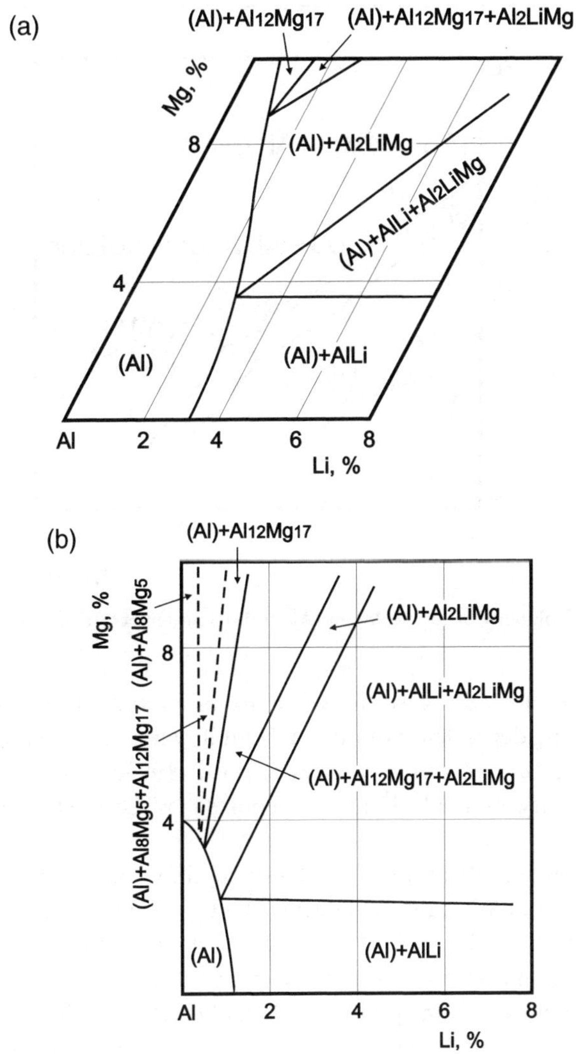

Figure 8.4. Isothermal sections of the Al–Li–Mg system at 500°C (a) and 200°C (b) (Drits et al., 1977).

Table 8.4. Mutual solid solubility of Li and Mg in (Al) at different temperatures (Mondolfo, 1976; Drits et al., 1977)

Three-phase phase field	470°C		430°C		200°C	
	Mg, %	Li, %	Mg, %	Li, %	Mg, %	Li, %
Solubility in binary systems	–	–	15.5	2.3	4.0	1.05
(Al) + Al_8Mg_5 + $Al_{12}Mg_{17}$	14.0	0.8	12.5	0.55	3.6	0.19
(Al) + $Al_{12}Mg_{17}$ + Al_2LiMg	9.3	1.4	7.2	1.72	3.4	0.32
(Al) + AlLi + Al_2LiMg	3.8	3.0	3.0	2.25	2.0	1.0

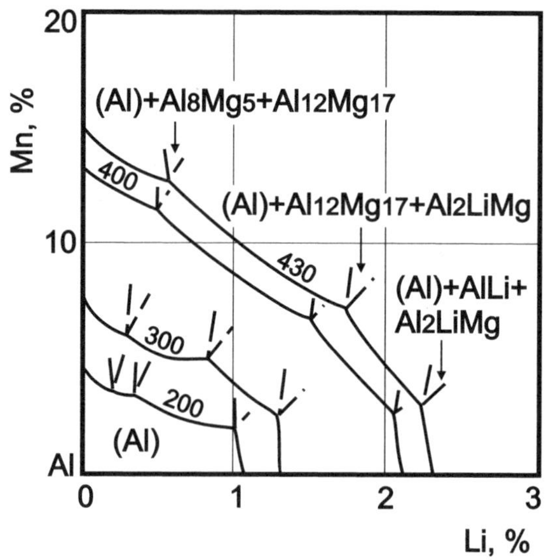

Figure 8.5. Solid solubility of Li and Mg in (Al) at different temperatures (after Ghosh, 1993c).

and Li to Al–Mg alloys decrease the solubility of Li and Mg in (Al), respectively. However, the strong dependence of mutual solubility in (Al) on temperature remains a characteristic feature of ternary alloys. Ghosh (1993c) noted that the data on magnesium solubility in solid aluminum reported by Drits et al. (1977) might be overestimated.

The equilibrium solidification of a 1420-type alloy (5.5% Mg, 2% Li) involves only the formation of (Al) grains. On decreasing the temperature, Al_2LiMg and then AlLi precipitate in the solid state. Therefore, the main excess phases in a 1420-type alloy are Al_2LiMg and AlLi in the form of precipitates.

Under more realistic, nonequilibrium solidification conditions the alloy may undergo reactions e_1 and P_1 (Figure 8.3, Table 8.3). The transition reaction P_1 may not complete. The final phase composition will be same as after equilibrium solidification, with the major difference that Al_2LiMg and AlLi particles are of the solidification (eutectic and peritectic) origin as well.

8.3. Al–Li–Mn PHASE DIAGRAM

Manganese is an alloying addition in some Al–Li alloys, e.g. VAD23. The Al–Li–Mn phase diagram is studied in the range of Al-rich alloys. No ternary phases are found in this region. Only binary AlLi and Al_6Mn phase are in equilibrium with (Al). According to Drits et al. (1977) Al_6Mn may dissolve up to 7% Li, whereas the solubility of Mn in AlLi does not exceed 0.1%. Mondolfo (1976) mentions that the solubility of the third component in either phase is very small.

Two ternary invariant reactions occur during solidification (Mondolfo,1976)

$$L + Al_4Mn \Rightarrow AlLi + Al_6Mn \text{ at } 640°C,\ 9.0\%\ Li,\ \text{and } 3.2\%\ Mn \text{ and}$$

$$L \Rightarrow (Al) + Al_6Mn + AlLi \text{ at } 597°C,\ 8.8\%\ Li,\ \text{and } 1.7\%\ Mn.$$

Figure 8.6 shows isothermal section of the Al–Li–Mn system at 590°C. The solid solubilities of Mn and Li in aluminum at different temperatures are presented in Table 8.5. Lithium and manganese considerably decrease the solubility of each other in solid aluminum. Mondolfo (1976) gives higher values of solubility for Mn and

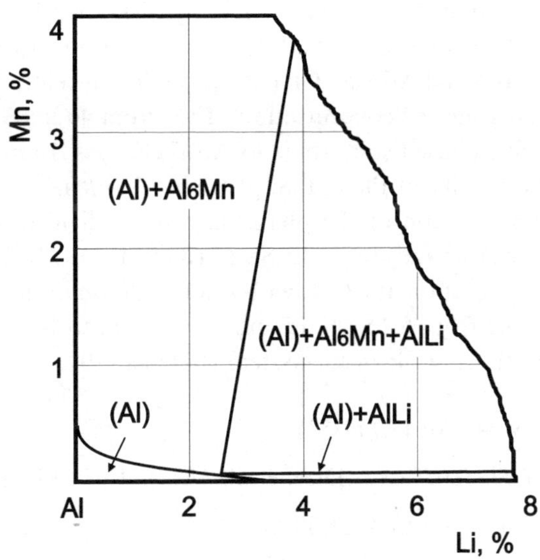

Figure 8.6. Isothermal section of the Al–Li–Mn system at 590°C (Drits et al., 1977).

Table 8.5. Solid solubilities of Li and Mn in aluminum (Drits et al., 1977)

Temperature, °C	590	500	400
Mn, %	0.05	0.03	0.01
Li, %	2.7	1.8	1.4

lower – for Li, e.g. 0.2% Mn and 1% Li at 597°C, as compared with data given in Table 8.5.

Addition of Mn to commercial Al–Li alloys results in the formation of Al_6Mn alongside T_x (AlLiCu) phases, in the presence of copper a ternary (AlCuMn) phase is formed (in VAD23).

8.4. Al–Li–Si PHASE DIAGRAM

Silicon is a common impurity in aluminum alloys and is present in commercial Al–Li alloys as well, up to 0.2%. Small additions of Li to casting Al–Si alloys are known to modify (refine) eutectics. The Al–Li–Si system is also interesting as a base for rapidly solidified alloys and composite materials with Al–Li matrix and SiC reinforcement.

In the aluminum corner of the Al–Li–Si phase diagram, (Si), AlLi, and ternary (AlLiSi) phases are in equilibrium with (Al). The composition and structure of the ternary phase are not clearly established. There is a possibility that this phase is an extension of the homogeneity range of AlLi by dissolving Si in the latter phase (Mondolfo, 1976). The composition of the ternary phase ranges from $Al_2Li_3Si_3$ to AlLiSi, through Al_2Li_2Si and $AlLi_2Si$. On increasing the concentration of Si (from 20 to 33 at.%) and decreasing the concentration of Li (from 40 to 33 at.%), the lattice parameter of this cubic phase (isomorphic to AlLi) changes from 0.612 to 0.593 nm (Bätzner, 1993). The density of the AlLiSi phase is 1.96 g/cm^3.

The aluminum-rich portion of the phase diagram is divided in two parts by a pseudo-binary section from (Al) to $Al_2Li_3Si_3$ (Drits et al., 1977). This section represents a simple eutectic reaction at 635°C with a composition of the eutectic point not yet established, ranging from 5.2 at.% Li and 3.5 at.% Si to 14 at.% Li and 4 at.% Si. Two eutectic reactions occur in the system (Bätzner, 1993):

$$L \Rightarrow (Al) + (Si) + (AlLiSi) \text{ at } 575°C$$

(11.5% Si, 0.05% Li (Drits et al., 1977), E_1 in Figure. 8.7) and

$$L \Rightarrow (Al) + AlLi + (AlLiSi) \text{ at } 595°C$$

(9.2% Li, 2.0% Si (Drits et al., 1977), E_2 in Figure 8.7)

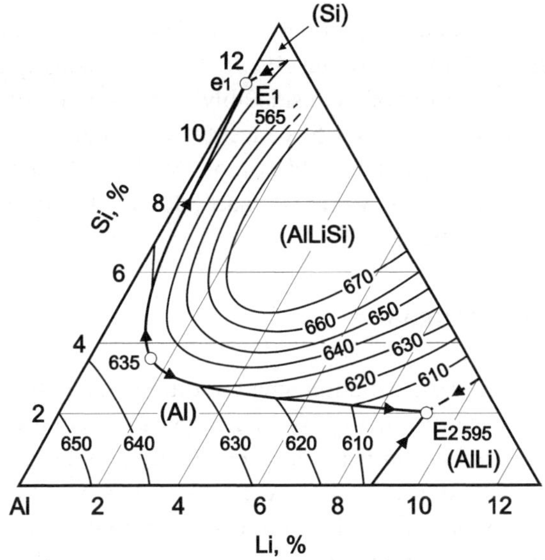

Figure 8.7. Projection of liquidus surface in the aluminum corner of the Al–Li–Si system (Drits et al., 1977).

on the Al–Si and Al–Li sides of the phase diagram, respectively. A lower temperature of 565°C has been reported for the first reaction by (Drits et al., 1977).

The solidification surface in the Al corner of the Al–Li–Si system is shown in Figure 8.7 following (Drits et al., 1977). It should be noted that there are still discrepancies between different sources regarding the positions of monovariant lines, temperatures and concentrations of the ternary eutectic points (Bätzner, 1993).

Addition of Si to Al–Li alloys results in the formation of (AlLiSi) phase of primary and eutectic origins. During rapid solidification, fine (AlLiSi) dispersoids are formed. These dispersoids have a beneficial effect on the homogeneity of plastic deformation and ductility of Al–Li alloys (Arcade et al., 1990). If an alloy contains less Si than Li (in at.%), all silicon is bound into (AlLiSi) particles that make plastic deformation more homogeneous, and lithium remaining in the solid solution (at a proportion higher than the equilibrium solubility due to large cooling rates) forms hardening Al_3Li precipitates during heat treatment (Champier and Samuel, 1986). Note that the concentration of lithium in an Al–Li–Si alloy required to retain sufficient amount of Li in the solid solution is higher than in binary Al–Li alloys (as part of Li is bound in the (AlLiSi) compound). Therefore, the resultant density of the ternary alloy is lower than that of the binary alloy with the same hardening ability. It is necessary to mention, though, that rapid solidification is required to achieve such an effect. Otherwise, the heterogeneous structure of the ternary alloy may be detrimental to the properties.

Addition of Li to Al–Si alloys modifies the Al–Si eutectics, although this effect is less pronounced than that of Na (Boom, 1963). The other consequence of adding Li to Al–Si alloys is the formation of a considerably larger amount of eutectics (ternary (Al) + (Si) + (AlLiSi)) at much lower Si concentrations than in binary Al–Si alloys (Figure 8.7). In an Al–2.5% Li alloy, addition of 4% Si results in the replacement of primary (Al) with primary (AlLiSi), more than 80% of the structure consisting of eutectics (Samuel et al., 1992).

8.5. Al–Li–Zr PHASE DIAGRAM

Zirconium is one of the most common small additions made to commercial Al–Li alloys. Most of the research on the Al–Li–Zr system is connected to metastable phases formed during rapid solidification (cubic Al_3Zr) and during decomposition of the supersaturated solid solution (cubic Al_3Zr, Al_3Li). We will consider the metastable phase selection of decomposition products in more detail later in this chapter.

In the aluminum corner of the system, only two binary phases are in equilibrium with (Al), namely, AlLi and Al_3Zr. The possibility of an invariant solidification reaction at 595.4°C (melt composition 24.65% Li and 5×10^{-5} at.% Zr) is mentioned by Saunders (1989) based on thermodynamic calculations. Although the nature of this reaction is not clear, we can suggest the following:

$$L + Al_3Zr \Rightarrow (Al) + AlLi$$

Al_3Zr may dissolve up to 1.3 at.% Li, whereas the solubility of Zr in AlLi is negligible (Saunders, 1989). Primary Al_3Zr particles are formed at trace amounts of Zr in Al–Li alloys, which makes zirconium a promising grain refiner. The addition of lithium to Al–Zr alloys considerably decreases the solubility of Zr in (Al).

Stiltz (1993) cites reports on the occurrence of a stable ternary compound in this system with the formula $Al_3(Li_xZr_{1-x})$ $(0.45 < x < 0.8)$. The formation and stability of this phase is questionable. On the other hand, the existence of the metastable $Al_3(Li_xZr_{1-x})$ phase that is formed during high-temperature (500–550°C) precipitation from the solid solution is likely to be true. This phase has a cubic structure of $L1_2$ type (similar to cubic Al_3Zr and Al_3Li) with a lattice parameter of 0.401 nm very close to those of the metastable cubic Al_3Zr and Al_3Li phases (Stiltz, 1993).

8.6. Al–Cu–Li–Mg PHASE DIAGRAM

The majority of commercial Al–Li alloys contain copper and magnesium within the compositional ranges 0.3–6% Mg, 1–5.5% Cu, and 1–3% Li. All three elements contribute to the formation of phases and structure during solidification, deformation, and aging. Therefore, the quaternary Al–Cu–Li–Mg phase diagram is very important.

Although the number of published works on the quaternary system is limited, the phase composition of aluminum-rich alloys is established. The following phases are in the equilibrium with aluminum: Al_2Cu (θ), AlLi (δ), Al_2CuMg (S), Al_2LiMg, $Al_{7.5}Cu_4Li$ (T_B), Al_2CuLi (T_1), and Al_6CuLi_3 (T_2) (Rokhlin et al., 1994a). In copper- and magnesium-rich alloys (>20% Cu, >11% Mg, 2% Li), the Al_4CuMg_5 phase was found in equilibrium with (Al) (Lawson-Jack et al., 1993). The T_1 phase dissolves substantial amounts of magnesium and its lattice parameters change, starting to resemble those of the R phase (Lawson-Jack et al., 1993). Magnesium may also dissolve in the T_2 phase (Rokhlin et al., 1994a).

Data on solidification reactions in the Al–Cu–Li–Mg system are limited. The analysis of experimental data shows that the following solidification invariant reactions are possible in aluminum-rich alloys (from Al–Cu–Mg towards Al–Cu–Li) (Fridlyander et al., 1993):

$$L + T_B \Rightarrow (Al) + \theta + S;$$
$$L \Rightarrow (Al) + \theta + S + T_B;$$
$$L + T_2 \Rightarrow (Al) + S + T_1;$$
$$L \Rightarrow (Al) + S + T_2 + Al_2LiMg\,(484^\circ C);$$
$$L + AlLi \Rightarrow (Al) + T_2 + Al_2LiMg.$$

The temperatures of these reactions (except one) are not determined due to the very small difference between them.

Figure 8.8 demonstrates isothermal sections of the Al–Cu–Li–Mg phase diagram at 400°C, showing the complex distribution of phase fields in the solid state. The sections are given for two copper concentrations of 1.5% (close to 8090 and 1441 alloy compositions, see Table 8.6) and 2.8% (close to 2090, CP276, and 1464 alloy compositions, see Table 8.6).

The nature of monovariant solidification reactions in this system is unclear. The comparison of phase compositions of 2090 and 8090-type alloy given by the Al–Cu–Li phase diagram (section 8.1) and by the Al–Cu–Li–Mg phase diagram shows that even small additions of Mg (0.25–1.0%) result in the formation of the S phase by a eutectic reaction and by precipitation from the solid solution.

Figure 8.9 gives approximate isotherms of solidus and solvus for Al–Cu–Li alloys with 0.5% Mg and 1.5% Mg. These approximations are based on the experimental work of Dorward (1988) who studied alloys containing 1.9–2.7% Li and 0.5–2.7% Cu and are accurate to ±5°C. The increase in the concentration of any given element results in a lower solidus and a higher solvus.

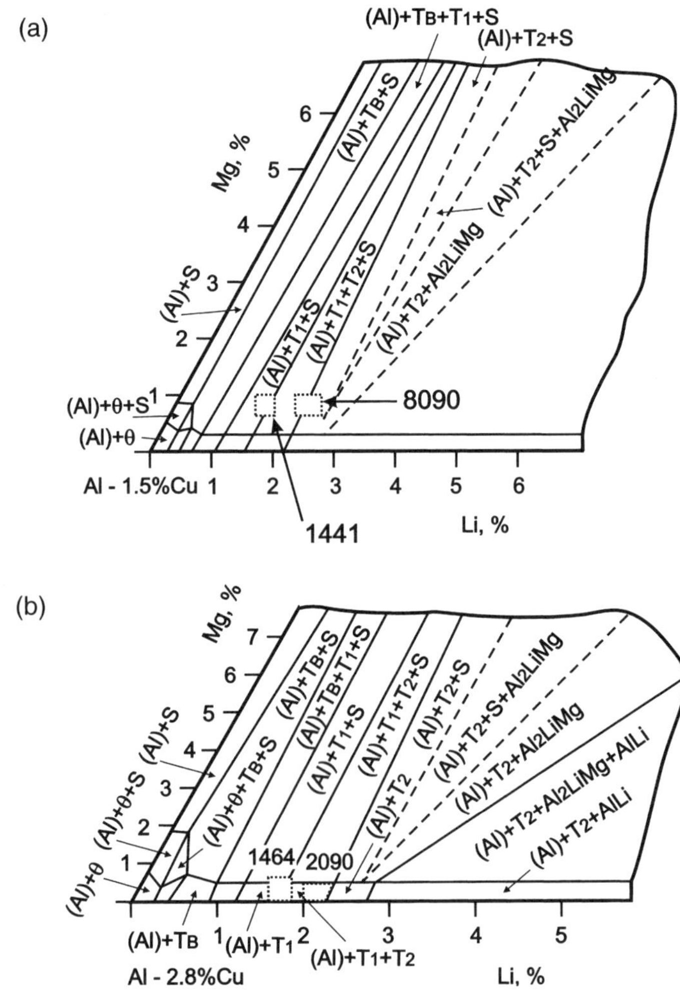

Figure 8.8. Isothermal sections of the Al–Cu–Li–Mg system at 400°C at a constant concentration of copper of 1.5% (a) and 2.8% (b) (after Fridlyander et al., 1993; Rokhlin et al., 1994a). T_B – $Al_{7.5}Cu_4Li$, T_1 – Al_2CuLi, T_2 – Al_6CuLi_3, S – Al_2CuMg, and θ – Al_2Cu.

These data can be used as a starting point for the right choice of solution heat treatment of such alloys as 2090, 8090, CP276, 1441, 1446, but not high-copper alloys, e.g. Weldalite049. The effects of alloying elements on the solidus temperature of Weldalite-type alloys (4–6.3% Cu, 0–2% Li, 0–0.8% Mg) were studied by Montoya et al. (1991). The variation in copper concentration above 4% has virtually no effect on the solidus temperature of the base alloy Al–1.3%Li–0.4%Mg, the solidus being at 512–513°C. Increasing the concentration of magnesium in the Al–(5–6)%Cu–1.3%Li alloys results in the continuously decreasing solidus

Table 8.6. Average chemical compositions of some commercial Al–Li alloys with solidus and liquidus temperatures

Grade	Chemical composition, %						T_{sol}, °C	T_{liq}, °C
	Li	Mg	Cu	Si	Fe	Other		
2090	2.25	<0.25	2.7	<0.10	<0.12	0.12Zr; 0.15Ti	560	650
2097	1.5	<0.35	2.8	<0.12	<0.15	0.35Mn; 0.12Zr; 0.15Ti	–	–
2020	1.1	–	4.5	–	–	0.5Mn	–	–
VAD23rus	1.15	–	5.15	–	–	0.6Mn; 0.18Cd	–	–
1420rus	2.0	5.5	–	–	–	0.12Zr	–	–
1421rus	2.0	5.0	–	–	–	0.2Mn; <0.2Sc; <0.15Zr	–	–
1424rus	1.7	5.0	0.7Zn	–	–	Sc, Zr	–	–
2091	2.0	1.5	2.15	<0.20	<0.30	0.1Zr; 0.1Ti	560	670
2094	1.1	0.4	4.8	<0.12	<0.15	0.1Zr; 0.1Ti; 0.25Mn; 0.43Ag	–	–
8090	2.45	0.95	1.3	<0.20	<0.30	0.12Zr; 0.1Ti	600	655
8091	2.6	0.85	2.0	<0.30	<0.50	0.12Zr	–	–
Weldalite049	1.3	0.4	5.4	–	–	0.14Zr; 0.4Ag	507–512	–
2095	1.1	0.53	4.4	<0.12	<0.15	0.1Zr; 0.1Ti; 0.43Ag	–	–
CP276	2.2	0.5	2.7	–	–	0.12Zr	–	–
1441rus	1.7	0.95	1.6	–	–	0.08Zr	–	–
1464rus	1.7	0.5	3.0	–	–	Sc, Zr	–	–

temperature, from 521 to 507°C. And the solidus temperature changes with the minimum at 1.3%Li (511°C) on increasing the amount of Li in the Al–(5–6)%Cu–0.4%Mg alloy. Additions of silver to Weldalite-type alloys do not have any appreciable effect on the solidus. These results are illustrated in Figure 8.10. A minimum solidus temperature of 507–513°C suggests the occurrence of a eutectic reaction at this temperature. The temperature is, however, about 10°C lower than that of the eutectic reaction $L \Rightarrow (Al) + T_1 + T_B$ of the Al–Li–Cu system (see Table 8.2). Mukhopadhyay et al. (2000) suggest that the eutectic reaction $L \Rightarrow (Al) + Al_2Cu + Al_2CuMg$ (S) that takes place at 507°C is responsible for the lowest melting

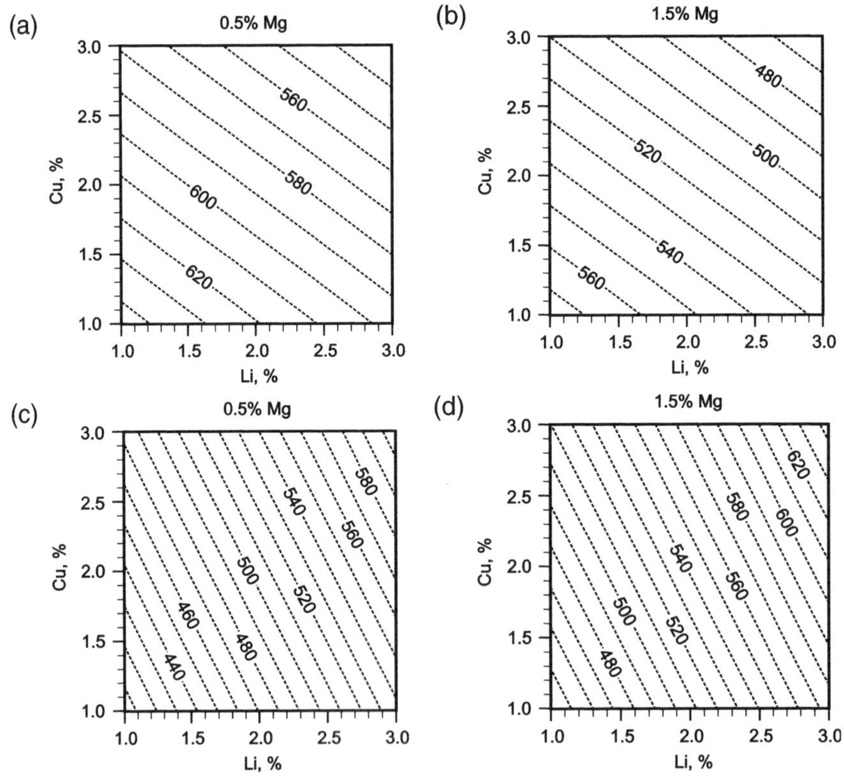

Figure 8.9. Solidus (a, b) and solvus (c, d) isotherms in Al–Cu–Li–Mg alloys containing 0.5% Mg (a, c) and 1.5% Mg (b, d) (after Dorward, 1988).

temperature in Weldalite-type alloys. They also observed the melting of the ternary eutectics (Al) + T_1 + T_B at 521°C. The phase composition of as-cast Weldalite-type alloys is (Al) + T_1 + T_B + Al_2Cu + Al_2CuMg (S), with the last phase being present only in small quantities (Mukhopadhyay et al., 2000). It should be noted that the possibility of the invariant eutectic reaction L ⇒ (Al) + T_1 + S that occurs at 505 ± 10°C cannot be excluded under nonequilibrium solidification conditions (Drits et al., 1977). The existence of low-melting eutectics in Weldalite-type alloys should be taken into account while choosing the correct regime of solution treatment.

8.7. Al–Li–Mg–Mn AND Al–Cu–Li–Mn PHASE DIAGRAMS

Some commercial Al–Li alloys contain small additions of manganese, e.g. 1421 and VAD23 (see Table 8.6). The effects of Mn on the phase equilibria in the Al–Li–Mg and Al–Li–Cu systems are reported for aluminum-rich alloys (Drits et al., 1977).

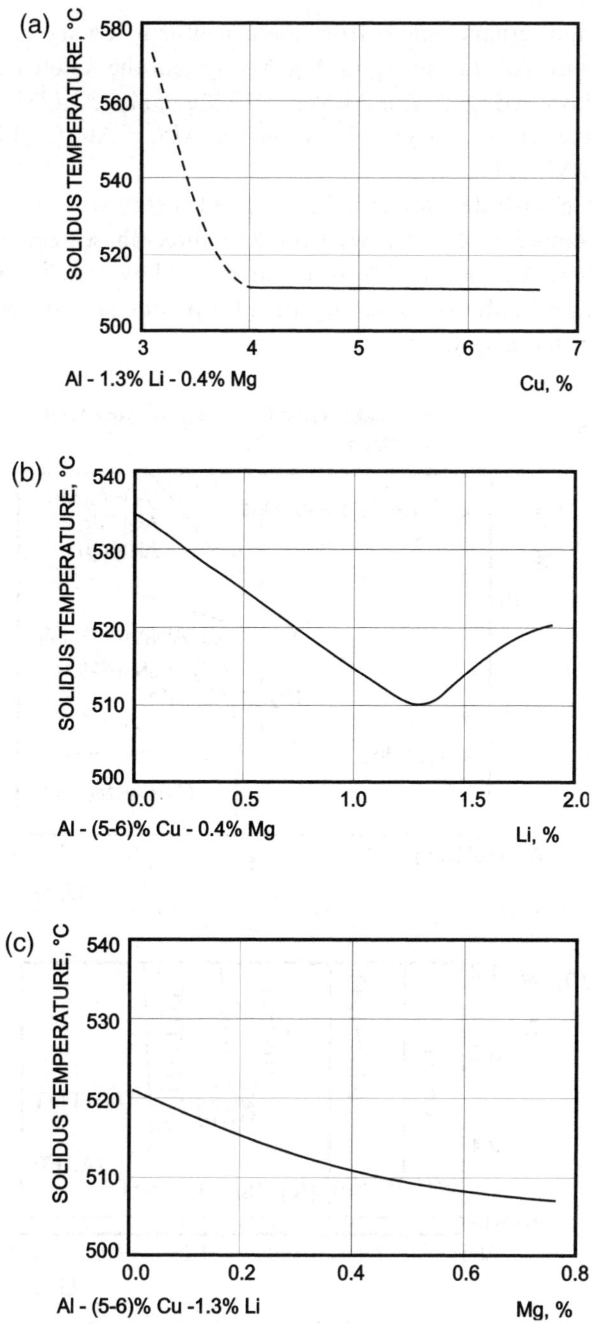

Figure 8.10. Effect of alloying elements (a, Cu; b, Li; and c, Mg) on the solidus temperature in Weldalite-type alloys (after Montoya et al., 1991).

Only binary and ternary phases from the constituent ternary systems are found in equilibrium with (Al). In the Al–Li–Mg–Mn system the following phases can be present in Al alloys: AlLi, Al_6Mn, Al_8Mg_5, $Al_{12}Mg_{17}$, and Al_3LiMg. In the Al–Li–Cu–Mn alloys, the set of the phases is as follows: Al_2Cu, $Al_{7.5}Cu_4Li$ (T_B), Al_2CuLi (T_1) and $Al_{20}Cu_2Mn_3$ (T).

Figure 8.11 demonstrates isothermal sections of these two systems. The decrease in temperature considerably narrows two- and three-phase regions and expands four-phase regions. Addition of Mn to commercial alloys results in the formation of Mn-containing (Al_6Mn and $Al_{20}Cu_2Mn_3$ (T)) phases in addition to the phases from the relevant ternary systems.

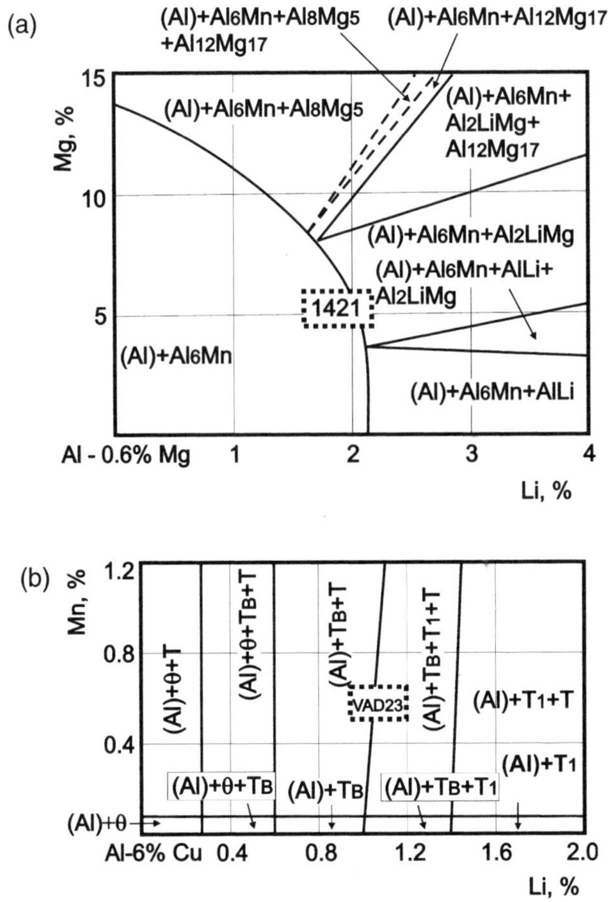

Figure 8.11. Isothermal sections of (a) Al–Li–Mg–Mn (430°C, 0.6% Mn) and (b) Al–Li–Cu–Mn (400°C, 6% Cu) systems (Drits et al., 1977). Compositional ranges of two commercial alloys are shown. T_B – $Al_{7.5}Cu_4Li$, T_1 – Al_2CuLi, and T – $Al_{20}Cu_2Mn_3$.

8.8. Al–Li–Mg–Si PHASE DIAGRAM

Silicon is always present in aluminum alloys as an impurity. However, silicon can be an alloying element in some Al–Li alloys, introduced either for further decrease of alloy density and/or for the improvement of casting properties. Lithium may also be added to Al–Mg–Si wrought alloys for changing hardening behavior or to Al–Si–Mg casting alloys for eutectics modification.

Figure 8.12 shows two isothermal sections of the Al–Li–Mg–Si system at 430 and 180°C. Addition of Si to Al–Li–Mg alloys containing 5–5.5% Mg (1420-type

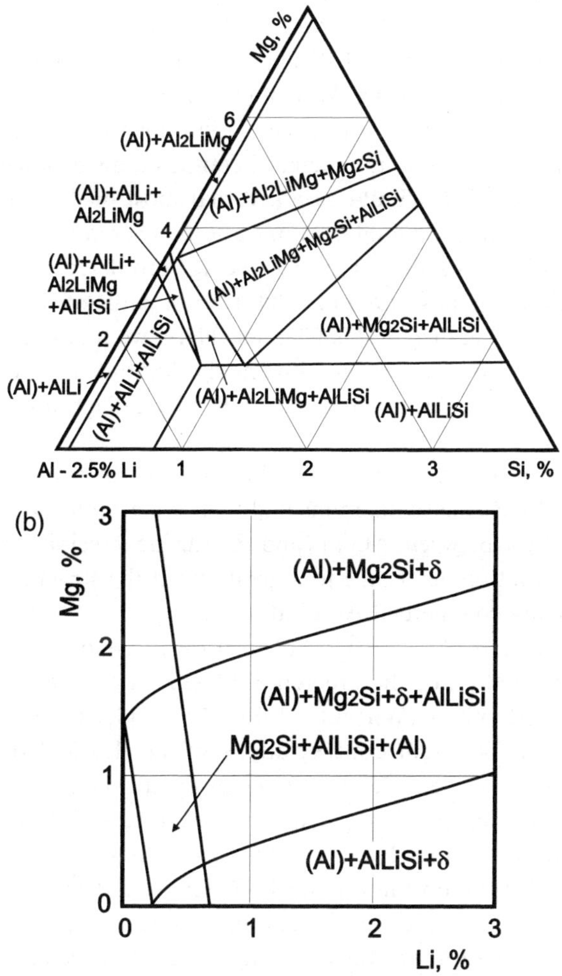

Figure 8.12. Isothermal sections of the Al–Li–Mg–Si system: (a) 430°C, 2.5% Li, experimental (after Rokhlin et al., 1994b) and (b) 180°C, 0.8% Si, calculated (after Chen et al., 2000).

alloys) causes formation of Mg_2Si phase in addition to Al_2LiMg and, possibly, AlLi. In Al–Mg–Si alloys (6XXX series), the effect of lithium depends on the Mg concentration. With increasing amount of Mg, the phase composition changes from (Al) + Mg_2Si + AlLi to (Al) + AlLi + (AlLiSi). The structure of Al–Si casting alloys containing Mg and Li exhibits the (AlLiSi) phase in addition to Si and Mg_2Si.

8.9. WROUGHT ALLOYS CONTAINING LITHIUM

Commercial Al–Li alloys can be conditionally divided into three groups: Al–Li–Cu (2090, 2020, VAD23rus), Al–Li–Mg (1420), and Al–Li–Cu–Mg (2091, 8090, Weldalite049, CP276, 1441rus, 1464rus). Most of Al–Li alloys contain Zr; some – other small additions such as Ag (Weldalite049), Sc (1421rus, 1424rus, 1464rus), or Cd (VAD23rus). The implications of these additions for the phase composition are discussed in this section. Average chemical compositions of some commercial alloys are shown in Table 8.6. It should be noted that most of these alloys are used in or developed for aerospace and military applications. Therefore, the exact and complete chemical composition is seldom reported, being considered confidential.

The equilibrium composition and solidification paths of some commercial Al–Li alloys were discussed earlier in this Chapter. Here, we focus on the phase composition of Al–Li alloys in connection with small additions and heat treatment. Some relevant sections of equilibrium and metastable phase diagrams will be given to assist the discussion.

Before considering multicomponent alloys, it is worth to look at the binary Al–Li and ternary Al–Li–Zr phase diagrams and their metastable variants. It is important because interaction between Al, Li, and Zr during precipitation from the supersaturated solid solution determine to a great extent the structure formation and the properties of many commercial Al–Li alloys.

The decomposition of the Al–Li supersaturated solid solution occurs with the homogeneous formation of the ordered δ' (Al_3Li) phase which later gives place to the equilibrium AlLi phase (Williams, 1981). The δ' phase has a cubic $L1_2$ structure with $a = 0.401$ nm. At usual concentrations of Li (1.5–2.5%), the solvus of the intermediate phase is about 150–250°C (Flower and Gregson, 1987) as shown in Figure 8.13. Very fine, spherical precipitates or clusters of Al_3Li are formed during quenching, and the artificial aging results only in their growth. Rapid lithium diffusion along dislocation lines, grain boundaries, and interfaces with dispersoids (including stable AlLi) causes the formation of precipitation-free zones and discontinuous precipitation of δ'. On further aging the coherent δ' phase transforms to the semicoherent and, finally, incoherent δ(AlLi) phase. The latter appears as plates and frequently forms onto grain boundaries. It is likely that the equilibrium AlLi

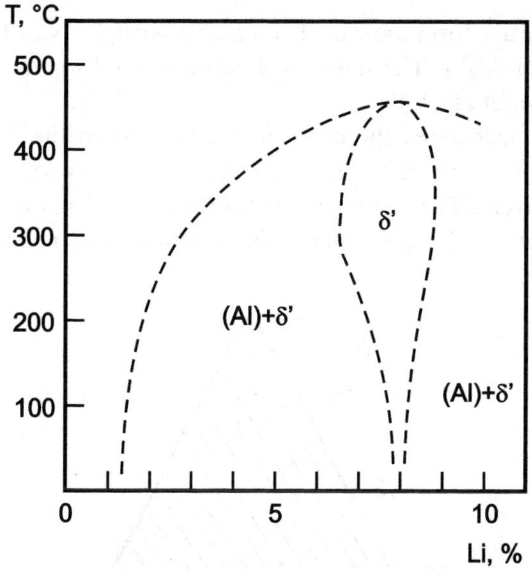

Figure 8.13. Solvus of the metastable δ′ (Al_3Li) phase in the binary Al–Li system (after Flower and Gregson, 1987).

phase nucleates and grows independently of its precursor, at the expense of dissolving δ′.

Al–Li alloys are quite unique with respect to the microstructure formed during aging. The δ′ phase once precipitated retains its coherency even after long aging times. These precipitates are also remarkable for their ability to retain coherency and morphology up to large sizes, 0.3 μm (Williams, 1981). The behavior of the δ′ phase is very similar to that of the Al_3Sc phase (Toropova et al., 1998).

However, the precipitation-free zones and grain-boundary precipitates alongside homogeneously distributed coherent particles inside grain bulk determine the limitation of Al–Li alloys – fast crack propagation and uneven dislocation slip with relevant stress accumulation and cracking.

In commercial alloys, the dislocation slip is homogenized and the precipitation-free zones are reduced by introducing dispersoids (e.g. Al_3Zr) and semicoherent/incoherent precipitates such as T_x, θ′ (Al_2Cu) and S′ (Al_2CuMg). Let us first look at the effect of Zr.

Zirconium is usually added to aluminum alloys in order to stabilize the substructure, thereby preventing the development of recrystallization. In Al–Li alloys, in addition to the retarded recrystallization, the introduction of transition metals, such as Zr, considerably slows down lithium diffusion and coarsening of δ′ particles (Miura et al., 1994). Moreover, the presence in the structure of coherent or semi-coherent Al_3Zr (cubic, $L1_2$) particles provides places for preferential nucleation of the δ′ phase. The latter forms so-called composite particles with the core of Al_3TM and the envelope of Al_3Li. The same mechanism is valid for Al–Li–Sc and Al–Li–Sc–Zr alloys (Miura et al., 1994).

Saunders (1989) calculated the metastable equilibria in the Al–Li–Zr system at high and low temperatures reflecting the conditions of homogenization annealing and aging, respectively. The results are demonstrated in Figure 8.14. During high-temperature annealing, the aluminum solid solution is in metastable equilibrium

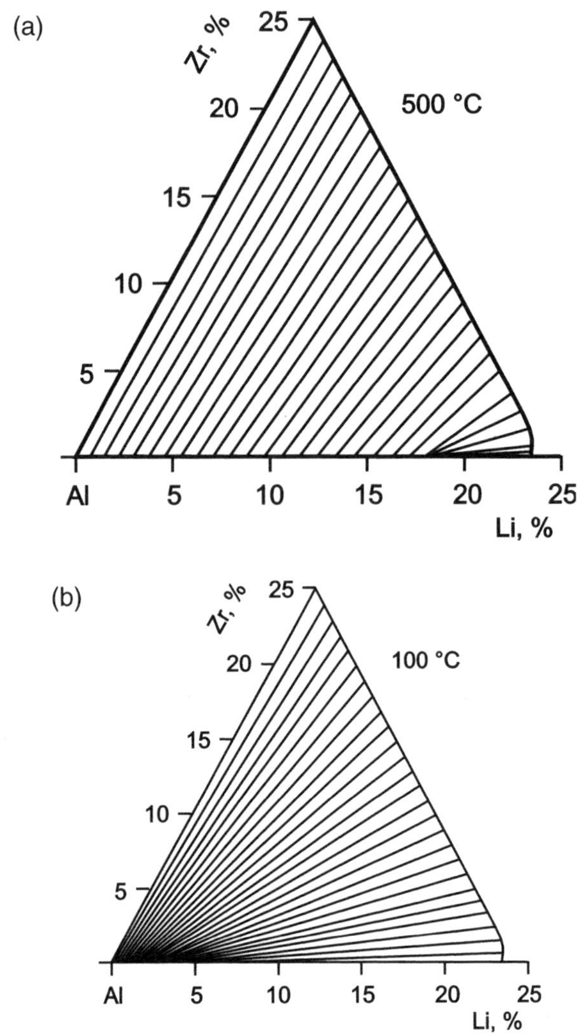

Figure 8.14. Tie lines for metastable phase boundaries between (Al) and $L1_2$ phases (Al_3Zr and Al_3Li, respectively) at 500°C (a) and 100°C (b) (after Saunders, 1989).

with cubic Al_3Zr that may dissolve substantial amount of Li, up to $Al_3Li_{0.4}Zr_{0.6}$. On decreasing the temperature, the metastable equilibrium turns to the metastable Al_3Li phase that contains almost no Zr. These calculations prove the experimentally observed fact that metastable Al_3Zr particles formed during solution treatment act as nucleation sites for the metastable Al_3Li phase, both phases being isomorphic.

Al–Cu–Li commercial alloys. Equilibrium and metastable phase compositions of commercial Al–Cu–Li alloys depend on the Cu:Li ratio. On increasing the ratio, the equilibrium phase composition shifts from $(Al)+T_1+T_2$ in the 2090 alloy to $(Al)+T_1+T_B$ in the VAD23 alloy. Zirconium in 2090 facilitates the uniform distribution of hardening phases and forms metastable Al_3Zr particles. In VAD23, manganese forms Al_6Mn and T (AlCuMn) particles and cadmium refines the hardening precipitates of θ′ (Al_2Cu).

The distribution of metastable phase fields in the Al–Cu–Li system is suggested by Riola and Ludwiczak (1986) and depicted in Figure 8.15.

The VAD23 alloy falls into the equilibrium phase field 3 and the 2090 alloy – in the equilibrium field 2 in Figure 8.15. The phase compositions reflecting the metastable equilibrium at the temperatures of aging are different. The hardening δ′ phase is always present at the hardening stage of precipitation in alloys with the compositions within phase fields 1 and 2 in Figure 8.15.

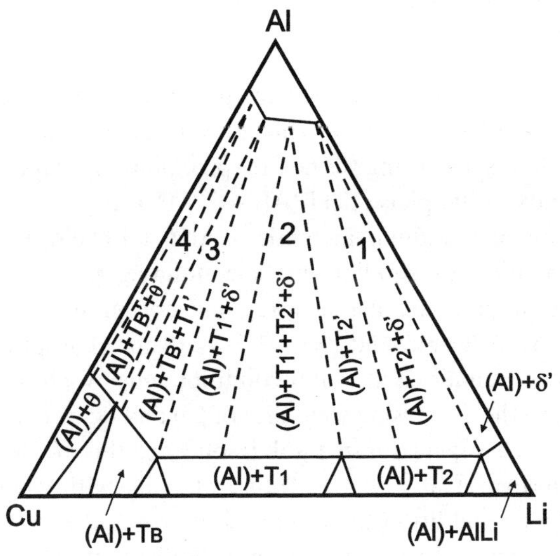

Figure 8.15. Equilibrium (solid) and metastable (dashed) phase fields in the Al–Cu–Li system (after Riola and Ludwiczak, 1986). Equilibrium three-phase fields at a low temperature are marked as (1) $(Al)+T_2+AlLi$ (δ); (2) $(Al)+T_1+T_2$; (3) $(Al)+T_1+T_B$; and (4) $(Al)+T_B+Al_2Cu$ (θ). T_B – $Al_{7.5}Cu_4Li$, T_1 – Al_2CuLi, and T_2 – Al_6CuLi_3.

Huang and Ardell (1986) demonstrated that the equilibrium phase diagram cannot predict the phase composition after aging. They studied two alloys, containing, respectively, 2.3% Li; 2.85% Cu (2090-type) and 2.90% Li; 0.99% Cu. Both alloys also contained 0.12% Zr. According to the equilibrium phase diagram, the first alloy has to contain T_1 and T_2 phases (phase field 2 in Figure 8.15) whereas the second one, δ and T_2 (phase field 2 in Figure 8.15). However, the TEM study revealed that both alloys contained after aging δ′ and T_1, and the first alloy in addition, θ′ (Huang and Ardell, 1986) or T_2' (Riola and Ludwiczak, 1986). The latter phase(s) frequently serves as a substrate for δ′ and is later transformed into stable T_2. The θ′ (T_2') phase is observed up to the peak hardness range, afterwards it dissolves giving place to the T_B and T_2 phases. The coherent δ′ phase dissolves and disappears completely upon long aging or at temperatures above 260°C. Precipitates of δ′ are spherical in shape and form composite particles with cubic Al_3Zr and tetragonal θ′. Particles of T_1 and θ′ phases are platelike in shape and may form composite particles with cubic Al_3Zr in high-copper alloys.

To summarize, the decomposition of the supersaturated solid solution in Al–Cu–Li alloys occurs as follows:

1. (2090-type alloys) $(Al)_{ss} \rightarrow (Al) + \delta'$ $(Al_3Li) + T_1 + \theta'$ $(T_2') \rightarrow (Al) + T_1$ $(Al_2CuLi) +$ T_2 (Al_6CuLi_2);
2. (VAD23-, 2020-type alloys) $(Al)_{ss} \rightarrow (Al) + \theta' + T_1 \rightarrow (Al) + T_B$ $(Al_{7.5}Cu_4Li) +$ T_1 (Al_2CuLi).

Figure 8.16 demonstrates a time–temperature diagram for phase distribution in a 1450 (Al–Li–Cu–Zr) Russian alloy. This alloy belongs to the same group as the 2090 alloys and behaves according to the first precipitation sequence. The maximum strength corresponds to the phase field $(Al) + \delta' + \theta' + T_1$.

The introduction of scandium in modern Al–Cu–Li alloys (1451rus, 1461rus) results in the formation of equilibrium Sc-containing phases, i.e. Al_3Sc and W $(Al_{6.5}Cu_{5.5}Sc)$, the latter phase has a tetragonal crystal structure with the lattice parameters $a = 0.855$–0.866 nm and $c = 0.505$–0.510 nm (Toropova et al.,1998) (see Chapter 9). However, under equilibrium conditions the W phase is formed only in the solid state due to the decreasing solubility of Cu and Sc in solid aluminum. Upon precipitation from the supersaturated solid solution, this phase is replaced by θ′ (Al_2Cu) (Kharakterova et al., 1994). Mutual alloying with Zr and Sc causes the appearance of Al_3Zr in addition to Al_3Sc and improves the structure of Al–Li–Cu alloys by strong grain refinement during solidification, sharp increase in the recrystallization temperature, and homogenizing of precipitation during aging.

Figure 8.17 gives a polythermal section of the Al–Li–Cu–Sc phase diagram at 1% Li and 0.6% Sc. This polythermal section reflects, apparently, nonequilibrium

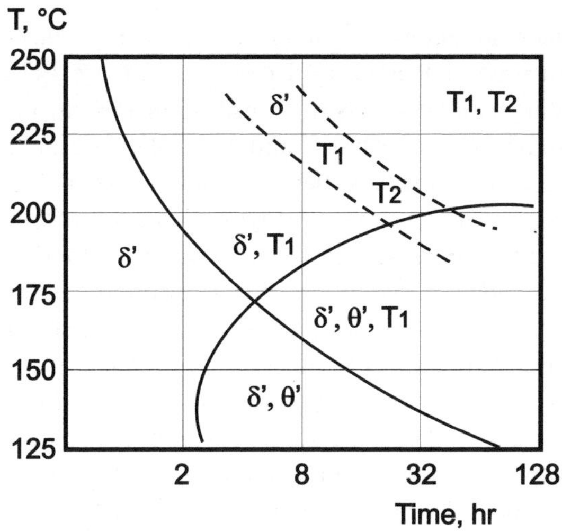

Figure 8.16. Time–temperature–transformation diagram for a 1450 (Al–Li–Cu–Zr) sheet alloy (after Davydov et al., 1996).

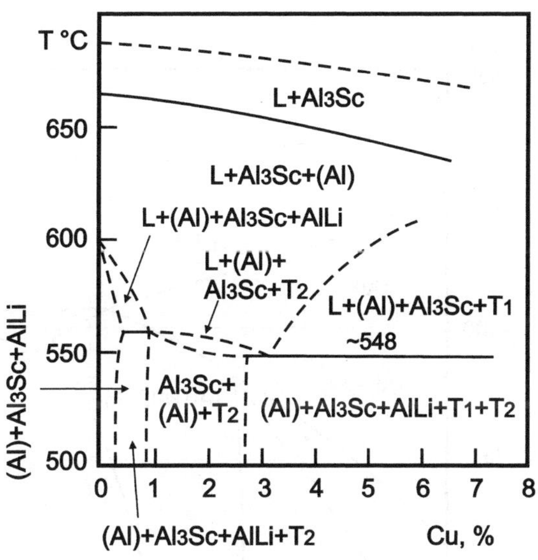

Figure 8.17. Polythermal section of the Al–Li–Cu–Sc phase diagram at 1% Li and 0.6% Sc (after Fridlyander et al., 2001).

conditions as it shows a five-phase field in the four-component system. The phase composition of commercial Al–Li–Cu–Sc–Zr alloys is (Al) + AlLi + Al_3Sc + Al_3Zr (tetragonal) + T_1 + T_2. The set of phases precipitating from the supersaturated solid solution appears to be as follows: θ′, δ′, T_1, Al_3Sc, Al_3Zr (cubic).

Al–Li–Mg commercial alloys. Commercial Al–Li–Mg alloys (without copper) were developed and used in Russia. In Table 8.6 these are 142X alloys. All alloys of this group contain small additions of transition metals such as Zr, Mn, and Sc.

The equilibrium phase composition and the solidification path of 142X-type alloys can be evaluated using polythermal and isothermal sections of ternary and quaternary phase diagrams shown in Figures 8.4 and 8.18. In ternary alloys, the equilibrium structure (at 4–6% Mg, 1.5–3% Li) consists of (Al) grains with secondary particles of AlLi and Al_2LiMg phases formed during cooling in the solid state. Upon decomposition of a supersaturated solid solution, δ′ (Al_3Li) phase forms coherent, hardening precipitates. The precipitation sequence in Al–Li–Mg alloys is as follows – supersaturated solid solution, δ′ (Al_3Li); Al_2LiMg. The only effect of magnesium at early stages of decomposition is the reduced solid solubility of lithium. Therefore, the precipitation density of δ′ increases. Later, magnesium and lithium form the ternary compound which is incoherent and nucleates on grain boundaries or dislocation networks during quenching or overaging.

Additional alloying with Zr and Sc results in more complex solidification behavior with primary solidification of Al_3Zr and Al_3Sc phases. There are indications that

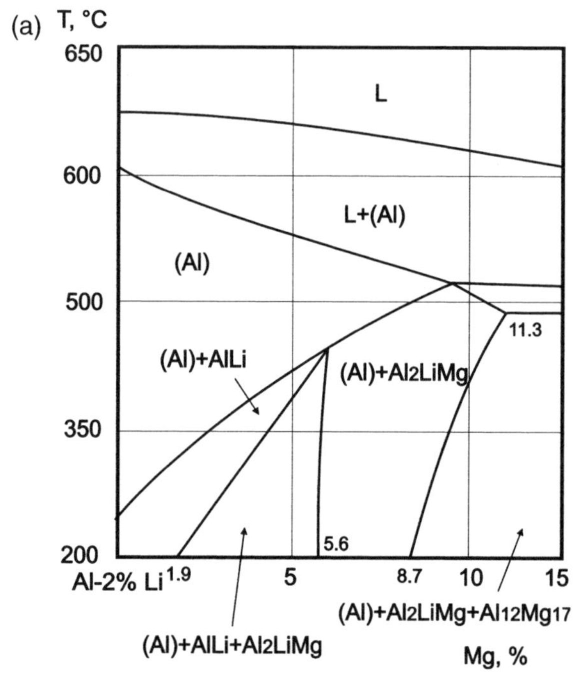

Figure 8.18. (a) Polythermal sections of Al–Li–Mg phase diagram at 2% Li; (b) Al–Li–Mg–Zr phase diagram at 4.5% Mg and 0.2% Zr (after Fridlyander et al., 2001); and (c) isothermal section of Al–Li–Mg–Sc phase diagram at 0.2% Sc and 400°C (after Toropova et al., 1998).

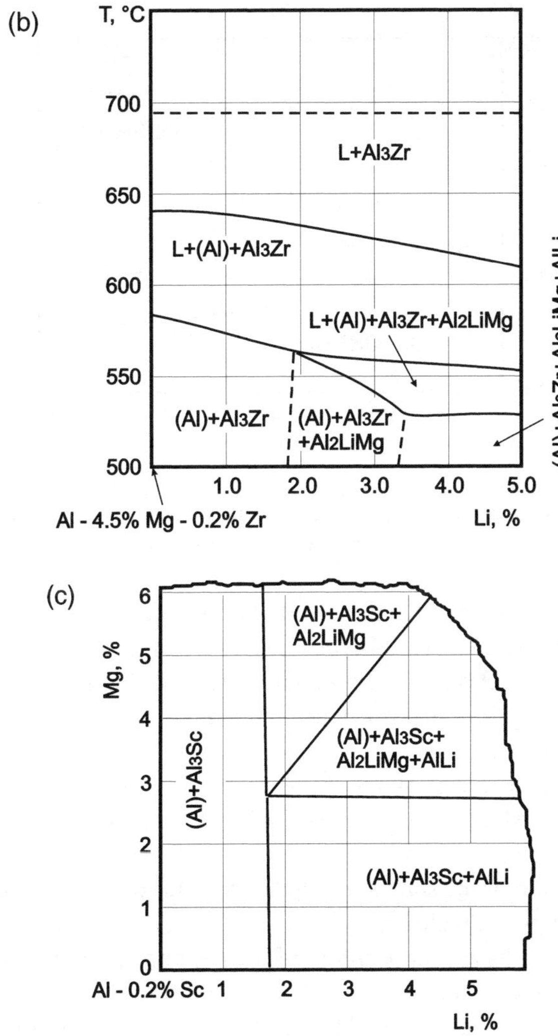

Figure 8.18 *(continued)*

Al_2LiMg phase may form peritectically at high Li concentrations (Fridlyander et al., 2001). The (nonequilibrium) solidus of Al–Li–Mg–Zr alloys is 530°C. Scandium and zirconium enter the aluminum solid solution during solidification and then precipitate upon high-temperature annealing to form coherent particles of stable Al_3Sc and metastable Al_3Zr phases. These particles facilitate homogeneous precipitation of δ' phase according to the mechanism of composite–particle formation discussed previously.

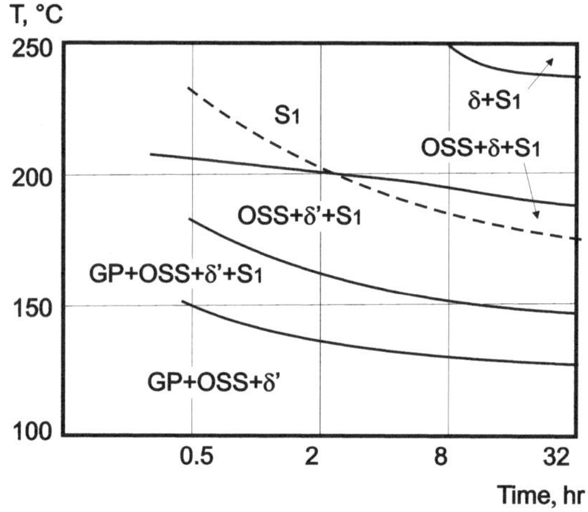

Figure 8.19. Temperature–time–transformation diagram for a 1424 alloy (Al–Li–Mg) in the range of precipitation hardening (after Davydov et al., 2000a). GP, Guinier–Preston zones; OSS, ordered solid solution; δ′ – Al_3Li; S_1 – Al_2LiMg.

A temperature–time–transformation diagram for the precipitation-hardening range of an Al–Li–Mg alloy (1424-type) is shown in Figure 8.19. This diagram can be used for the correct choice of aging regime. The maximum hardening is observed after aging at 150–175°C, 16–32 h with coherent δ′ and relatively fine incoherent Al_2LiMg precipitates present in the structure.

Al–Li–Cu–Mg commercial alloys. Magnesium and copper are the most widely used additions to Al–Li alloys. They improve strength by solid-solution and precipitation hardening and minimize the formation of precipitation-free zones during decomposition of the supersaturated solid solution, thereby increasing fatigue endurance and fracture toughness.

The equilibrium phase composition of this group of alloys was discussed in Section 8.6. The as-cast and annealed alloys contain the following phases: (Al), T_1 (Al_2CuLi), T_2 (Al_6CuLi_3), and S (Al_2CuMg). Figure 8.20 demonstrates the polythermal section of the quaternary phase diagram at 3% Cu and 2.5% Li, which is relevant for the analysis of such commercial alloys as CP276, 2091, 8091 (up to 3% Mg). One can easily see that the solidification for these alloys starts with the formation of the aluminum solid solution, then T_2 and S phases are formed through eutectic reactions. The T_1 phase is formed by precipitation from the solid solution.

The precipitation in Al–Li–Cu–Mg alloys depends on the ratio of all three elements (Tiryakioglu and Staley, 2003). On increasing the Cu:Li ratio, the set of main precipitating phases changes from T_1 to $T_1 + \delta'$ and then to $T_1 + S'$ (Al_2CuMg) + θ′ (Al_2Cu). Magnesium additions to Al–Li–Cu alloys favor the precipitation of θ′ and S′ phases with the amount of the latter phase increasing with the Mg

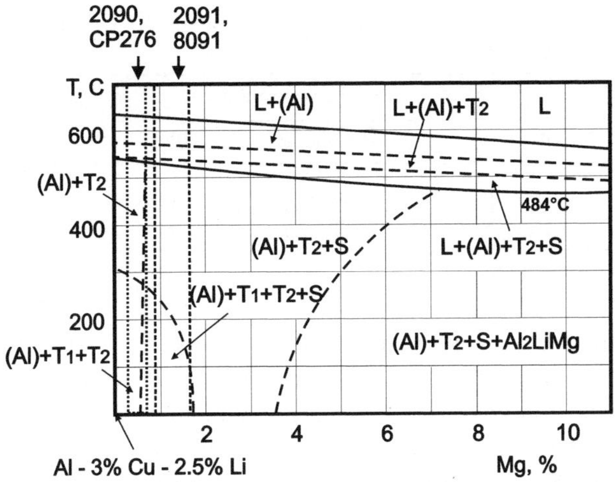

Figure 8.20. Polythermal section of the Al–Li–Cu–Mg phase diagram at 3% Cu and 2.5% Li (after Fridlyander et al., 1993). T_1 – Al_2CuLi, T_2 – Al_6CuLi_3, and S – Al_2CuMg.

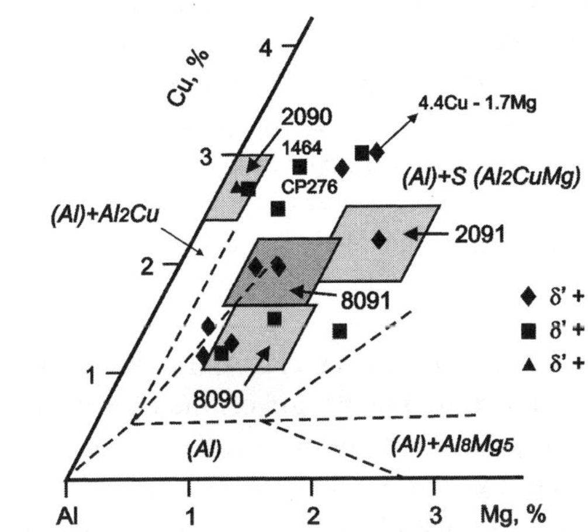

Figure 8.21. Metastable phase composition of aged commercial Al–Li–Cu–Mg alloys containing 2–3% Li superimposed on the isothermal section of the Al–Cu–Mg phase diagram at 190°C (after Flower and Gregson, 1987). Equilibrium phase fields for Al–Cu–Mg alloys are shown in italics.

concentration. The Al_2LiMg phase appears when the alloys contain more than 2% Mg. The occurrence of δ′ is suppressed in the alloys containing more than 4.5% Cu and 1% Li (Weldalite049). Even small amounts of Mg in Weldalite-type alloys may result in the formation of the S′ phase during precipitation (Lee et al., 1999). The effect of Cu:Mg ratio on the metastable phase composition of Al–Cu–Mg–(2–3)%Li alloys is illustrated in Figure 8.21.

Table 8.7. Effect of lithium concentration on the phase composition of an annealed Al–4%Cu–0.3% Mg–0.4%Ag alloy (after Lee, 1998)

	Phase composition			
Li, %	0	0.5	0.8	1.0
Peak hardness	Ω, θ′, S′	Ω, θ′, S′	T_1, θ′, S′	T_1, θ′, S′
Overaging	Ω, θ′, S	Ω, T_2, T_B	T_2, T_B, T_1, R	T_1, T_2, T_B, R

The hardening is associated with δ′ (which precipitates first) and homogeneously or heterogeneously nucleated S′. The precipitation of θ′ phase is likely to occur at low Mg concentrations, e.g. in 2090 and Weldalite049 alloys. The S′ particles decorate subboundaries and prevent or delay the development of dynamic recovery, thus improving high-temperature stability of the alloy up to 250°C. The T_1 phase nucleates heterogeneously on dislocations and subgrain boundaries, forming laths. The overall structure provides for more uniform precipitations and favors cross-slip of dislocations. In addition to these phases formed in the grain bulk, the precipitation of the icosahedral T_2 phase occurs at grain boundaries in the peak-age regime. The effect of zirconium and scandium is similar to that in other Al–Li alloys.

Some of the commercial Al–Li–Cu–Mg alloys contain silver that is also under scrutiny as a promising alloying element in Al–Cu and Al–Cu–Mg alloys. Lee (1998) traced the influence of increasing Li concentration on the phase composition of an Al–4% Cu–0.3%Mg–0.4%Ag alloy annealed after quenching. The results are shown in Table 8.7. The difference of phase compositions at peak hardness and in overaged alloys is obvious.

However, the addition of silver in commercial Al–Li alloys does not change the equilibrium phase composition and arguably affects the metastable phase composition during aging. There are few reports on the occurrence of additional phases in Ag-containing Al–Li alloys, e.g. Lee et al. (1999), Chen et al. (2004), but most authors agree that there are no new phases formed in the concentration range of commercial alloys (Ω is a metastable phase based on Al_2Cu). At the same time, it is well acknowledged that silver considerably increases the hardening effect in Al–Li–Cu–Mg alloys by influencing nucleation and distribution of hardening δ′, θ′, and S′ precipitates.

In the conclusion of this chapter, it is worth to note one more time the limitations of equilibrium phase diagrams. The phase diagrams are very useful for the analysis of solidification (though with reservations regarding incomplete phase transformations and changing positions of phase fields on increasing cooling rate) but should be used with extreme caution when it comes to the prediction of metastable phase selection after decomposition of a supersaturated solid solution.

Chapter 9

Alloys with Transition Metals

Transition metals (TM) are traditional additions to commercial aluminum alloys. Manganese, chromium, and titanium (frequently together with boron and carbon) were used as minor alloying elements for decades, serving as grain refiners and anti-recrystallizing agents. Starting from the 1970s, zirconium attracted attention as a powerful anti-recrystallizer, especially in high-strength alloys. Later on, zirconium and scandium joint additions proved to be very efficient for both the grain refining and the recrystallization control. Some transition metals such as nickel and iron are used as major alloying additions to aluminum alloys in order to improve elevated-temperature properties (these alloys are considered in more detail in Chapter 7). Manganese is a main alloying element in 3XXX series alloys (see Chapter 1). In recent years, transition metals are used in rapidly solidified aluminum materials where the formation of supersaturated solid solutions and fine as-cast structures produces new qualities of the material. Amorphous and quasicrystalline alloy is yet another and most recent application of transition metals in aluminum-based materials.

In this chapter, we first consider some phase diagrams of aluminum with transition metals, including scandium and rare-earth elements. After that, general features of stable and metastable interaction between aluminum and transition metals alongside some metastable phase diagrams are discussed. Finally, some advanced alloys are considered.

9.1. PHASE DIAGRAMS OF SOME Al-BASED SYSTEMS WITH TRANSITION METALS

This part is dedicated to equilibrium and non-equilibrium phase diagrams of some Al-based systems relevant to commercial and emerging alloys produced by conventional or rapid solidification routes. First, phase diagrams of alloys containing titanium, scandium, and zirconium are discussed. These systems are important for grain refining and recrystallization control. Then, promising alloying systems for conventional, rapidly solidified, and amorphous materials are considered.

9.1.1. Phase diagrams of Al–Ti–B and Al–Ti–C systems

Two alloying systems, namely Al–Ti–B and Al–Ti–C, are very important for understanding the grain refining effect of Ti, Ti + B, and Ti + C master alloys on commercial aluminum alloys.

The Al–Ti–B phase diagram has been a subject of investigation since the early 1970s. Under equilibrium solidification conditions, the following phases can be found together with aluminum in the solid state: Al_3Ti, AlB_2, and TiB_2. The Al_3Ti phase has a tetragonal crystal structure (space group *I4/mmm*) with $a = 0.385$ nm and $c = 0.86$ nm (Mondolfo, 1976). The open question is the existence of continuous solid solution between AlB_2 and TiB_2. Both compounds have a hexagonal crystal structure (space group *P6/mmm*) with close lattice parameters: $a = 0.3003$ nm and $c = 0.3251$ nm for AlB_2 and $a = 0.3032$ nm and $c = 0.3231$ nm for TiB_2 (Fjellstedt et al., 1999). Thermodynamical calculations show the possible formation of $(AlTi)B_2$ solid solution. However, numerous experimental results attest for the formation of separate, well-distinguished borides.

The liquid–solid equilibria in Al-rich Al–Ti–B alloys was studied in detail (Abdel-Hamid and Durnad, 1985). The invariant and monovariant solidification reactions are shown in Table 9.1 and the schematic projection of the solidification surface is given in Figure 9.1.

Alloys in compositional ranges 1 and 2 (Figure 9.1) start solidification with the formation of TiB_2, then Al_3Ti and TiB_2 are formed through the eutectic reaction ($S–P_1$). Small TiB_2 particles are observed inside bulk Al_3Ti crystals. The equilibrium crystallization finishes with the invariant reaction P_1 when the aluminum solid solution is formed. Under nonequilibrium conditions, the peritectic reaction P_1 is unfinished and the remaining liquid solidifies as divorced eutectics along line $P_1–P_2$. The described solidification path is typical of Al–Ti–B master alloys.

In real casting situation, rims of Al_3Ti onto TiB_2 and rims of TiB_2 onto AlB_2 are frequently observed suggesting complex incomplete solidification reactions and possible mechanisms of grain refinement (Fjellstedt et al., 1999). The aluminum solid solution nucleates on Al_3Ti that is dispersed in the melt by primary solidified borides, and much less titanium is required for the same refining effect.

Table 9.1. Invariant and monovariant reactions in the aluminum-rich Al–Ti–B alloys (Abdel-Hamid and Durnad, 1985)

Reaction	*T*, °C	Point/line in Figure 9.1
$L + Al_3Ti \Rightarrow (Al)$	665	p_1
$L + Al_3Ti \Rightarrow (Al) + TiB_2$	~665	P_1
$L \Rightarrow (Al) + TiB_2 + AlB_2$	659	e_1
$L + TiB_2 \Rightarrow (Al) + AlB_2$	~659	P_2
$L \Rightarrow TiB_2 + Al_3Ti$	T > 665	$S–P_1$
$L + Al_3Ti \Rightarrow (Al)$	–	$p_1–P_1$
$L \Rightarrow TiB_2 + AlB_2$	T > 659	$F–P_2$
$L \Rightarrow (Al) + AlB_2$	–	$P_2–e_1$
$L + TiB_2 + (Al)$ either peritectic or eutectic	659 < T < 665	$P_1–P_2$

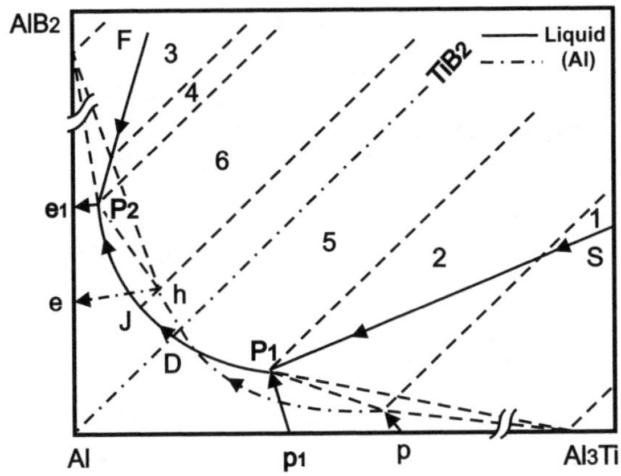

Figure 9.1. Schematic phase diagram of the Al–Ti–B system (after Abdel-Hamid and Durnad, 1985).

In compositional ranges 3 and 4, the primary phase is TiB_2. The solidification continues along line F–P_2 (formation of $TiB_2 + AlB_2$ eutectics) to point P_2 where the invariant peritectic reaction occurs with the formation of AlB_2 and (Al). Under real casting conditions, some liquid may remain at temperatures below the temperature of the peritectic reaction. This liquid solidifies according to line P_2–e_1 forming the (Al) + AlB_2 eutectics.

Alloys with intermediate compositions 5 and 6 also start to solidify with the formation of TiB_2 as the primary phase. Then the aluminum solid solution is formed as a result of the peritectic reaction $L + TiB_2 \Rightarrow (Al)$ (P_1–P_2), this reaction may transform to the eutectic reaction $L \Rightarrow (Al) + TiB_2$ until all liquid vanishes at point P_2. In compositional range 6, the peritectic reaction $L + TiB_2 \Rightarrow (Al) + AlB_2$ is likely to occur.

The alloys with the Ti:B ratio less or equal to unity are not considered here as they do not possess the grain refining ability. Relevant information can be found elsewhere (Zupanič et al., 1998; Fjellstedt and Jarfors, 2001).

Another system that is important for grain refining is Al–Ti–C. In the aluminum corner of this system three phases are in equilibrium with (Al): Al_3Ti, Al_4C_3, and TiC_x ($0.48 < x < 0.98$). The Al_4C_3 carbide (25.3% C) has a rhombohedral structure with $a = 0.855$ nm and $\beta = 22°28'$ (Mondolfo, 1976) or a hexagonal structure with $a = 0.33328$ nm and $c = 2.5026$ nm (Villars and Calvert, 1985). And TiC_x has a cubic structure (space group *Fm*3*m*) with $a = 0.43176$ (Villars and Calvert, 1985).

The invariant equilibrium $L + TiC \Rightarrow Al_3Ti + Al_4C_3$ occurs in the aluminum corner of the phase diagram at 693°C (0.53 at.% Ti, 7×10^{-6} at.% C) (Frage et al., 1998), other temperatures of 700 or 812°C are also reported for the same equilibrium.

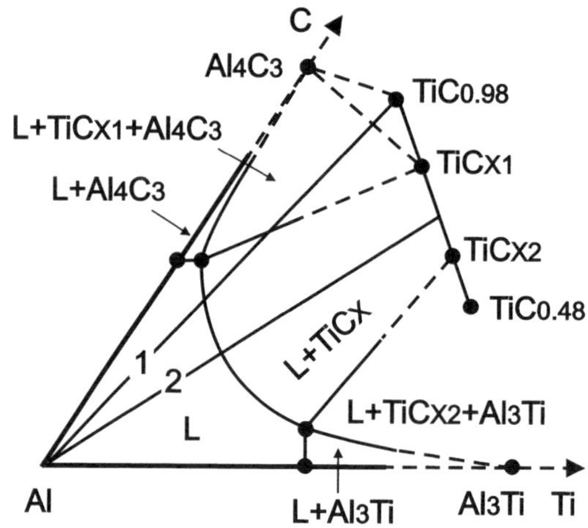

Figure 9.2. Schematic isothermal section of the Al–Ti–C system relevant to 680–1130°C (after Frage et al., 1998).

In the structure of solidified samples, the association of Al_3Ti with TiC and TiC with Al_4C_3 is frequently observed, suggesting the complex character of solidification reactions in this system (Svendsen and Jarfors, 1993). However, the full sequence of solidification is not clear yet. A schematic isothermal cross section is shown in Figure 9.2. Line 1 connects Al with the stoichiometric composition of TiC, whereas line 2 reflects the equilibrium with the non-stoichiometric titanium carbide. The latter line does not cross phase fields with Al_4C_3 present and, therefore the Al–TiC_x (off-stoichiometric) system can be considered as quasi-binary (Frage et al., 1998).

9.1.2. Phase diagrams of alloys containing Sc and Zr

Aluminum alloys containing scandium and zirconium use several mechanisms to control structure and properties. Primary particles of aluminides act as nucleants for the aluminum solid solution during solidification. A considerable amount of scandium and zirconium is retained in the solid solution after solidification. This supersaturated solid solution decomposes at relatively high temperatures with the formation of coherent and semi-coherent particles that can either harden the alloy (this effect is used in Al–Mg alloys) or retard recrystallization (this is used in heat-treatable aluminum alloys).

Phase compositions of Al–Li–Zr, Al–Li–Mg–Sc–(Zr) and Al–Li–Cu–Sc alloys are discussed in Chapter 8. Here, we focus on other alloying systems.

Al–Sc–Zr phase diagram. In the aluminum corner of this system only binary Al_3Sc and Al_3Zr phases are in equilibrium with (Al). The Al_3Sc phase is formed at 1320°C during a peritectic reaction, has a cubic ordered structure of $L1_2$ type with $a = 0.4104$ nm and can dissolve Zr up to the composition $Al_3Sc_{0.6}Zr_{0.4}$ (35% Zr) (Toropova et al., 1998). The Al_3Zr phase melts congruently at 1577°C, has a tetragonal structure of $D0_{23}$ type with $a = 0.4006$–0.4014 nm and $c = 1.727$–1.732 nm, and dissolves Sc up to the composition $Al_3Zr_{0.8}Sc_{0.2}$ (5% Sc) (Toropova et al., 1998; Villars and Calvert, 1985). These phases participate in the invariant reaction $L + Al_3Zr \Rightarrow (Al) + Al_3Sc$, at 659°C. Mutual equilibrium solubility of Zr and Sc in solid (Al) is 0.06% Zr, 0.03% Sc, and 0.09% Zr, 0.06% Sc at 550 and 600°C, respectively (Toropova et al., 1998).

The typical concentration of scandium and zirconium in commercial aluminum alloys is lower than 0.3% Sc and 0.15% Zr or <0.45% in total (Davydov et al., 2000b). Figure 9.3 shows the polythermal and isothermal sections of the Al–Sc–Zr phase diagram (Toropova et al., 1998). Depending on the Zr:Sc ratio, either Al_3Sc (Zr:Sc < 1) or Al_3Zr (Zr:Sc > 1) solidifies as the primary phase. The efficiency of Sc as a grain refiner is much improved in the presence of Zr. The reason for that is under discussion. Some authors suggest the formation of a ternary $Al_3(ScZr)$ phase with the crystal structure similar to that of stable Al_3Sc and metastable Al_3Zr. However, there is no evidence in favor of the formation of a new phase in the Al–Sc–Zr system. Metallographic examination of primary particles in Al–Sc–Zr alloys show that Al_3Sc phase (as a result of the peritectic reaction) forms a rim on primary Al_3Zr particles (Figure 9.4). This surface layer possesses very good refining ability of Al_3Sc and "activates" Al_3Zr, allowing strong grain refinement at relatively low Sc concentrations.

Al–Cu–Sc phase diagram. The aluminum solid solution in the Al–Cu–Sc system can be in equilibrium with the Al_2Cu and Al_3Sc phases from the constituent systems and the ternary W phase. The ternary phase exists in the range of compositions and has the following formula $Al_{5-8}Cu_{7-4}Sc$ with the average composition 31.8% Al, 58.9% Cu, and 9.3% Sc (Toropova et al., 1998). This phase has a tetragonal crystal structure (space group *4mmm*) with lattice parameters $a = 0.8546$–0.8621 nm and $c = 0.5036$–0.5091 nm (Toropova et al., 1998).

In the aluminum corner of the system, two invariant reactions take place:

$L + Al_3Sc \Rightarrow (Al) + W$ at 572°C, 25% Cu, 0.22% Sc (point P in Figure 9.5a);

$L \Rightarrow (Al) + Al_2Cu + W$ at 546°C, 31.2% Cu and 0.07% Sc (point E in Figure 9.5a).

Figure 9.5 shows the projection of the solidification surface, and isothermal and polythermal sections of the Al–Cu–Sc system. The narrow solidification range of primary (Al) is very close to the Al–Cu side. Even small additions of Sc cause

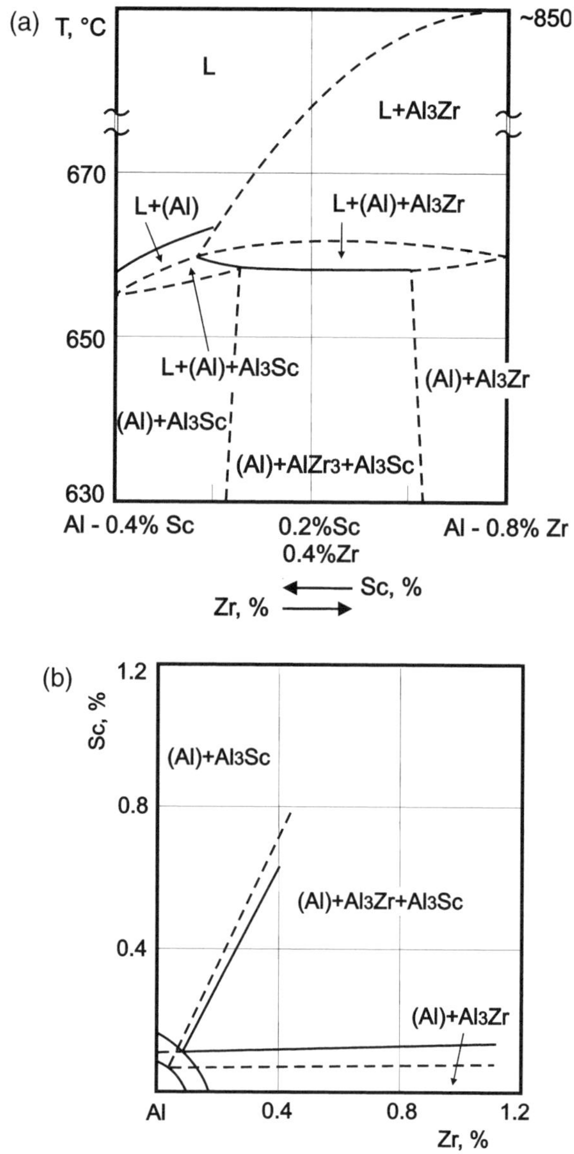

Figure 9.3. Section of the Al–Sc–Zr phase diagram: (a) polythermal section from Al–0.4% Sc to Al–0.8% Zr and (b) isothermal sections at 550°C (dashed lines) and 600°C (solid lines) (after Toropova et al., 1998).

the formation of primary Al_3Sc or W crystals. Norman et al. (1998) observed the formation of Al_3Sc (alongside Al_2Cu and (Al)) in an Al–4.5%Cu–0.8%Sc alloy, with a very strong grain refining. No W phase was found in this alloy under non-equilibrium solidification conditions. Alloying of Al–Sc alloys with as little as 1%

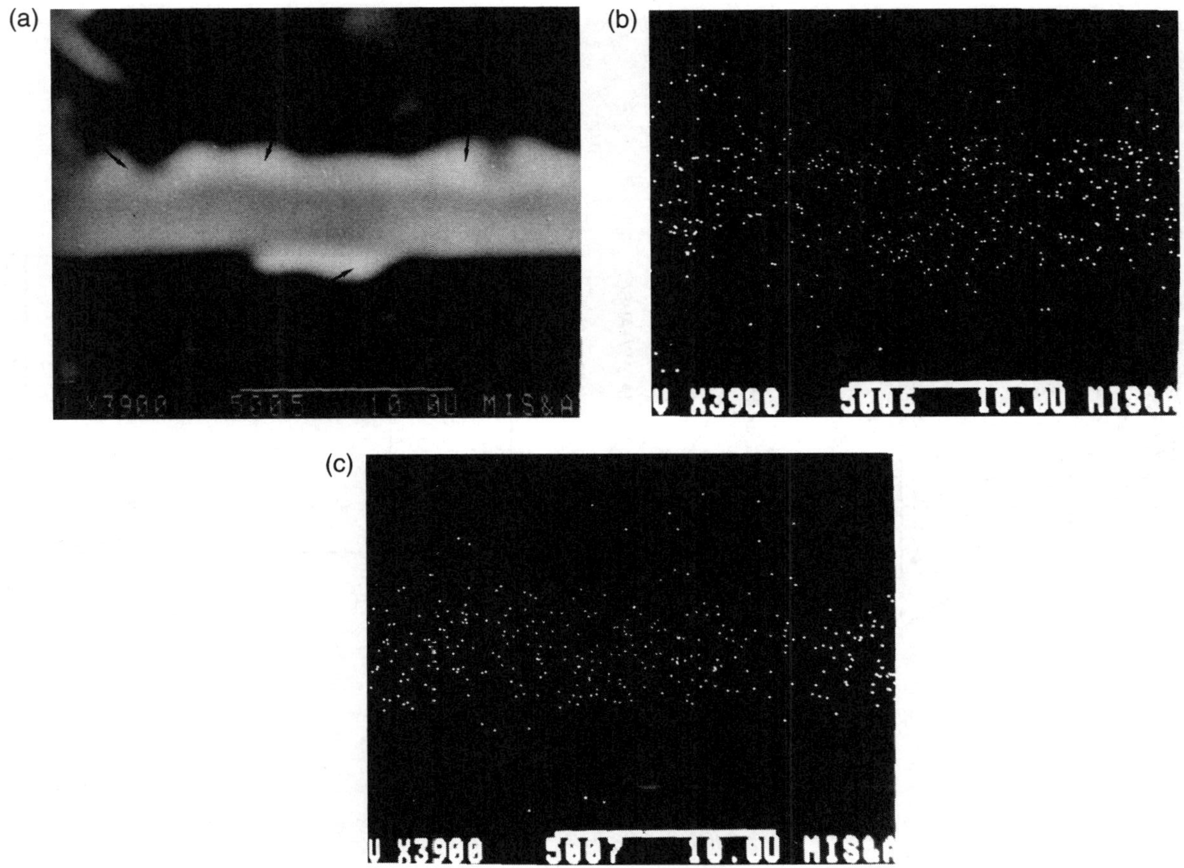

Figure 9.4. Precipitation of Sc-rich phase onto primary Zr-rich particles during solidification: (a) secondary-electron SEM image, arrows point at the surface Sc-rich layer; (b) Sc characteristic radiation; and (c) Zr characteristic radiation.

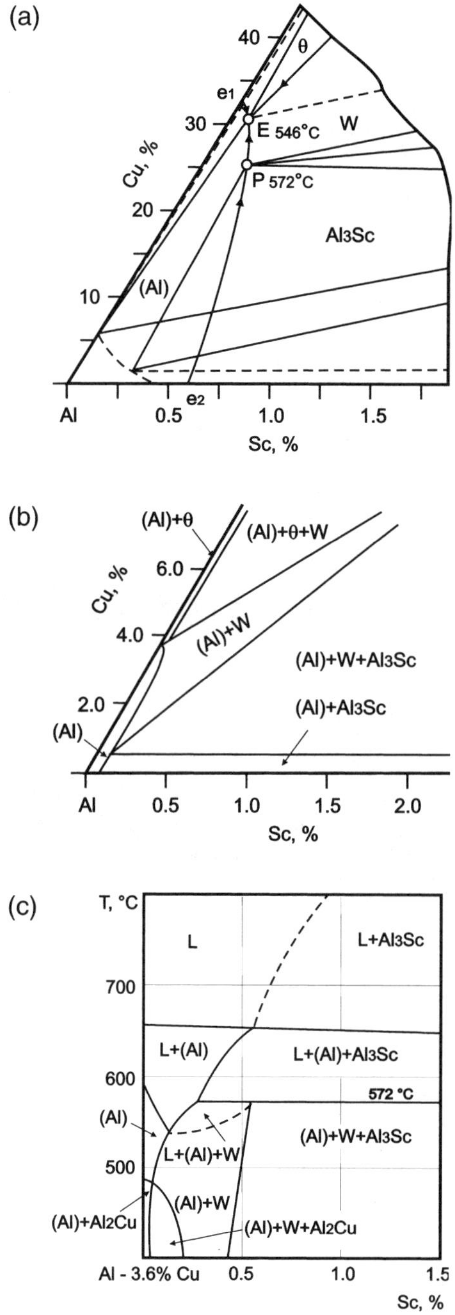

Figure 9.5. Projection of the solidification surface (a); isothermal section at 500°C (b) (after Toropova et al., 1998); and polythermal section at 3.6% Cu (after Rokhlin et al., 1998) (c) of the Al–Cu–Sc system.

Cu results in the solidification of the ternary eutectics and significant decrease in the solidus temperature.

Line e_1E in Figure 9.5a represents the formation of the binary (Al) + Al_2Cu eutectics and runs from 548°C to 546°C separating fields of primary solidification of (Al) and Al_2Cu phases. Line e_2P reflects the eutectic reaction $L \Rightarrow (Al) + Al_3Sc$ and runs from 655°C to 572°C, bordering fields of primary solidification of (Al) and Al_3Sc. The primary solidification fields of (Al) and W are separated by line EP of the formation of the binary eutectics (Al) + W, temperature decreasing from 572°C to 546°C.

Copper slightly dissolves in Al_3Sc, but Sc does not dissolve in Al_2Cu. The solid solubility of Sc and Cu in (Al) is shown in the following table. Alloying with copper has virtually no effect on the solubility of Sc in (Al), whereas scandium somewhat decreases the solubility of copper in solid aluminum.

Phase field (Figure 9.5b)	450°C		500°C	
	Sc, %	Cu, %	Sc, %	Cu, %
(Al) $\Longleftrightarrow$ $Al_2Cu + W$	0.02	2.40	0.025	3.75
(Al) $\Longleftrightarrow$ $Al_3Sc + W$	0.04	0.45	0.07	0.50

The equilibrium phase diagram shows great stability of the W phase, it appears in a wide range of Sc concentrations at typical concentrations of Cu (3–5%), Figure 9.5b, c. However, under nonequilibrium conditions of decomposition of a supersaturated solid solution the Al_3Sc phase precipitates (alongside θ (Al_2Cu)) instead of the W phase in the alloys falling into (Al) + Al_2Cu + W and (Al) + W equilibrium phase fields (Kharakterova et al., 1994).

Al–Mg–Sc–(Zr) phase diagram. Thorough studies of the Al–Mg–Sc system revealed no new phases to be in equilibrium with (Al), except those known from the constituent binary systems, i.e. Al_8Mg_5 and Al_3Sc. The invariant eutectic reaction $L \Rightarrow (Al) + Al_3Sc + Al_8Mg_5$ occurs in the aluminum corner of the system at 447 ± 3°C and 0.1–0.5% Sc (Toropova et al., 1998).

The liquidus surface, polythermal, and isothermal sections of the Al-rich portion of the Al–Mg–Sc phase diagram are shown in Figure 9.6.

Under equilibrium solidification conditions, an alloy containing 3–5% Mg and 0.2% Sc starts solidification with the formation of primary (Al) grains, then the (Al) + Al_3Sc eutectic forms and the solidification ceases somewhere between 600 and 610°C (Figure 9.6a). Al_3Sc and Al_8Mg_5 precipitate from the solid solution on further cooling. Under nonequilibrium (real) casting conditions, the solidification continues down to 447–450°C when the ternary (Al) + Al_3Sc + Al_8Mg_5 eutectics is formed.

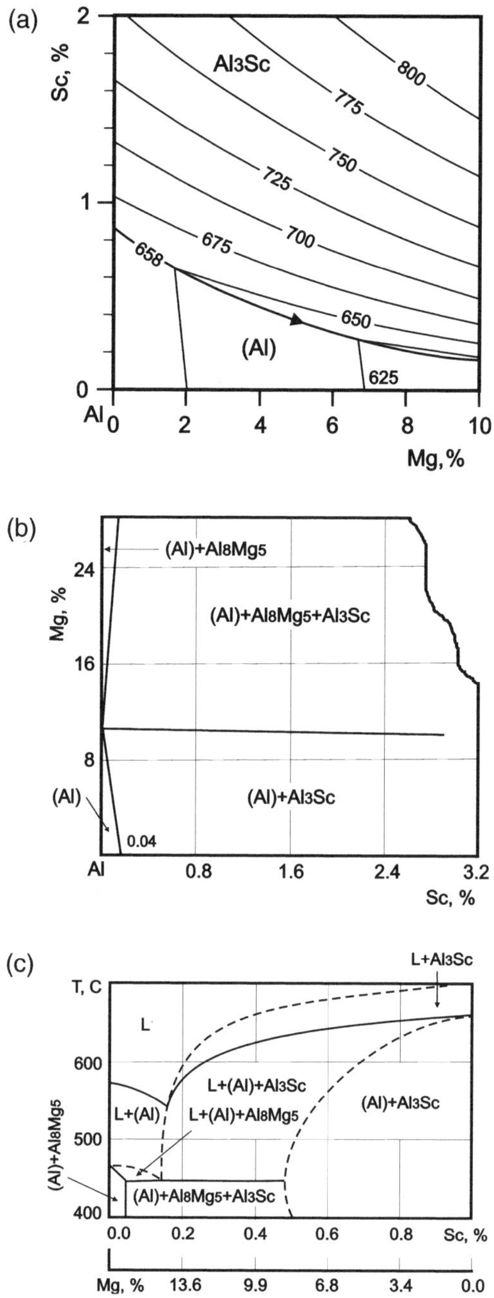

Figure 9.6. Liquidus surface (a) (after Pisch et al., 2000); isothermal section at 430°C (b) (after Toropova et al., 1998); and polythermal section from Al–17% Mg to Al–1% Sc (c) (after Toropova et al., 1998) of the Al–Mg–Sc phase diagram.

Commercial alloys of 1570-type (5–6% Mg) usually contain less than 0.3% Sc (typical concentrations of Sc and Zr are less than 0.15% each). In this compositional range, the ternary Al–Mg–Sc alloy is on the border between primary solidification of either (Al) or Al_3Sc (Figure 9.6a). However, if we take into account the joint presence of Sc and Zr in commercial alloys and the mechanism of their formation (Figures 9.3 and 9.4), it becomes clear that the typical concentration of Sc and Zr is sufficient for very good grain refining of Al–Mg alloy upon casting. The temperature of liquidus increases in the presence of Zr as shown in Figure 9.7a. The addition of Zr in Al–Mg–Sc alloys results in the formation of Al_3Zr phase in addition to Al_3Sc as shown in Figure 9.7b (compare to Figure 9.6b).

In addition to grain refinement, the main purpose of introduction of Sc to Al–Mg alloys is to assure the precipitation of coherent and semi-coherent Al_3Sc particles from the supersaturated solid solution. The precipitation hardening and structural hardening (due to the retarded recrystallization) significantly add to the solid-solution hardening effect, which is the typical mechanism in strengthening Al–Mg alloys. Therefore, the solubility of Mg and Sc are of great importance in these alloys. The solubility of magnesium in solid aluminum considerably decreases in the presence of scandium. The limit solubility of Mg and Sc in solid aluminum at the temperature of the ternary eutectics is 10.5% Mg and 0.007% Sc as compared to 13.5–14% Mg in the binary system (Toropova et al., 1998). The solubility of Mg and Sc decreases with temperature as shown in Figure 9.8. These data are useful for the correct choice of annealing temperatures for homogenization or precipitation.

Another implication of adding Mg to Al–Sc alloys is the fact that magnesium increases the lattice parameter of the aluminum solid solution and, hence decreases the dimensional misfit between Al_3Sc and the matrix (from 0.012 in a binary Al–Sc alloy to 0.00054 in a ternary Al–6.5%Mg–Sc alloy (Toropova et al., 1998)). As a result, the coherency of Al_3Sc is retained at higher temperatures and on longer exposures, providing for a higher thermal stability of Al–Mg–Sc alloys.

Al–Sc–Si phase diagram. Aluminum alloys with silicon are used mostly as foundry materials for shape casting. Rare-earth metals are known to refine the structure of the Al–Si eutectics. Additions of scandium to hypoeutectic alloys promote grain refinement, improve thermal stability, and provide for some precipitation hardening upon further heat treatment.

Three phases are found to be in equilibrium with (Al): (Si), Al_3Sc, and V (AlScSi). The ternary V phase appears as gray particles without etching and darkens upon electrolytic polishing. Experimental studies show that the V phase does not have considerable homogeneity range and has a composition that fits the formula $AlSc_2Si_2$ (Toropova et al., 1998). This ternary phase has a tetragonal structure (space group *P4/mbm*) with $a = 0.6597$ nm and $c = 0.3994$ nm (Toropova et al., 1998).

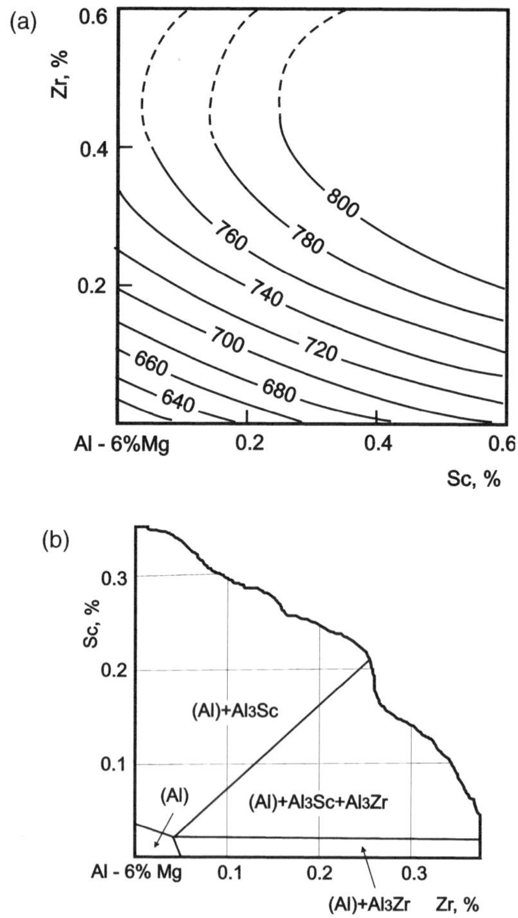

Figure 9.7. Liquidus surface (a) and isothermal section at 500°C (b) of the Al–Mg–Sc–Zr system at 6% Mg (after Toropova et al., 1998).

As compared to other Al–Si–TM systems, scandium behaves like Hf and Zr (IV group) rather than like rare earth metals of the yttrium subgroup.

The following solidification reactions occur in the aluminum corner of the Al–Si–Sc system (Toropova et al., 1998):

$L + Al_3Sc \Rightarrow (Al) + V(AlScSi)$ at 617°C, 4.2% Si and 0.52% Sc (point P in Figure 9.9a);
$L \Rightarrow (Al) + (Si) + V(AlScSi)$ at 574°C, 11.2% Si and 0.18% Sc (point E in Figure 9.9a).

Figure 9.9 demonstrates the solidification surface, isothermal, and polythermal sections of the Al–Si–Sc system. The important feature of the Al–Si–Sc phase diagram is the narrow solidification range of primary (Al) that lies along the Al–Si

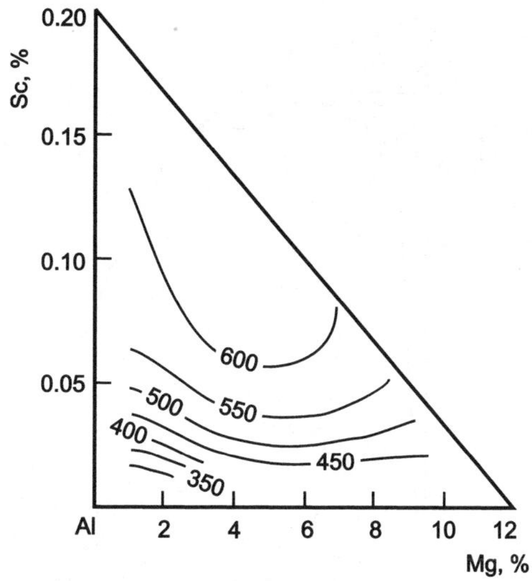

Figure 9.8. Solvus isotherms in the Al–Mg–Sc system (after Toropova et al., 1998).

side of the system (Figure 9.9a). Small additions of scandium cause the formation of primary crystals of either Al_3Sc or V (AlScSi) (Figure 9.9a, c, d). The solidus, however, decreases only slightly as the temperature of the ternary eutectics is only 3 K lower than that of the binary (Al) + (Si) eutectics. Alloying of Al–Sc alloys with silicon, on the contrary, considerably decreases the solidus temperature and broadens the solidification range due to the formation of the (Al) + Al_3Sc + V eutectics. During nonequilibrium solidification, Al_3Sc particles were found in Al–0.4% Sc alloys containing as much as 0.78% Si (Røyset et al., 2002). Although according to the phase diagram such alloys should contain only the V phase (Figure 9.9b), the hindered peritectic reaction P (Figure 9.9a) prevents the reaction of Al_3Sc with liquid to form V upon solidification, and Al_3Sc remains the dominant phase.

The solid solubilities of scandium and silicon in aluminum are given below according to the results of electron microprobe analysis and the geometry of the phase diagram (Toropova et al., 1998). Silicon has a negligible effect on the solubility of scandium, whereas scandium markedly decreases the solubility of silicon in solid (Al).

Phase field (Figure 9.9b)	500°C		550°C	
	Sc, %	Si, %	Sc, %	Si, %
(Al) $\Longleftrightarrow$ (Si) + V	0.03–0.05	0.62–0.74	0.08–0.12	0.97–1.05
(Al) $\Longleftrightarrow$ Al_3Sc + V	0.05–0.08	0.26	0.08–0.10	0.29

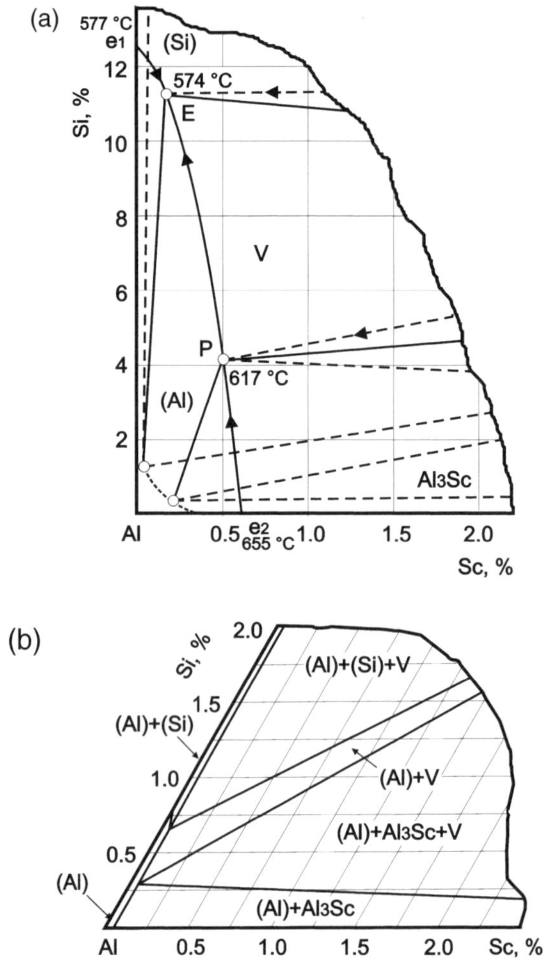

Figure 9.9. Projection of solidification surface (a), isothermal section at 500°C (b) and polythermal sections at 6% Si (c) and 11% Si (d) of the Al–Si–Sc system (after Toropova et al., 1998; Rokhlin et al., 1998).

The supersaturated solid solution formed in Al–Sc–Si alloys during solidification contains about 1% Si and 0.1% Sc. This composition falls into the (Al) + (Si) + V phase field of the equilibrium phase diagram. During annealing in the temperature range 100–450°C, precipitates of (Si) and V phase (probably metastable modification) are formed (Kharakterova et al., 1994).

Al–Cu–Mg–Sc–Zn–Zr phase diagram. The Al–Cu–Mg–Zn system is a base for high-strength aluminum alloys of the 7XXX series. Many of these alloys contain small additions of Zr as a grain refiner and anti-recrystallization agent. Introduction

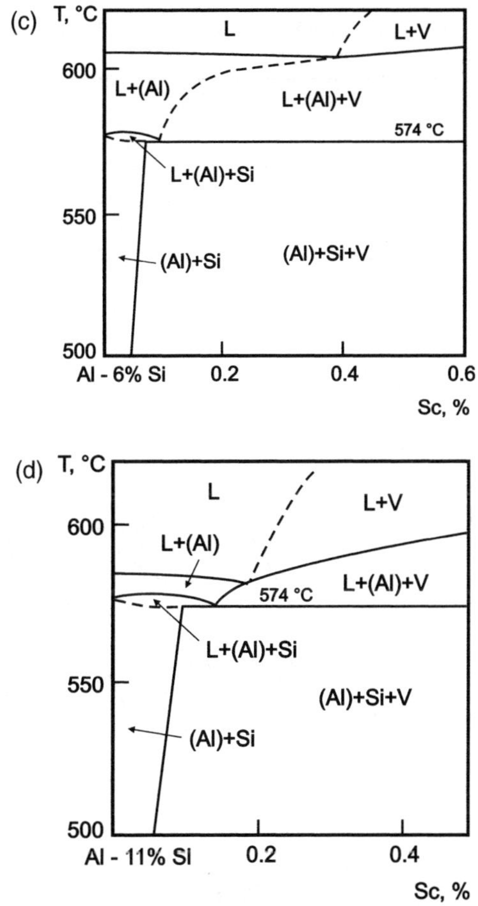

Figure 9.9 (*continued*)

of scandium to these alloys can be very beneficial for improving casting properties, weldability, and resource characteristics.

Due to a very complex character of this system, many phases can be in equilibrium with (Al). The complete list is given in Table 9.2. Phases T and M are quaternary solid solutions formed by the isomorphous compounds $Al_2Mg_3Zn_3$ and Al_6CuMg_4 (T) and $Mg(AlCu)_2$ and $MgZn_2$ (M), respectively. Some of the constitutive phase diagrams, i.e. Al–Zn–Mg, Al–Zn–Mg–Cu, and Al–Cu–Mg are considered elsewhere in this book.

In the Al–Cu–Zn–Zr subsystem containing up to 0.3% Zr, only binary phases are found in equilibrium with (Al) at 500°C, i.e. Al_2Cu and Al_3Zr as shown in Figure 9.10a. Alloys rich in Zn and Al are still in the semi-liquid state at 500°C as can be seen from this isothermal section. The ternary $Al_3Cu_5Zn_2$ (T) phase should

Table 9.2. Crystal structure of phases in equilibrium with aluminum in commercial Al–Zn–Mg–Cu–Sc–Zr alloys (Rokhlin et al., 2004; Toropova et al., 1998)

Phase	Crystal structure	Lattice parameters, nm	
		a	*c*
Al_2Cu	Tetragonal	0.6063	0.4872
Al_3Zr	Tetragonal	0.4014	1.732
Al_3Sc	Cubic	0.4104	–
W ($Al_{5-8}Cu_{7-4}Sc$)	Tetragonal	0.8546–0.8621	0.5036–0.5091
T ($(AlCuZn)_{49}Mg_{32}$)	Cubic	1.428–1.435	–
T ($Al_2Mg_3Zn_3$)	Cubic	1.416–1.422	–
M ($Mg(AlCuZn)_2$)	Hexagonal	0.5221	0.8567
η ($MgZn_2$)	Hexagonal	0.5221	0.8567
M ($Mg(AlCu)_2$)	Hexagonal	0.507–0.512	0.829–0.839

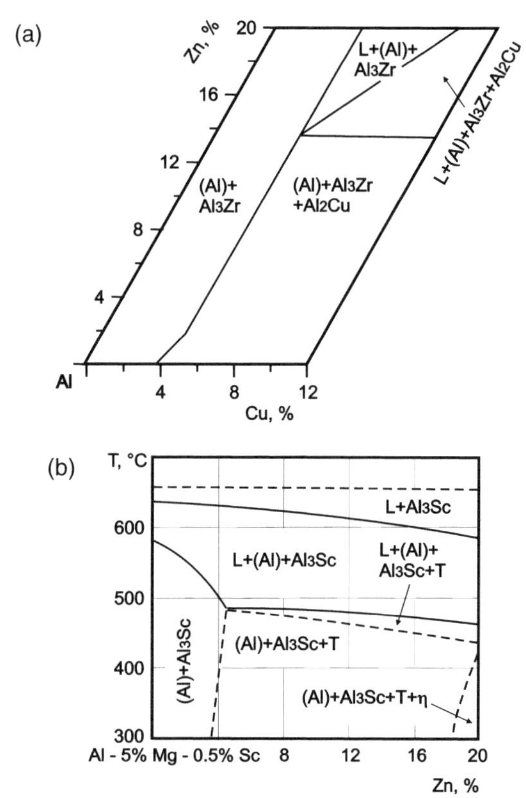

Figure 9.10. Isothermal section at 500°C of the Al–Zn–Cu–Zr phase diagram at 0.3% Zn (a) and polythermal sections of the Al–Zn–Mg–Sc (b); Al–Zn–Cu–Mg–Zr (c); Al–Zn–Cu–Mg–Sc (d); and Al–Zn–Cu–Mg–Sc–Zr (e) phase diagrams (after Rokhlin et al., 2004; Fridlyander et al., 2001). Compositions for isopleths are given in the diagrams. T – $(AlCuZn)_{49}Mg_{32}$ and M – $Mg(AlCuZn)_2$.

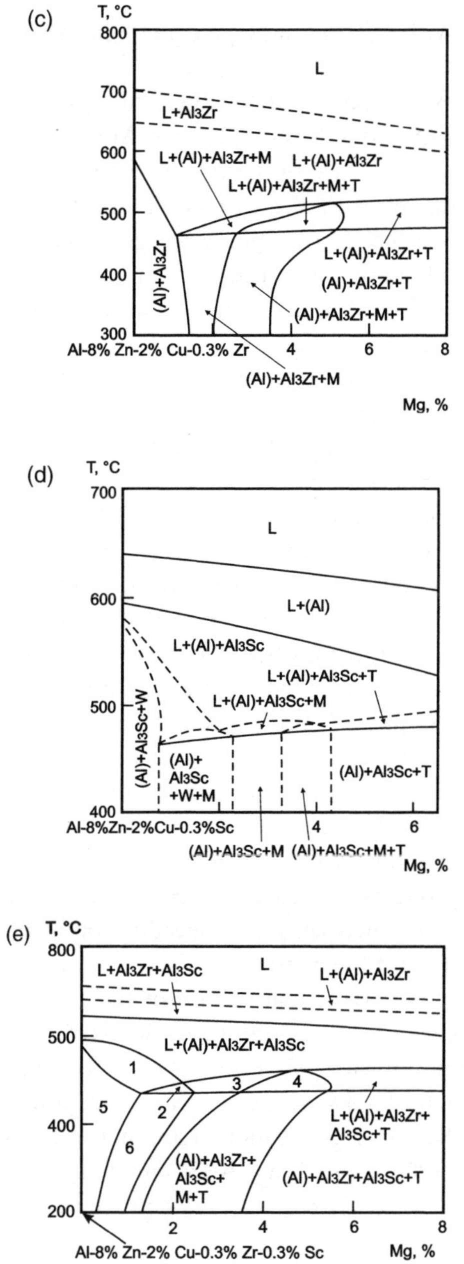

1 - L+(Al)+Al3Zr+Al3Sc+W   4 - L+(Al)+Al3Zr+Al3Sc+M+T
2 - L+(Al)+Al3Zr+Al3Sc+M+W   5 - (Al)+Al3Zr+Al3Sc+W
3 - L+(Al)+Al3Zr+Al3Sc+M   6 - (Al)+Al3Zr+Al3Sc+W+M

Figure 9.10 (*continued*)

also be in equilibrium with (Al) and may precipitate in the solid state upon cooling (see also Section 6.2).

The following phases can be found together with (Al) in Al–Mg–Zn–Sc alloys (another subsystem): Al_3Sc, $Al_2Mg_3Zn_3$ (T), and $MgZn_2$ (η). Figure 9.10b shows that alloys containing 0.5% Sc are in the range of primary solidification of Al_3Sc. The solidification continues with the formation of (Al) and $Al_2Mg_3Zn_3$ phases through eutectic reactions. The $MgZn_2$ phase precipitates during cooling in the solid state. The effect of scandium on phase transformations during solidification of such alloys as 7046 (7% Zn, 1.3% Mg) and 7076 (7.5% Zn, 1.6% Mg, 0.3% Cu) can be analyzed using this section.

The strongest Al–Zn–Mg–Cu alloys, e.g. 7001 and V96ts (1960rus), contain 7–9% Zn, 1.5–2.5% Cu, and 2.5–3.5% Mg, and the effect of Zr and Sc on their phase composition can be traced using polythermal sections in Figures 9.10c, d. Addition of up to 1% Mg to Al–Zn–Cu–Zr alloys does not change the phase composition. On further increasing the concentration of Mg the M and T phases appear as it should be according to the Al–Cu–Mg–Zn phase diagram (see Chapter 6). Rokhlin et al. (2004) report that alloys containing 3% Mg melt at 480°C.

A similar pattern is observed on addition of magnesium to Al–Zn–Cu–Sc alloys. However, there are some distinctions between these two subsystems, i.e. (1) the (Al) solid solution is a primary phase (in the given compositional range in the Al–Cu–Sc–Zn system), (2) the Al_3Sc phase is present in all phase fields instead of Al_3Zr, and (3) the ternary W phase is formed in alloys containing less than 3% Mg. Cast samples also show particles of Al_2Cu phase that disappear during annealing, which attests for the nonequilibrium character of Al_2Cu formation (Rokhlin et al., 2004). The Al_3Sc phase is formed during the eutectic reaction $L \Rightarrow (Al) + Al_3Sc$. The M and T phases are also products of eutectic reactions. The liquidus temperature continuously decreases with increasing the concentration of Mg, whereas the solidus temperature has a tendency to increase.

Figure 9.10e shows the polythermal section of the Al–Cu–Mg–Sc–Zn–Zr phase diagram. In the chosen compositional range, Al_3Sc, Al_3Zr, W, M, and T phases are detected to be in equilibrium with (Al). Particles of Al_3Zr and Al_3Sc cannot be easily separated under an optical microscope, so X-ray identification is required. We already know that Al_3Sc dissolves substantial amount of Zr and Al_3Zr can dissolve some Sc. Rokhlin et al. (2004) found that Al_3Sc contained 27% Zr and 10% Zn, and Al_3Zr contained 3% Sc and 6% Zn in an alloy containing 6% Zn, 2% Cu, 7% Mg, 0.3% Sc, and 0.3% Zr.

9.1.3. Phase diagrams of aluminum alloys with transition and rare-earth metals

Al–Mg–REM systems. We already discussed Al–Mg–Sc–(Zr) alloys and their phase composition. Scandium found application in commercial Al–Mg alloys by greatly improving their mechanical and technological properties. Commercial alloys containing magnesium and transition metals and relevant phase diagrams are considered in Chapter 2 and 4. Rare-earth metals (REM) are among promising alloying elements and their interaction with Al–Mg alloys is the subject of this section.

In the Al–Mg–Ce system, three phases are in equilibrium with (Al), i.e. Al_8Mg_5 (β). $Al_{11}Ce_3$ (Al_4Ce), and $Al_2Ce_{0.15}Mg_{0.85}$ or $Al_{13}CeMg_6$ (τ) (Drits et al., 1977; Odinaev et al., 1996; Gröbner et al., 2002). The $Al_{11}Ce_3$ phase has an orthorhombic structure (space group *Immm*) with $a = 0.4395$ nm, $b = 1.3025$ nm and $c = 1.0092$ nm (Mondolfo, 1976). The ternary phase has a hexagonal crystal structure of $MgZn_2$ type (space group $P6_3/mmc$) with $a = 0.552$ or 0.531 nm and $c = 0.889$ or 0.894 nm (Drits et al., 1977; Odinaev et al., 1996). These phases participate in the following solidification reactions in Al-rich alloys (Gröbner et al., 2002):

$L + Al_{11}Ce_3 \Rightarrow (Al) + Al_{13}CeMg_6(\tau)$ at 453°C (Gröbner et al., 2002) or 445°C (Odinaev et al., 1996);

$L \Rightarrow (Al) + Al_8Mg_5(\beta) + Al_{13}CeMg_6(\tau)$ at 450°C (Gröbner et al., 2002) or

$L \Rightarrow (Al) + Al_8Mg_5(\beta) + Al_{11}Ce_3$ at 441°C (Odinaev et al., 1996).

In Al–Mg alloys containing small additions of cerium the ternary τ phase is more stable than the binary β phase. So the equilibrium composition of solid alloys would be either $(Al) + Al_{11}Ce_3$ or, on increasing amount of Mg, $(Al) + Al_{11}Ce_3 + \tau$.

The solubility of Ce in solid aluminum is 0.05% at 635°C and 0.01 at 525°C (Mondolfo, 1976). However, this solubility increases dramatically with increasing the cooling rate in solidification.

In Al-rich Al–Mg–Gd alloys, the Al_8Mg_5, Al_3Gd, and $Al_2Mg_{0.5}Gd_{0.5}$ phases are in equilibrium with (Al). The $AlGd_3$ phase is formed (in Al-rich alloys) by a eutectic reaction at 650°C, has a hexagonal crystal structure (space group $P6_3/mmc$) with lattice parameters $a = 0.6308$–0.6323 nm, $c = 0.4589$–0.4598 nm and density 4.96 g/cm^3 (Mondolfo, 1976). The $Al_2Mg_{0.5}Gd_{0.5}$ phase is formed at 800°C (Rokhlin et al., 1997) or 761°C (Gröbner et al., 2001) through a peritectic reaction. This ternary phase has a hexagonal structure (space group $P6_3/mmc$) with lattice parameters $a = 0.552$ nm, $c = 0.888$ nm (Rokhlin et al., 1997) or $a = 0.5525$ nm, $c = 1.777$ nm (De Negri, 2003). The following solidification reactions are reported in the Al-corner of the Al–Mg–Gd system:

$L + Al_3Gd \Rightarrow (Al) + Al_2Mg_{0.5}Gd_{0.5}$ at 565°C (Rokhlin et al., 1997) or at 515 °C (Gröbner et al., 2001);

$L \Rightarrow (Al) + Al_8Mg_5 + Al_2Mg_{0.5}Gd_{0.5}$ at 448–450°C (Rokhlin et al., 1997; Gröbner et al., 2001).

Gröbner et al. (2001) also report, based on thermodynamical calculations, some solid-state reaction including the peritectoid reaction

$$(Al) + Al_2Mg_{0.5}Gd_{0.5} \Rightarrow Al_8Mg_5 + Al_3Gd \text{ at } 314°C$$

and the eutectoid reaction

$$Al_2Mg_{0.5}Gd_{0.5} \Rightarrow Al_8Mg_5 + Al_3Gd + Al_{12}Mg_{17} \text{ at } 243°C.$$

Similar phase composition is observed in Al–Mg alloys with Dy and Ho (De Negri, 2003; Cacciamani et al., 2003). The ternary phases found in these systems are $Al_2Dy_{0.36}Mg_{0.64}$ (hexagonal, $MgNi_2$ type, $a = 0.5490$ nm, $c = 1.7697$) and $Al_2Ho_{0.39}$ $Mg_{0.61}$ (same structure, $a = 0.5471$ nm, $c = 1.7671$ nm) (De Negri, 2003).

Phase diagrams of Al alloys relevant to rapid solidification processing. Aluminum alloys containing transition metals are promising for rapid solidification processing including the formation of RS/PM, quasicrystalline, and amorphous materials. In this subsection we consider some alloying systems. It should be noted that the phase equilibria in the aluminum corner of these systems are seldom well established, let alone the complete solidification paths. That is why we give only phases that are in equilibrium with the aluminum solid solution and, in some cases, the phases which are next (contain less aluminum) on the phase diagram. The latter phases may occur in alloys solidified under nonequilibrium conditions of rapid solidification processing.

Tables 9.3–9.5 summarize the available information on the alloying systems that are important for rapid solidification processing with the formation of supersaturated, quasicrystalline or amorphous materials. For Al–Cu–TM alloys (Table 9.3), ternary phases exist in equilibrium with (Al) in the following systems: Al–Cu–Ce, Al–Cu–Co, Al–Cu–Fe, and Al–Cu–La. In other systems given in Table 9.3 the ternary phases, though existing, are not in equilibrium with (Al). Similar in composition and structure ternary phases, Al_7Cu_2TM, are formed in Al–Cu–TM systems where TM = Co, Fe, Mo. Cubic Al_5CuTM_2 phases exist (not in equilibrium with (Al)) in Al–Cu–TM alloys where TM = Hf, Ti, Zr. In Al–Ni–TM alloys (Table 9.4), ternary phases are found in equilibrium with (Al) in the following systems Al–Ni–Fe, Al–Ni–Gd, Al–Ni–Nd, and Al–Ni–Ru. An invariant eutectic reaction with the formation of (Al), Al_3Ni, and either binary AlTM or ternary AlNiTM phases takes place in the aluminum corner of Al–Ni–TM alloys. Ternary phases are not commonly found in equilibrium with (Al) in the systems listed in Table 9.5. However, $Al_{10}Fe_2Me$ phases exist in the systems with Me = Y, La, Ce, Pr Nd, Sm, Gd, Tb, Dy, Ho, Er, Tm, Yb, Lu (Thiede et al., 1998). Alloys of the corresponding systems are prone to glass formation upon solidification, and

Table 9.3. Characteristics of phases and known data on phase equilibria in Al–Cu–TM systems

System	Phase	Crystal structure	Lattice parameters, nm			Comments	Ref.
			a	*b*	*c*		
Al–Cu–Ce	$Al_{11}Ce_3$	Orthorhombic *Immm*	0.4395	1.302	1.009		Rogl (1991a), Mondolfo (1976)
	Al_2Cu*	Tetragonal *I4/mcm*	0.6063	–	0.4872	Density 4.34 g/cm^3	
	Al_8CeCu_4	Tetragonal *I4/mmm*	0.883–0.885	–	0.516–0.518		
	Al_4CeCu	Unknown	–	–	–	Homogeneity range 43.7–47.2% Ce; 19–23.9% Cu	
	$Al_{3.25\text{-}3}CeCu_{0.75–1.0}$	Tetragonal *I4/mmm*	0.425–0.428	–	1.065–0.1080	Not in equilibrium with (Al)	
Al–Cu–Co	Al_9Co_2	Monoclinic $P2_1/c$	0.8556	0.6290	0.6213; $\gamma = 94.76°$	Density 3.67 g/cm^3 $L \Rightarrow (Al) + Al_2Cu + Al_9Co_2$ at 536–540°C, 0.5% Co or $L \Rightarrow (Al) + Al_2Cu + Al_7CoCu_2$ at 536–544.5°C, 32% Cu, 0.33% Co; $L + Al_9Co_2 \Rightarrow (Al) + Al_7CoCu_2$ at 559 or 568°C	Ran (1991a), Mondolfo (1976), Zhang and Gille (2004)
	Al_7CoCu_2	Tetragonal *P4/mnc*	0.63047	–	1.4756	Density 4.0–4.4 g/cm^3	
	$Al_{65.3}Ci_{8.7}Cu_{26}$	Decagonal	–	–	–		
Al–Cu–Cr	$Al_{45}Cr_7$ ($Al_{13}Cr_2$; Al_7Cr)	Monoclinic *C2/m*	2.5196	0.7574	1.0949; $\gamma = 128.7°$	Density 3.05–3.18 g/cm^3 $L \Rightarrow (Al) + Al_2Cu + Al_{45}Cr_7$ at 545°C, 28.5% Cu, 1.5% Cr	Ghosh (1991)
	$Al_{85-x}Cr_{15}Cu_x$	Icosahedral	–	–	–	Not in equilibrium with (Al) $0 < x < 5$ and $x = 20$ phase is icosahedral; at $x = 10–15$ mixture of i-phase and Al_2Cu	
	$Al_{67}Cr_{15}Cu_{18}$	Hexagonal	1.2820	–	1.2710	Equilibrium phase, not in equilibrium with (Al)	

(*continued*)

Table 9.3. Continued

System	Phase	Crystal structure	Lattice parameters, nm			Comments	Ref.
			a	*b*	*c*		
Al–Cu–Fe	Al_3Fe	Monoclinic *C2/m*	1.5487	0.80831	1.2476; $\gamma = 107°43'$	Density 3.896 g/cm^3 $L \Rightarrow Al_3Fe + Al_{23}CuFe_4$ at 626°C; $L + Al_3Fe \Rightarrow Al_{23}CuFe_4 + Al_7Cu_2Fe$ at 622°C; $L + Al_3Fe \Rightarrow (Al) + Al_{23}CuFe_4$ at 620°C	Legendre and Harmelin (1991), Mondolfo (1976)
	$Al_{23}CuFe_4$ ($Al_6(FeCu)$)	Orthorhombic *Cmcm*	0.64343	0.74604	0.87769	Density 3.45 g/cm^3, stabilized Al_6Fe $L + Al_{23}CuFe_4 \Rightarrow (Al) + Al_7Cu_2Fe$ at 590°C	
	Al_7Cu_2Fe	Tetragonal *P4/mnc*	0.6336	–	1.4870	Density 4.3 g/cm^3, homogeneity range 12–20% Fe, 29–39% Cu $L \Rightarrow (Al) + Al_2Cu + Al_7Cu_2Fe$ at 542°C	
	$Al_{62}Cu_{25.5}Fe_{12.5}$ $Al_{64-70}Cu_{10-23}Fe_{10-20}$	Icosahedral	–	–	–	Not in equilibrium with (Al)	
	$Al_{65}Cu_{20}F_{12}$	Cubic	1.407	–	–	Equilibrium phase, not in equilibrium with (Al)	
Al–Cu–Hf	Al_3Hf^{HT}	Tetragonal *I4/mmm*	0.389–0.393	–	0.893–0.889	High-temperature modification > 727°C	Lysova (1991)
	Al_3Hf^{LT}	Tetragonal *I4/mmm*	0.3980–0.4010	–	1.7140–1.7130	Low-temperature modification < 727°C	
	Al_5CuHf_2	Cubic $Pm\bar{3}m$	0.4013	–	–	Not in equilibrium with (Al)	
	$(Al_{1-x}Cu_x)_{92.3}Hf_{7.7}$	Tetragonal *I4/mmm*	–	–	–	Not in equilibrium with (Al)	

Al–Cu–La	$Al_{11}La_3^{LT}$	Orthorhombic *Immm*	0.4431	1.3142	1.0132	$L \Rightarrow (Al)+(Al_xCu_{1-x})_4La$ at 610°C; $L \Rightarrow (Al)+Al_{11}La_3^{LT}+(Al_xCu_{1-x})_4La$ at 590°C; $L \Rightarrow (Al)+Al_{10}Cu_2La$ at 580°C; $L \Rightarrow (Al)+Al_{10}Cu_2La+(Al_xCu_{1-x})_4La$ at 550°C; $L \Rightarrow (Al)+Al_2Cu+Al_{10}Cu_2La$ at 542°C	Ran (1991b)
	$Al_{10}Cu_2La$	Tetragonal *I4/mmm*	0.880	–	0.517		
	$(Al_xCu_{1-x})_4La$	Tetragonal *I4/mmm*	0.4294–0.4326	–	1.053–1.088	$0.75 < x < 0.875$	
Al–Cu–Mo	$Al_{12}Mo$	Cubic $Im\bar{3}$	0.75815	–	–		Harmelin and Legendre (1991b),
	Al_7Cu_2Mo	Orthorhombic	0.5052	0.8413	1.9685	Not in equilibrium with (Al)	
Al–Cu–Ti	Al_3Ti	Tetragonal *I4/mmm*	0.385	–	0.8596	Density 3.37 g/cm^3 $L \Rightarrow (Al)+Al_2Cu+Al_3Ti$ at 540°C	Ran and Stadelmaier (1992b), Mondolfo (1976)
	Al_5CuTi_2	Cubic $Pm\bar{3}m$	0.3927	–	–	Not in equilibrium with (Al) $L+Al_5CuTi_2 \Rightarrow Al_2Cu+Al_3Ti$ at 570°C	
	$Al_{11}CuTi_4$	Tetragonal					
		Low-temp	0.3901	–	1.660	Not in equilibrium with (Al)	
		High-temp	0.3917	–	1.6524	$L+Al_{11}CuTi_4 \Rightarrow Al_3Ti+Al_5CuTi_2$ at 1280°C	
Al–Cu–Zr	Al_3Zr	Tetragonal *I4/mmm*	0.4016	–	1.7320	Density 4.1 g/cm^3	Tretyachenko (1992), Mondolfo, (1976)
	Al_3Zr	Cubic $Pm\bar{3}m$	0.405	–	–	Metastable	
	τ_7	Tetragonal *I4/mmm*	0.512–0.490	–	0.850–0.856	Not in equilibrium with (Al) Exists at 42–51% Cu and 7.7% Zr	
	$Al_{2.5}Cu_{0.5}Zr$	Cubic $Pm\bar{3}m$	0.402–0.404	–	–	Not in equilibrium with (Al) Observed at 12.5% Cu and 25% Zr	
	τ_9	–	–	–	–	Not in equilibrium with (Al) Observed at 24% Cu and 12% Zr	

* The Al_2Cu phase is in equilibrium with (Al) in all given systems (not shown for other given systems to avoid repetition)

Table 9.4. Characteristics of phases and known data on phase equilibria in Al–Ni–TM systems

System	Phase	Crystal structure	Lattice parameters, nm			Comments	Ref.
			a	*b*	*c*		
Al–Ni–Ce	$Al_{11}Ce_3$	Orthorhombic *Immm*	0.4395	1.302	1.009	L⇒(Al)+Al_3Ni+$Al_{11}Ce_3$ at 627°C	Ferro et al. (1991b), Belov et al. (1999)
	Al_3Ni*	Orthorhombic *Pnma*	0.6611	0.7366	0.4811	Density 3.950 g/cm^3	
	Al_7CeNi_2	Unknown	–	–	–	Not in equilibrium with (Al)	
Al–Ni–Cr	$Al_{45}Cr_7$ (Al_7Cr)	Monoclinic *C*2/*m*	2.5196/ 2.5256	0.7574/ 0.7582	1.0949; $\gamma = 128.7°$/ 1.0955; $\gamma = 128.68°$	Density 3.05–3.18 g/cm^3 L⇒(Al)+Al_3Ni+$Al_{45}Cr_7$ at 635°C, 5–6% Ni, 1.3% Cr	Rogl (1991b), Rosell-Laclau et al. (1996)
	$Al_{11}Cr_2$	Monoclinic	1.2880	0.7652	1.0639; $\gamma = 122.2°$	Not in equilibrium with (Al)	
		Orthorhombic *Cmcm*	1.24	3.46	2.0		
	Al_3Ni_2	Hexagonal *P*$\bar{3}$*m*1	0.40363	–	0.49005	Not in equilibrium with (Al)	
	λ (AlNiCr)	Orthorhombic *Immm*	1.255	1.255 2.510	3.075	Not in equilibrium with (Al)	
		Triclinic	1.255		$\alpha = 1.775$; $\beta = 110.7°$; $\gamma = 90°$		
Al–Ni–Fe	Al_3Fe	Monoclinic *C*2/*m*	1.5487	0.80831	1.2476; $\gamma = 107°43'$	L+Al_3Fe+Al_3Ni⇒Al_9FeNi at 809°C, 3.8% Fe, 20.5% Ni; L+Al_3Fe⇒(Al)+Al_9FeNi at 649–650°C, 1.8% Fe, 1.2% Ni; L⇒(Al)+Al_3Ni+Al_9FeNi at 638–640°C, 0.28% Fe, 6.27% Ni;	Budberg and Price (1992), Drits et al. (1977)
	Al_9FeNi	Monoclinic *P*2$_1$/*c*	0.8598	0.6271	0.6207; $\gamma = 94.66°$	Homogeneity range 4.5–14% Fe, 18–28% Ni	
	$Al_{10}Fe_3Ni$	Hexagonal	0.7688/ 0.7703	–	0.7668/ 0.7698	Not in equilibrium with (Al)	

Al–Ni–Gd	Al_3Gd	Hexagonal *$P6_3/mmc$*	0.6331	–	0.4600	Density 4.96 g/cm^3	Mondolfo (1976),
	$Al_{15}Gd_2Ni_3$ ($Al_{16}GdNi_3$)	Orthorhombic	0.407	1.601	2.724	L⇒(Al)+Al_3Gd+$Al_{15}Gd_2Ni_3$ at 633°C; L+Al_3Ni_2⇒Al_3Ni+$Al_{15}Gd_2Ni_3$ at 843°C; L⇒(Al)+Al_3Ni+$Al_{15}Gd_2Ni_3$ at 644°C	Gao et al. (2003), Ferro et al. (1992)
	Al_4GdNi	Orthorhombic *Aba2*	0.4087	0.669	1.530	Not in equilibrium with (Al)	
Al–Ni–Hf	Al_3Hf	Tetragonal *I4/mmm*	0.3980–0.4010	–	1.7140–1.7130	Low-temperature modification <727°C	Ghosh (1993a)
	$Al_{16}Hf_3Ni_6$	Tetragonal *I4/mmm*	0.4010	–	1.4120	Not in equilibrium with (Al)	
Al–Ni–La	$Al_{11}La_3^{LT}$	Orthorhombic *Immm*	0.4433	1.315	1.013	Exists below 807 or 915°C L→(Al)+Al_3Ni+$Al_{11}La_3^{LT}$ at 625°C	Gödecke et al. (2001b), Mondolfo (1976), Ferro et al. (1993a)
Al–Ni–Mo	$Al_{12}Mo$	Cubic *$Im\bar{3}$*	0.75815	–	–	Exists below 712°C	Kubaschewski (1993),
	Al_5Mo	Hexagonal *$P\bar{3}1c$*	0.4957	–	2.622	Not in equilibrium with (Al), exists below 650°C	Grushko et al. (2002)
	(Al, Ni)$_3$Mo	Tetragonal *I4/mmm*	0.3702–0.3732	–	0.8361–0.8430	Close to Al_8Mo_3, not in equilibrium with (Al)	
	$Al_{75-77}Mo_5Ni_{20-18}$	Unknown	–	–	–	Not in equilibrium with (Al)	
	or $Al_{75}Mo_{11}Ni_{14}$	Orthorhombic	1.0054	1.5288	0.8519		
Al–Ni–Nb	Al_3Nb	Tetragonal *I4/mmm*	0.3841	–	0.8609	Density 4.54 g/cm^3 Al-corner is not studied	Saunders (1993)
Al–Ni–Nd	$Al_{11}Nd_3^{LT}$	Orthorhombic *Imma*	0.4359	1.2924	1.0017	Exists below 934 or 950°C	Gödecke et al. (2001a, b), Ferro et al. (1993b)

(continued)

Table 9.4. Continued

System	Phase	Crystal structure	Lattice parameters, nm			Comments	Ref.
			a	*b*	*c*		
	Al_7NdNi_2	Unknown	–	–	–	$L+Al_{11}Nd_3^{HT}\Rightarrow Al_{11}Nd_3^{LT}+Al_7NdNi_2$ at 860°C; $L+Al_3Ni\Rightarrow(Al)+Al_7NdNi_2$ at 629°C; $L+Al_7NdNi_2\Rightarrow Al_3Ni+Al_{11}Nd_3^{LT}$ (metastable); $L\Rightarrow(Al)+Al_{11}Nd_3^{LT}+Al_7NdNi_2$ at 626°C; $L\Rightarrow(Al)+Al_3Ni+Al_{11}Nd_3^{LT}$ (metastable)	
	Al_3Nd	Hexagonal *P*6$_3$*/mmc*	0.6471	–	0.4605	Not in equilibrium with (Al)	
	Al_4NdNi	Orthorhombic	0.4110	1.548	0.6669	Not in equilibrium with (Al)	
Al–Ni–Re	$Al_{12}Re$	Cubic *Im*3	0.7528	–	–	Dissolves <3 at.% Ni $L+Al_6Re+Al_3Ni\Rightarrow(Al)$ (5 at.% Ni, 0.5 at.% Re); $L+Al_6Re\Rightarrow(Al)+Al_{12}Re$;	Mondolfo (1976), Cornish and Witcomb (1999)
	Al_6Re	Orthorhombic *Cmcm*	0.659	0.761	0.902	Dissolves 4.5 at.% Ni, appears under nonequilibrium solidification conditions alongside (Al)	
	Al_4Re	Triclinic?	–	–	–	Dissolves <1.5 at.% Ni, appears under nonequilibrium solidification conditions alongside (Al)	
	Al_3Re	Unknown	–	–	–	Dissolves 5.5 at.% Ni, appears under nonequilibrium solidification conditions alongside (Al)	
Al–Ni–Ru	Al_6Ru	Orthorhombic *Cmcm*	0.65559	0.74882	0.89605	Isomorphic with Al_6Fe, exists <732°C $L+Al_6Ru\Rightarrow(Al)+Al_{13}Ru_4$	Mondolfo (1976), Hohls et al. (2000), Tretyachenko et al. (1993)
	$Al_{13}Ru_4$	Monoclinic *C*2/*m*	1.586	0.8188	1.274; γ=107°80′	Equilibrium, isomorphic with Al_3Fe; $L+Al_{13}Ru_4\Rightarrow(Al)+Al_{14}Ni_2Ru$;	
	$Al_{14}Ni_2Ru$ (Hohls et al., 2000)	Unknown	–	–	–	$L+Al_{14}Ni_2Ru\Rightarrow(Al)+Al_3Ni$ Another version of phase diagram suggests no ternary	

						phases, then: $L+Al_{13}Ru_4 \Rightarrow Al_3Ni+Al_6Ru$; $L \Rightarrow (Al)+Al_3Ni+Al_6Ru$ (<640°C)	
Al–Ni–Ta	Al_3Ta	Tetragonal *I4/mmm*	0.384	–	0.8537–0.8553	Al-corner is not studied	Mondolfo (1976)
Al–Ni–Ti	Al_3Ti	Tetragonal *I4/mmm*	0.385	–	0.8596	Density 3.37 g/cm^3 $L+Al_3Ni_2 \Rightarrow Al_3Ni+Al_3Ti$ at 820°C; $L+Al_3Ti \Rightarrow (Al)+Al_3Ni$ at 645–650°C	Budberg (1993)
	$Al_{6.5}NiTi_{2.5}$, $Al_{67}Ni_{7.5}Ti_{25.5}$, Al_8NiTi_3	Cubic $Pm\bar{3}m$	0.393–0.394	–	–	Not in equilibrium with (Al); density 3.75 g/cm^3	
	X	Orthorhombic *Pmmm*	1.805	1.405	20.205	Metastable, in RS alloy $Al_{97.2}Ni_{0.5}Ti_{2.3}$	Ding et al. (2001)
	$Al_{83.1}Ni_{4.3}Ti_{12.6}$	Cubic $Fm\bar{3}c$	2.421	–	–	Metastable, in RS alloys	
	Al_9Ni_2	Monoclinic $P2_1/c$	0.854	0.617	0.612	Metastable, in RS alloys	
	Y	Tetragonal *I4/mmm*	0.3890	–	1.6822	Metastable, in RS alloy $Al_{95.8}Ni_{0.7}Ti_{3.5}$	
Al–Ni–V	$Al_{10}V$	Cubic *Fd3m*	1.4492–1.4586	–	–	No ternary phases found in the system. A succession of peritectic reactions in the Al corner. Homogeneity range 15.1–15.9% V	Mondolfo (1976)
	$Al_{45}V_7$ (Al_7V)	Monoclinic *C2m*	2.54	0.759	1.10; $\gamma=127°$	Not in equilibrium with (Al)	
	$Al_{23}V_4$ (Al_6V)	Hexagonal $P6_3/mmc$	0.769–0.7718	–	1.704–1.715	Not in equilibrium with (Al)	
Al–Ni–Y	Al_3Y	Hexagonal	0.619–0.622	–	2.1128–2.13	Exists above 627°C; density 3.6 g/cm^3	Mondolfo (1976), Ferro et al. (1993c)
	Al_3Y	Hexagonal $P6_3/mmc$	0.6299	–	0.4584	Exists below 645°C	
	$Al_{16}Ni_3Y$	Orthorhombic	0.408	1.604	2.729	Not in equilibrium with (Al)	
	Al_4NiY	Orthorhombic *Aba2*	0.408	1.544	0.662	Not in equilibrium with (Al)	
Al–Ni–Zr	Al_3Zr	Tetragonal *I4/mmm*	0.4016	–	1.7320	Density 4.1 g/cm^3	Ghosh (1993b)
	Al_5Ni_2Zr	Tetragonal	0.401/0.4023	–	1.441/1.444	Not in equilibrium with (Al)	

* The Al_3Ni phase is in equilibrium with (Al) in all given systems (not shown for other given systems to avoid repetition)

Table 9.5. Characteristics of phases and known data on phase equilibria in Al-based systems with Ce, Co, Fe, and other TM

System	Phase	Crystal structure	Lattice parameters, nm *a*	*b*	*c*	Comments	Ref.
Al–Ce–Co	$Al_{11}Ce_3$	Orthorhombic *Immm*	0.4395	1.302	1.009		Grieb (1991a)
	Al_9Co_2	Monoclinic $P2_1/c$	0.8556	0.6290	0.6213; $\gamma=94.76°$		
	Al_8CeCo_2	Orthorhombic	1.241	1.430	0.412	Not in equilibrium with (Al)	
Al–Ce–Fe*	Al_3Fe	Monoclinic *C2/m*	1.5487	0.80831	1.2476; $\gamma=107°43'$	$L+Al_{11}Ce_3 \Rightarrow Al_{10}CeFe_2$ at 940°C	Grieb (1991b)
	$Al_{10}CeFe_2$	Orthorhombic	0.894	1.022	0.906	Not observed in quenched alloys	
Al–Co–Hf**	Al_3Hf	Tetragonal *I4/mmm*	0.3980–0.4010	–	1.7140–1.7130	Low-temperature modification <727°C	Kubaschewski (1991)
	$Al_{13}Co_4$ ($Al_{19}Co_6$)	Monoclinic	1.5183	0.8122	1.2340; $\gamma=107.9°$	Not in equilibrium with (Al)	
	Ternary phases					Not in equilibrium with (Al)	
Al–Co–Ti**	Al_3Ti	Tetragonal *I4/mmm*	0.3849	–	0.861		Schmid (1991)
	$Al_{1-2}CoTi_{0-1}$	Cubic	1.193	–	–	Not in equilibrium with (Al)	
Al–Co–Zr**	Al_3Zr	Tetragonal *I4/mmm*	0.4014	–	1.732		Ferro et al. (1991a)
	$Al_{19}Co_6$	Monoclinic	1.5183	0.8122	1.2340; $\gamma=107.9°$	Not in equilibrium with (Al)	
	Al_4CoZr	Hexagonal	0.726	–	0.909	Not in equilibrium with (Al)	
Al–Cr–Mn	Al_7Cr	Monoclinic *C2/m*	2.5196	0.7574	1.0949; $\gamma=128.7°$	Dissolves up to 11.55% Mn $L \Rightarrow (Al)+Al_7Cr+Al_6Mn$ at 657°C, 0.3% Cr, 2.0% Mn	Mondolfo (1976), Grieb (1991c)
	Al_6Mn	Orthorhombic *Cmcm*	0.649–0.651	0.754–0.757	0.886–0.887	Dissolves <1% Cr; density 3.09–3.27 g/cm^3	

	$Al_{12}(CrMn)$	Cubic *Im*3	0.7507–0.7509	–	–	Stable modification of metastable $Al_{12}Mn$ (2–4% Cr, 12–10% Mn); density 2.92 g/cm^3; (Al)+Al_7Cr+Al_6Mn⇒$Al_{12}(CrMn)$ at 590°C	
	$Al_{11}Cr_2$	Monoclinic	1.2880	0.7652	1.0639; $\gamma = 122.2°$	Not in equilibrium with (Al)	
Al–Cr–Zr***	Al_3Zr	Tetragonal *I4/mmm*	0.4014	–	1.732		Guzei (1991)
Al–Fe–Cr***	Al_3Fe	Monoclinic *C2/m*	1.5487	0.80831	1.2476; $\gamma = 107°43'$	L+Al_5Cr⇒$Al_{45}Cr_7$ + Al_3Fe at 709°C, 4.2% Fe, 2.1% Cr; L⇒(Al)+Al_3Fe+$Al_{45}Cr_7$ at 640°C, 1.7% Fe, 0.3% Cr	Mondolfo (1976), Belov et al. (2002a),
	Al_6(Fe, Cr)	Orthorhombic *Cmcm*	–	–	–	Isomorphic to Al_6Mn. Metastable, in RS alloys	Ding et al. (2001)
Al–Fe–Gd****	Al_3Gd	Hexagonal *P6₃/mmc*	0.6308–0.6323	–	0.4589–0.4598	Density 4.96 g/cm^3; dissolves up to 1 at.% Fe	Hackenberg et al. (2002),
	$Al_{10}Fe_2Gd$	Orthorhombic	0.8970	1.0162	0.9023	L+Al_3Fe+Al_3Gd⇒$Al_{10}Fe_2Gd$ at 1040°C, 8.15 at.% Fe, 9.58 at.% Gd; L+Al_3Fe⇒(Al)+$Al_{10}Fe_2Gd$ at 647°C, 0.8 at.% Fe, 0.6 at.% Gd; L⇒(Al)+Al_3Gd+$Al_{10}Fe_2Gd$ at 638°C, 0.2 at.% Fe, 3.84 at.% Gd	Thiede et al. (1998)
Al–Fe–Hf****	Al_3Hf	Tetragonal *I4/mmm*	0.3980–0.4010	–	1.7140–1.7130	Low-temperature modification <727°C	Schmid (1992)
	$Al_{66}Fe_{20}Hf_{14}$	Unknown	–	–	–	Not in equilibrium with (Al)	
	Al_6Fe_6Hf	Tetragonal *I4/mmm*	0.850–0.865	–	0.492	Not in equilibrium with (Al). Exists at 7.7 at.% Hf and 35–61 at.% Al	
Al–Fe–La****	$Al_{11}La_3{}^{LT}$	Orthorhombic *Immm*	0.4433	1.315	1.013		Grieb (1992), Thiede et al. (1998)
	$Al_{10}Fe_2La$	Orthorhombic	0.9051	1.0249	0.9122	Equilibrium with (Al) not clear	
	Al_7Fe_2La	Unknown	–	–	–	Not in equilibrium with (Al)	
	Al_8Fe_4La	Tetragonal *I4/mmm*	0.882/0.890	–	0.519/0.5075	Not in equilibrium with (Al)	

(continued)

Table 9.5. Continued

System	Phase	Crystal structure	Lattice parameters, nm			Comments	Ref.
			a	*b*	*c*		
Al–Fe–Nd****	$Al_{11}Nd_3^{LT}$	Orthorhombic *Immm*	0.4359	1.2924	1.0017	Exists below 927°C	Mondolfo (1976), Thiede et al. (1998)
	$Al_{10}Fe_2Nd$	Orthorhombic	0.9006	1.0206	0.9069		
Al–Fe–V****	Al_6Fe	Orthorhombic *Cmcm*	0.7440	0.6464	0.8779	Al_6Fe is formed at high cooling rates	Ghosh (1992b), He et al. (2002), Ramanan et al. (1991)
	$Al_{10}V$ ($Al_{11}V$)	Cubic $Fd3m$	1.4492	–	–	15.1–15.9% V	
	$Al_{80}Fe_{7.5}V_{12.5}$	Unknown	–	–	–	Stable phase observed after rapid solidification	
	$Al_{87}Fe_xV_y$	Icosahedral	0.508–0.526	–	–	Observed in melt-spun alloys	
Al–Fe–Zr****	Al_3Zr	Tetragonal $I4/mmm$	0.4016	–	1.7320	L⇒(Al)+Al_3Fe+Al_3Zr at ~660°C	Belov et al. (2002a), Alekseeva (1992)
	Al_6Fe	Orthorhombic *Cmcm*	0.7440	0.6464	0.8779	In metastable equilibrium with (Al)	
	$Al_{0.92-x}Fe_xZr_{0.07}$	Tetragonal $I4/mmm$	0.849–0.8595	–	0.4890–0.4966	$0.31 < x < 0.54$. Not in equilibrium with (Al)	
	$(Al_xFe_{1-x})_2Zr$	Cubic $Fd\bar{3}m$	0.7430–0.7461	–	–	$0.15 < x < 0.175$. Not in equilibrium with (Al)	

* Contains also $Al_{11}Ce_3$
** Contains also Al_9Co_2
*** Contains also $Al_{45}Cr_7$ (Al_7Cr)
**** Contains also Al_3Fe

specifically in the range of concentrations where the ternary phase occurs in the equilibrium phase diagram (Hackenberg et al., 2002). These ternary phases have an orthorhombic crystal structure of the $Al_{10}Fe_2Yb$ type (space group *Cmcm*) with lattice parameters in the following ranges $a = 0.894$–0.9205 nm, $b = 1.0114$–1.0300 nm, and $c = 0.8980$–0.9169 nm (Thiede et al., 1998). Lack of data does not allow one to make the general conclusion about the equilibrium of the ternary phase with (Al) in all the systems. However, at least in one system, Al–Fe–Gd, the ternary $Al_{10}Fe_2Gd$ phase is shown to be in equilibrium with (Al) and form a ternary eutectics with the corresponding binary phases (Hackenberg et al., 2002) (see Table 9.5).

9.2. GENERAL FEATURES OF INTERACTION BETWEEN ALUMINUM AND TRANSITION METALS

The main peculiar feature of transition metals in Al-based alloys is the formation of supersaturated solid solutions and, at some specific compositions, quasicrystalline and amorphous materials during solidification.

Let us first focus on the formation of supersaturated solid solutions as this phenomenon is most widely used in modern commercial alloys. The higher the solidification (cooling) rate, the higher the solubility of transition metals in solid aluminum. Supersaturated solid solutions formed in this way are characterized by high stability, with the decomposition temperature being between 300 and 650°C, depending on the alloying element. The development of metallurgical methods involving high cooling rates, e.g. granule, powder, laser metallurgy, has led to a large number of investigations into Al–TM alloys. Modern ingot metallurgy operates with cooling rates of 10^0 to 10^2 K/s. At such cooling rates, the resultant aluminum solid solution can be saturated with transition metals to the equilibrium concentration at a temperature of the three-phase equilibrium in eutectic systems and to the composition of the liquid phase for peritectic systems.

The following features are typical of the equilibrium interaction between aluminum and transition metals (Toropova et al., 1998).

- The high-temperature solidus of Al-based alloys, close to the melting temperature of aluminum, determines a narrow temperature range of solidification for Al–TM solid solutions.
- The low solubility of TM in solid aluminum (the solubility limit from a fraction of one percent to 1.8%).
- The solubility of TM in solid aluminum sharply decreases with temperature.
- Low diffusion coefficients of TM metals in solid aluminum (by 3–4 orders of magnitude lower than those of Zn, Mg, Cu, Si, and Ag).

These peculiarities of the physico-chemical interaction of TM and aluminum determine the as-cast structure formed upon nonequilibrium solidification. The most typical and important effects of TM on as-cast material are:

- The formation of supersaturated solid solutions with aluminum, the maximum content of TM exceeding the terminal solubility on increasing the cooling rate.
- The refinement of the as-cast grain structure.

The ability of transition metals to form supersaturated solid solutions in solidification is a base for their positive effect on the properties of deformed semi-finished products. Secondary precipitates of TM aluminides which are formed during solid–solution treatment (homogenization) and hot deformation of ingots affect the dislocation structure of deformed semi-finished products and the precipitation hardening of heat treating Al-based alloys. The refinement of the as-cast structure is frequently related to the formation of stable and metastable phases during solidification. This is considered in more detail later in this chapter.

The supersaturation of the solid solution can be characterized by the ratio of the alloying element's concentration in the solid solution at a given cooling rate (C_v) to the limit equilibrium concentration of the alloying element in aluminum (C_0). The available information is summarized in Table 9.6.

The data shown in Table 9.6 suggests that in the case of peritectic systems, there is a threshold cooling rate above which the aluminum solid solution is supersaturated with an alloying element ($C_v/C_0 > 1$). This threshold cooling rate is about 10^2 K/s for aluminum alloys. No such threshold exists for alloys of eutectic systems. The schematic dependence of the supersaturation on the cooling rate for eutectic and peritectic alloys is given in Figure 9.11. Thus, the minimum undercooling thermodynamically required for the solidification of a metastable solid solution in eutectic systems is considerably lower than that for peritectic alloys.

Taking into account not only the C_v/C_0 ratio but also the actual solubility of TM in aluminum, one can highlight manganese (a limit solubility of 1.8%), scandium (0.4%) (both have eutectic phase diagrams with aluminum), and chromium (solubility in the liquid phase 0.4% and in the solid phase 0.8% at the peritectic equilibrium temperature) as additions which form a wide range of solid solutions at a relatively low undercooling. The supersaturation of solid solutions is strongly affected by the solubility of alloying elements in liquid aluminum, the position of the liquidus, the melt overheating, and the temperature range of solidification. The lower the liquidus temperature and the higher the melt overheating, the higher the solid solution supersaturation at a given cooling rate.

Figure 9.12 shows the data on the solubility of some transition metals in liquid aluminum. The most suitable for supersaturation liquidus is exhibited by Al–Mn, Al–Fe, and Al–Sc alloys, followed by Al–Cr alloys.

Table 9.6. Effect of cooling rate on the supersaturation of Al-based solid solutions with transition metals (Toropova et al., 1998)*

V_c, K/s	Mn	Fe	Cr	Zr	Mo	W	V	Ti	Sc
	C_v E	C_v E	C_v P	C_v P	C_v P	C_v P	C_v P	C_v P	C_v E
10^{-1} (equilibrium)	1.8	0.04	0.82	0.28	0.25	0.16	0.37	0.28	0.4
10^1	2.5	0.04	0.4	0.1**	0.1	0.06	0.1	0.1	0.4
10^2	2.8	0.1–0.2	0.72–1.0	0.15–0.17	–	–	0.25	0.2	0.5
10^3	3.0–5.0	0.1–0.2	1.5–3.5	0.5–0.7	0.7	–	1.3	0.41–1.0	1.1
10^4	5.0–7.0	0.15–0.2	2.0–5.7	0.9–1.7	–	–	–	1.4	–
10^5	6.1–10.0	–	5.35–5.7	1.2–2.0	1.0	1.0	1.3	–	3.2
10^6	8.6–14.0	0.35–8.0	5.8–7.75	2.5–2.8	2.7	6.7	2.6	3.7–5.0	–
10^7	–	9.6	–	3.78	3.5	11.5	2.6–3.7	–	–
V_c, K/s	Mn	Fe	Cr	Zr	Mo	W	V	Ti	Sc
	C_v/C_0	C_v/C_0	C_v/C_0	C_v/C_0	C_v/C_0	C_v/C_0	C_v/C_0	C_v/C_0	C_v/C_0
10^{-1}	1.0	1.0	1.0	1.0	1.0	1.0	1.0	1.0	1.0
10^1	1.3	1.0	0.49	0.36	0.4	0.37	0.11	0.35	1.0
10^2	1.5	3.3–5.0	0.9–1.2	0.53–0.60	–	–	0.67	0.7	1.38
10^3	1.7–2.8	3.3–5.0	1.9–4.4	1.8–2.5	2.8	–	3.5	1.45–3.5	2.7
10^4	2.8–3.9	3.7–5.0	2.5–7.1	3.2–6.05	–	–	–	5.0	–
10^5	3.4–5.5	–	6.7–7.1	4.3–7.1	4.0	6.2	3.5	–	7.5
10^6	4.8–7.8	8.7–20.0	7.1–9.4	8.9–9.5	10.8	45	7.0	13.3–16.1	–
10^7	–	240	–	13.5	14.0	72	7.0–10.0	–	–

* C_v is the limit solubility at a cooling rate V_c; C_0 is the equilibrium limit solubility (C_v at $V_c = 10^{-1}$ K/s); E denotes the eutectic and P, the peritectic equilibrium in the Al corner of the binary phase diagram

** Up to 0.8% Zr can be introduced in solid (Al) upon casting with high melt overheating (Belov, 1994, 1996; Belov et al., 2004)

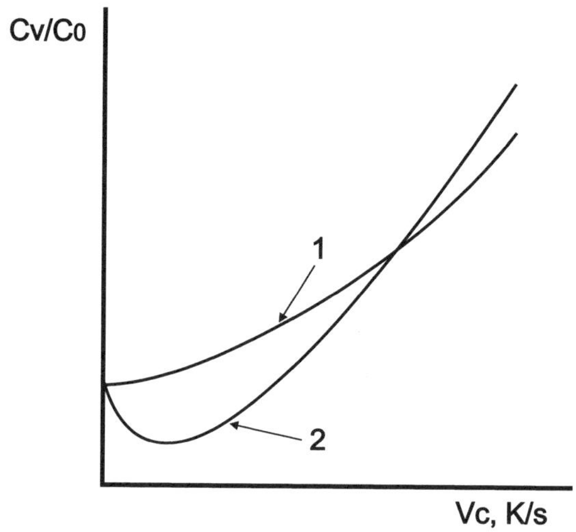

Figure 9.11. Schematic dependence of the supersaturation of (Al) solid solution on the cooling rate during solidification: 1 – eutectic systems and 2 – peritectic systems.

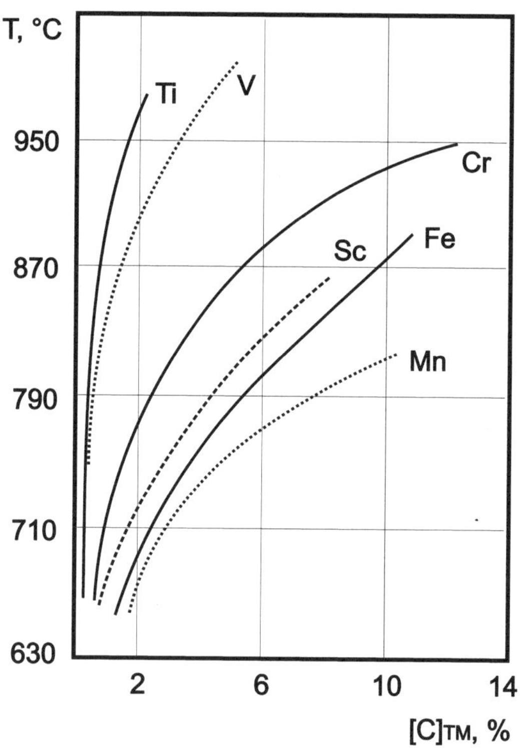

Figure 9.12. Solubility of some transition metals in liquid aluminum.

In principle, any alloy that is homogeneous in the liquid state can solidify as a solid solution. But the realization of such a process depends on the properties of the alloy and primarily on the following factors:

- The diffusion mobility of dissolved atoms: the higher the activation energy for the diffusion of the alloying element in liquid aluminum, the more stable the liquid;
- The nature of the phase precipitating from the liquid upon equilibrium solidification: the higher the difference in free energies of liquid and solid phases, the less stable the liquid.

The thermodynamic stability of the system can be estimated by the diffusion coefficient of the alloying element in solid aluminum and the decomposition temperature of a supersaturated solid solution. The transformation rate V (rate) for the decomposition of the supersaturated solid solution can be written in a form of the Arrhenius equation as follows:

$$V(\text{rate}) = \text{A}\exp(-\Delta U/\text{k}T).$$

Accordingly, the logarithm of the transformation rate depends linearly on the absolute temperature. Solvus from equilibrium phase diagrams appears as almost straight lines on $\log(C_{at})$ $-1/T$ axes (where C_{at} is the concentration in at. % and T is the absolute temperature in K), especially at high temperatures (Toropova, 1987) (Figure 9.13). The slope of solvus lines obtained in this way allows one to determine the enthalpy of mixing. The stability of metastable phases (in our case, supersaturated solid solutions) can be estimated using reference data on diffusion coefficients, decomposition temperatures of supersaturated solid solutions, and calculated enthalpy of mixing (Toropova et al., 1998). The distinction between eutectic and peritectic alloys is obvious. In the case of eutectic systems, the temperature dependence of the reaction rate is clearly pronounced, the decomposition temperature ranges from 300 to 350°C, the enthalpy of mixing is 36–80 kJ/kg, and the diffusion coefficient is one order of magnitude higher that that of peritectic alloys. The decomposition temperature of peritectic alloys falls in the range of 400 to 600°C and the enthalpy of mixing is considerably lower, 0.6–12 kJ/kg.

9.3. METASTABLE AND NONEQUILIBRIUM PHASE INTERACTIONS IN SOME Al–TM SYSTEMS

Apart from the formation of supersaturated solid solution, rapid solidification is known to result in metastable and nonequilibrium solidification structures and the

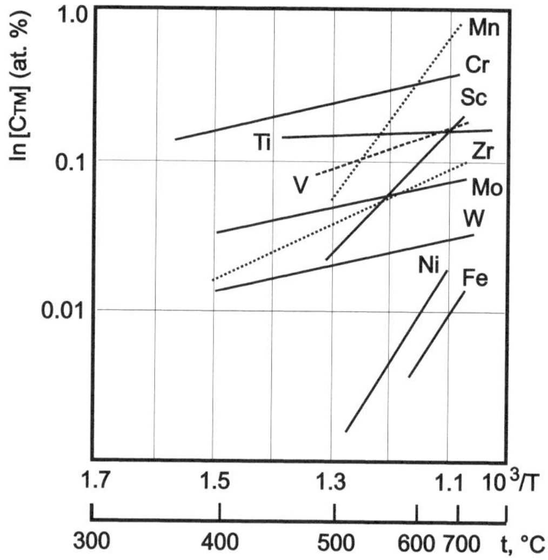

Figure 9.13. Temperature dependence of solubility of some transition metals in solid aluminum.

precipitation of metastable aluminides of transition metals in Al-rich alloys. In this section we consider some practically important examples.

Inoue (1998) formulated some rules for the formation of nonequilibrium phases in Al–RE (ETM)–LTM alloys produced by rapid solidification (ETM stands for Early Transition Metals of IV–VI groups and LTM – for Late Transition Metals of VII–VIII groups). The obtained structures systematically change from (Al) + compound → (Al) + quasicrystals → (Al) + amorphous phase → amorphous phase with the decreasing group number of RE, ETM, LTM in the periodic system. The decreasing group number of the constituent elements facilitates the formation of (meta)stable supercooled liquid as a result of increased atomic radius ratios and negative heats of mixing.

9.3.1. Peritectic systems

Al–Cr alloys. Chromium is present as a small addition in a number of aluminum alloys. The peritectic reaction in the Al corner of this system is suppressed at high cooling rates, and the formation of supersaturated solid solutions and nonequilibrium $Al_{11}Cr_2$ phase (instead of Al_7Cr) is observed. The latter phase also precipitates during decomposition of the supersaturated Al–Cr solid solution. The effect of cooling rate on the supersaturation of (Al) is shown in Table 9.6. Figure 9.14a demonstrates a structural stability diagram for the Al–Cr system. Dendrites of Al_7Cr can be easily substituted for $Al_{11}Cr_2$.

Al–Ti alloys. Most of the aluminum alloys contain titanium (alone or in combination with B, C, and Zr) as a powerful grain refiner. Although the formation

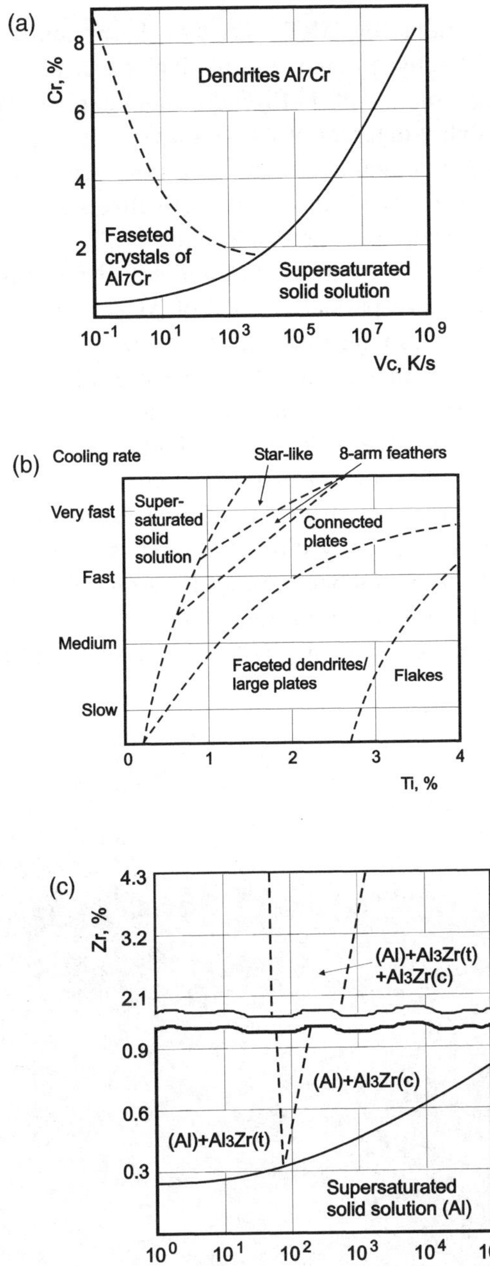

Figure 9.14. Structural stability diagrams for (a) Al–Cr (after Dobatkin et al., 1995); (b) Al–Cr (Blake after Svendsen and Jafors, 1993); and (c) Al–Zr alloys (after Toropova et al., 1998). Al_3Zr (t) is the equilibrium phase with $D0_{23}$ tetragonal crystal structure and Al_3Zr (c) is the metastable phase with $L1_2$ cubic crystal structure.

of metastable modifications of Al_3Ti has not been confirmed during rapid solidification, some distinctly different morphologies can be observed in relation to the cooling rate and amount of Ti (Svendsen and Jarfors, 1993). Figure 9.14b gives the structural stability diagram of Al–Ti alloys.

Al–Zr alloys. Modern wrought aluminum alloys frequently contain zirconium as a grain-refining and anti-recrystallizing agent. Both effects are related to the formation of metastable modification of the equilibrium Al_3Zr phase. Depending on the cooling rate, the phases which are in equilibrium with the aluminum solid solution appear either as needle-shaped particles typical of Al_3Zr or as fine star-like precipitates. With respect to the cooling rate and zirconium concentration, the volume proportion of these particles changes: needle-shaped precipitates dominate at cooling rates less than 80 K/s and in alloys with more than 0.3% Zr, but on increasing the cooling rate above 80 K/s the Al_3Zr phase solidifies mainly in the form of small stars. Increasing the zirconium content leads to the formation of a three-phase region where needle- and star-shaped particles are observed simultaneously as shown in Figure 9.15. According to an X-ray diffraction analysis, the star-like precipitates observed under an optical microscope have an fcc structure of the $L1_2$ type (space group $Pm\bar{3}m$) with the lattice parameter 0.405 nm as distinct from the equilibrium Al_3Zr phase that is tetragonal ($D0_{23}$ type, space group $I4/mmm$) with

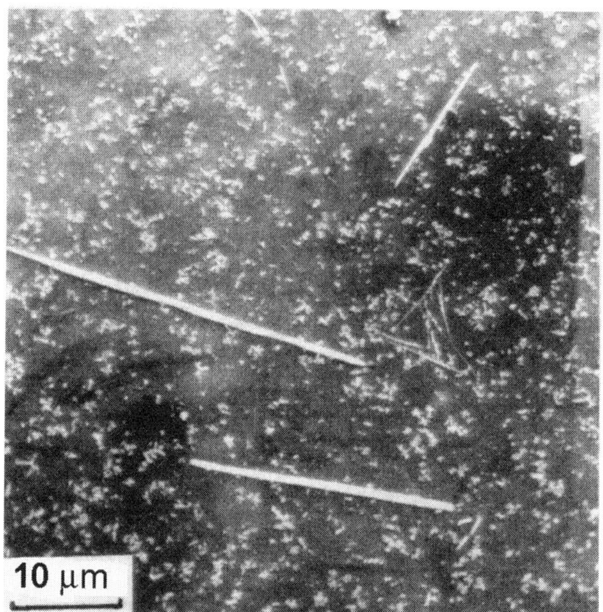

Figure 9.15. Microstructure an Al–2.2% Zr alloy solidified at 10^2 K/s, SEM.

$a = 0.4016$ nm and $c = 1.7320$ nm. The metastable cubic Al_3Zr phase precipitates also upon decomposition of a supersaturated solid solution of Zr in (Al), its coherent and semi-coherent particles efficiently hindering the recrystallization (Toropova et al., 1998).

The metastable Al_3Zr phase appears at cooling rates above 10^2 K/s. The simultaneous existence of the equilibrium and metastable Al_3Zr phases is probably due to the considerable broadening of the solidification range on increasing the zirconium concentration, which markedly affects the undercooling in solidification (Toropova et al., 1998). Therefore, at specific cooling rates the undercooling achieved under given temperature–concentration conditions promotes first the solidification of the metastable phase and then the formation of the equilibrium phase. Figure 9.14c shows the structural stability diagram for Al–Zr alloys (Toropova et al., 1998).

The metastable solidus is experimentally determined at cooling rates of 10^2 and 10^3 K/s (Figure 9.16). Thermal analysis shows that the temperature of the three-phase equilibrium in all given alloys is higher than that of the primary solidification of aluminum. The distribution of zirconium over a dendritic cell of a solid-solution alloy confirms that the Al–Zr phase diagram is of the peritectic type under given cooling and compositional conditions. One can suppose that the transformations

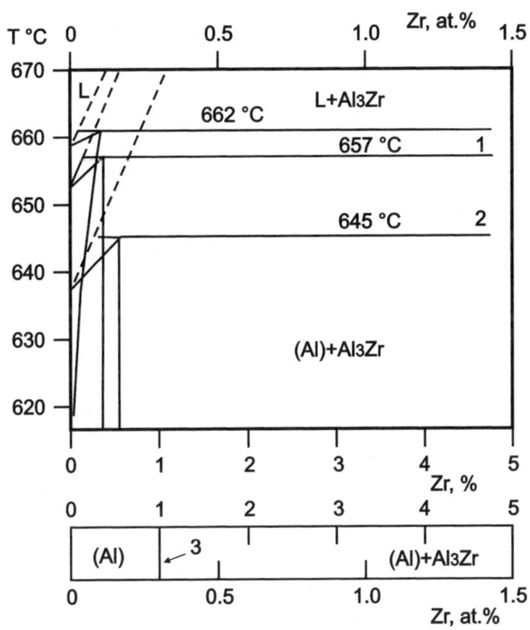

Figure 9.16. Metastable solidus lines for Al–Zr alloys at different cooling rates: (1) 5 K/s, (2) 10^2 K/s, and (3) 10^3 K/s (after Toropova et al., 1998).

based on interphase diffusion are suppressed or incomplete under strong volume undercooling when the diffusion in the liquid phase is severely hampered. In dilute alloys of the peritectic type, the solid-solution solidification occurs along the metastable solidus and involves all the phenomena of nonequilibrium solidification (microsegregation). With a higher zirconium content, solidification starts with the precipitation of Al_3Zr (tetragonal) crystals and should finish with the diffusion interaction between the remaining liquid and the solid phase. However, strong undercooling suppresses the peritectic reaction and the alloy solidifies like a solid solution even below the peritectic temperature. If the cooling rate is not high enough to assure the solid-solution solidification, the rest of the liquid can be transformed according to the eutectic reaction $L \Rightarrow (Al)_m + Al_3Zr$ (cubic) (both phases are metastable), the temperature of the invariant transformation being higher than the melting point of aluminum but below that of the equilibrium peritectic reaction. One can consider this as the "superimposition" of a eutectic reaction on the peritectic phase diagram.

Sigli (2004) estimated the metastable solubility of Zr in solid (Al) in binary Al–Zr and some commercial alloys. First of all, the metastable limit solubility of Zr in (Al) in Al–Zr alloys is confirmed to be significantly higher than that in the equilibrium, 0.87% Zr (cubic Al_3Zr in metastable equilibrium with (Al)) versus 0.28% Zr (tetragonal Al_3Zr in stable equilibrium with (Al)) at 660°C. Alloying considerably decreases the metastable solubility, most significantly in Al–Li and Al–Mg alloys as shown here. As a result the precipitation density of Al_3Zr in these alloys is significantly higher than in binary or 7XXX-series alloys.

Alloy composition	Metastable solubility of Zr in (Al) at 480°C, %
Al–1%Zr	0.147
Al–1%Zr–2%Zn	0.134
Al–1%Zr–2%Cu	0.122
Al–1%Zr–2%Mg	0.087
Al–1%Zr–2%Li	0.011

9.3.2. Eutectic systems

Al–Fe Alloys. Aluminum alloys containing iron as an alloying element are among promising RS/PM materials. The Al_3Fe phase is in equilibrium with (Al) forming a eutectic at 652°C and 1.8% Fe. During rapid solidification processing several metastable phases are formed as listed in Table 9.7.

On increasing the cooling rate, the eutectic point shifts towards higher iron concentrations and the eutectic temperature lowers. The range of primary (Al)

Table 9.7. Metastable phases observed in Al–Fe alloys (Hollingsworth et al., 1962; Kosuge and Mizukami, 1972; Mondolfo, 1976; Simensen and Vellasamy, 1977; Young and Clyne, 1981; Belov et al., 2002a)

Phase	Crystal structure	Lattice parameters, nm			Comments
		a	*b*	*c*	
Al_6Fe	Orthorhombic *Cmcm*	0.6492	0.7437	0.8788	Density 3.45 g/cm^3 25.6% Fe
Al_xFe	Monoclinic	2.16	0.93	0.905; $\beta = 94°$	Composition close to Al_5Fe Observed at 10^{-1}–10^1 K/s
Al_mFe	Tetragonal	0.884	–	2.160	Composition close to Al_9Fe_2 Observed at >20 K/s
Al_9Fe_2	Monoclinic $P2_1/c$	0.869	0.635	0.632; $\beta = 93.4°$	

solidification expands from 1.8% Fe (equilibrium) to 2.3% Fe at 10^3 K/s. In the concentration range 2.3 to 9.2% Fe two primary phases are formed, i.e. Al_3Fe and Al_6Fe. The effects of concentration and cooling rate on the phase diagram and structure are summarized in Figure 9.17.

Al–Mn alloys. Manganese is one of the first transition metals that has been introduced in aluminum alloys. It is still used as a small addition in numerous commercial alloys and as a main alloying element in 3XXX-series alloys. In the Al-corner of the Al–Mn phase diagram, (Al) is in the eutectic equilibrium (657°C, 1.9% Mn) with Al_6Mn which is formed peritectically at 710°C by the reaction of liquid with Al_4Mn.

The Al_6Mn phase (25.34% Mn) has an orthorhombic crystals structure (space group *Cmcm*) with $a = 0.649$–0.651 nm, $b = 0.754$–0.757 nm, and $c = 0.886$–0.887 nm, density 3.09–3.27 g/cm^3. The Al_4Mn phase (30–33% Mn) has a hexagonal structure with $a = 2.84$ nm and $c = 1.24$ nm (Mondolfo, 1976).

During fast cooling upon solidification, the peritectic reaction is partially or completely suppressed and primary crystals of Al_4Mn can be found in the structure alongside (Al) + Al_6Mn eutectics. The concentration and temperature of the eutectic equilibrium are also shifted as shown in Figure 9.18a. The structural stability diagram is given in Figure 9.18b.

Supersaturated manganese solid solution decomposes with the precipitation of several metastable phases. $Al_{12}Mn$ phase is the dominant phase and it has a cubic structure (space group *Im*3) with $a = 0.748$–0.758 nm. This phase forms semi-coherent precipitates and eventually transforms to Al_6Mn. Chromium stabilizes the $Al_{12}Mn$ phase in the form of $Al_{12}(CrMn)$ and makes it in equilibrium with Al–Cr–Mn alloys (Mondolfo, 1976). Several other metastable phases are mentioned by Mondolfo, with hexagonal, cubic, and rhombohedral structures (Mondolfo, 1976). Their presence is probably influenced by impurities.

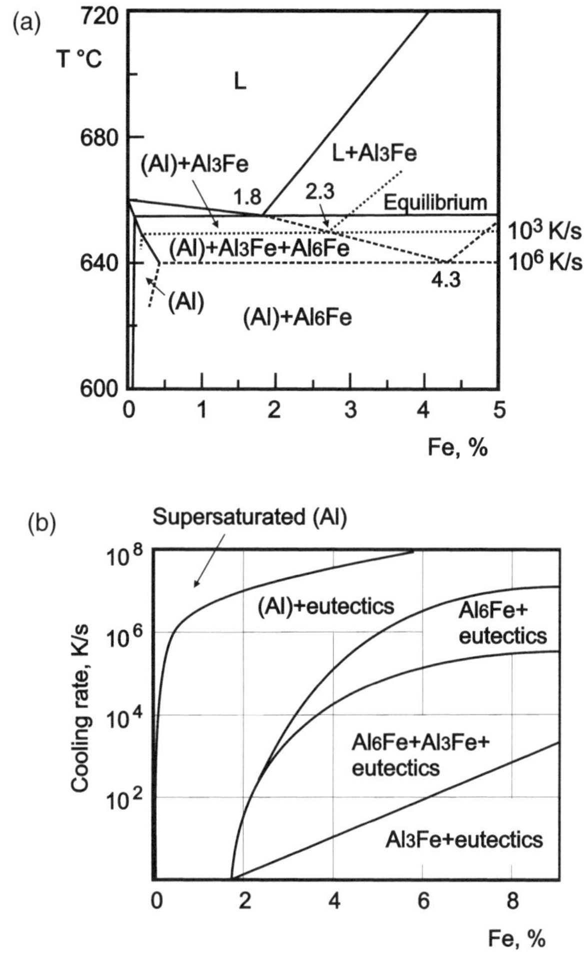

Figure 9.17. Metastable phase diagram (a) and structural stability diagram (b) of the Al–Fe system (Belov et al., 2002a).

Al–Sc alloys. Scandium attracts much attention as a promising addition to commercial aluminum alloys (Toropova et al., 1998). Table 9.6 shows that scandium forms supersaturated solid solutions with aluminum during solidification. The limit solubility of Sc in (Al) increases with the cooling rate and is equal to 0.5, 1.5, and 3.2% at 10^2, 10^3, and 10^5 K/s, respectively (Toropova et al., 1998). The eutectic concentration shifts to higher Sc concentrations and is found to be 1.5 and 4.5% Sc at a cooling rate of 10^3 and 10^5 K/s, respectively (compare with 0.6% Sc in the equilibrium diagram). The eutectic temperature decreases to 623–625°C at 10^3 K/s as compared to 655°C in the equilibrium (Toropova et al., 1998).

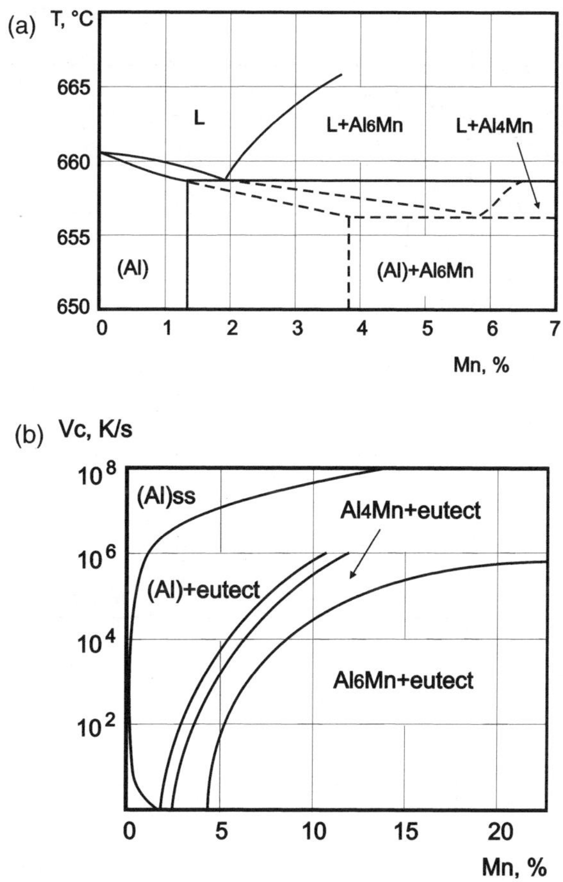

Figure 9.18. Metastable phase diagram (a) and structural stability diagram (b) for the Al–Mn system (after Dobatkin et al., 1995). Eutectics is (Al) + Al_6Mn, "ss" stands for supersaturated solid solution.

No metastable phases were found in the Al–Sc system. The equilibrium Al_3Sc phase (cubic structure of $L1_2$ type with $a = 0.4104$ nm) remains in the metastable equilibrium with the supersaturated solid solution and forms coherent and semi-coherent precipitates during its decomposition.

Figure 9.19 demonstrates the metastable phase diagram and the structural stability diagram for the Al–Sc system.

Based on the given information one can conclude that the general features of metastable (and non-equilibrium) phase diagrams of eutectic systems are as follows:

- Increased compositional range of primary (Al) solidification;
- Formation of supersaturated solid solutions with the degree of supersaturation increasing with the cooling rate;

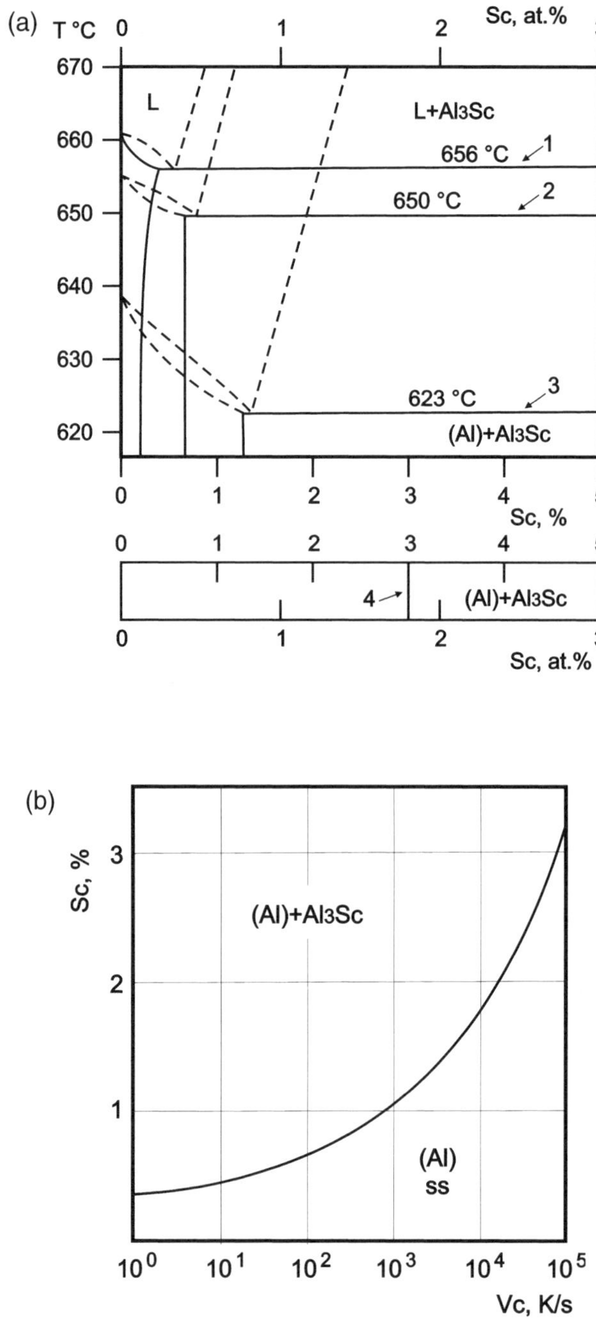

Figure 9.19. Metastable phase diagram ((a) 1 – 5 K/s; 2 – 10^2 K/s; 3 – 10^3 K/s; and 4 – 10^5 K/s) and structural stability diagram (b) for the Al–Sc system (after Toropova et al., 1998).

– Decreased eutectic temperature and shift of the eutectics concentration towards higher concentrations of the alloying element.

In many cases, the formation of metastable (Al_6Fe, Al_mFe) or non-equilibrium (Al_4Mn) phases is observed at high cooling rates. However, this is not the general case as exemplified by the Al–Sc system. Sometimes, the metastable phase forms only during decomposition of the supersaturated solid solution ($Al_{12}Mn$).

9.4. ALLOYS WITH TRANSITION METALS

9.4.1. Conventional aluminum alloys with transition metals

Although transition metals are used as major alloying elements mainly in highly specialized alloys produced by special techniques, there are some commercial and promising alloys that contain transition metals and are manufactured using the conventional process routine. Some of these alloys are already discussed in Chapters 1 and 7. Here, we consider Al–Ce–Ni alloys as an example of microstructural approach to alloy design. These alloys can be produced by conventional casting techniques at cooling rates of 1 to 10 K/s but possess the fine, thermally stable microstructure typical of rapidly solidified alloys. The base of these alloys is the formation of thermally stable and fine-in-constitution complex eutectics.

The aluminum corner of the Al–Ce–Ni phase diagram has a relatively simple constitution with only binary phases $Al_{11}Ce_3$ (Al_4Ce) and Al_3Ni phases in the equilibrium with (Al), and the invariant eutectic reaction between these phases at 627°C (Table 9.4, Figure 9.20). The aluminides have very narrow homogeneity ranges and dissolve less than 1% of the third element (Belov et al., 1999). The matrix remains as nearly pure aluminum. In the range of concentrations 5–6% Ni and 11–13% Ce, the structure of the alloys cast at 20 K/s consists almost completely of the ternary eutectics. However, the structure is inhomogeneous and exhibits zones with very fine ternary eutectics alongside regions with hypoeutectic structure (primary (Al) dendrites). An example of such two-zone structure is shown in Figure 9.21a while Figure 9.21b demonstrates a fine internal structure of the ternary eutectics (with particles less than 0.3 μm in size). The binary Al–Ce eutectics gains the same fineness of internal structure only at cooling rates above 10^4 K/s. Inhomogeneous structures like that shown in Figure 9.21a are typical of rapidly solidified alloys, e.g. Al–Fe granules produced at 10^6 K/s (Demarkar, 1986).

The analysis of volume fractions of structure constituents shows that the eutectic composition depends on the alloy composition (the eutectic is richer in Ce and Ni in hypoeutectic alloys) and the amount of eutectics is lower than it should be according to the lever rule and the equilibrium phase diagram (Belov et al., 1999).

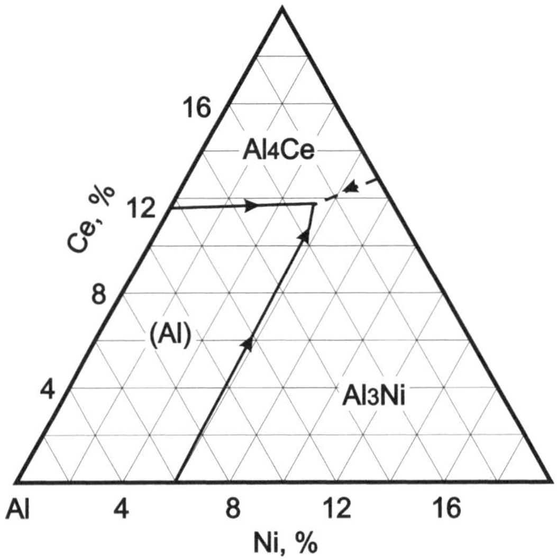

Figure 9.20. Aluminum corner of the Al–Ce–Ni phase diagram (Belov et al., 1999). Primary phases are shown in respective phase fields.

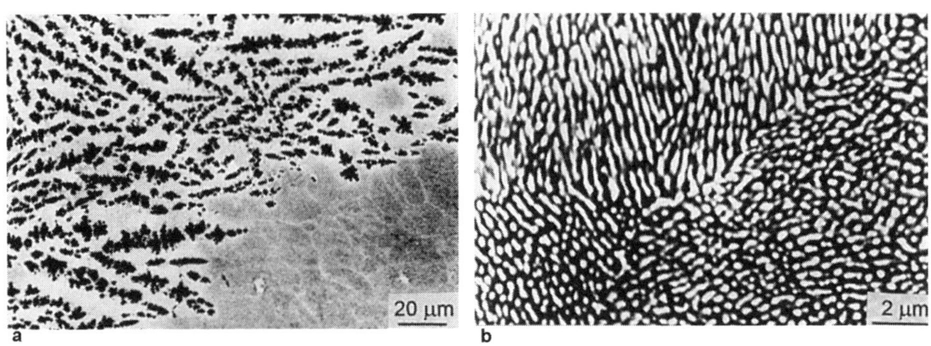

Figure 9.21. Microstructure of an Al–12% Ce–4% Ni alloy cast at 20 K/s (SEM, backscattered electrons): (a) two-zone structure with hypoeutectic region on the left and the eutectic region on the right and (b) ternary (Al) + Al_4Ce + Al_3Ni eutectics.

Apparently, even moderate cooling rates (10–20 K/s) cause large deviations from the equilibrium.

The fine structure of the ternary eutectics and thermal stability of the constituent phases (no visible structure changes at temperatures up to 400°C, spheroidization, and coarsening of eutectics particles to 0.5–1 μm at 450°C and to 5 μm at 600°C) assures good mechanical properties of eutectic alloys at room and high temperatures, alongside very good casting characteristics (Belov et al., 1999).

The same approach to alloy design is used for Al–Ce–Fe–Ni alloys with the complex (Al) + $Al_{10}CeFe_2$ + Al_9FeNi + Al_4Ce eutectics formed at 3% Fe and for Al–Fe–Mn–Ni–Si alloys with a complex eutectics formed at 1.5% Fe (Belov et al., 2002a).

9.4.2. Rapidly solidified aluminum alloys with transition metals

Rapid solidification of aluminum-based materials is an effective way of achieving a unique combination of properties, which cannot be obtained using traditional technologies. Aluminum alloys with transition metals are prone to the formation of supersaturated solid solutions and metastable phases during solidification, as well as to the precipitation of stable and metastable dispersoids upon decomposition of the supersaturated solid solutions, as has been discussed in Sections 9.2 and 9.3 of this chapter. Table 9.8 shows some compositions of the alloys produced by rapid solidification.

The effectiveness of the RS/PM technology is largely dependent on the cooling rate V_c in solidification, which may vary from 10^3 to 10^6 K/s for different alloys and technologies. One of the most widely known methods for achieving such a rate is the melt spinning technique – sputtering of the melt on a rapidly rotating copper disk or pouring between cooled copper rollers. Among other ways of obtaining rapidly cooled specimens one can mention the atomization of liquid metal by gas or ultrasonic oscillations, the extraction of the melt from a suspended drop or the melt surface, and the flattening of a drop with the help of electromagnetic field or by impact.

Table 9.8. Chemical compositions of some RS/PM alloys (Dobatkin et al., 1995; Belov et al., 2002a)

Alloy	TM1, %	TM2, %	TM3, %	Other elements, %
AlFeV	12Fe	2V	–	–
FVS0812 (8009)	8.5Fe	1.3V	–	1.7Si
FVS1212	12.4Fe	1.15V	–	2.3Si
FVS0512	5.5Fe	0.5V	–	1.1Si
FVS0611 (8022)	6.5Fe	0.6V	–	1.3Si
AlFeYZr	7Fe	4.6Y	2Zr	–
CU78	8.3Fe	4Ce	–	–
AlFeCe	2–4Fe	1Ce	1–6Ti	1Zr, 1W, 2–6Cu
AlFeMo	8Fe	2Mo	–	–
AlFeNi	5.9Fe	6.2Ni	–	–
AlFeCo	8Fe	2Co	–	–
01489rus	8.5Fe	1.1Cr	1.1Zr	1.1Mo
AlCrZr	5Cr	2Zr	–	–
01435rus	2.5–3.0Cr	2.5–3.0Zr	–	–
01419rus	1.2Cr	2Mn	0.5Zr	0.5Ti, 0.5V

Let us first consider rapidly solidified alloys based on the Al–Fe system. In binary Al–Fe alloys obtained at cooling rates above 10^6 K/s, a formed ultrafine structure consists of the metastable Al_6Fe (Al_mFe) phase and the supersaturated solid solution of iron in aluminum (>0.4% Fe) (see Section 9.3.2, Figure 9.17). Such alloys show a very high hardness at room temperature. However, coarsening of the structure and a sharp fall in strength occurs in heating above 300–350°C. This is associated mainly with the transition of metastable phases (Al_6Fe, Al_mFe, etc.) to the stable Al_3Fe phase and with the coagulation of fine intermetallic particles, accelerated by their extremely developed interface. As a result, the production of massive details from granules and flakes of binary Al–Fe alloys seems to be inefficient, because it is impossible to preserve a complex of properties inherent in rapidly cooled alloys.

A search for alloying systems, which would permit the stabilization of the structure and retention of high properties of RS/PM Al–Fe alloys, is based on the following principles (Dobatkin et al., 1995):

1. Changing and complicating the phase composition;
2. Adding soluble transition metals that enter the solid solution during solidification and inhibit diffusion of iron in aluminum.
3. Adding of soluble transition metals that enter the solid solution during solidification and form thermally stable precipitates upon decomposition of the supersaturated solid solution.

Examples of the first approach are alloys of the Al–Fe–Ce and Al–Fe–V–Si systems, in which complex phases stable to high-temperature heating are formed. As a result of the second approach, Al–Fe–Mo alloys are developed, where molybdenum plays a role of iron-diffusion inhibitor. The third approach is realized with additions of Zr and Cr to RS/PM alloys.

Studies aimed at developing heat-resistant Al–Fe–TM alloys obtained by RS/PM technology are based on the Al–Fe structure-stability diagram shown in Figure 9.17, which predicts the shift of the eutectic point towards the region of higher iron concentrations (up to 4–5%) on increasing the solidification rate up to 10^6 K/s.

Since binary Al–Fe alloys exhibit low thermal stability at elevated temperatures, the multicomponent compositions are developed with additions of transition metals (Co, Ni, Mo, V, etc.) and lanthanides (La, Ce, Nd, etc.), which are low-soluble in the aluminum solid solution (Dobatkin et al., 1995). Certain alloys contain additionally a small amount of silicon entering into the composition of complex compounds.

The choice of optimal compositions of aluminum alloys with iron additives is determined, firstly, by the necessity to provide the full binding of alloying elements in highly dispersed eutectics and, secondly, by obtaining phases with the required combination of properties, in particular, high thermal stability. It is customary to describe the composition of heat-resistant alloys obtained by RS/PM technology as Al–A% Fe–B% X, where X is at least one element of the group Co, Ni, Cr, Mo, V, Zr, Ti, Y, Ce; A varies from 7 to 15%; and B varies in the range 1–10% (Belov et al., 2002a). The cooling rate upon solidification of such alloys must be at least 10^6 K/s, while the very fine eutectics must be the main structural constituent of the alloy (its relative amount must be no less than 70%, or ideally, 100%). The size of iron-containing and other particles in the eutectics must not exceed 100 nm, whereas the volume fraction of these particles should be in the range from 25 to 45 vol.%. Among the elements denoted by X, zirconium, vanadium, and yttrium most effectively affect the structure, being added either separately or in combinations, e.g. Al–12% Fe–2% V and Al–7% Fe–4.6% Y–2% Zr.

In an Alcoa's CU78 alloy (Table 9.8) two phases, Al_3Fe (Al_6Fe) and $Al_{11}Ce_4$, are present in the structure obtained at cooling rates up to 10^3 K/s. However, when the cooling rate increases to 10^6 K/s, the ternary $Al_{10}Fe_2Ce$ phase is formed, which enters into the eutectic composition and stabilizes the properties of alloys at elevated temperatures. A noticeable coagulation of eutectic particles is observed only after heating for 2 h at 400°C. It is only heating at 500°C that results in the sharply changed morphology of intermetallic particles (Dobatkin et al., 1995). The quasicrystalline $Al_{20}Fe_5Ce$ phase is also reported in Al–Fe–Ce alloys of similar composition. Same regularities are observed after adding neodymium, lanthanum, and other REMs into Al–Fe alloys (Table 9.5), but the high cost of these metals limits their use. Different results are reported by Zhang et al. (2002) for an atomized and extruded Al–8.3% Fe–3.4% Ce alloy. After extrusion, this alloy contains metastable Al_mFe and Al_8Ce (rhombohedral, $a = 1.4$ nm, $c = 0.7$ nm) phases alongside equilibrium Al_3Fe. On increasing the annealing temperature to 315°C, metastable phases are substituted with equilibrium Al_3Fe and non-equilibrium $Al_{13}Fe_3Ce$ and $Al_{16}Fe_3Ce$ phases. And, after annealing at 400°C, the structure comprises Al_3Fe and $Al_{13}Fe_3Ce$. The latter phase has an orthorhombic structure and can be a modification of the stable $Al_{10}Fe_2Ce$ phase.

After extensive studies of Al–Fe alloys with Zr, Mo, Hf, Nb, and V additions, the Pratt and Whitney Corporation developed an Al–Fe–Mo alloy (Table 9.8). In the as-cast condition, the structure of this alloy (in flakes) consists of fine intermetallic Al_x(FeMo) particles, while extrusion at 290°C leads to the formation of globular inclusions of different iron-containing phases.

Heat-resistant alloys based on the Al–Fe–V–Si system are developed by Allied Signal (Table 9.8). These alloys, while exhibiting high ductility, fracture toughness, and fatigue resistance at room temperature, also display high heat resistance at temperatures up to 375°C. The hardening phase in these alloys, $Al_{12}(FeV)_3Si$ (cubic structure, $a = 1.26$ nm), is formed during rapid solidification and as a result of transformation from an icosahedral quasicrystalline phase during annealing (at ~270°C) (Srivastava and Ranganathan, 2001). This phase exhibits an extremely high resistance to coalescence. The advantage of Al–Fe–V–Si alloys over Al–Fe–Ce, Al–Fe–Mo, and Al–Fe–Mo–V alloys is kept up to 450°C.

Wang et al. (1998) studied the effect of mishmetal (MM) on the structure of an $Al_{93.3}Fe_{4.3}V_{0.7}Si_{1.7}$ alloy and showed that an increase in the MM concentration leads to the grain refinement of the aluminum matrix and the reduction in size of excess particles. In the presence of mishmetal, the nonequilibrium Al_8Fe_4MM and quasi-crystalline $Al_{20}Fe_5MM$ phases are formed at the expense of $Al_{12}(FeV)_3Si$. The formation of the latter phase is completely suppressed at 1% MM. In heating at 400°C, the Al_8Fe_4MM phase transforms to the more stable $Al_{12}(FeV)_3Si$ phase, raising the thermal stability of the alloy.

Rapidly solidified Al–Fe–Ni and Al–Fe–Co alloys gain their high-temperature resistance due to the fine particles of Al_9FeNi and $Al_9(CoFe)_2$ (Staley, 1985; Stéfaniay et al., 1996). These particles are quite stable up to 250–300°C. At higher temperatures coarsening occurs, and $Al_9(CoFe)_2$ transforms to $Al_3(CoFe)$.

RS/PM aluminum alloys containing Fe, Cr, Zr, and Mo (e.g. 01489 in Table 9.8) use a combination of stability and strengthening mechanisms and contain elements that are soluble and insoluble in solid aluminum at high cooling rates. The iron content corresponds to the eutectic point at a cooling rate of 10^6 K/s (Dobatkin et al., 1995). Additions of molybdenum and chromium form complex phases with iron and aluminum and retard diffusion of iron at high temperatures. Zirconium enters into the solid solution during solidification, and Al_3Zr particles precipitated in heating are responsible for additional strengthening of the alloy.

However, all the Al–Fe alloys considered here can work for a long time only at temperatures up to 290–300°C, which is their main disadvantage in comparison with RS/PM Al–Cr–Mn–Zr alloys having a higher thermal stability (Dobatkin et al., 1995). An introduction of zirconium and other soluble transition metals in Al–Fe alloys, while raising their strength characteristics at room and elevated temperatures, does not affect significantly the long-term thermal stability. The high-temperature stability is controlled by the fragmentation and coagulation of particles of iron-containing phases. In very fine eutectics formed in RS/PM alloys, these processes proceed intensely starting from 300–350°C and degenerate the structure.

Rapidly solidified alloys based on the Al–Cr system (AlCrZr, 01419 and 01435 in Table 9.8) contain fine primary particles of the Al_7Cr or $Al_{11}Cr_2$ phase (in alloys containing more than 2.5% Cr) and very fine (about 10 nm) and thermally stable dispersoids of Al_7Cr and Al_3Zr that precipitate from the supersaturated solid solution (Dobatkin et al., 1995; Bouchaud et al., 1990). Figure 9.22 shows how the range of an Al-based solid solution in Al–Cr–Zr alloys extends on increasing the cooling rate. Addition of manganese (alongside Ti and V) produces a thermally

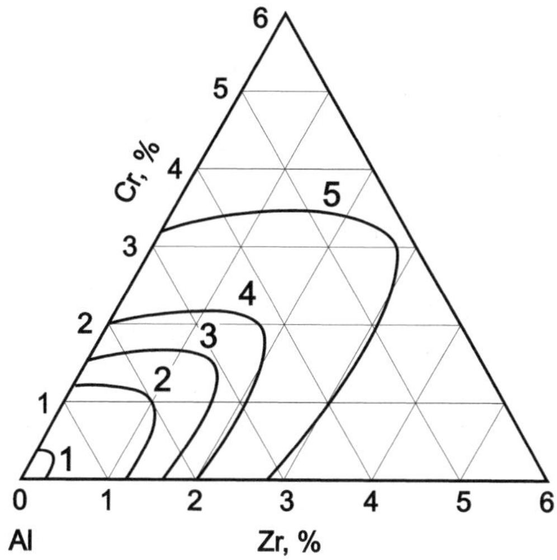

Figure 9.22. Solubility of Cr and Zr in solid (Al) under different solidification conditions: 1, equilibrium solubility at 640°C; 2, 3, 4, and 5, solubility at 20°C after solidification at 10^1, 10^2, 10^3, and 10^4 K/s, respectively (after Guzei, 1991; Dobatkin et al., 1995).

stable, complex X phase isomorphic to Al_6Mn that may dissolve Cr, Ti, and V (Dobatkin et al., 1995).

9.4.3. Quasicrystalline aluminum alloys with transition metals

In some cases, ultra-rapid solidification of aluminum alloys or annealing of metallic glasses is accompanied by the formation of intermediate, so-called quasicrystalline structures, in which the long-range order exists in the atomic arrangement. This atomic order corresponds, however, to the odd (5-fold) symmetry. The corresponding phases most often belong to the icosahedral symmetry. It should be noted that quasicrystals do not exhibit the same ordering and periodicity as classical crystals. Their description calls for the six-dimensional space. For this reason, the so-called lattice parameter, which is often used for icosahedral structures is only an averaged parameter derived from X-ray diffraction data, which does not have the same physical sense as the lattice parameter of true crystals.

Icosahedral phases are found in a number of aluminum alloys containing transition metals, e.g. Al–Cu–Co, Al–Cu–Cr, Al–Cu–Fe, Al–Fe–V (Tables 9.3 and 9.5).

Quasicrystals are formed either as metastable phases during rapid solidification or as stable phases always present in some compositional range.

Metastable quasicrystals are observed in all binary aluminum alloys with transition metals from V to Ni, from Mo to Pd, and from W to Pt (Grushko and

Velikanova, 2004). Stable quasicrystalline phases appear in the ternary systems based on one of the binary systems mentioned earlier. The third element can be also one of the mentioned transition metals or Cu. The stable icosahedral and decagonal phases can be considered as the extension of the homogeneity range of binary quasicrystals (Grushko and Velikanova, 2004). All stable Al-based quasicrystalline phases are found in the following systems: Al–Cu (or Ni, or Pd)–TM, where TM = Co, Fe, Mn, Rh, Ru. The element forming the quasicrystalline phase is usually the earlier TM and the dissolved element stabilizing the quasicrystal is the later transition metal (Grushko and Velikanova, 2004). Stable quasicrystals are formed during solidification by peritectic or transition reactions and are thermally stable, e.g. in the Al–Fe–Ni system the decagonal phase decomposes at about 100°C below its melting point into three neighboring phases which, in turn, transform during further heating to a high-temperature stable decagonal phase (Grushko and Velikanova, 2004). Figure 9.23 shows tentative compositional ranges of stable quasicrystalline phases in aluminum systems with transition metals.

9.4.4. Amorphous aluminum alloys with transition metals

The development of amorphous and nanocrystalline structures in aluminum alloys is one of the most promising trends in modern physical metallurgy. The results achieved on experimental specimens suggest that in the nearest future aluminum materials with such structures will be actually competitive with titanium alloys in mechanical properties at working temperatures up to 400°C.

Amorphous alloys, or metallic glasses, represent ideally a solid phase without a long-range atomic order. A short-range order exists, however, in such phases. It should be noted that TEM and X-ray diffraction examination, the absence of reflections from a crystalline phase and the formation of the so-called halo are often interpreted as evidence of the amorphous state. In this situation it is customary to say about an X-ray amorphous state. In other words, the diffraction-forming regions are comparable in size to the X -ray wavelength. The existence of such micro- or nanocrystalline regions determines the formation of the so-called nanocrystalline structure.

To produce an amorphous structure, it is necessary to satisfy the following conditions (criteria):

1. A thermodynamic criterion based on the assumption that the composition-dependent temperature T_0 exists, at which the solid and liquid phases have the same free energy;
2. A morphological criterion determining the velocity of interface motion at which this surface is morphologically stable, i.e. no grains are formed;

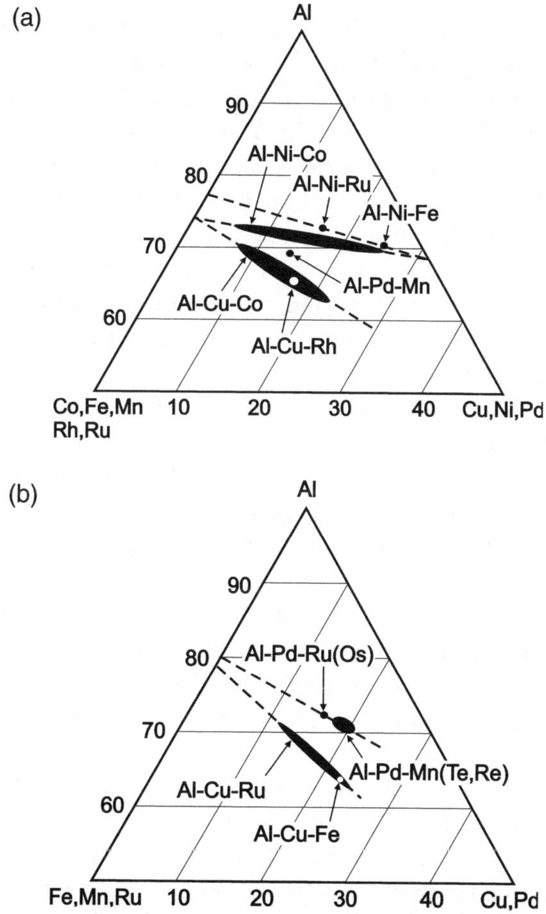

Figure 9.23. Schematic representation of compositional ranges of stable decagonal (a) and icosahedral (b) phases in ternary Al–TM1–TM2 alloys (axes in at.%) (after Grushko and Velikanova, 2004).

3. A thermal criterion, determining the supercooling of the liquid at which solidification of the entire melt is possible, even in the absence of the external heat removal;
4. A kinetic criterion, which determines the cooling rate required to prevent the formation of nuclei of the crystalline phase;
5. Structural criteria, for example, the necessity of a certain mismatch (>15%) between atomic radii of the components.

Eutectic systems between aluminum and transition metals is a source of Al-based amorphous and quasicrystalline materials. Although amorphous phase can be produced by melt quenching in Al–Fe–B, Al–Co–B, Al–Fe–Si, Al–Fe–Ge and

Al–Mn–Si alloys, they are extremely brittle. Much more promising alloying systems appeared to be Al–Early TM (ETM, IV–VI groups)–Late TM (LTM, VII–VIII groups and Cu), such as Al–Zr–Cu, Al–Zr–Ni, Al–Nb–Ni; Al–REM–Late TM where rare earth (REM) is represented by Y, La, and Ce, and late TM, by Fe, Co, Ni, and Cu (Inoue, 1998). The data on the equilibrium phases in these systems are given in Tables 9.3 to 9.5. A significant factor affecting the glass-forming ability of alloys is the interaction between constitutive atoms that results in the increased viscosity of the supercooled liquid and its stronger temperature dependence. In combination with the decreased melting temperature due to the existence of low-temperature eutectic reactions, this makes the ternary systems most attractive as a base of amorphous materials.

The heating of amorphous materials leads to their crystallization, which may occur without changing the composition, through the mechanism of primary crystallization and by the eutectic mechanism. The crystallization, which usually occurs during compacting and processing amorphous powders, is accompanied by the formation of a very fine and stable crystalline or quasicrystalline structure with unique properties.

Amorphous phases can be obtained during rapid solidification of alloys of the following general composition $Al_{70}LTM_{20}ETM_{10}$, where LTM is represented by Fe, Co, Ni, or Cu and ETM by Ti, Zr, Hf, V, Nb, Ta, Cr, and Mo (Inoue, 1998). The greatest ability to form metallic glasses is exhibited by Zr and Hf followed by Ti and V, especially in combination with Ni and Cu. In Al–Ni–Zr alloys the amorphization is observed in the compositional ranges 8 to 32% Ni and 3 to 18% Zr.

The formation of metallic glass is observed in melt-spun binary Al–REM alloys, where REM represents Y, La, Ce, Pr, Nd, Sm, Gd, Tb, Dy, Er, or Yb. Except for the Al–Nd system, the range of glass formation lies between the eutectic point and $Al_{11}REM_3$ or Al_3REM compound, roughly between 8 and 11–16 at.% REM (Inoue, 1998). It should be noted that the supersaturated solid solution is formed in the compositional range lower than the glass-formation range, up to the eutectic point. The reason why the eutectic composition in binary alloys does not produce the amorphous phase may be connected to the very low eutectic compositions in Al–REM systems (Inoue, 1998).

Wide compositional ranges of glass formation are observed in ternary Al–REM–TM alloys, where REM is Y, La or Ce and TM is Fe, Co, Ni or Cu. The widest formation ranges (from 0 to more than 20 at.% TM) are for Al–Y–Ni, Al–Ce–Ni, and Al–La–Fe(Co or Ni) systems (Inoue, 1998).

For more information on Al-based amorphous alloys, we address the reader to a very good review by A. Inoue (1998).

Chapter 10

Composite Materials with SiC, Al_2O_3, and SiO_2

This chapter considers phase composition of metal–matrix composite materials (MCMs) based on the Al–C–Si and Al–O–Si systems. In other words, these MCM have a matrix of aluminum or its alloys reinforced with fibers or particles of SiO_2, Al_2O_3, or SiC.

Interaction at the matrix–reinforcement interface in the presence of the liquid phase or upon heat treatment is one of the essential processes accompanying MCM production. For such an interaction to occur, a reinforcing element (fiber or particle) should be in direct contact with the matrix (liquid or solid). Under such conditions, active chemical reactions with the formation of various phases proceed at the interface, which may result in deterioration of mechanical and other properties of the final composite material.

In our view, the understanding of the interaction processes in aluminum-based MCMs requires the analysis of corresponding ternary and more complex phase diagrams.

10.1. Al–C–Si PHASE DIAGRAM

The Al–C–Si system is the basis for aluminum–matrix composites reinforced with SiC. The analysis of phase interactions in this system is very important, especially by taking into account that SiC is known to actively react with aluminum melt forming Al_4C_3 (or Al_4C_4Si) and free silicon.

The SiC, Al_4C_3, (Si), and (C) phases from respective binary systems can be in equilibrium with (Al) in the aluminum corner of the Al–C–Si system. In addition, two ternary compounds, Al_8C_7Si and Al_4C_4Si, can be in equilibrium with (Al). Data on crystal structure and density of these phases are given in Table 10.1.

The SiC phase forms in binary Si–C alloys by the peritectic reaction (Doboleg, 1963; Elliott, 1965):

$$L + C \Rightarrow SiC\,(25\text{ at.\% C}, 2545°C).$$

Then the following eutectic reaction occurs in the Si–C system:

$$L \Rightarrow Si + SiC\,(0.25\text{ at.\% C}, 1404°C).$$

Table 10.1. Crystal structure and density of phases of Al–C–Si system (Kotelnikov et al., 1968; Oden and McCune, 1987; Lukas, 1990; Drits, 1997)

Phase	Crystal structure	Lattice parameters			Density, g/cm^3
		a, nm	*c*, nm	β	
β-SiC	Cubic, $F\bar{4}3m$	0.43596	–	–	3.2–3.8
α-SiC	Hexagonal $P6_3mc$	0.3078	n·0.2518*	–	
Al_4C_3	Rhombohedral	0.855	–	22°28′	2.93–2.96
(C)	Hexagonal $P6_3/mmc$	0.2464	0.6711	–	2.66
(Si)	Cubic $Fd3m$	0.54285	–	–	2.33
Al_8C_7Si	Hexagonal $P6_3/mmc$	0.33128	1.92424	–	2.98
Al_4C_4Si	Hexagonal	0.32771	2.1676	–	3.03

* $n=4$ to 15 is the number of layers per unit cell

According to Kotelnikov et al. (1968) the solubility of carbon in liquid silicon is rather small as shown here:

T, °C	1725	1600	1520
C, at.%	0.43	0.12	0.05

Both crystal forms of SiC are thermodynamically close (Doboleg, 1963) and, therefore even minor variation in process conditions can be sufficient for either β-SiC or α-SiC to appear. For the same reason, it is difficult to determine exactly which of the modifications is high-temperature and which is a low-temperature form. Consequently, the temperature position of the modifications is not reflected in the phase diagram. SiC decomposes at atmospheric pressure, failing to melt up to 2700°C. The hardness of SiC at room temperature is HV3600, the ultimate strength in tensile tests at room temperature is 180 MPa (Drits, 1997).

The Al_4C_3 phase (42.86 at.% [25.03%] C) forms in binary Al–C alloys by the following peritectic reaction (Elliott, 1965; Schuster, 1991):

$$C_{solid} + L \Rightarrow Al_4C_3 \text{ (at 25\% C and 2156°C or 1990°C).}$$

It can also form at a temperature of ≈660°C by the eutectic reaction:

$$L \Rightarrow (Al) + Al_4C_3 (<0.01\% \text{ C in the liquid phase}).$$

Upon heating, the crystalline carbide remains solid up to 2027°C. In the amorphous state, this compound is less stable and decomposes at 1227°C into (Al) and (C).

There are reports on the existence of the Al_3C carbide (12.9% C) that is in equilibrium with (Al) and Al_2O_3, and of the Al_2C_6 carbide (57.2% C) forming in Fe-rich alloys. Carbides AlC_2 (47.5% C) and AlC (30.7% C) are also known, but their formation in aluminum alloys is doubtful.

The solubility of carbon in liquid aluminum is very small as shown here (Shunk, 1969):

T, °C	1200	1100	1000	800
C, % (at.%)	0.32 (0.71)	0.16 (0.35)	0.14 (0.31)	0.1 (0.22)

The maximal solubility of carbon in solid (Al) is approximately 0.015% (0.03 at.%).

The Al–C–Si phase diagram was studied in detail (Oden and McCune, 1987; Viala et al., 1990) with quasi-binary Al_4C_3–SiC and isothermal sections given by Oden and McCune (1987).

Figure 10.1a presents a projection of monovariant liquidus lines and ternary invariant planes on the Al–C–Si concentration triangle as suggested by Oden and McCune (1987). The possible invariant reactions are listed in Table 10.2 according to different reference sources.

Many authors note that high-temperature phase equilibria given by Oden and McCune (1987) do not explain a number of experimental observations on the interaction of silicon carbide with aluminum at temperatures below 2000°C.

Viala et al. (1990) proposed a model for describing the Al–C–Si system using three separate phase diagrams:

1. A stable phase diagram in which equilibrium is achieved at temperatures above 1930°C;
2. A metastable phase diagram where equilibrium is achieved at temperatures from 1630 down to 1400°C;
3. A metastable phase diagram where equilibrium is achieved at temperatures below 1400°C.

At temperatures between 1630 and 2000°C there is a possibility of reaching the equilibrium at which the Al_4C_4Si and Al_8C_7Si phases can coexist with (Al) (Viala et al., 1990; Aksenov et al., 1995, 2001a).

Within the temperature range from 1400 up to ~1630°C, the interaction between aluminum and SiC produces the ternary carbide Al_4C_4Si rather than Al_8C_7Si (Viala et al., 1990). As a result, the line separating the liquidus surfaces of the SiC and Al_4C_3 phases in Figure 10.1b vanishes. Instead, two other lines appear in the phase diagram, reflecting the decomposition of two liquid phases according to the following monovariant reactions:

$$L_1 \Rightarrow Al_4C_4Si + SiC \text{ and}$$

$$L_2 \Rightarrow Al_4C_4Si + Al_4C_3.$$

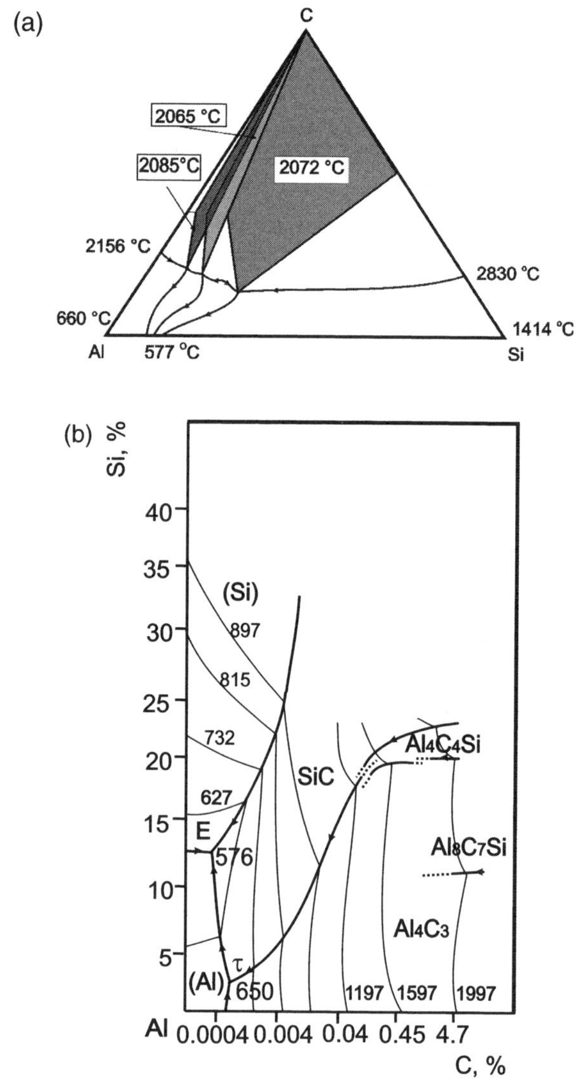

Figure 10.1. (a) Projection of monovariant lines and invariant planes (after Oden and McCune, 1987) and (b) projection of the liquidus surface of the Al–C–Si phase diagram (after Viala et al., 1990).

And, finally, the following monovariant eutectic reaction possibly occurs in the temperature range 650–1400°C:

$$L \Rightarrow Al_4C_3 + SiC.$$

The composition of the liquid phase during this reaction changes along the line separating the liquidus surfaces of the SiC and Al_4C_3 phases, with the concentration

Table 10.2. Invariant reactions in Al–C–Si system (Shunk, 1969; Oden and McCune, 1987; Viala et al., 1990; Lukas, 1990)

No	Reaction	Temperature, °C	Concentrations in liquid phase	
			Si, at.%	C, at.%
1	$L + Al_4C_3 + (C) \Rightarrow Al_8C_7Si$	2085	10	17
2	$L + (C) \Rightarrow Al_8C_7Si + SiC$	2072	27	10
2′	or $L + (C) \Rightarrow Al_4C_4Si + SiC$*			
3	$L + (C) \Rightarrow Al_8C_7Si + Al_4C_4Si$	2065	18	16
4	$L + Al_4C_3 \Rightarrow (Al) + SiC$	650	1.5	<0.001
4′	or $L + Al_4C_3 \Rightarrow (Al) + Al_8C_7Si$*	645*		
5	$L + Al_8C_7Si \Rightarrow (Al) + Al_4C_4Si$*	620*	–	–
6	$L + Al_4C_4Si \Rightarrow (Al) + SiC$*	~582*	–	–
7	$L \Rightarrow (Al) + (Si) + SiC$	576	12.3	<0.001

* Invariant reaction given by Lukas (1990)

of C in the liquid remaining rather low and the concentration of Si decreasing from 16 at.% at 1300°C down to 1.5 at.% at 650°C (Viala et al., 1990).

At 650°C, SiC and (Al) are formed from Al_4C_3 and the liquid, i.e. through invariant peritectic reaction 4 in Table 10.2 (point τ in Figure 10.1b). Then the remaining liquid undergoes a monovariant eutectic transformation within the temperature range 650–576°C:

$$L \Rightarrow (Al) + SiC.$$

And solidification completes at a temperature of 576 ± 1°C by invariant eutectic reaction 7 (point E in Figure 10.1b) to form (Al), (Si), and SiC.

10.2. Al–O–Si PHASE DIAGRAM

The Al–O–Si phase diagram is required for understanding the interactions between aluminum on one side and alumina (Al_2O_3), silica (SiO_2), and mullite (AlSiO) on the other. This system is also basic for ceramic technologies, e.g. for interpreting physico-chemical processes that occur upon annealing, melting, and crystallization of various alumo-silicate refractory mixtures and upon their interaction with various media.

From the analysis of the Al–O binary system it follows that in the absence of water a stable compound, aluminum oxide Al_2O_3 (47.1% O) is in equilibrium with aluminum solid solution. The compound can exist in different forms, some of which are listed in Table 10.3. These forms are not true polymorphs but rather transition

Table 10.3. Crystal structure and lattice parameters of Al_2O_3 (Mondolfo, 1976; Morrissey et al., 1985)

Phase modification	Crystal structure	Lattice parameters				Density, g/cm^3
		a, nm	b, nm	c, nm	β	
α-Al_2O_3	Rhombohedral ($R\bar{3}c$)	0.5129	–	–	55°17′	3.96–4.02 at
(stable, >1200°C)	Hexagonal	0.476	–	1.299		20°C
γ-Al_2O_3 (>670°C)	Cubic ($Fd3m$)	0.790	–	–	–	3.6
θ-Al_2O_3	Hexagonal	0.84	–	1.365	–	–
	Monoclinic	1.124	0.572	1.174	103°20′	
δ-Al_2O_3 (>800°C)	Tetragonal $c/a = 2.9$	0.794	–	2.304	–	–

phases from amorphous alumina to stable α-Al_2O_3. Therefore, the transition occurs only on increasing the temperature.

The aluminum oxidation proceeds by the formation of amorphous oxide and its subsequent transition to α-Al_2O_3 on heating. This transition is possible by two schemes:

1. The main route: γ-Al_2O_3 ⇒ δ-Al_2O_3 ⇒ α-Al_2O_3; and
2. A parallel route, where the formation of the θ-Al_2O_3 phase from γ-Al_2O_3 is possible.

It is important to note that air moisture should be taken into account while considering the oxidation of aluminum under natural conditions. Several types of hydroxides are known to be formed by the interaction of aluminum with moist atmosphere: γ-$Al(OH)_3$, α-$Al(OH)_3$, γ-AlO(OH), and α-AlO(OH). The most stable form under common conditions is γ-$Al(OH)_3$. Numerous studies have shown that in air a film of γ-$Al(OH)_3$ covers aluminum, e.g. aluminum powder (Gopienko and Smagorinskii, 1993).

The melting temperature of Al_2O_3 varies within 2037 to 2072°C.

The combined heating of Al and Al_2O_3 leads to the formation of two more oxides – Al_2O and AlO (Gopienko and Smagorinskii, 1993). The Al_2O oxide forms within the temperature range 1100–1500°C. It has a cubic structure with lattice parameter $a = 0.498$ nm. The compound AlO is observed at temperatures between 1500°C and 1900°C. It is also cubic with lattice parameter $a = 0.567$ nm.

There is a miscibility gap between liquid aluminum and liquid alumina (Gopienko and Smagorinskii, 1993). The monotectic temperature is close to the melting temperature of Al_2O_3 (2046.5°C), and the temperature of eutectic transformation almost coincides with the melting temperature of pure aluminum.

The solubility of oxygen in (Al) is negligibly small and does not exceed 0.067 at.%.

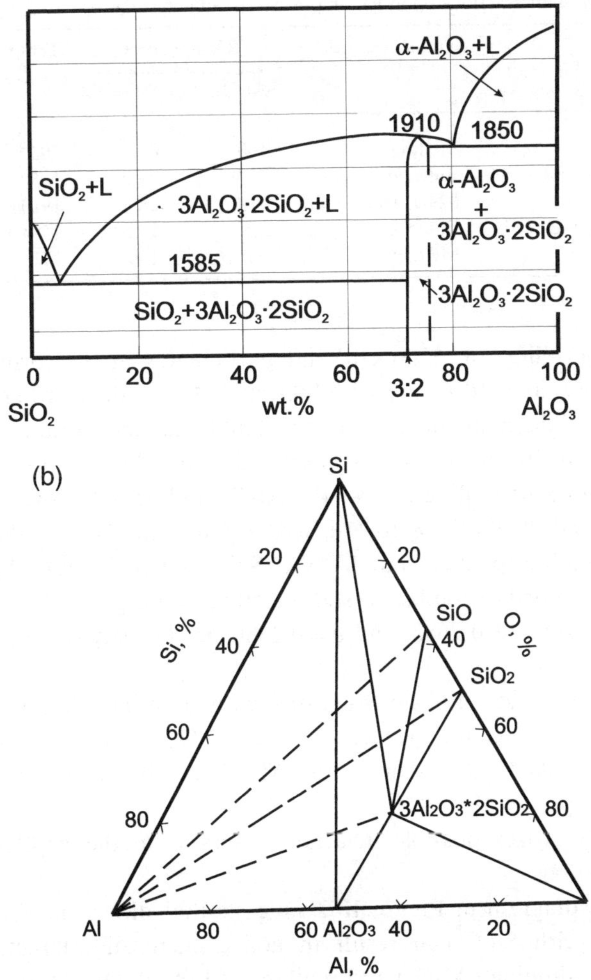

Figure 10.2. Al–O–Si phase diagram: (a) section Al_2O_3–SiO_2 and (b) triangulation of the system in the region Al–Al_2O_3–SiO_2–Si.

In the Si–O system, the following oxides can form: SiO_2, SiO, Si_2O_3, and Si_3O_4. The SiO phase melts at 1710 ± 10°C and decomposes into a mixture of Si and SiO_2 at a temperature of ~1100°C (Toropov et al., 1969).

The phase equilibria in the aluminum corner of the Al–O–Si system is less studied, the Al_2O_3–SiO_2 section being mostly investigated and is shown in Figure 10.2a.

The quasi-binary Al_2O_3–SiO_2 section contains a ternary compound called mullite, the formation and composition of which is a matter of discussion. In one version of the phase diagram, it forms directly from the liquid phase at 1860°C, has the

Table 10.4. Invariant reactions upon heating in Al_2O_3–SiO_2 system

Reaction	Transformation	Composition		Temperature, °C
		Al_2O_3, %	SiO_2, %	
$Al_2O_3 \Rightarrow L$	Melting	100	0	2050
$Al_2O_3 + 3Al_2O_3{\cdot}2SiO_2 \Rightarrow L$	Eutectic	79.0	21.0	1850 ± 10
$3Al_2O_3{\cdot}2SiO_2 \Rightarrow L$	Congruent melting	71.8	28.2	1910 ± 10
$3Al_2O_3{\cdot}2SiO_2 + SiO_2 \Rightarrow L$	Eutectic	5.5	94.5	1584 ± 10
$SiO_2 \Rightarrow L$	Melting	0	100	1713

composition $2Al_2O_3{\cdot}SiO_2$ or $Al_4Si_2O_{10}$ and participates in the formation of two eutectics, with Al_2O_3 at 1810°C and with SiO_2 at 1640°C (Shepherd et al., 1909). In the other version of the quasi-binary section, mullite has the formula $3Al_2O_3{\cdot}2SiO_2$ or $Al_6Si_2O_{13}$ and melts incongruently. Depending on the accepted formula, the chemical composition of mullite is: 64.84% Al_2O_3 and 35.41% SiO_2 for $Al_4Si_2O_{10}$ or 71.8% Al_2O_3 and 28.2% SiO_2 for $Al_6Si_2O_{13}$. One can also say that the homogeneity range of mullite spreads from $3Al_2O_3{\cdot}2SiO_2$ to $2Al_2O_3{\cdot}SiO_2$ (Toropov et al., 1969). Mullite has an orthorhombic crystal structure (space group *Pbam*) with lattice parameters $a = 0.7585$, $b = 0.7682$, and $c = 0.2886$ nm. Density of this compound is 3.11–3.26 g/cm^3.

Mullite forms a range of solid solutions with alumina. A possible, not yet confirmed quasi-binary section between Si and Al_2O_3 is suggested as a result of triangulation of the Al–O–Si system in the Al–Al_2O_3–SiO_2–Si region as shown in Figure 10.2b (Toropov et al., 1969).

Table 10.4 summarizes possible invariant reactions in the Al_2O_3–SiO_2 system (Toropov et al., 1969).

Analysis of the diagram in Figure 10.2 suggests that direct and long contact of liquid aluminum with SiO_2 can result in active chemical interaction with the formation of both alumina (Al_2O_3) and mullite ($3Al_2O_3{\cdot}2SiO_2$). However, the most accepted reaction is (Toropov et al., 1969):

$$4Al + 3SiO_2 \Rightarrow 3Si + 2Al_2O_3.$$

The same reaction is reported to occur in the solid state (at temperatures as low as 440–550°C) at the interface between aluminum and silica (Aksenov et al., 1991). It is noteworthy that one of the reaction products is free silicon. This gives an opportunity to control the extent of interaction at the interface between the aluminum matrix and ceramic fibers or particles by monitoring the composition of the matrix.

10.3. Al–C–Si PHASE DIAGRAM FOR THE ANALYSIS OF INTERFACIAL PROCESSES IN Al–SiC AND Al–Si–SiC METAL–MATRIX COMPOSITES

The correct analysis of phase transformations and reactions occurring in the solid state in Al-based composite materials requires the knowledge of metastable equilibrium and nonequilibrium phase selection. In this section we consider the interaction between aluminum matrix and SiC reinforcement and suggest some metastable and nonequilibrium section of the Al–C–Si phase diagram as applicable to the composite materials.

Analysis of the literature data and our own results shows that the harmful Al_4C_3 phase forms according to the following chemical reaction at temperatures below 1400°C:

$$4Al + 3SiC \Rightarrow 3Si + Al_4C_3.$$

10.3.1. Experimental study of the matrix–reinforcement interaction

Two types of MCMs were selected for the examination of the interaction between the matrix and the reinforcing phases:

1. MCMs containing 20% SiC particles, so-called MCMp (compositions are given in Table 10.5); and
2. MCMs containing 10 vol.% SiC fibers, so-called MCMf (compositions are given in Table 10.6).

The matrix alloys were prepared using 99.99% pure aluminum and 99.999% pure silicon by melting in an electrical furnace in alumina crucibles.

Table 10.5. Compositions of MCMp reinforced with SiC particles

MCMp	1	2	3	4
Si in the matrix, %	–	5	7	12
Fraction of SiC, wt%	20			

Table 10.6. Compositions of MCMf reinforced with SiC fibers

MCMf	5	6	7	8	9	10
Si in the matrix, %	–	1	3	5	7	12
Volume fraction of SiC, vol.% (%)	10 (10.6)					

Particles were mechanically mixed with the melt in the semi-solid state as described elsewhere (Polkin et al., 1993; Aksenov et al., 1994). The time of contact of the aluminum melt with SiC at a given temperature did not exceed 15 min. As a result MCMp castings with uniform spatial distribution of SiC particles were obtained. The average size of SiC particles was 10 μm, and the crystal structure, according to X-ray analysis, was α-SiC.

Fiber-reinforced MCMf were obtained by vacuum impregnation of a melt of a bundle of long coreless SiC fibers according to the method described elsewhere (Aksenov et al., 1995). This method ensures the longitudinal arrangement of the fibers in an MCMf specimen with their sufficiently uniform transverse distribution, i.e. without noticeable clusters. SiC fibers had the crystalline β-SiC structure.

The use of particles and fibers in our investigation gave us an opportunity to reveal the difference in interaction behaviors of α- and β-SiC and, in addition, to clarify the effect of impurities on the kinetics and phase composition of interaction products (fibers contained rather high concentrations of free carbon and oxygen).

To study the interaction processes, specimens of all MCM were held for various times at temperatures of 700, 800, and 900°C, i.e. above the liquidus of the matrix alloys. Anneals were performed in alumina crucibles either under pure Ar atmosphere or in air. The temperature was maintained accurate to ±5°C. The slurry was subsequently cooled at a rate of 10 K/s, which made it possible to model the real conditions of MCM production and casting.

Interaction in particle-reinforced materials (MCMp). Figure 10.3 shows the initial structure of MCMp 1 (Table 10.5) and the kinetic dependences of the SiC, Al_4C_3, and (Si) mass fractions (assessed by X-ray analysis) on the holding time at 700, 800, and 900°C. Apparently, Al_4C_3 and (Si) are already present in the initial state, immediately after the MCMp was obtained. This implies that reaction $4Al + 3SiC \Rightarrow 3Si + Al_4C_3$ already begins in the preparation stage. On holding at a high temperature, the amount of Al_4C_3 and (Si) phases rapidly rises during first 2–3 h and then virtually does not change upon holding for as long as 22 h. Simultaneously, the amount of SiC decreases by a similar law. It is important to note that the phase composition changes as a result of intensive diffusion of Al and Si in opposite directions across the interface. The observed dependences are general for all tested temperatures. However, the reaction rate (amount of reaction products) increases with the temperature.

Interaction in fiber-reinforced materials (MCMf). The initial structure of MCMf 5 (Table 10.6) differs significantly from the initial structure of MCMp 1 (Table 10.5) and consists of only (Al) and β-SiC fibers (Figure 10.4a). The X-ray analysis does not reveal any interaction products. Possibly, this is due to a very small time (less than 30 s) of contact between fibers and the melt during the production stage. At all temperatures studied, a latent period of interaction is observed in MCMf

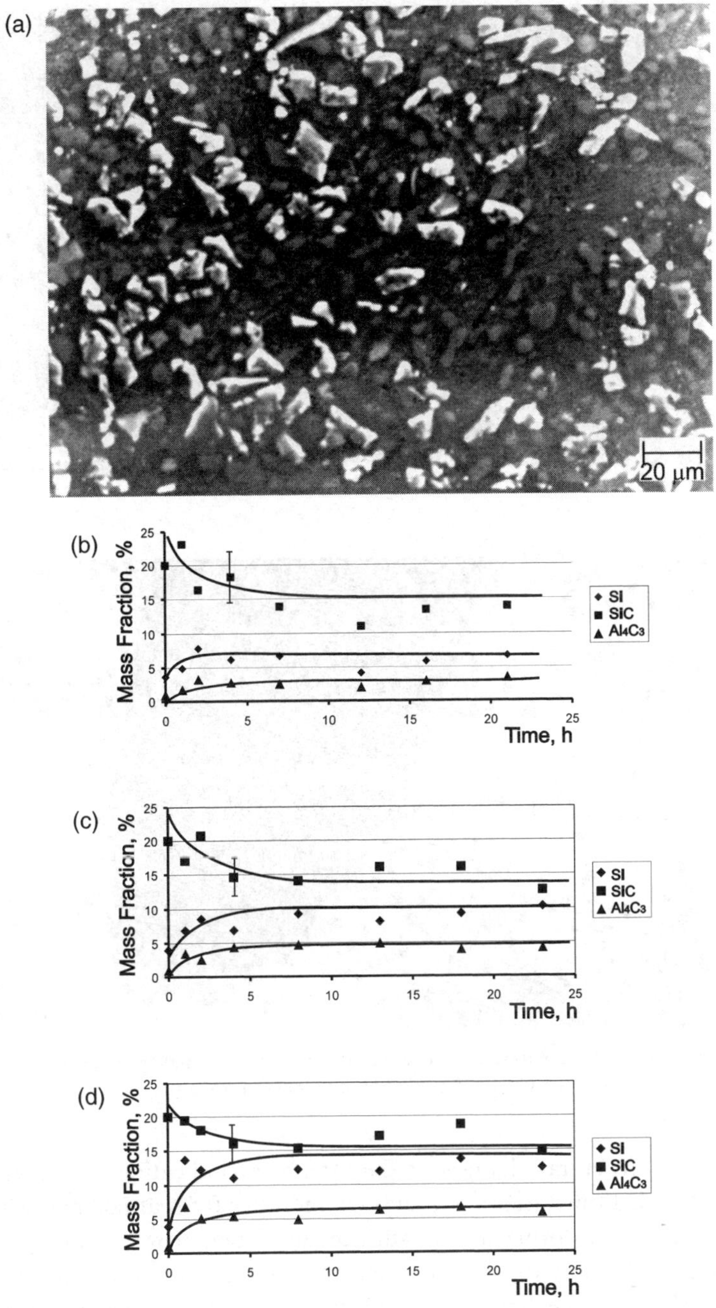

Figure 10.3. Initial structure (a) and mass fractions of SiC, Al_4C_3, and Si in the composite material Al–20%SiC as a function of holding time at temperatures of 700 (b), 800 (c), and 900°C (d).

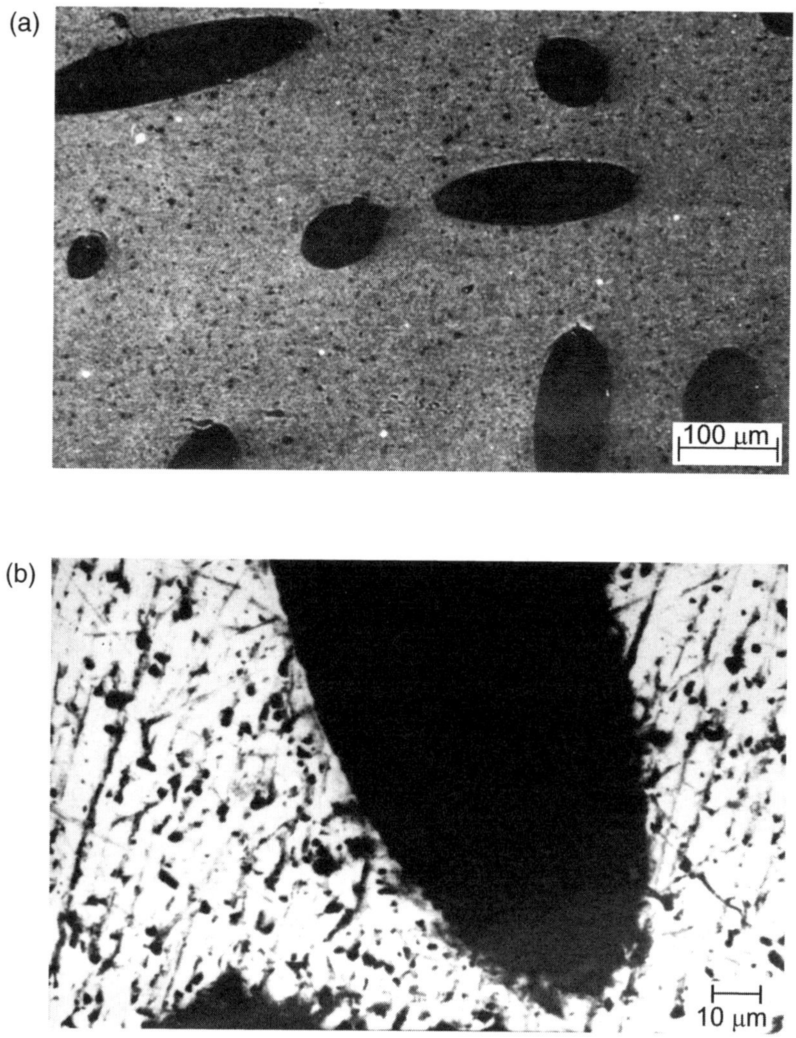

Figure 10.4. Structure of the composite material Al–10%SiC in the initial state (a) and after holding at 800°C for 0.5 h (b, c), and at 900°C for 4 h (d).

materials. First structural changes at the interfaces, indicating the onset of interaction, are revealed only after 1 h at 700°C; and after 0.5 h at 800 and 900°C.

The reaction zone forms at the surface and then grows inwards SiC fibers as clearly seen in Figures 10.4b–d. The extent of interaction of the matrix melt with β-SiC fibers during annealing is assessed by the size of the reaction zone (*t*), the thickness of the unaffected part of a fiber (*d*), and the size of the conglomerate (fiber + reaction zone, *D*) as average of 100 measurements (Figure 10.5). Figure 10.6

(c)

(d)

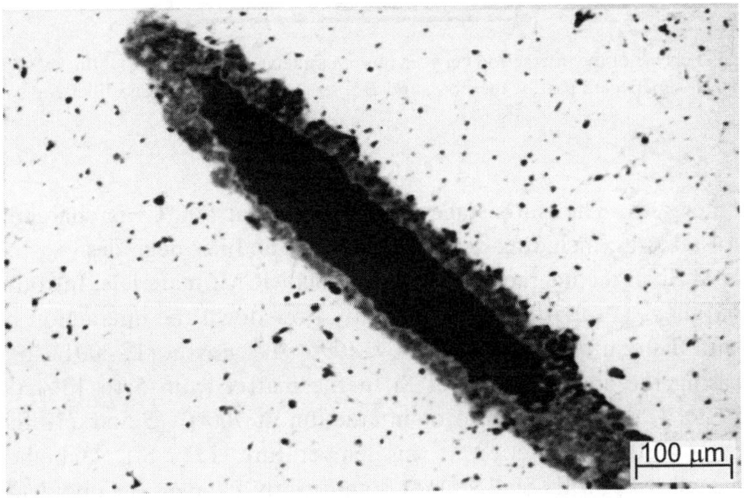

Figure 10.4 *(continued)*

shows that the reaction zone t extends during interaction eating the fiber d, with the size of the agglomerate D remaining virtually the same. This indicates that the interaction zone spreads into the fiber. Three stages of the process can be clearly distinguished from the data in Figure 10.6. At the first stage, the reaction zone rapidly grows up to a thickness of 3–7 μm, the growth spreading deep inside the fiber. Then the reaction slows down, due to the formation of a barrier layer of Al_4C_3. At the second stage, the thickness of the reaction zone remains constant or only

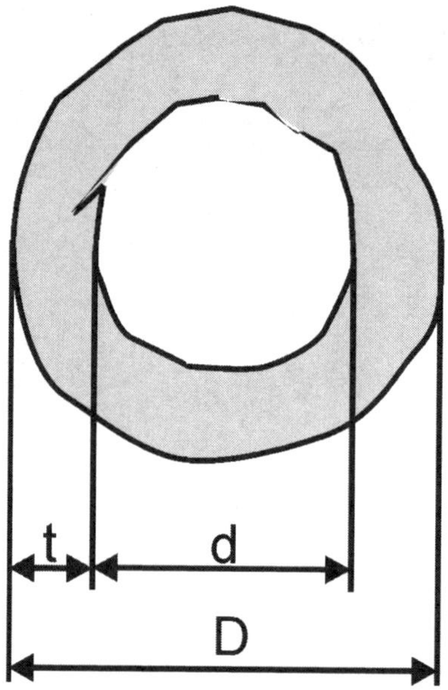

Figure 10.5. Scheme of the interaction between aluminum melt and SiC fiber: *t* – thickness of the reaction zone; *d* – unaffected part of the fiber; and *D* – size of the reaction zone–fiber conglomerate.

slightly increases. The third stage (which occurs at 900°C) is characterized by a change of all sizes, including decreasing *D*, i.e. the fiber degrades.

Effect of Si on the interaction in MCMp and MCMf materials. Introduction of Si to matrix alloys is reported to significantly slow down the interaction between the matrix and the reinforcement (Aksenov, 1996; Aksenov et al., 2001a, b).

Increasing the concentration of Si in the matrix from 5 to 12% (Table 10.5, MCMp 2 to 4) delays the onset of interaction at 700°C (5 and 7% Si) and even totally suppresses the reaction at this temperature (12% Si). At higher temperatures, the alloying with Si does not significantly change the interaction kinetics (Figure 10.7). Similar results were observed in MCMf materials based on Al–Si alloy matrices (Table 10.6, MCMf 6 to 10).

10.3.2. Refinement of Al–C–Si phase diagram

Metastable equilibria in Al–C–Si system. The obtained experimental results and available reference data allow us to suggest the metastable Al–C–Si phase diagram that adequately describes the phase transformations and composition in aluminum-based composite materials at temperatures and compositions relevant to the industrial practice.

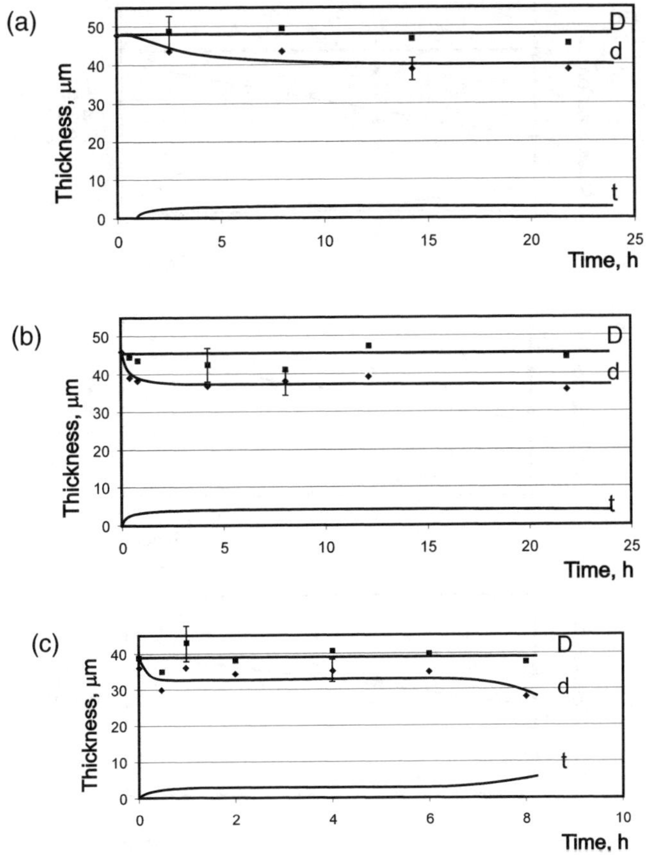

Figure 10.6. Dependence of structural parameters of Al–β-SiC interaction on holding time at temperatures of 700 (a), 800 (b), and 900°C (c). *D*, *d*, and *t* are explained in Figure 10.5.

The polythermal sections shown in Figure 10.8 were constructed using the generally accepted rules, data on liquidus (Figure 10.1b; Viala et al., 1990), invariant reactions (Table 10.2), and results reported elsewhere (Schuster, 1991). These sections go through the compositions of MCMs given in Tables 10.5 and 10.6.

The concentrations of the components at temperatures of invariant transformations are calculated as described elsewhere (Belov, 1998; Aksenov et al., 2001a, b), the compositions of all phases being assumed constant.

Figure 10.8a presents the polythermal section Al–SiC (Aksenov et al., 2001a, b). It shows that a monovariant peritectic reaction with the formation of Al_4C_3 occurs in alloys containing 0.00*X*–2.3% SiC. At the SiC concentration exceeding 3.6%, a monovariant eutectic reaction produces the $Al_4C_3 + SiC$ eutectics. The invariant peritectic reaction $L + Al_4C_3 \Rightarrow (Al) + SiC$ at 650°C (Table 10.2) proceeds in

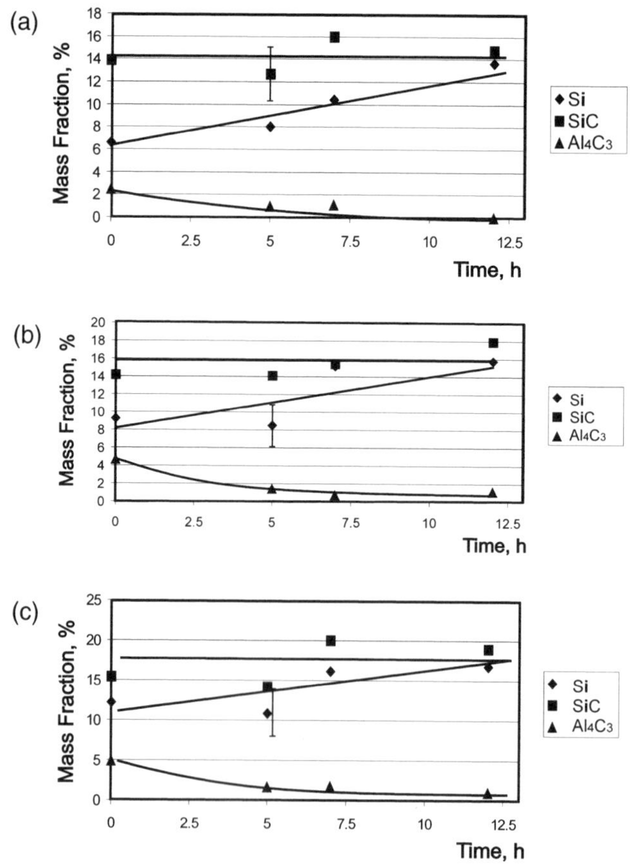

Figure 10.7. Dependence of mass fraction of SiC, Al_4C_3, and (Si) on Si concentration in the matrix of a composite material Al–Si–(α-SiC) at (a) 700, (b) 800, and (c) 900°C (holding time 7 h).

Al-based MCMs. No further phase transformations occur on decreasing temperature below 650°C. The phases (Al), Al_4C_3, and SiC are present in the final structure. This explains the possibility of the interaction reaction $4Al + 3SiC \Rightarrow 3Si + Al_4C_3$, which does not proceed to the end.

The polythermal sections given in Figure 10.8 show that the studied composite materials (>10% SiC) in the temperature range from 700 to 900°C fall into the three-phase region $L + Al_4C_3 + SiC$.

Figures 10.8b–f demonstrate the effect of silicon on the phase transformations upon solidification of Al–Si–SiC materials. The invariant peritectic reaction $L + Al_4C_3 \Rightarrow (Al) + SiC$ occurs only in alloys containing less than 3% Si, e.g. at 1% Si and at >0.9% SiC. Silicon carbide also forms as a primary phase. On increasing the concentration of silicon, the invariant peritectic transformation is suppressed.

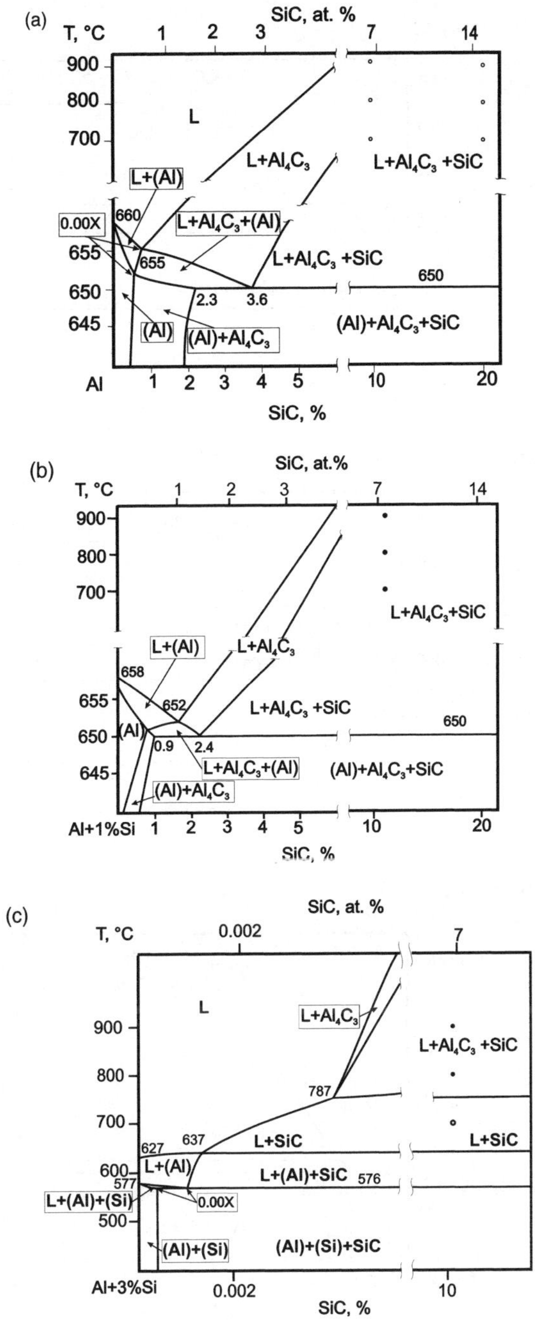

Figure 10.8. Polythermal sections Al–SiC (a), Al–1%Si–SiC (b), Al–3%Si–SiC (c), Al–5%Si–SiC (d), Al–7%Si–SiC (e), and Al–12%Si–SiC (f) of the Al–C–Si phase diagram.

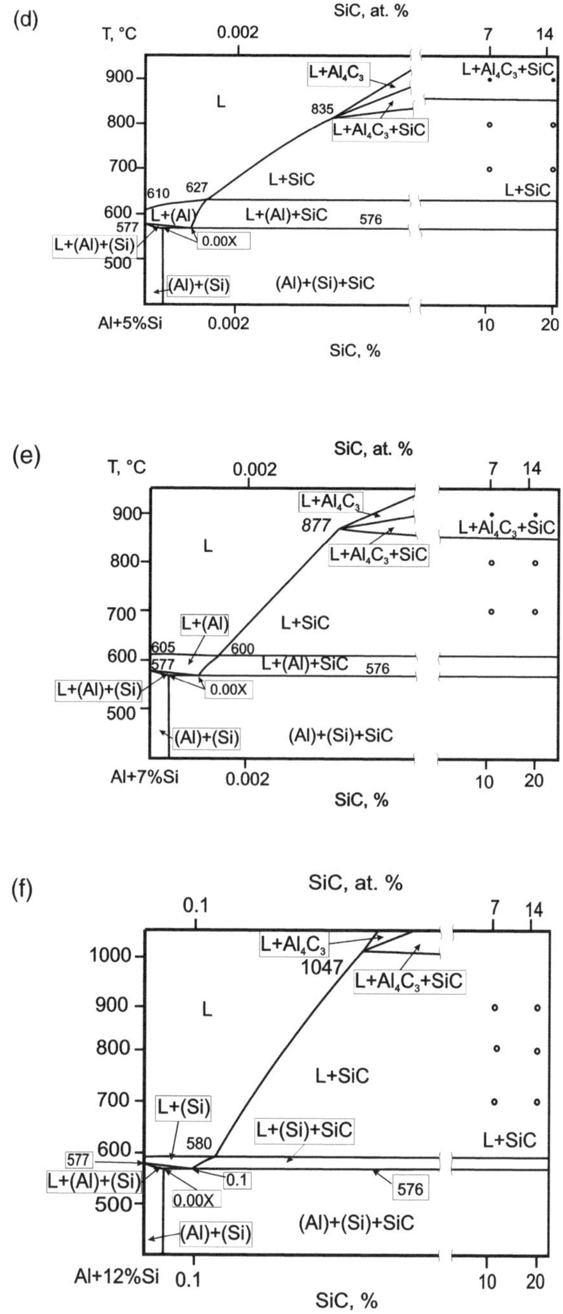

Figure 10.8 *(continued)*

At low concentrations of SiC, primary (Al) forms in alloys with 3–7% Si and primary (Si) – in alloys with 12% Si (Figures 10.8, b–f). In composite materials (hence, at considerable amount of SiC) the Al_4C_3 phase is formed as a primary phase. Then the monovariant peritectic reaction $L + Al_4C_3 \Rightarrow SiC$ occurs. Under equilibrium conditions, this reaction results in the disappearance of aluminum carbide. On further cooling, the monovariant eutectic reaction $L \Rightarrow (Al) + SiC$ proceeds at almost constant onset temperature, the temperature range of this reaction narrowing with increasing amount of Si in the material. And the equilibrium solidification ceases with the invariant eutectic reaction at 576°C with the formation of (Al), SiC, and (Si) (Table 10.2). According to these polythermal sections, the studied MCMs fall into the (L + SiC) phase region at a lower temperature and into the ($L + Al_4C_3 + SiC$) phase region at a higher temperature.

However, under real processing conditions we observe the following deviations from the metastable equilibrium diagram presented in Figure 10.8 (that may correctly describe the high-temperature phase composition of composite materials).

1. In MCMs based on Al and Al–Si alloys containing up to 3% Si, only (Al), SiC, and Al_4C_3 should be present in the structure at room temperature (Figures 10.8a–c). However, free (Si) is observed experimentally as well.
2. In MCMs based on Al–Si alloys containing 5 to 12% Si, only the phases (Al), (Si), and SiC should be present in the structure at room temperature (Figures 10.8d–f). However, the Al_4C_3 phase is often observed in annealed composite materials.

These phenomena can occur only if some solidification reactions do not complete and, therefore nonequilibrium conditions have to be applied.

Polythermal sections of Al–C–Si system for nonequilibrium conditions of MCM processing. When plotting the nonequilibrium polythermal sections, the following deviations from equilibrium were taken into account (Belov, 1998):

- Lower concentrations of alloying elements dissolved in (Al);
- Formation of nonequilibrium eutectic phases;
- Extension of the region of (Al) primary crystallization;
- Lower temperatures of eutectic reactions upon faster cooling;
- Partial or complete suppression of peritectic reactions.

Note that some of the general rules of phase equilibrium may not be observed in nonequilibrium diagrams, e.g. the number of phases after nonequilibrium solidification can be more than three (for a ternary system) and the rules of geometrical thermodynamics can be violated (Belov, 1998).

As applied to the Al–C–Si system, the major factor affecting the real phase composition is the suppression (partial or complete) of the invariant four-phase peritectic reaction $L + Al_4C_3 \Rightarrow (Al) + SiC$ (reaction 4 in Table 10.2, point τ in Figure 10.1b). As a result, the Al_4C_3 phase is retained after the end of solidification.

The methodology allows one to use the experimentally determined mass fraction (Q_M) of phases obtained at room temperature for the analysis of phase equilibria at elevated temperatures. Experimental values in Table 10.7 were obtained by X-ray diffraction analysis of MCMs annealed for more than 20 h at temperatures of 700 to 900°C. When calculating the mass fractions that are also given in Table 10.7, we used the composition of the liquid phase taken from the phase diagram depicted in Figure 10.1b (Viala et al., 1990). One can see good agreement between the experimental and calculated values, some discrepancy being attributed to the inaccurate monovariant line in Figure 10.1b at high temperatures.

Figure 10.9 shows nonequilibrium polythermal sections. If one compares these nonequilibrium sections with the metastable equilibrium sections given in Figure 10.8, two clear differences can be observed.

At low concentrations of Si (Figures 10.8b and 10.9b), a region of (Al) + (Si) eutectics appears in the Al corner of the section. This is a result of the following. During nonequilibrium solidification, the peritectic reaction $L + Al_4C_3 \Rightarrow (Al) + SiC$

Table 10.7. Calculated and experimental mass fractions of phases in the three-phase region $L + Al_4C_3 + SiC$ in Al–C–Si system

Alloying system	T, °C	Phase	Mass fraction Q_M, %	
			Experiment	Calculation
	700	SiC	10.3	17.05
		Si	6.0	–
		Al_4C_3	4.0	3.53
Al–SiC	800	SiC	12.7	14.14
		Si	9.1	–
		Al_4C_3	4.2	7.02
	900	SiC	13.5	7.48
		Si	12.2	–
		Al_4C_3	5.8	14.94
Al–5% Si–SiC	900	SiC	15.5	14.39
		Si	13.8	–
		Al_4C_3	5.0	6.63
Al–7% Si–SiC	900	SiC	17.8	17.16
		Si	16.3	–
		Al_4C_3	2.5	3.31

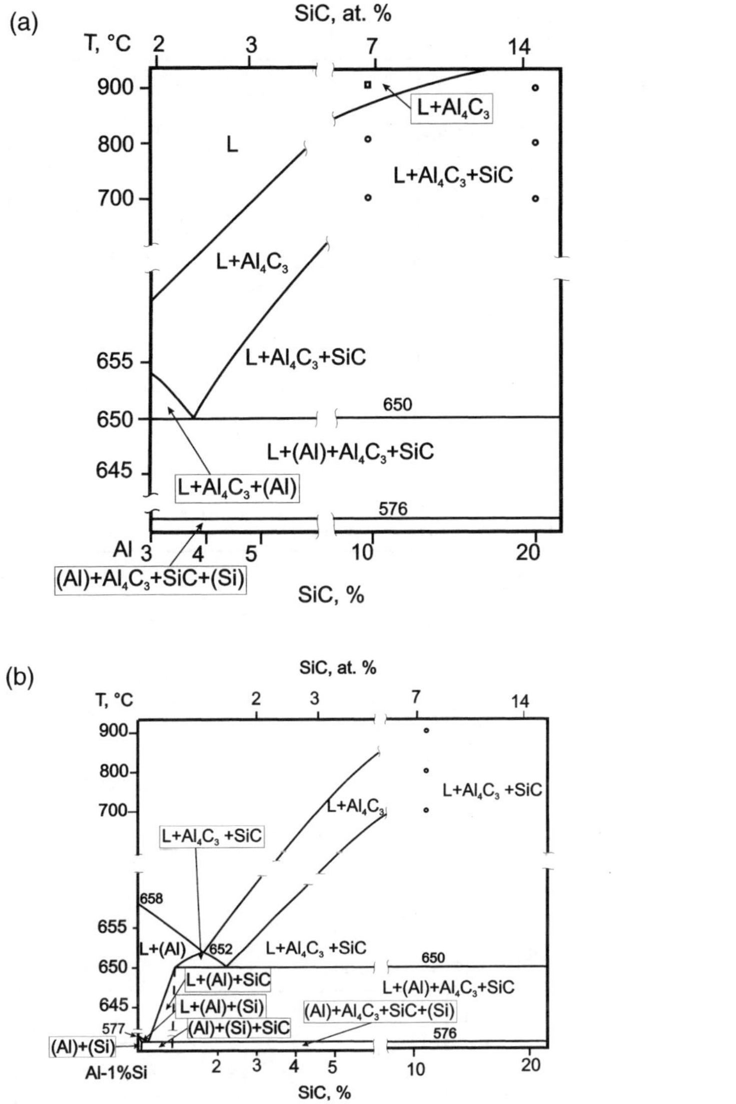

Figure 10.9. Nonequilibrium polythermal sections Al–SiC (a), Al–1%Si–SiC (b), Al–3%Si–SiC (c), Al–5%Si–SiC (d), and Al–7%Si–SiC (e) of the Al–C–Si phase diagram.

does not proceed to the end, and the silicon-enriched liquid after the monovariant eutectic reaction $L \Rightarrow (Al) + SiC$ or $L \Rightarrow (Al) + (Si)$ decomposes by the invariant eutectic reaction $L \Rightarrow (Al) + (Si) + SiC$. The final phase composition in the solid state is $(Al) + Al_4C_3 + SiC + (Si)$ that is in good agreement with experimental observations (Table 10.7).

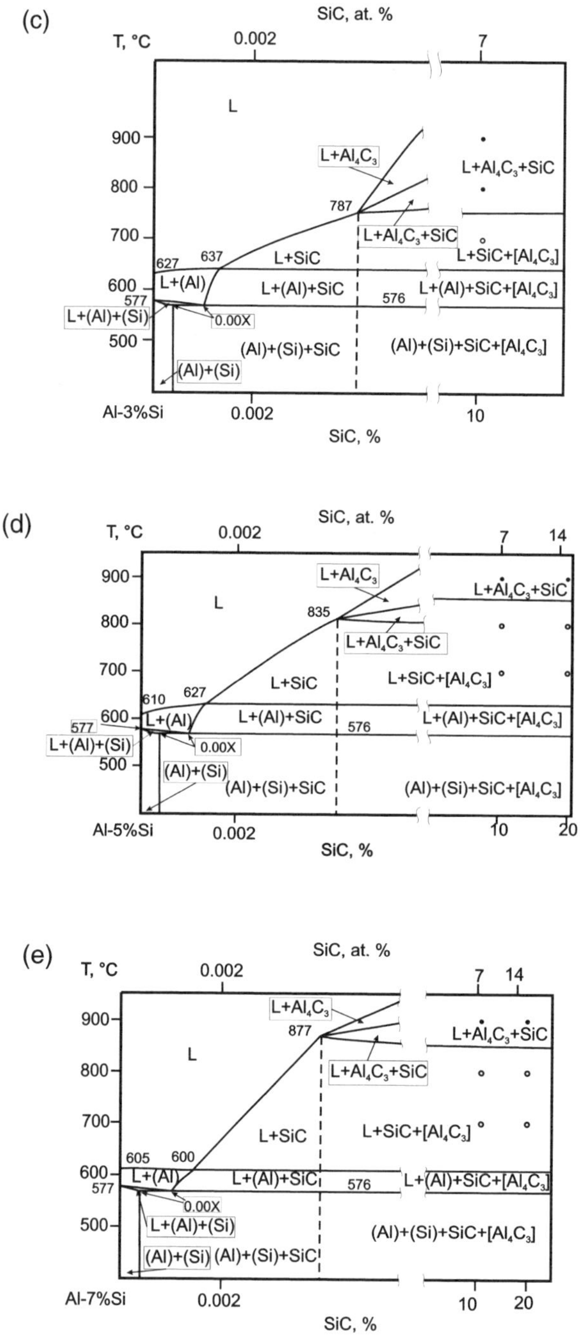

Figure 10.9 *(continued)*

The polythermal sections at higher concentrations of silicon shown in Figures 10.9c–e do not differ much from the equilibrium sections in Figures 10.8c–e only at SiC concentrations lower than 0.9%. At a larger SiC content, the Al_4C_3 phase does not vanish during the peritectic reaction $L + Al_4C_3 \Rightarrow (Al) + SiC$ and is retained in the completely solid state.

In MCMs with 12% Si, the interaction of the melt with SiC is totally suppressed at all temperatures studied and the Al_4C_3 phase is absent, i.e. the reaction opposite to $L + Al_4C_3 \Rightarrow (Al) + SiC$ does not occur. The experimentally obtained phase composition of such an MCM is consistent with the equilibrium phase composition (Figure 10.8f). Therefore, all transformations described by the phase diagram given in Figure 10.1b do occur under nonequilibrium conditions as well.

Construction of isothermal sections of Al–C–Si system for metastable equilibrium conditions. The isothermal sections shown in Figure 10.10 are constructed using our experimental data, liquidus isotherms reported by Viala et al. (1990) (Figure 10.1b), and some general rules (Belov, 1998; Aksenov et al., 2001a, b). The nominal compositions of MCMs from Tables 10.5 and 10.6 are also given in these isopleths.

There are two main observations that can be made based on these isothermal sections. Firstly, on increasing the temperature the phase regions L + SiC and $L + Al_3C_3$ widen and shift towards higher Si concentrations, effectively causing the appearance of Al_4C_3 in MCMs with higher concentrations of silicon. For example, materials with 5 to 12% Si are in the phase region L + SiC at 700°C and only an MCM containing 12% Si remains in this region at 900°C. Secondly, decrease of Si content in the matrix alloy shifts MCMs to the three-phase region $L + Al_4C_3 + SiC$. This suggests that prolonged holding at 900°C may lead to the complete degradation of SiC reinforcing elements, which is observed in practice.

The results on the refinement of the Al–C–Si phase diagram for equilibrium, metastable and nonequilibrium conditions are summarized in the flow chart given in Figure 10.11. Invariant reactions with the corresponding temperatures are shown in frames, monovariant and bivariant reactions are given without indication of temperature range, phases that are formed as a result of reactions are underlined, and the nonequilibrium Al_4C_3 phase is shown in brackets.

In commercial MCMs, brittle layers of Al_4C_3 at SiC–Al interfaces decrease the strength of composite materials. It is also known that aluminum carbide is extremely unstable in water and in some other corrosive media, impairing therefore the corrosion resistance of the composite (Lide, 1992; Aksenov, 1996). In addition, uncontrollable release of silicon into the melt as an interaction product causes the formation of the Al–Si eutectics both at the matrix–reinforcement interface and in the matrix bulk at dendrite boundaries, which often has a negative effect on the mechanical properties of the material.

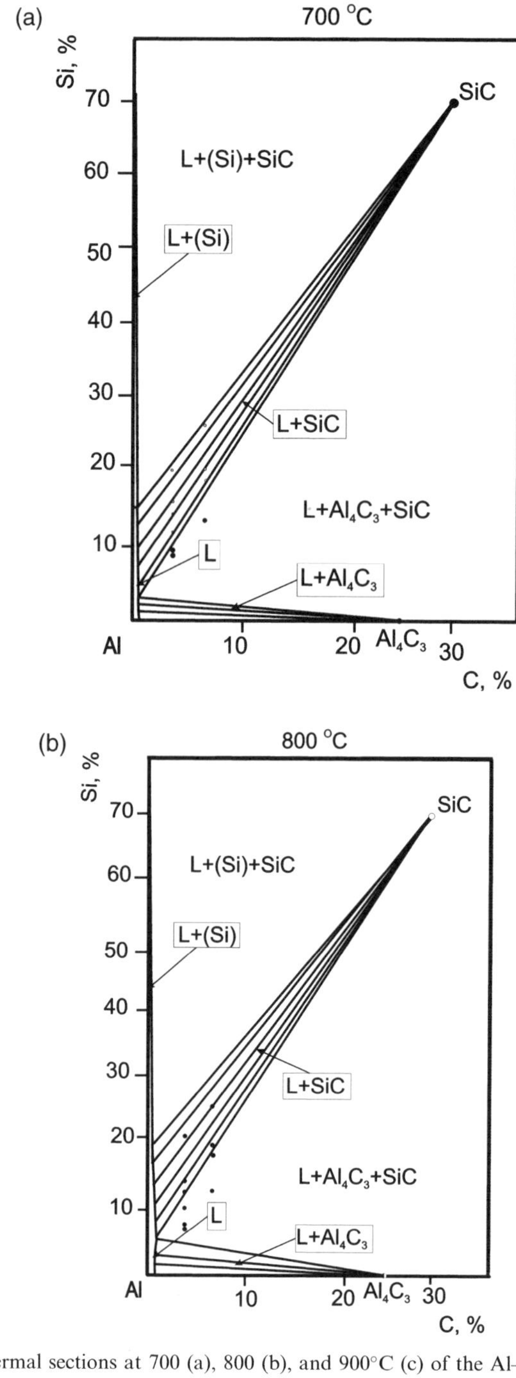

Figure 10.10. Isothermal sections at 700 (a), 800 (b), and 900°C (c) of the Al–C–Si phase diagram.

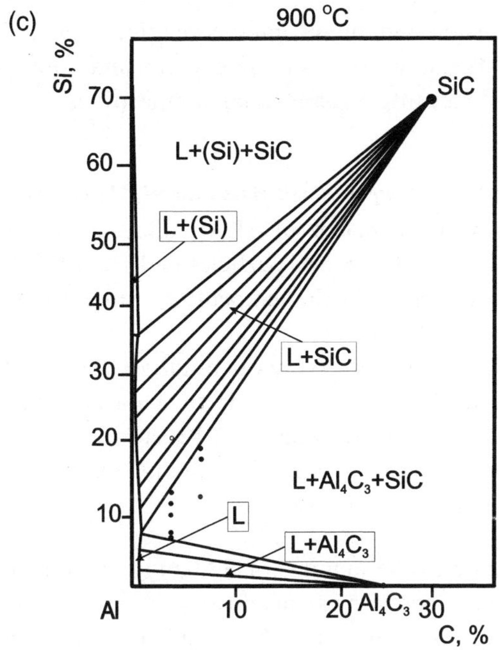

Figure 10.10 (*continued*)

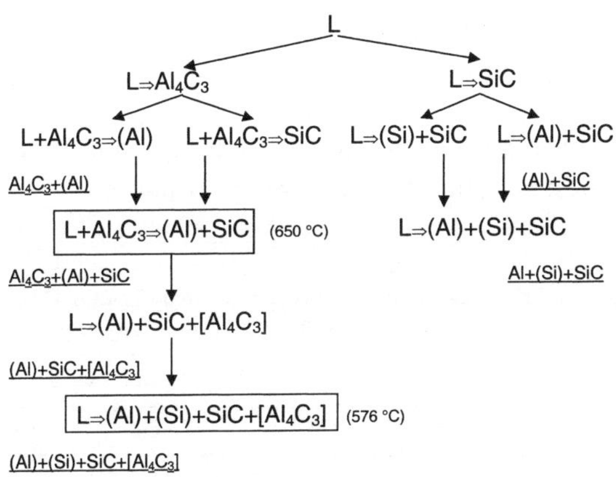

Figure 10.11. Flow chart of the Al–C–Si system.

10.4. Al–C–Mg–Si PHASE DIAGRAM FOR THE ANALYSIS OF INTERFACIAL PROCESSES IN Al–Mg–SiC AND Al–Si–Mg–SiC COMPOSITE MATERIALS

There are only few data available in the literature on the interaction of SiC reinforcement with Mg-containing aluminum alloys. In this section, we briefly discuss the effect of magnesium on the interaction kinetics and phase composition of SiC-reinforced MCMs. As a result of the experimental studies, we suggest some polythermal sections relevant to commercial compositions.

10.4.1. Interaction in MCM composite materials with Al–Mg matrix

Composite materials with the matrix of Al–Mg alloys containing 1 to 6% Mg and the reinforcement of either 20% α-SiC particles or 10% β-SiC fibers were examined after high temperature anneals in the temperature range from 700 to 900°C in air and under protective Ar atmosphere.

Table 10.8 shows the results of X-ray quantitative analysis of phase composition of α-SiC-reinforced MCMs annealed in the temperature range 700 to 900°C for 25 h under protective atmosphere. Figure 10.12 shows the structure of an Al–6% Mg–10%SiC composite material with the reaction zone. The following general features of the interaction are observed:

- A gradual decrease in the amount of SiC and the simultaneous emergence of the Al_4C_3 phase on increasing temperature;
- The Al_4C_3 phase is formed within the volume of former SiC particles/fibers;
- During interaction Si diffuses from SiC reinforcing element to the matrix, forming the Mg_2Si or (Si) phases during solidification of an MCM;
- The interaction zone contains Al, Mg, and Si;
- Increase in the Mg concentration decreases the amount of (Si) phase formed in MCM;
- MgO and $MgAl_2O_4$ can be present in small amounts (not more than 3%) if the interaction occurs without protective atmosphere;

Table 10.8. Results of the quantitative X-ray phase analysis of Al–Mg–SiC composite materials annealed under Ar atmosphere for 25 h at various temperatures

Mg, %	*T*, °C	Amount of phases, %				
		(Al)	(Si)	Al_4C_3	Mg_2Si	SiC
1		73.7	4.8	9.7	0.3	11.5
2	700	66.7	4.6	12.8	0.5	15.4
6		80.3	1.7	3.2	2.3	12.5
1		81.7	5.3	5.7	0.7	6.7
2	800	71.7	3.7	12.1	3.4	9.2
6		63.2	3.2	11.9	6.3	15.4
1		73.0	6.5	13.1	1.9	5.6
2	900	72.6	8.3	11.9	3.0	4.1
6		81.0	9.3	0.6	2.0	7.1

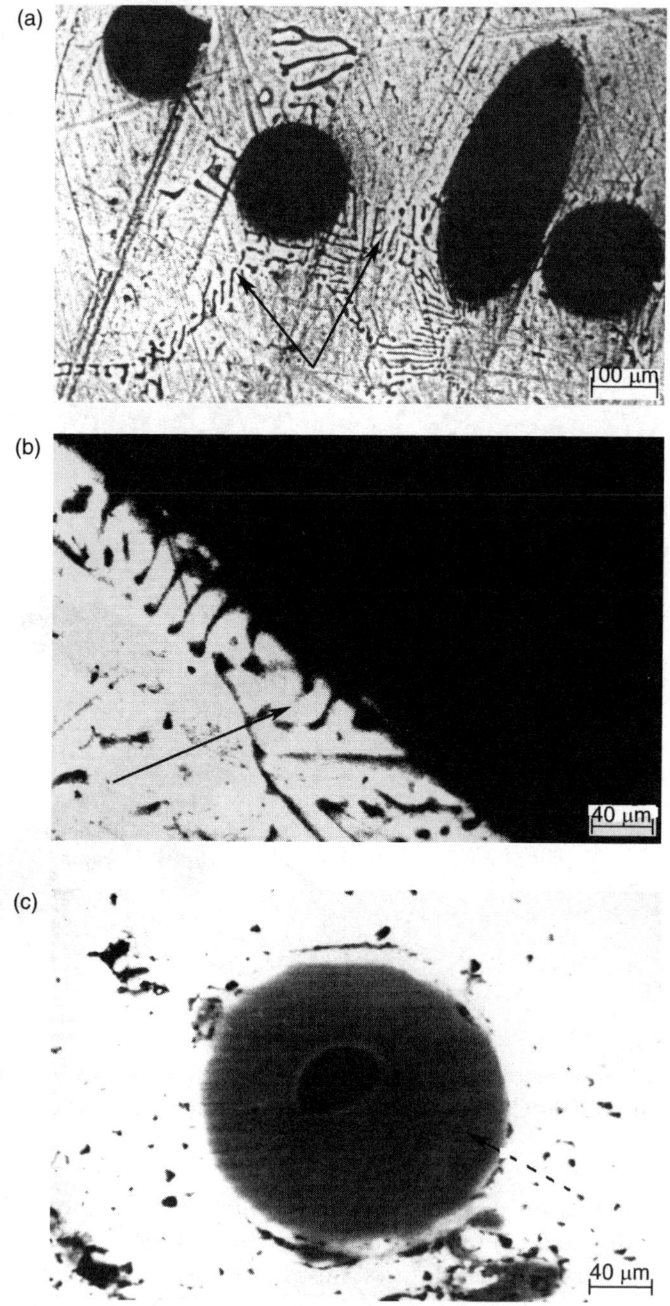

Figure 10.12. Structure of an MCM Al–6%Mg–10%SiC after annealing at 700°C for 0.5 h (a, b) and at 900°C for 3 h (c–e). Dashed arrows show the reaction zone enriched with magnesium. Solid arrows show the Mg_2Si phase.

(d)

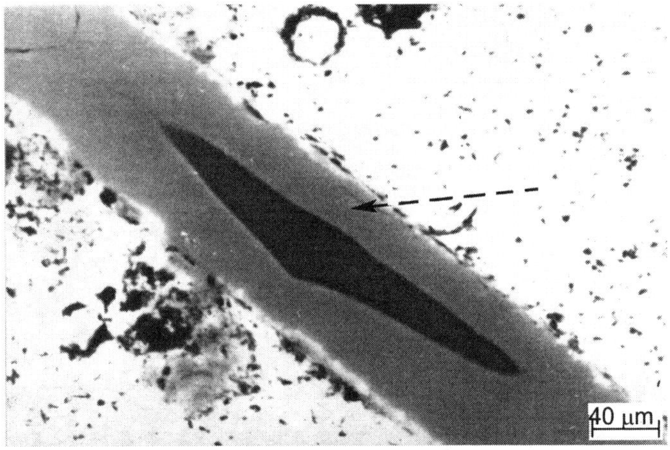

(e)

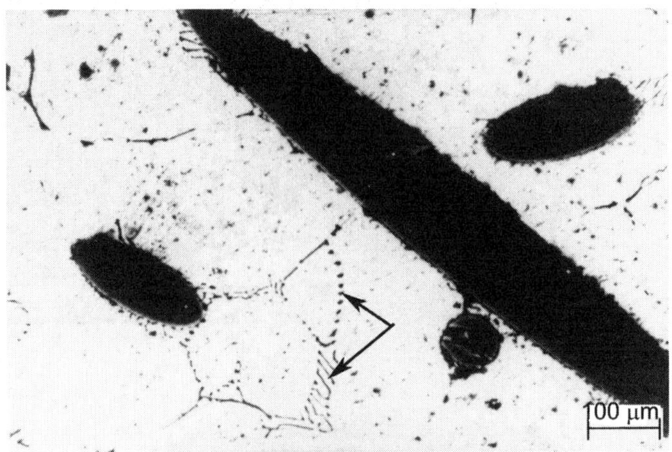

Figure 10.12 *(continued)*

- Protective atmosphere contributes to some slowdown of SiC degradation at a temperature of 900°C and has no effect on the interaction processes at 700 and 800°C.

Based on these results, we can suggest the following mechanism of interaction between the matrix and reinforcement in the presence of Mg.

Initially, Mg from the melt diffuses into the reinforcing element, and Si simultaneously diffuses from the reinforcing element. In the melt, Si forms Mg_2Si during subsequent solidification. After the diffusion is completed, conditions are created for chemical interaction of the matrix with SiC reinforcement. In contrast to composite materials with the aluminum matrix, reaction $4Al + 3SiC \Rightarrow 3Si + Al_4C_3$ is inhibited in the presence of Mg in the matrix and, instead the following reaction occurs:

$$4Al + 6Mg + 3SiC \Rightarrow Al_4C_3 + 3Mg_2Si$$

Magnesium also interacts with oxygen in the fiber to form Mg oxides (if interaction occurs in air). The reaction products precipitate at the matrix–reinforcement interface and slow down the reaction.

To avoid the active diffusion of Mg into SiC particles or fibers, the temperature of MCM preparation should not exceed 700°C.

10.4.2. Al–C–Mg–Si phase diagram

Figure 10.13 shows a part of the Al–C–Mg–Si phase diagram that is relevant to metal–matrix composites reinforced with SiC, i.e. the tetrahedron Al–Al_8Mg_5–(Si)–Al_4C_3. Table 10.9 gives the invariant reactions occurring in Al–C–Mg–Si alloys.

As the solubility of C in (Al) is negligibly small, we made the following assumptions:

- In the four-component phase diagram the invariant reactions involving carbon-containing phases have the same type as in the constituent ternary systems;
- Temperatures and concentrations of these invariant points are also close to those in the constituent ternary systems.

In addition, based on our experimental data we assumed that the crystal structure of SiC does not affect the interaction processes.

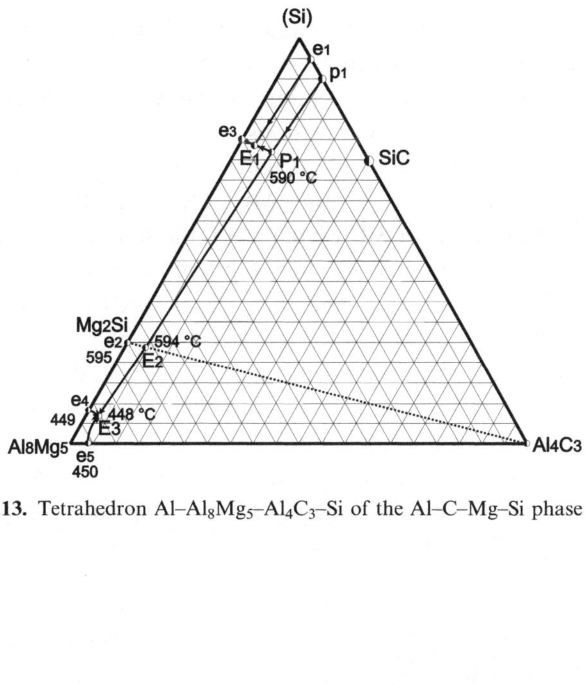

Figure 10.13. Tetrahedron Al–Al_8Mg_5–Al_4C_3–Si of the Al–C–Mg–Si phase diagram.

Table 10.9. Invariant reactions in Al–C–Mg–Si alloys

Point in Figure 10.13	Reaction	Phase	Concentration			*T*,°C
			Si, %	Mg, %	C, %	
	Binary					
–	$L \Rightarrow (Al) + (Si)$	L	12.5	–	–	577
–	$L \Rightarrow (Al) + Al_8Mg_5$	L	–	34	–	450
–	$L \Rightarrow (Al) + Al_4C_3$	L	–	–	<0.001	≈660
	Ternary					
p_1	$L + Al_4C_3 \Rightarrow (Al) + SiC$	L	1.5	–	<0.001	650
e_1	$L \Rightarrow (Al) + (Si) + SiC$	L	12.3	–	<0.001	576
e_2	$L \Rightarrow (Al) + Mg_2Si$	L	7.75	8.15	–	595
	(quasi-binary section)	(Al)	0.68	1.17	–	
e_3	$L \Rightarrow (Al) + (Si) + Mg_2Si$	L	12.95	4.96	–	555
		(Al)	1.10	0.85	–	
e_4	$L \Rightarrow (Al) + Mg_2Si + Al_8Mg_5$	L	0.37	32.2	–	449
		(Al)	0.05	15.3	–	
e_5	$L \Rightarrow (Al) + Al_8Mg_5 + Al_4C_3$	L	–	34	<0.001	450
		(Al)	–	17.4	<0.001	
	Quaternary					
P_1	$L + Al_4C_3 \Rightarrow (Al) + SiC + Mg_2Si$*	L	12.5	4.95	<0.001	590
E_1	$L \Rightarrow (Al) + (Si) + SiC + Mg_2Si$*	L	12.3	4.96	<0.001	550
E_3	$L \Rightarrow (Al) + Al_8Mg_5 + Mg_2Si + Al_4C_3$*	L	0.37	32.2	<0.001	448
E_2	$L \Rightarrow (Al) + Mg_2Si + Al_4C_3$* (quasi-ternary section)	L	7.75	8.15	<0.001	594

* The phases appear in the quaternary phase diagram from the corresponding ternary phase diagram

In order to give a more complete picture of solidification, Table 10.10 shows monovariant reactions proceeding in Al–C–Mg–Si alloys.

Figure 10.14 presents polythermal sections of the Al–C–Mg–Si phase diagram for matrix Al–Si alloys of different compositions.

At a low Mg concentration of 0.3% (Figure 10.14a), the primary phase changes on increasing the amount of SiC from (Al) to SiC and then to Al_4C_3. Only at extremely low concentrations of SiC (<0.001%), the solidification finishes with the formation of the (Al) + (Si) eutectics. In the compositional range of SiC primary solidification, the binary (Al) + SiC and then ternary (Al) + (Si) + SiC eutectics are formed at a virtually constant temperature. If the Al_4C_3 phase solidifies as a primary phase, it reacts through a peritectic reaction to form SiC and then the solidification sequence is the same as at a lower SiC concentration. Magnesium in these quantities remains completely dissolved in aluminum.

According to the Al–Mg–Si phase diagram (Section 2.1) the Mg_2Si phase should appear in the structure of alloys containing more than 0.3% Mg. Figure 10.14b

Table 10.10. Monovariant reactions in Al–C–Mg–Si alloys

Temperature range,°C	Reaction
For tetrahedron Al–Al_8Mg_5–Mg_2Si–Al_4C_3	
450–448	$L \Rightarrow (Al) + Mg_5Al_8 + Al_4C_3$
449–448	$L \Rightarrow (Al) + Mg_5Al_8 + Mg_2Si$
594–448*	$L \Rightarrow (Al) + Al_4C_3 + Mg_2Si$
For tetrahedron Al–(Si)–Mg_2Si–Al_4C_3	
594–590*	$L \Rightarrow (Al) + Al_4C_3 + Mg_2Si$
650–590	$L + Al_4C_3 \Rightarrow (Al) + SiC$
590–540	$L \Rightarrow (Al) + SiC + Mg_2Si$
576–540	$L \Rightarrow (Al) + (Si) + SiC$
555–540	$L \Rightarrow (Al) + (Si) + Mg_2Si$

* A degenerated monovariant reaction, because its character cannot be determined exactly

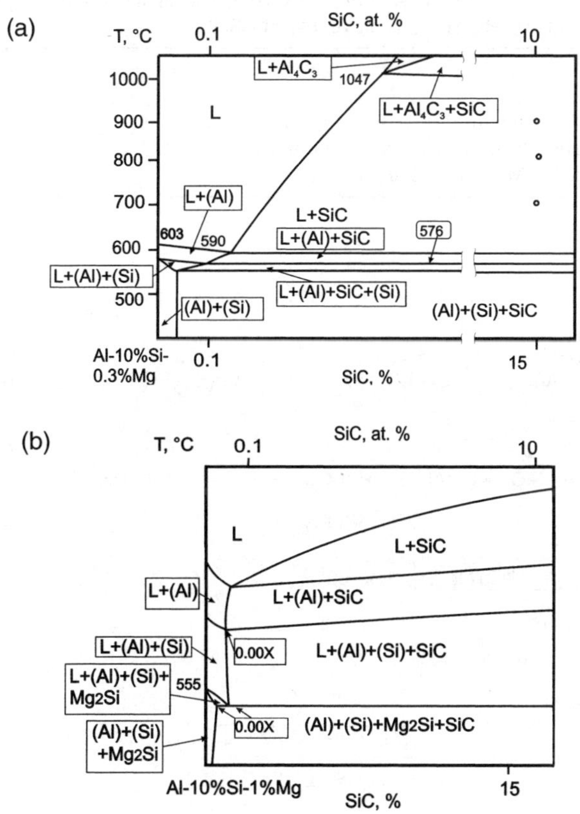

Figure 10.14. Polythermal sections Al–10%Si–0.3%Mg–SiC (a), Al–10%Si–1%Mg–SiC (b), and Al–12%Si–1%Mg–SiC (c) of the Al–C–Mg–Si phase diagram.

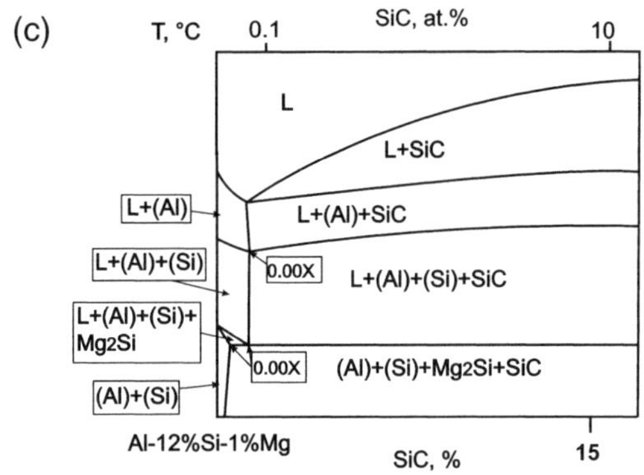

Figure 10.14 (*continued*)

Table 10.11. Solidification reactions in Al–Mg–Si–SiC composite materials (Al–1% Mg–SiC, Al–(2–10)% Mg–SiC, Al–0.5% Si–0.5% Mg–SiC, Al–1% Si–1% Mg–SiC)

$L \Rightarrow Al_4C_3$
$L + Al_4C_3 \Rightarrow SiC$
$L + Al_4C_3 \Rightarrow (Al) + SiC$ (only for Al–(2–10)% Mg–SiC);
$L + Al_4C_3 \Rightarrow SiC + Mg_2Si$ (for other given compositions);
$L + Al_4C_3 \Rightarrow (Al) + SiC + Mg_2Si$ ($T = 590°C$)

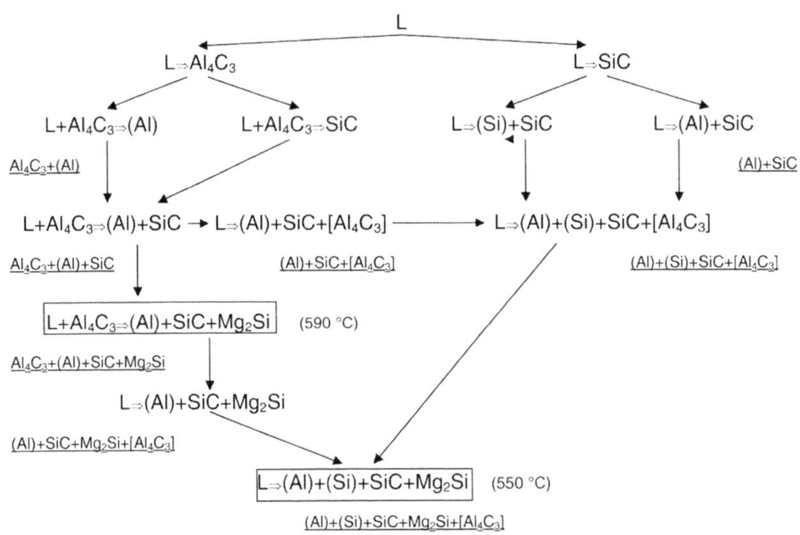

Figure 10.15. Flow chart for the tetrahedron Al–Mg_2Si–Si–Al_4C_3 of the Al–C–Mg–Si system.

shows the polythermal section for an Al–10%Si–1%Mg matrix alloy. At very low concentrations of SiC, the solidification occurs as in ternary Al–Mg–Si alloys with the formation of binary (Al) + (Si) and ternary (Al) + (Si) + Mg_2Si eutectics. On increasing the SiC concentration, silicon carbide solidifies first, then the binary (Al) + SiC and ternary (Al) + (Si) + SiC eutectics are formed. The solidification ends at 550°C with the invariant eutectic reaction L ⇒ (Al) + (Si) + SiC + Mg_2Si (point E_1 in Figure 10.13 and Table 10.9). As a result, the phases (Al), (Si), Mg_2Si, and SiC should be present at room temperature. The same phase composition is observed in SiC-reinforced alloys with 12% Si and 1% Mg (Figure 10.14c).

Table 10.11 summarizes the solidification reactions occurring in SiC-reinforced composite materials with the matrix of Al–Mg and Al–Mg–Si alloys.

A flow chart of solidification reactions given in Figure 10.15 is a convenient way of understanding solidification and interaction in MCMs. Invariant reactions with the corresponding temperatures are shown in frames, monovariant and bivariant reactions are given without indication of temperature range, phases that are formed as a result of reactions are underlined, and the nonequilibrium Al_4C_3 phase is shown in brackets.

10.5. Al–C–Cu–Si, Al–C–Si–Zn, AND Al–C–Cu–Mg–Zn PHASE DIAGRAMS FOR THE ANALYSIS OF INTERFACIAL PROCESSES IN Al–Cu–SiC AND Al–Zn–SiC COMPOSITE MATERIALS

Zinc and copper are major alloying components in high-strength aluminum alloys. Therefore, the knowledge of the interaction between reinforcing elements and high-strength matrix is important from both fundamental and practical points of view.

Experimental studies similar to those described in previous sections showed that copper and zinc do not change the nature of interaction between SiC and the matrix as compared to that of MCMs with aluminum or Al–Mg matrices, respectively.

Figure 10.16a shows the polythermal section Al–(1–6)%Zn–SiC. Zinc is completely dissolved in (Al) and does not form own phases. The solidification (providing sufficient amount of SiC) starts with the formation of Al_4C_3 phase followed by the bivariant transformation with a tentative reaction L ⇒ (Al) + Al_4C_3. The exact nature of this transformation is unclear. At room temperature, the structure of the MCMs comprises (Al) and Al_4C_3. On further increasing the concentration of SiC, bivariant Al_4C_3 + SiC and monovariant (Al) + Al_4C_3 + SiC eutectics are formed, ultimately defining the structure at room temperature. The only effect of Zn (compare Figures 10.16a and 10.8a) is in the changed type of the reaction L + Al_4C_3 ⇒ (Al) + SiC that becomes monovariant and proceeds in a temperature range as reflected in Figure 10.16a by the region L + (Al) + Al_4C_3 + SiC.

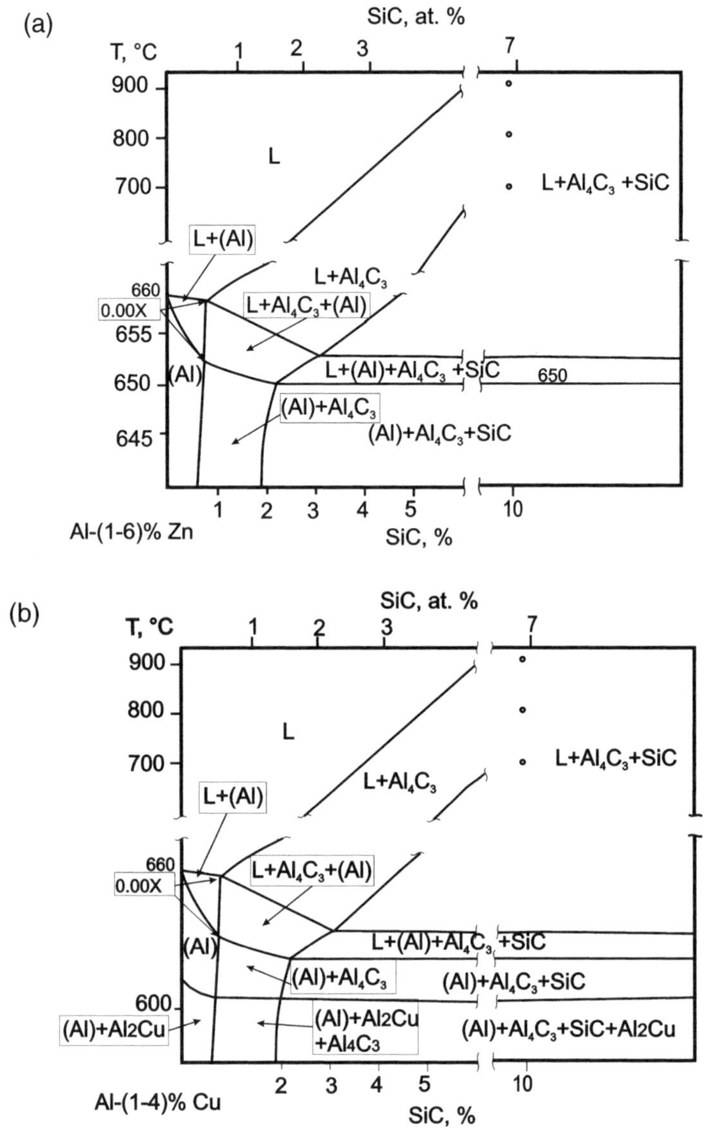

Figure 10.16. Polythermal sections Al–(1–6)%Zn–SiC (a) and Al–(1–4)%Cu–SiC (b).

In the case of an Al–(1–4)% Cu matrix, the general sequence of phase transformation on the increasing SiC concentration is the same (Figure 10.16b). In addition, Al_2Cu phase precipitates from the aluminum solid solution during cooling in the solid state, following the solvus line. A polythermal line reflecting the formation of Al_2Cu during solidification also appears in the polythermal sections.

10.6. Al–O–Si PHASE DIAGRAM FOR THE ANALYSIS OF Al–SiO_2 AND Al–MULLITE COMPOSITE MATERIALS

Let us consider the interaction between SiO_2 and mullite fibers on one side and the matrix of a 332.0-type piston alloy (12% Si, Cu, Mg, Ni) on the other side.

Experiments on annealing of SiO_2-reinforced MCMs in the temperature range 720 to 800°C showed that SiO_2 fibers degraded within several minutes of contact with a liquid Al-alloy. At a lower temperature of 620°C, it is possible to follow the interaction kinetics. After a latent period (about 10 min), the interaction starts by advancement of the interaction zone into the fiber with complete degradation of the latter within 3 h (Figure 10.17a). Simultaneously with the change of the fiber structure, the matrix structure undergoes dramatic changes. As the holding time increases, the matrix becomes enriched with Si and the structure changes from hypoeutectic to hypereutectic (Figures 10.17b–d, primary crystals of Si are visible). This result can be expected from the following reaction:

$$4Al + 3SiO_2 \Rightarrow 3Si + 2Al_2O_3.$$

Similar results are obtained for the Al–Mullite system, though the interaction in this system is slower.

The following sequence of interaction between aluminum melt and silica-containing ceramic fibers can be suggested based on our experimental results.

After complete wetting of the fiber with the melt, initially at the sites of best contact at the interface, aluminum diffuses into the fiber. The moving force is a

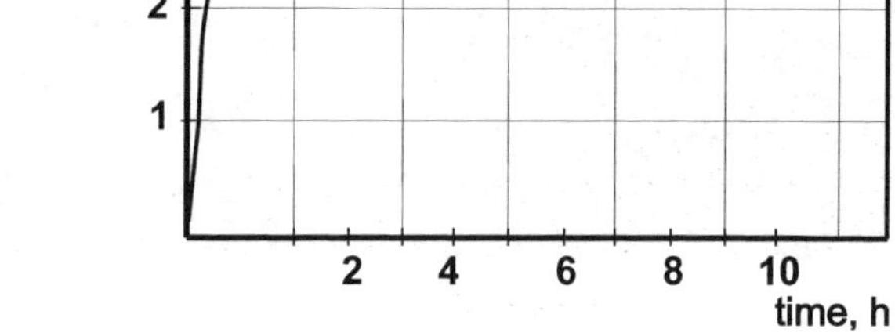

Figure 10.17. Dependence of the thickness of the interaction zone on holding time at 620°C (a) and the structure of composite material Alloy 332.0–11% SiO_2, obtained by impregnation under pressure at a melt temperature of 620°C followed by holding at the same temperature for 1 (b), 3 (c), and 7 h (d).

(b)

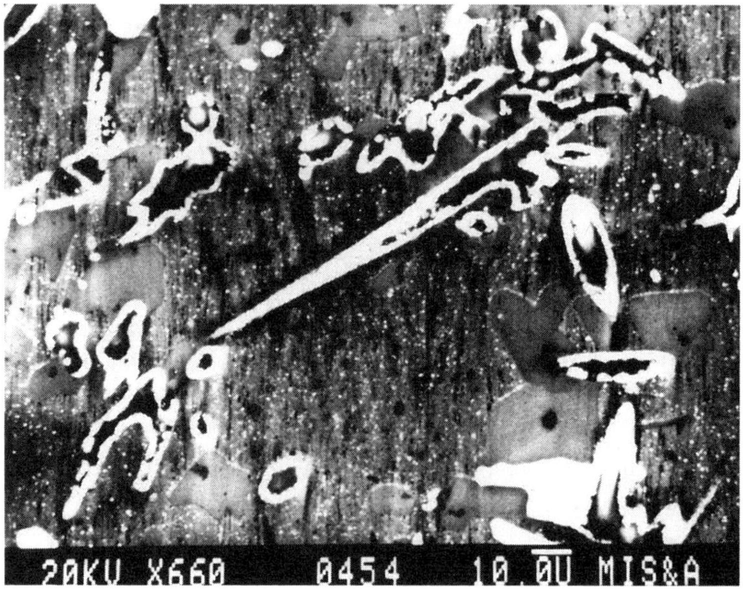

(c)

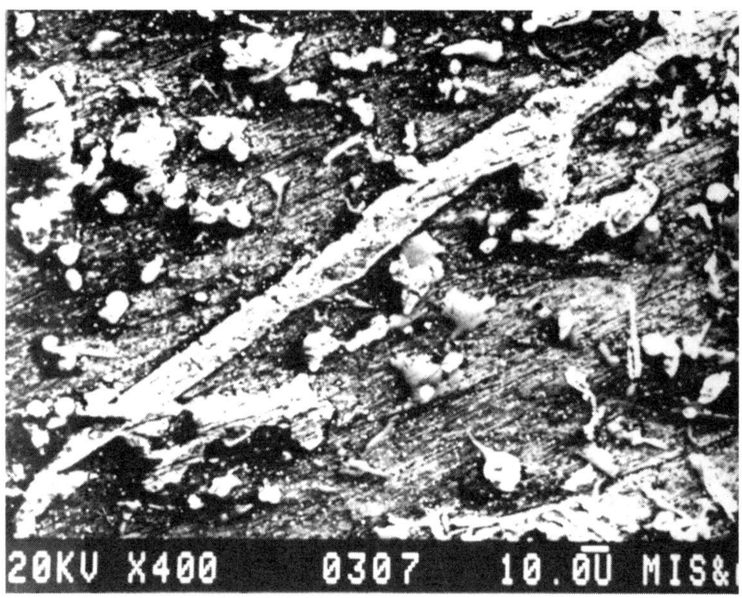

Figure 10.17 (*continued*)

(d)

Figure 10.17 (*continued*)

larger affinity of aluminum to oxygen than that of silicon. It is possible that the formation of alumina goes through an intermediate stage of mullite formation as follows:

$$Al + SiO_2 \Rightarrow Al + Si + 3Al_2O_3 \cdot 2SiO_2 \Rightarrow Al + 3Si + 2Al_2O_3.$$

In the case of the Al–Mullite system, the interaction can proceed in accordance with the following tentative reaction:

$$Al + 3Al_2O_3 \cdot 2SiO_2 \Rightarrow Al + Si + Al_2O_3.$$

Simultaneously with "substitution" of the oxides, free Si is transferred into the melt and upon subsequent solidification, precipitates as eutectic or primary crystals.

Appendix 1

CHARACTERISTICS OF BINARY PHASE DIAGRAMS

Table A1.1. Main characteristics of binary phase diagrams on the Al-side of the systems (Mondolfo, 1976; Drits et al., 1977)

Element	Reaction*	*a*, %**	*e*; *p*, %**	$T_{e;\,p}$, °C	Second phase β (%)
Ag	E	55.6	70.5	566	$AlAg_2$ (85–93 Ag)
As	E	<1	<1	~657	AlAs (73.5 As)
Au	E	0.35	5.0	642	Al_2Au (77.8–79 Au)
B	E	0.01	0.022	659.5	AlB_2 (44.5 B)
Ba	E	0.1	3.5 or 4.89	651–652	Al_4Ba (56 Ba)
Be	E	0.05 or 0.08	0.5 or 0.8	645	(Be) (>99 Be)
Bi	E (M)	<0.01 (<0.01)	99.7 (3.5)	269 (657)	(Bi) (>99.7 Bi)
C	E	<0.01	<0.01	~660	Al_4C_3 (25.3 C)
Ca	E	~0.01	7.6	617	Al_4Ca (27 Ca)
Cd	E (M)	0.005 (0.4)	99.9 (7.2)	321 (649)	Cd (>99.9 Cd)
Ce	E	0.05	10 or 12	637 or 638	Al_4Ce (56.5 Ce)
Co	E	<0.02	~1	657	Al_9Co_2 (32.7 Co)
Cr	P	0.72	0.4	661	Al_7Cr (21.6 Cr)
Cs	E (M)	<0.01	n/a	28.6 (660)	(Cs)
Cu	E	5.7	33.2	548	Al_2Cu (53.2–53.9 Cu)
Dy	E	0.05	35	637	Al_3Dy (69.8 Dy)
Er	E	0.05	~6	655	Al_3Er (67.3 Er)
Fe	E	0.05	1.8	655	Al_3Fe (37.3–40.7 Fe)
Ga	E	~20	99	26.3	(Ga) (>99.95 Ga)
Gd	E (P′)	0.2	10 or 25	643 or 652 (402)	Al_3Gd (68.9 Gd)
Ge	E	7.2	54	424	(Ge) (>99.5 Ge)
Hf	P	1.19	0.48	~662.2	Al_3Hf (68.8 Hf)
Hg	E	<0.01	99.9977	-39.3	(Hg) (>99.999)
Ho	E	0.05	14	647	Al_3Ho (67.1 Ho)
In	E (M)	<0.01 (0.15)	99.95 (17.3)	155 (637)	(In) (>99.955 In)
K	E (M)	n/a	n/a	63.5 (657)	(K)
La	E	0.05	11.5 or 12	642	Al_4La (56.3–58.4 La)
Li	E	4.2	7.5	601 or 602	AlLi (17.8–25 Li)
Mg	E	17.4	35	450	Al_8Mg_5 (34.8–37.1 Mg)
Mn	E	1.8	1.9	658.5	Al_6Mn (25.3 Mn)
Mo	P	0.25	~0.1	660.3	$Al_{12}Mo$ (22.6 Mo)
Na	E (M)	<0.002 (0.0025)	~100 (0.14)	97.5 (659)	(Na) (~100 Na)
Nb	P	0.22	0.15	661.4	Al_3Nb (53.5 Nb)
Nd	E	0.05	16.24 or 30	640 or 637	$Al_{11}Nd_3$ (57.2–59.3 Nd)
Ni	E	0.04	6.0	640	Al_3Ni (42 Ni)

(continued)

Table A1.1. Continued

Element	Reaction*	*a*, % **	*e*; *p*, %**	$T_{e;\ p}$, °C	Second phase β (%**)
O	n/a	<0.001	n/a	~660	Al_2O_3 (47.1 O)
P	E	<0.001	<0.015	~660 or 900	AlP (53.5 P)
Pb	E (M)	<0.002 (0.17)	99.997 (1.5)	327 (658.5)	(Pb) (>99.997 Pb)
Pd	E	<0.002	23 or 24.2	605 or 615	Al_3Pd (56–56.9 Pd)
Pr	E	<0.1	~12	654	Al_4Pr (56.6 Pr)
Pt	E	<0.1	6.8	657	Al_4Pt (64.4 Pt)
Pu	E	~0.06	13 or 15.58	652	Al_4Pu (68.9–69.2 Pu)
Re	E	~0.1	~2	~597	$Al_{12}Re$ (36.4 Re)
S	E	<0.001	~0.05	659	Al_2S_3 (63.2 S)
Sb	E	<0.1	~1	658	AlSb (81.9 Sb)
Sc	E	0.3	0.6	655	Al_3Sc (36 Sc)
Se	E	<0.001	<0.1	~657 or 648	Al_2Se_3 (81.6 Se)
Si	E	1.65	12.7	577	(Si) (>99.5 Si)
Sm	E	<0.1	~15	633	Al_3Sm (64.9 Sm)
Sn	E	~0.02 or 0.06	99.5	228.3	Sn (>99.6 Sn)
Sr	E	<0.01	<1	~659	Al_4Sr (44.8 Sr)
Ta	P	0.24	0.12	668	Al_3Ta (69.1 Ta)
Tb	E (P′)	0.1	10 (59.5)	644 (397)	Al_3Tb (66.3 Tb)
Th	E	~0.8	~24	627	Al_3Th (74.2 Th)
Ti	P	1.3	0.12	665	Al_3Ti (37.2 Ti)
U	E	0.06	13	~640–647	Al_4U (64.2–66.3 U)
V	P	~0.1	~0.37	661.9	$Al_{10}V$ (15.1–15.9 V)
W	P	0.06	0.16	660	$Al_{12}W$ (36.2 W)
Y	E	0.17	9.5 or 8.96	627 or 635–645	Al_3Y (52.34 Y)
Yb	E	0.2	~21	625	Al_3Yb (68.13 Yb)
Zn	E (M)	82.2 (31.6)	94.9 (78)	382 (275)	(Zn) (>99.4 Zn)
Zr	P	0.28	0.11	660.5	Al_3Zr (53 Zr)

* E, eutectic; P, peritectic; M, monotectic; and P′, peritectoid reactions; temperatures and concentrations are given only for the first two type of reactions (see in Figure A1.1)
** Concentration of the second element

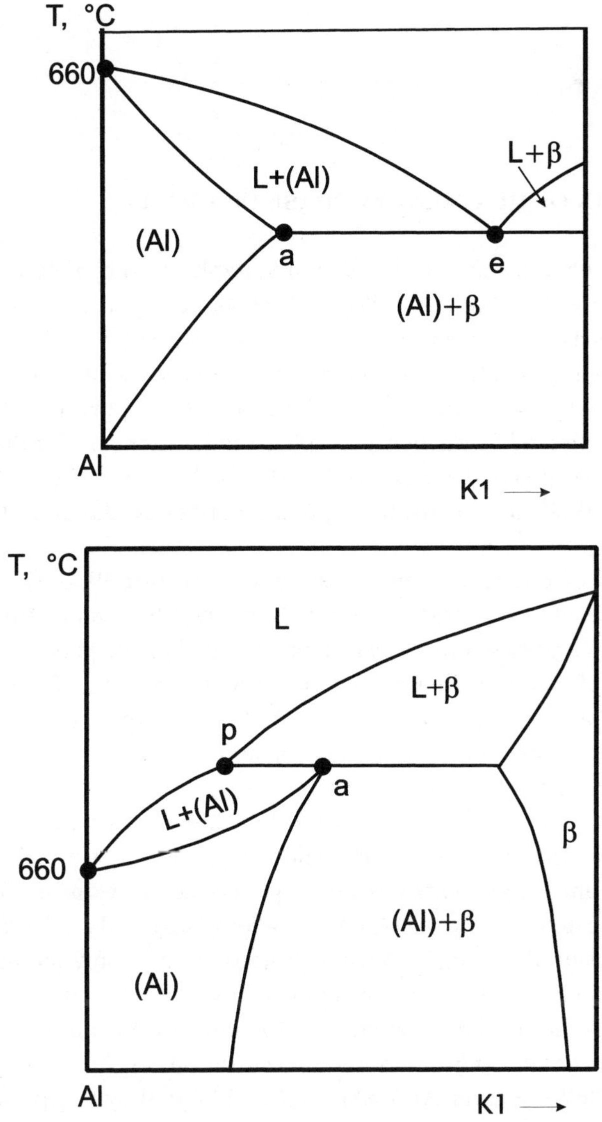

Figure A1.1. Schemes of eutectic (top) and peritectic (bottom) phase diagrams of the second element K_1 with Al.

Appendix 2

VISUALIZATION OF QUATERNARY PHASE DIAGRAMS

Ternary phase diagrams are, as a rule, not suitable for the analysis of phase composition of commercial aluminum alloys. This is especially true for the systems where a quaternary phase is formed, e.g. widely used casting 356 (357)-type alloys and wrought 6XXX-series alloys. Although these alloys contain only two alloying components – magnesium and silicon, the unavoidable presence of iron impurity causes the formation of the $Al_8FeMg_3Si_6$ phase and, hence calls for the consideration of the quaternary phase diagram Al–Fe–Mg–Si. We have chosen this system to illustrate the methods of visualization of quaternary phase diagrams that are used in this book.

The analysis of quaternary phase diagram is intrinsically difficult as the composition requires a concentration tetrahedron and the temperature, yet another dimension. As we possess only a two-dimensional sheet of paper, a true graphical representation of the four-dimensional construction demands for very elaborative calculations and diagrams (Zakharov, 1964; Zakharov, 1985; Petrov, 1991). Therefore, some concessions are usually made and, nowadays two main methods of visualization are used in practice.

The first method involves the projection of the aluminum corner of the isothermal tetrahedron onto the opposite concentration plane (Mondolfo, 1976). In this case, the phase diagram is represented as a triangle projection with the compositions of binary alloys located in the vertices, ternary alloys lying at the sides, and quaternary alloys falling inside the triangle. Such a projection gives only tentative impression of the phase equilibria. The concentration is given as the ratio of three alloying components, not in absolute percentages. In order to put an alloy onto the projection, one needs first to normalize the concentrations of three components (except aluminum) to 100%, e.g. an Al–7%Si–2%Mg–1%Fe alloy is represented by point S in Figure 2.4a (70 rel.% Si, 20 rel.% Mg, 10 rel.% Fe). Another assumption used in this visualization approach is neglecting the solid solubility of the alloying element in aluminum. Hence, the phase composition of an alloy can be assessed provided the following two conditions hold: (i) the aluminum solid solution (Al) is present at all temperatures and (ii) concentrations of all alloying components exceed their limit solubility in (Al).

A quaternary phase diagram can be graphically represented either as the distribution of phase fields in the solid state (Figure 2.4a) or as the polythermal diagrams (Drits et al., 1977) or liquidus projection (Mondolfo, 1976) (Figure 2.4b).

In the case of the solid-state phase distribution, the diagram shows two-, three-, and four-phase regions, the vertices of the concentration triangle reflecting binary alloys of the corresponding binary systems (see Appendix 1). As applicable to the Al–Fe–Mg–Si system, the vertices of the triangle contain (Si), Al_8Mg_5, and Al_3Fe phases from the Al–Si, Al–Mg, and Al–Fe systems, respectively (Figure 2.4a). The sides contain phases from the constituent ternary systems, i.e. Mg_2Si from the Al–Mg–Si system and Al_8Fe_2Si and Al_5FeSi from the Al–Fe–Si system. And the quaternary $Al_8FeMg_3Si_6$ compound is located inside the triangle. Such a diagram allows one to assess the phase composition of quaternary alloys in the solid state. For example, Figure 2.4a shows that 6XXX-series alloys can contain all listed phases except Al_8Mg_5, and that iron and silicon in 5XX.0, 5XXX-series alloys can form only Al_3Fe and Mg_2Si phases, respectively.

For the analysis of solidification, the projection of polythermal planes, lines, and points, reflecting bi-, mono-, and invariant reactions is used. Such a projection for the Al–Fe–Mg–Si systems is shown in Figure 2.4b. It is assumed that (Al) is solidified in all alloys as the primary phase. After that, bivariant eutectic reactions occur depending on the concentration range of the alloy as follows: (Si)–e_3–E_2–P_1–e_4 , L ⇒ (Al) + (Si); e_1–E_1–e_6–P_4–P_3–P_2–E_2–e_3–e_2, L ⇒ (Al) + Mg_2Si; Al_8Mg_5–e_1–E_1–e_5, L ⇒ (Al) + Al_8Mg_5; Al_3Fe–p_2–P_4–e_6–E_1–e_5, L ⇒ (Al) + Al_3Fe; p_2–p_1–P_3–P_4, L ⇒ (Al) + Al_8Fe_2Si; p_1–e_4–P_1–P_2–P_3, L ⇒ (Al) + Al_5FeSi; and P_2–P_1–E_2, L ⇒ (Al) + $Al_8FeMg_3Si_6$. For example, high-pure 356 (357)-type alloys contain the binary Al–Si eutectics, whereas an increased concentration of iron would result in the formation of the (Al) + Al_5FeSi eutectics in the same type of alloys. The change in the liquid composition can be roughly estimated by the projection of the line connecting the alloy composition with the second eutectic phase. After the solidification of the binary eutectics, the liquid composition falls onto the polythermal lines reflecting monovariant eutectics and peritectic reactions: e_3–E_2, L ⇒ (Al) + (Si) + Mg_2Si; e_1–E_1, L ⇒ (Al) + Mg_2Si + Al_8Mg_5; e_5–E_1, L ⇒ (Al) + Al_3Fe + Al_8Mg_5; e_4–P_1, L ⇒ (Al) + (Si) + Al_5FeSi; e_6–E_1 and e_6–P_4, L ⇒ (Al) + Al_3Fe + Mg_2Si; P_4–P_3, L ⇒ (Al) + Al_8Fe_2Si + Mg_2Si; P_3–P_2, L ⇒ (Al) + Al_5FeSi + Mg_2Si; P_1–E_2, L ⇒ (Al) + $Al_8FeMg_3Si_6$ + (Si); P_2–E_2, L ⇒ (Al) + $Al_8FeMg_3Si_6$ + Mg_2Si; p_2–P_4, L + Al_3Fe ⇒ (Al) + Al_8Fe_2Si; p_1–P_3, L + Al_8Fe_2Si ⇒ (Al) + Al_5FeSi; and P_2–P_1, L + Al_5FeSi ⇒ (Al) + $Al_8FeMg_3Si_6$. The solidification ends with invariant eutectic and peritectic reactions, the liquid composition being given by points. The knowledge of temperatures and concentrations of invariant reactions is very important for the correct analysis of solidification in quaternary alloys. Table 2.8 gives these values for the Al–Fe–Mg–Si system.

In some cases, the distribution of phase fields in the solid state and the polythermal diagram can be assessed using the data on ternary diagrams, with two main assumptions: (i) no quaternary phases are formed and (ii) the compositions of the ternary phases do not change significantly on addition of the fourth component. Examples of such assessed diagrams in this book are Al–Mg–Ni–Zn (Figure 7.10), Al–Fe–Mg–Zn (Figure 6.4), and Al–Cu–Ni–Si (Figure 7.9).

The second method of visualization of quaternary phase diagrams is the construction of isothermal sections at a constant concentration of one alloying element or the construction of polythermal sections or isopleths providing the constant concentrations of two alloying components. This method is rather visual for semiquantitative estimates of the phase composition. However, as distinct from the first method, it does not describe the phase equilibria in the entire system. Moreover, the experimental construction of such sections is very laborious, and the availability of sections as applicable to commercial alloys is limited. The evaluation of such sections using the data on ternary systems and the information on the general constitution of the quaternary phase diagram can be very useful. This can be done using calculation techniques described in Appendices 4 and 5.

Appendix 3

VISUALIZATION OF QUINARY PHASE DIAGRAMS

Although quaternary phase diagrams adequately describe the phase composition of some commercial alloys, for many commercial compositions they are not sufficient. The presence of five and more phases in the structure of commercial aluminum alloys demands the construction and analysis of quinary phase diagrams.

Present data on aluminum-based phase diagrams with five and more components are very few and relate only to some commercially important alloys. These data are limited, as a rule, by the detected phases and the assumed invariant reactions, the identification of the latter being based on one or two studies or being purely hypothetical (Mondolfo, 1976). Since the available equilibria in five-component systems are often inconsistent with ternary and quaternary diagrams, the authors sought to uncover detected discrepancies and to suggest a more suitable variant by using their own method of prediction and construction of five-component phase diagrams, as applied to the most important groups of commercial aluminum alloys (Belov et al., 1998, 2002a, b).

This method uses well-known reference books by Mondolfo (1976), Drits et al. (1977), Red Book (1955–1994), and available published articles directly or indirectly containing information on the phase composition of quinary aluminum alloys, and original studies performed by the authors of this book. The method is explained here.

Commercial aluminum alloys based on the binary Al–Si, Al–Cu, and Al–Mg systems contain the main alloying component in the amount that significantly exceeds the content of other elements. As a consequence, irrespective of other additive concentrations, the main phases, i.e. (Si), Al_2Cu, and Al_8Mg_5, must be in equilibrium with (Al), unless the main alloying element enters completely in (Al). This condition enables one to substantially simplify analysis of quinary systems as illustrated by the example of Al–Si alloys (silumins) in which the second basic element (in addition to aluminum) is silicon.

The Al–Cu–Fe–Mg–Si system attracts our attention because most casting alloys of the 3XX.0 series contain as alloying elements magnesium and copper, as well as iron as impurity (see Section 3.8). These elements, depending on their concentrations, can enter in the composition of various phases, and therefore the prediction of the phase composition of a particular alloy directly from ternary or quaternary systems becomes incorrect.

To the best of our knowledge, all phases found in silumins are formed on the basis of compounds containing no more than four components. This fact is used in the suggested method of predicting five-component phase diagrams, because the absence of quinary phases considerably facilitates the polyhedration of the system.

Appendix 2 demonstrates that the analysis of quaternary systems can be performed on the concentration triangle, which allows one to study the effect of all three alloying elements. We suggest extending this method of plotting phase diagrams to five-component systems using the most evident analogies.

When analyzing the phase composition of 3XX.0-type alloys, it is natural to assume that all phase regions in the solid state contain both (Al) and (Si). This enables us to consider the effect of the three remaining elements using the concentration triangle, because the maximum number of the remaining equilibrium phases is three. Hence, the polyhedration of the five-component system is reduced to "triangulation", i.e. the analogy exists with the method of plotting the phase regions of quaternary systems. The difference is that, instead of projecting the aluminum corner, we project an Al–X% Si alloy onto the base of the concentration triangle with the vertices (i) Al–X% Si–Y_1% K_1, (ii) Al–X% Si–Y_2% K_2, and (iii) Al–X% Si–Y_3% K_3 (where K_i denotes one of the other three elements of the quinary system and Y_i is the corresponding concentration). In this case, the vertices of the concentration triangle represent the projections of (Al)–(Si)–Φ_i phase regions of the corresponding ternary Al–Si–K_i systems, while the sides of this triangle are the projections of the three- and four-phase regions (Al)–(Si)–Φ_i–Φ_j belonging to the corresponding quaternary Al–Si–K_i–K_j systems. It is evident that the concentrations Y_i at any concentration ratio must be lower than the concentrations at which (Si) may disappear. In this case, the phase boundaries on the concentration triangle are constant irrespective of silicon contents. To perform polyhedration, one has to separate the five-phase ("quasi-three-phase") regions. This can be made a priori by using the triangulation rules (Zakharov, 1985, 1990), or by invoking the experimental data from extensive literature.

After the polyhedration ("triangulation") the concentration triangle of the quinary system shows three-, four-, and five-phase fields similar to single-, two-, and three-phase regions of a ternary system or two-, three,- and four-phase regions of a quaternary system.

The concentration is written as a relative fraction of three alloying components K_i, e.g. an Al–5% Si–3% Cu–1% Mg–1% Fe alloy corresponds to point Z in Figure 3.8a (60 rel.% Cu, 20 rel.% Mg and 20 rel.% Fe).

The distribution of phase fields in the solid state in the Al–Cu–Fe–Mg–Si system is shown in Figure 3.8a, as applied to 3XX-series alloys. The following phases can be in equilibrium with (Al) and (Si): Al_2Cu, Mg_2Si, Al_5FeSi, $Al_5Cu_2Mg_8Si_6$, and $Al_8FeMg_3Si_6$. Under equilibrium conditions, the maximum number of phases in a quinary alloy cannot exceed three (in addition to (Al) and (Si)).

The analysis of solidification of quinary alloys requires a polythermal diagram. An additional assumption should be made in order to make the visualization possible: only alloys in which the primary solidification of (Al) is followed by the formation of the (Al) + (Si) eutectics are considered. With this limitation, four-, five-, and six-phase reactions with the obligatory participation of (Al) and (Si) proceed in the system. On the concentration triangle, two-dimensional regions reflect the bivariant eutectic reactions and lines correspond to the invariant transformations. In such a way, alloys containing 9–11% Si and up to 0.6% Fe, i.e. to the concentrations below the line connecting the compositions of (Al) and the eutectic Al–11% Si–0.8% Fe alloy in the Al–Fe–Si phase diagram (Point E in Figure 1.1a, b), can be analyzed. For Al–Si alloys of other compositions, this method can be used for the analysis of solidification below the temperatures at which the reactions without (Si) are already completed.

The knowledge of temperatures and compositions of invariant reactions is very important for the correct analysis of solidification in quinary alloys. For the Al–Cu–Fe–Mg–Si system, these data are given in Table 3.19.

Appendix 4

MATHEMATICAL APPROXIMATION OF PHASE DIAGRAMS

In practice, the most convenient application of multicomponent phase diagrams to the analysis of a specific alloy is the use of polythermal and isothermal sections that enable one to quantitatively determine critical temperatures (T_i) and, for ternary systems, the relative mass fractions of the phases (Q_i). Not much of such sections are available in the literature and most researchers need to evaluate these sections themselves, using known rules (Zakharov, 1990). However, the correct construction of curved lines requires meticulous processing of primary graphical information (liquidus, solidus, solvus etc.) that, in turn may be not accurate, which degrades the accuracy of T_i and Q_i estimation.

Another approach includes the use of thermodynamical calculations. As a result all the surfaces are described analytically, and T_i and Q_i can be calculated exactly. Some of the results obtained in such a way are summarized in Appendix 5. However, the software packages based on thermodynamic calculations, e.g. Thermocalc, are specialized and rather expensive. Frequently, the application of such a software is limited by available thermodynamic functions or databases, especially in the case of ternary and quaternary phases.

We suggest some principles and methods of quantitative analysis of multicomponent phase diagrams using a widely available Excel spreadsheet software that allows one to create convenient calculation forms as applicable to a specific problem.

The proposed method is based on the calculation of mass fractions of phases in any N-phase alloy and the determination of the boundaries of N-phase regions in N-component systems. We are using the rule that the compositions of all phases (at a constant temperature) in an N-phase region of an N-component phase diagram are constant, irrespective of the alloy composition. The relative mass fractions of the phases can be calculated by solving a system of N linear balance equations, where the concentration of each alloy component (X_i) is determined as the sum of products $Y_{ij}Q_{Mj}$ (where Y_{ij} is the concentration of the i-component in the j-phase and Q_{Mj} is the mass fraction of the j-phase):

$$\begin{aligned} Y_{11}Q_{M1} + Y_{12}Q_{M2} + \cdots + Y_{1N}Q_{MN} &= X_1 \\ Y_{21}Q_{M1} + Y_{22}Q_{M2} + \cdots + Y_{2N}Q_{MN} &= X_2 \\ &\cdots \\ Y_{N1}Q_{M1} + Y_{N2}Q_{M2} + \cdots + Y_{NN}Q_{MN} &= X_N \end{aligned} \tag{1}$$

As only $N-1$ concentrations are independent, and the sum of all mass fractions of N phases equals unity (or 100%), system (1) can be modified as follows:

$$
\begin{aligned}
&Y_{11}Q_{M1} + Y_{12}Q_{M2} + \cdots + Y_{1N}Q_{MN} = X_1 \\
&Y_{21}Q_{M2} + Y_{22}Q_{M2} + \cdots + Y_{2N}Q_{MN} = X_2 \\
&\qquad\qquad \ldots \\
&Q_{M1} + Q_{M2} + \cdots + Q_{MN} = 1
\end{aligned} \tag{2}
$$

The system of linear equations with N unknowns is usually solved through an N-degree determinant, e.g. for system (2) the matrix of principal determinant reads as follows:

$$
\Delta = \begin{vmatrix} Y_{11} & Y_{12} & \cdots & Y_{1N} \\ Y_{21} & Y_{22} & \cdots & Y_{2N} \\ \cdots & \cdots & \cdots & \cdots \\ 1 & 1 & \cdots & 1 \end{vmatrix} \tag{3}
$$

Partial determinants $\Delta_1, \Delta_2 \ldots \Delta_N$ can be obtained by substitution of a column Y_{ij} for free members (X_1, X_2, …, X_{N-1}, 1). After that, the values of Q_{Mi} can be obtained as $Q_{Mi} = \Delta_i / \Delta$.

The volume fractions Q_{Vi}, which are very useful for evaluation of many alloy properties, can be calculated using the phase density ρ_i:

$$
Q_{Vi} = (Q_{Mi}/\rho_1)/(Q_{M1}/\rho_1 + Q_{M2}/\rho_2 + \cdots Q_{MN}/\rho_N) \tag{4}
$$

Example 1: Ternary system. Let us consider three-phase equilibria in the aluminum-corner of the Al–Fe–Si system (Section 1.1) (Belov and Matveeva, 2001). Three four-phase solidification reactions involving (Al) occur in the system (Table 1.2), i.e.

$$\mathrm{L} + \mathrm{Al_3Fe} \Rightarrow \mathrm{(Al)} + \mathrm{Al_8Fe_2Si};$$

$$\mathrm{L} + \mathrm{Al_8Fe_2Si} \Rightarrow \mathrm{(Al)} + \mathrm{Al_5FeSi}; \text{ and}$$

$$\mathrm{L} \Rightarrow \mathrm{(Al)} + \mathrm{Al_5FeSi} + \mathrm{(Si)}$$

Correspondingly, the following three three-phase regions are formed in the solid state (Figure 1.1c), i.e.

$$\mathrm{(Al)} + \mathrm{Al_3Fe} + \mathrm{Al_8Fe_2Si};$$

$$\mathrm{(Al)} + \mathrm{Al_8Fe_2Si} + \mathrm{Al_5FeSi}; \text{ and}$$

$$\mathrm{(Al)} + \mathrm{Al_5FeSi} + \mathrm{(Si)}$$

Despite numerous experimental data that are available on this system, the complexity of this phase diagram hinders the analysis of alloy phase compositions without adequate sections. The demand for the efficient analysis of Al–Fe–Si alloys facilitated the development of the calculation procedure described earlier. The following set of balance equations was solved:

$$\begin{aligned} Q_{M1}C_{Fe1} + Q_{M2}C_{Fe2} + Q_{M3}C_{Fe3} &= C_{FeX} \\ Q_{M1}C_{Si1} + Q_{M2}C_{Si2} + Q_{M3}C_{Si3} &= C_{SiX} \\ Q_{M1}C_{Al1} + Q_{M2}C_{Al2} + Q_{M3}C_{Al3} &= C_{AlX}, \end{aligned} \qquad (5)$$

where Q_{M1}, Q_{M2}, and Q_{M3} are the mass fractions of the respective phases, and C_{Fei}, C_{Sii}, C_{Ali}, and C_{FeX}, C_{SiX}, C_{AlX} are the mass concentrations of Fe, Si, and Al in the i-phase and in the analyzed alloy X, respectively. Evidently, the last equation can be substituted for

$$Q_{M1} + Q_{M2} + Q_{M3} = 1.$$

Here we consider the calculation for the phase region (Al) + Al_3Fe + Al_8Fe_2Si. By taking into account that Al_3Fe and Al_8Fe_2Si have virtually constant composition in the entire temperature range, 20 to 660°C, the calculation of Q_{Mi} requires only the concentrations of iron and silicon in the aluminum solid solution, C_{FeAl} and C_{SiAl}, respectively. Using data in Table 1.4, one can calculate temperature dependences of C_{FeAl} and C_{SiAl}. These dependences are introduced in the Excel spreadsheet form. Now, the user fills only tree fields in the form, i.e. the alloy composition (Fe, Si) and temperature (shown in bold in Figure A4.1), and receives the sought parameters Q_{Mi} and Q_{Vi} as an output of calculations. This form allows one to calculate the boundaries of the phase field (Al) + Al_3Fe + Al_8Fe_2Si at any temperature and composition in the solid state, by assigning a zero value to either Q_{Mi} or Q_{Vi} of the phase that disappears at the boundary, i.e. to calculate solvus surfaces. The volume fraction of phases can be calculated provided the values of density are known (Figure A4.1).

Similarly, the calculation forms for other three-phase regions – (Al) + Al_8Fe_2Si + Al_5FeSi, (Al) + Al_5FeSi + (Si), L + (Al) + Al_3Fe, L + (Al) + Al_8Fe_2Si, L + (Al) + Al_5FeSi, and L + (Al) + (Si) – can be developed.

Such forms can be used for the calculation of invariant reactions (Table 1.2) that allows one to determine the limit points at the horizontal lines of polythermal sections as shown in Figure 1.7. The liquidus polytherms can be constructed using approximating functions derived from liquidus isotherms (Figure 1.1b). The construction of boundaries of single- and two-phase regions requires additional data, but these regions are usually so narrow that their precise position is not practically important.

Construction of isothermal sections like those shown in Figure 1.6 is based on the calculation of canode triangles. If phases have narrow homogeneity ranges and the composition of (Al) is known, then the position of boundaries can be determined quite accurately.

Same forms can be used for the calculation of the dependences of phase volume fractions on the concentration of one of the components (Figure 1.9) or on the temperature. These dependences can be used in the optimization of alloy composition and heat treatment.

And, of course, all these calculations can be embedded in a single file/spreadsheet.

System Al–Fe–Si	$(Al) + Al_3Fe + Al_8Fe_2Si$				
COMPONENTS	(Al) F1	Al_3Fe F2	Al_8Fe_2Si F3	**ALLOY**	*T*, °C
Si, %	0.21	0.00	7.00	**0.30**	**550**
Fe, %	0.017	37.00	33.00	**2.00**	
	1	1	1	1	
			RESULT		
		(Al)	Q_{M1},% = *94.49*	Q_{V1},% = *96.03*	
		Al_3Fe	Q_{M2},% = *4.12*	Q_{V2},% = *2.90*	
		Al_8Fe_2Si	Q_{M3},% = *1.39*	Q_{V3},% = *1.07*	
			SUM = *100.00*	SUM = *100.00*	
Phase density, g/cm^3					
(Al)	2.7				
Al_3Fe	3.89				
Al_8Fe_2Si	3.58				

Figure A4.1. Excel spreadsheet for calculation of mass (Q_M) and volume (Q_V) fractions of phases in Al–Fe–Si alloys in the $(Al) + Al_3Fe + Al_8F3_2Si$ phase field.

Example 2: Quaternary system. Here we consider the aluminum corner of the Al–Fe–Mg–Si system (Section 2.4) (Belov, 2005). The following four-phase regions exist in this portion of the system in the solid state (Figure 2.4a):

$(Al) + (Si) + Mg_2Si + Al_8FeMg_3Si_6$;
$(Al) + (Si) + Al_5FeSi + Al_8FeMg_3Si_6$;
$(Al) + Mg_2Si + Al_5FeSi + Al_8FeMg_3Si_6$;
$(Al) + Mg_2Si + Al_5FeSi + Al_8Fe_2Si$;
$(Al) + Mg_2Si + Al_8Fe_2Si + Al_3Fe$; and
$(Al) + Mg_2Si + Al_8FeSi + Al_3Fe + Al_8Mg_5$

The composition of all phases, except (Al), is constant. The composition of (Al) can be calculated (by approximating functions) using the data on Al–Mg–Si alloys (Table 2.4) because the solubility of Fe in (Al) is very small.

The following system of four linear equations should be solved in order to calculate the mass fractions of phases:

$$
\begin{aligned}
&Q_{M1}C_{Si1} + Q_{M2}C_{Si2} + Q_{M3}C_{Si3} + Q_{M4}C_{Si4} = C_{SiX} \\
&Q_{M1}C_{Mg1} + Q_{M2}C_{Mg2} + Q_{M3}C_{Mg3} + Q_{M4}C_{Mg4} = C_{MgX} \\
&Q_{M1}C_{Fe1} + Q_{M2}C_{Fe2} + Q_{M3}C_{Fe3} + Q_{M4}C_{Fe4} = C_{FeX} \\
&Q_{M1} + Q_{M2} + Q_{M3} + Q_{M4} = 1,
\end{aligned}
\qquad (6)
$$

where Q_{M1}, Q_{M2}, Q_{M3}, and Q_{M4} are the mass fractions of the respective phases, and C_{Sii}, C_{Mgi}, C_{Fei}, and C_{SiX}, C_{MgX}, C_{FeX} are the mass concentrations of Si, Mg, and Fe in the *i*-phase and the analyzed alloy *X*, respectively.

Figure A4.2 shows the spreadsheet form for the phase region (Al) + (Si) + $Mg_2Si + Al_8FeMg_3Si_6$. The negative volume fraction (in this case – Mg_2Si) means that the equilibrium alloy at this particular temperature does not fall into the four-phase region containing this phase. The dependences of the volume fractions of individual phases on the concentration of an alloying element can reveal some tendencies that cannot be found without such a calculation, see Figure 2.8 as an example.

Let us look in more detail at the construction of a polythermal section of this system at 7% Si and 0.2% Fe as shown in Figure 2.13a. The liquidus polytherm for the primary solidification of (Al) can be calculated using data on the liquidus isotherms of the constituent ternary systems. Similarly, one can calculate the polytherm for the solidification of the binary L ⇒ (Al) + (Si) eutectics that occurs in all alloys of the given polythermal section. Then the calculation shows that two invariant horizontals should be present in the considered compositional range, one for the eutectic reaction L ⇒ (Al) + (Si) + $Mg_2Si + Al_8FeMg_3Si_6$ at 554°C and the other the peritectic reaction L + Al_5FeSi ⇒ (Al) + (Si) + $Al_8FeMg_3Si_6$ at 567°C (Table 2.8). The points at these horizontals from which the lines bordering the three- and four-phase regions start can also be determined. The calculation of these phase fields at different temperatures allows one to draw their borders.

Isothermal sections at 7% Si shown in Figure 2.12 are constructed based on the calculation of canode tetrahedrons. As the boundaries of four-phase fields in the isothermal plane should appear as straight lines, their position can be determined by two points.

In the case of 6XXX-series alloys, the phase field boundaries are so close to each other that only calculation allows one to distinguish these boundaries (see Figure 2.7).

System Al–Fe–Si–Mg $(Al) + Al_3Fe + Mg_2Si + Al_8FeMg_3Si_6$						
COMPONENTS	(Al) F1	(Si) F2	Mg_2Si F3	$Al_8FeMg_3Si_6$ F4	ALLOY	*T*, °C
Si, %	0.11	99.5	36.7	32.9	**7**	**200**
Mg, %	0.05	0	63.2	14.1	**0.3**	
Fe, %	0.01	0	0	10.9	**0.5**	
	1	1	1	1	1	
			RESULT			
			Q_M, wt%	Q_V, vol.%		
		(Al)	*90.38*	*89.97*		
		(Si)	*5.72*	*6.60*		
		Mg_2Si	*−0.60*	*−0.86*		
		$Al_8FeMg_3Si_6$	*4.50*	*4.29*		
		SUM=	*100*	*100*		
	Phase density, g/cm^3					
	(Al)	2.7				
	(Si)	2.33				
	Mg_2Si	1.88				
	$Al_8FeMg_3Si_6$	2.82				

Figure A4.2. Excel spreadsheet for calculation of mass (Q_M) and volume (Q_V) fractions of phases in Al–Fe–Mg–Si alloys in the $(Al) + (Si) + Mg_2Si + Al_8FeMg_3Si_6$ phase field.

Example 3: Quinary system. The algorithm considered here for ternary and quaternary systems does not change much on transition to more complex systems. Moreover, the calculation of multicomponent phase diagram is, in many cases, the only way to extract the required information.

Take as an example the Al–Cu–Fe–Mg–Si system (Section 3.7) in the range of Al–Si alloys. The distribution of phase fields in the solid state of this system is shown in Figure 3.8a. By taking into account that the compositions of excess phases are constant, and the composition of (Al), due to a small solubility of iron, can be calculated from the data on the Al–Cu–Mg–Si system (Table 3.13), we have all necessary information for the calculation spreadsheet forms for all five-phase regions. The calculation of mass fractions is based on the following system of five linear equations:

$$\begin{aligned}
&Q_{M1}C_{Si1} + Q_{M2}C_{Si2} + Q_{M3}C_{Si3} + Q_{M4}C_{Si4} + Q_{M5}C_{Si5} = C_{SiX}\\
&Q_{M1}C_{Cu1} + Q_{M2}C_{Cu2} + Q_{M3}C_{Cu3} + Q_{M4}C_{Cu4} + Q_{M5}C_{Cu5} = C_{CuX}\\
&Q_{M1}C_{Mg1} + Q_{M2}C_{Mg2} + Q_{M3}C_{Mg3} + Q_{M4}C_{Mg4} + Q_{M5}C_{Mg5} = C_{MgX}\\
&Q_{M1}C_{Fe1} + Q_{M2}C_{Fe2} + Q_{M3}C_{Fe3} + Q_{M4}C_{Fe4} + Q_{M5}C_{Fe5} = C_{FeX}\\
&Q_{M1}C_{Al1} + Q_{M2}C_{Al2} + Q_{M3}C_{Al3} + Q_{M4}C_{Al4} + Q_{M5}C_{Al5} = C_{AlX},
\end{aligned} \quad (7)$$

where Q_{M1}, Q_{M2}, Q_{M3}, Q_{M4}, Q_{M5} are the mass fractions of the respective phases, and C_{Sii}, C_{Cui}, C_{Mgi}, C_{Fei}, C_{Ali}, C_{SiX} etc. are the mass concentrations of Si, Cu, Mg, Fe, and Al in the *i*-phase and the analyzed alloy (X), respectively.

Figure A4.3 shows the spreadsheet for the calculation of Al–Si–Cu–Mg–Fe phase diagram. The user enters element concentrations for the analyzed alloy in the corresponding fields of the spreadsheet, and receives as a result the selection of phases with their mass and volume fractions. This spreadsheet can also be programmed for the calculation of (Al) composition at different temperatures. Figure 3.8a exhibits three possible sets of phases in the range of Al–Si alloys.

The properties of an alloy are to a great extent determined by the selection of phases and their amount. The suggested calculations can help in assessment of the effect of any given element on the values of Q_M and Q_V of all phases. An example of such a calculation is demonstrated in Figure 3.19 for the effect of silicon on the amount of excess phases in a 222.1 casting alloy. One can see a profound influence of this element.

The polythermal section of the Al–Cu–Fe–Mg–Si phase diagram at 10% Si, 5% Cu, and 0.5% Mg shown Figure 3.15 was constructed as follows. First, the position of liquidus lines and the polytherm of the beginning of the binary eutectic reaction L $\Rightarrow$ (Al) + (Si) are assessed. Then the calculation shows that this section contains two invariant horizontals (Table 3.19), for the eutectic reaction L $\Rightarrow$ (Al) + (Si) + Al_2Cu + $Al_5Cu_2Mg_8Si_6$ + $Al_8FeMg_3Si_6$ at ~505°C and for the peritectic reaction

	ALLOY AK5M7	F1 (Al)	F2 (Si)	F3 Al_2Cu	F4 Mg_2Si	F5 Al_5FeSi	F6 Q	F7 π
Si, %	9	0.08	99.5	0	36.8	13.5	32.1	32.9
Cu, %	1.8	0.05	0	52.5	0	0	19.2	0
Mg, %	0.5	0.05	0	0	63.2	0	33	14.1
Fe, %	1	0	0	0	0	27.5	0	10.9
	1	1	1	1	1	1	1	1
Al, %	87.7	99.82	0.5	47.5	0	59	15.7	42.1
				RESULT	wt%		vol. %	
Phase density	g/cm^3		(Al)	Q_{M1} =	*83.47*	Q_{V1} =	*84.79*	
R1 =	2.7		(Si)	Q_{M2} =	*7.58*	Q_{V2} =	*8.00*	
R2 =	2.6		Al_2Cu	Q_{M3} =	*3.349*	Q_{V3} =	*2.11*	
R3 =	4.35		Al_5FeSi	Q_{M5} =	*2.35*	Q_{V5} =	*1.93*	
R4 =	1.88		π	Q_{M7} =	*3.25*	Q_{V7} =	*3.16*	
R5 =	3.33			sum =	*100*		*100*	
R6 =	2.79							
R7 =	2.82							

Figure A4.3. Excel spreadsheet for calculation of mass (Q_M) and volume (Q_V) fractions of phases in the Al–Cu–Fe–Mg–Si system in the range of Al–Si alloys.

$L + Al_5FeSi \Rightarrow (Al) + (Si) + Al_2Cu + Al_8FeMg_3Si_6$ at ~554°C. And finally, the calculation of five-phase regions determines their boundaries.

It should be noted that the sections constructed using the described technique are quite similar to those produced with a thermodynamical software, provided the same starting data are used. Some of the sections of quaternary diagrams given in this book, e.g. Al–Cu–Mg–Si and Al–Fe–Ni–Si, were obtained by the authors using Thermocalc.

Appendix 5

REFERENCES ON SOME CALCULATED AL-BASED SYSTEMS

In recent years, considerable success has been achieved in the quantitative analysis of multicomponent phase diagrams using thermodynamics-based software. Vast literature has been published on this subject including those of Kaufman and Bernstein (1970) and articles in CALPHAD and Journal of Phase Equilibria. This technique could be quite useful for the analysis of aluminum alloys that are, as a rule, multicomponent and multiphase. However, the thermodynamic calculations of complex systems require vast experimental data both for the assessment of chemical potential functions and for valiadation of the calculation results. Thermodynamic cacluation of little studied systems can be quite inadequate. A combination of experimental work (for qualitative study of phase equilibria) with thermodynamic calculation seems to be an appropriate way for establishing an adequate model with subsequent solution of practical problems, i.e. construction of isothermal and polythermal sections, calculation of critical temperatures and phase ratios in a specific alloy. Table A5.1 give some references to the articles where the results of thermodynamic calculation of Al-based systems have been published.

Table A5.1. References to some calculated multicomponent phase diagrams in the range of aluminum alloys

System	Reference
Al–Ca–Mg	J. Gröbner, D. Kevorkov, I. Chumak, R. Schmid-Fetzer. Experimental investigation and thermodynamic calculation of ternary Al–Ca–Mg phase equilibria, Z. Metallkde., 2003, Vol. 94, no. 9, pp. 976–982.
Al–Ce–Mg	J. Gröbner, D. Kevorkov, R. Schmid-Fetzer. Thermodynamic modeling of Al–Ce–Mg phase equilibria coupled with key experiments, Intermetallics, 2002, Vol. 10, no. 5, pp. 415–422.
Al–Cr–Ni	N. Dupin, I. Ansara, B. Sundman. Thermodynamic re-assessment of the ternary system Al–Cr–Ni. Calphad, 2001, Vol. 25, no. 2, pp. 279–298.
Al–Cu–Fe–Mg–Mn–Si	Y.A. Chang, X.-Y. Yan, F.-Y. Xie, S.-L. Chen. Phase formation and solidification path analysis of multicomponent aluminum alloys. Solidification of Aluminum Alloys as held at the 2004 TMS Annual Meeting; Charlotte, NC; USA; 14–18 Mar. 2004. Warendale: TMS, 2004, pp. 3–18.

(*continued*)

Table A5.1. Continued

System	Reference
Al–Cu–Fe–Mg–Mn–Si–Zn	X.-Y. Yan, Y.A. Chang, F.-Y. Xie et al. Calculated phase diagrams of aluminum alloys from Al–Cu to multicomponent commercial alloys. J. Alloys Comp., 2001, Vol. 320, pp. 151–160.
Al–Cu–Mg	A. Zhu, B.M. Gable, G.J. Shiflet, E.A. Starke Jr. Trace element effects on precipitation in Al–Cu–Mg–(Ag, Si) alloys: a computational analysis, Acta Mater., 2004, Vol. 52, no. 12, pp. 3671–3679 (also considers small additions of Si and Ag).
	A.M. Camacho, H.V. Atkinson, P. Kapranos, B.B. Argent. Thermodynamic predictions of wrought alloy compositions amenable to semi-solid processing, Acta. Mater., 2003, Vol. 51, pp. 2319–2330.
	S.L. Chen, Y. Zuo, H. Liang, Y.A. Chang. A thermodynamic description for the ternary Al–Mg–Cu system, Metall. Mater. Trans. A, 1997, Vol. 28A, no. 2, pp. 435–446.
Al–Cu–Mg–Si	Y.A. Chang, X.-Y. Yan, F.-Y. Xie, S.-L. Chen. Phase formation and solidification path analysis of multicomponent aluminum alloys. Solidification of Aluminum Alloys as held at the 2004 TMS Annual Meeting; Charlotte, NC; USA; 14–18 Mar. 2004. Warendale: TMS, 2004, pp. 3–18.
	X.-Y. Yan, Y.A. Chang, F.-Y. Xie et al. Calculated phase diagrams of aluminum alloys from Al–Cu to multicomponent commercial alloys. J. Alloys Comp., 2001, Vol. 320, pp. 151–160.
Al–Cu–Mg–Zn	Y.A. Chang, X.-Y. Yanm F.-Y. Xie, S.-L. Chen. Phase formation and solidification path analysis of multicomponent aluminum alloys. Solidification of Aluminum Alloys as held at the 2004 TMS Annual Meeting; Charlotte, NC; USA; 14–18 Mar. 2004. Warendale: TMS, 2004, pp. 3–18.
	A.M. Camacho, H.V. Atkinson, P. Kapranos, B.B. Argent. Thermodynamic predictions of wrought alloy compositions amenable to semi-solid processing, Acta. Mater., 2003, Vol. 51, pp. 2319–2330.
	J.L. Murray. Thermodynamic modeling of the Al–Cu–Mg–Zn system in the aluminum corner, CALPHAD and Alloy Thermodynamics as held at the 2002 TMS Annual Meeting; Seattle, WA; USA; 17–21 Feb. 2002. Warrendale: TMS, pp. 165–175.
	H.J. Seifert, P. Liang, H.L. Lukas, F. Aldinger, S.G. Fries, M.G. Harmelin, F. Faudot, T. Jantzen. Computational phase studies in commercial aluminium and magnesium alloys, Mater. Sci. Technol., 2000, Vol. 16, no. 11–12, pp. 1429–1433.
Al–Cu–Zn	A.M. Camacho, H.V. Atkinson, P. Kapranos, B.B. Argent. Thermodynamic predictions of wrought alloy compositions amenable to semi-solid processing, Acta. Mater., 2003, Vol. 51, pp. 2319–2330.
Al–Fe–Mg–Mn–Si	X.-Y. Yan, Y.A. Chang, F.-Y. Xie et al. Calculated phase diagrams of aluminum alloys from Al–Cu to multicomponent commercial alloys. J. Alloys Comp., 2001, Vol. 320, pp. 151–160.
Al–Fe–Mn–Si	E. Balitchev, T. Jantzen, I. Hurtado, D. Neuschütz. Thermodynamic assessment of the quaternary system Al–Fe–Mn–Si in the Al-rich corner, Calphad, 2003, Vol. 27, no. 3, pp. 275–278.

(*continued*)

Table A5.1. Continued

System	Reference
Al–Mg–Si	X. Dore, H. Combeau, M. Rappaz. Modelling of microsegregation in ternary alloys: application to the solidification of Al–Mg–Si, Acta Mater., 2000, Vol. 48, no. 15, pp. 3951–3962.
Al–Mg—Zn	A.M. Camacho, H.V. Atkinson, P. Kapranos, B.B. Argent. Thermodynamic predictions of wrought alloy compositions amenable to semi-solid processing, Acta. Mater., 2003, Vol. 51, pp. 2319–2330.
	S.L. Chen, S. Daniel, F. Zhang, et al. The PANDAT software package and its applications, Calphad, 2002, Vol. 26, no. 2, pp. 175–188.
	P. Liang, T. Tarfa, J.A. Robinson et al. Experimetal investigation and thermodynamic calculation of the Al–Mg–Zn system, Themochim. Acta, 1998, Vol. 314, pp. 87–110.
Al–Mn–Si	Y. Du, J.C. Schuster, F. Weitzer et al. A thermodynamic description of the Al–Mn–Si system over the entire composition and temperature ranges. Metall. Mater. Trans. A, 2004, Vol. 35A, pp. 1613–1628.
Al–Si–Ce	J. Gröbner, D. Mirkovic, R. Schmid-Fetzer. Thermodynamic aspects of the constitution, grain refining, and solidification enthalpies of Al–Ce–Si alloys, Metall. Mater. Trans. A., 2004, Vol. 35A, no. 11, pp. 3349–3362.
Al–Si–Sb	S.H. Hansen, C.R. Loper. Jr. Effect of antimony on the phase equilibrium of binary Al–Si alloys. Calphad, 2000, Vol. 24, no. 3, pp. 339–352.

REFERENCES

Abdel-Hamid, A. & Durnad, F. (1985) *Z. Metallkde.*, **76**, pp 739–746.

Aksenov, A., Belov, N.A. & Zolotorevskii, V.S. (2003) Patent US 6,585,392 B1 Jul. 1, Aluminum-based material and a method for manufacturing products from aluminum-based material.

Aksenov, A.A. (1996) *Izv. Vyssh. Uchebn. Zaved., Tsvetn. Metall.*, no. 2, pp 34–46.

Aksenov, A.A., Zolotorevskii, V.S., Bendovskii, E.B. & Hayurov, S.S. (1991) *Izv. Vyssh. Uchebn. Zaved., Tsvetn. Metall.*, no. 4, pp 104–107.

Aksenov, A.A., Zolotorevskii, V.S., Kuznetsov, G.M. & Kiselev, A.N. (1992) in *Metallovedenie i obraborka tsvetnykh splavov* (Physical Metallurgy and Processing of Nonferrous Alloys), Nauka, Moscow, pp.171–180.

Aksenov, A.A., Churmukov, E.A. & Romanova, V.S. (1994) *Izv. Vyssh. Uchebn. Zaved., Tsvetn. Metall.*, no. 4–6, pp 145–150.

Aksenov, A.A., Churmukov, E.A., Zolotorevskii, V.S. & Indenbaum, G.V. (1995) *Isv. RAN. Metally*, no. 1, pp 21–26.

Aksenov, A.A., Belov, N.A. & Medvedeva, S.V. (2001a) *Z. Metallkde.*, **92**, pp 1103–1110.

Aksenov, A.A., Belov, N.A. & Medvedeva, S.V. (2001b) *Izv. Vyssh. Uchebn. Zaved., Tsvetn. Metall.*, no. 1, pp 18–25.

Alekseeva, Z.A. (1992) in *Ternary Alloys*, Eds. Petzow, G. & Effenberg, G., **Vol. 5**, VCH, Weinheim, pp 524–528.

Arcade, P., He, X., Adawy, Z. & Champier, G. (1990) Aluminum Alloys'90. *Proc. 2nd Inter. Conf. Aluminum Alloys*, Intern. Acad. Publishers, Beijing, China, pp 134–139.

Bäckerud, L., Król, E. & Tamminen, J. (1986) Solidification Characteristics of Aluminium Alloys. Vol. 1: Wrought Alloys, SkanAluminium/Universitetsforlaget AS, Oslo.

Bäckerud, L., Chai, G. & Tamminen, J. (1990) Solidification Characteristics of Aluminum Alloys. Vol. 2: Foundry Alloys, Des AFS/SkanAluminium, Plaines.

Balitchev, E., Jantzen, T., Hurtado, I. & Neuschbtz, D. (2003) *Computer Coupling of Phase Diagrams and Thermochemistry (CALPHAD)*, **27**, pp 275–278.

Bätzner, C. (1993) in *Ternary Alloys*, Eds. Petzow, G. & Effenberg, G., **Vol. 6**, VCH, Weinheim, pp 394–397.

Belov, N.A. (1993a) *Izv. RAN. Metally*, no. 5, pp 133–138.

Belov, N.A., (1993b) Patent RUS 2001145, 15.11.1993. *Casting aluminum alloy.*

Belov, N.A., (1993c) Patent RUS 2001147, 15.11.1993. *Casting aluminum alloy.*

Belov, N.A. (1994) *J. Adv. Mater.*, **1**, pp 321–329.

Belov, N.A. (1996) *Mater. Sci. Forum*, pp 217–222, 293.

Belov, N.A. (1998) *Organizatsiya eksperimenta* (Planning of Experiments), MIS&A, Moscow.

Belov, N.A. (2005) *Izv. Vyssh. Uchebn. Zaved., Tsvetn. Metall.*, no. 2, pp 71–76.

Belov, N.A. & Koltsov, A.V. (2002) *Izv. Vyssh. Uchebn. Zaved., Tsvetn. Metall.*, no. 2, pp 37–48.

Belov, N.A. & Matveeva, I.A. (2001) *Izv. Vyssh. Uchebn. Zaved., Tsvetn. Metall.*, no. 1, pp 34–39

Belov, N.A. & Zolotorevskii, V.S. (1995) in *Metallovedenie, lit'e i obrabotka splavov* (Physical Metallurgy, Casting and Treatment of Alloys), VILS, Moscow, pp 29–37.

Belov, N.A. & Zolotorevskii, V.S. (2001) *J. Adv. Mater.*, **5**(3), pp 1–8.

Belov, N.A. & Zolotorevskii, V.S. (2002) *Mater. Sci. Forum.*, **396–402**, pt. 2, pp 935–940.

Belov, N.A. & Zolotorevskii, V.S. (2003) *Tsvetn. Met.*, no. 2, pp 99–105.

Belov, N.A., Zolotorevsky, V.S. & Tagiev, E.E. (1992) *Izv. RAN. Metally*, no. 1, pp 146–151.

Belov, N.A., Zolotorevskii, V.S. & Luzgin, D.V. (1994) *Izv. RAN, Metally*, no. 3, pp 70–76.

Belov, N.A., Zolotorevskii, V.S. & Luzgin, D.V. (1996) *J. Adv. Mater.*, **3**, pp 228–238.

Belov, N.A., Yu Gusev, A. & Eskin, D.G. (1998) *Z. Metallkde.*, **89**(9), pp 618–622.

Belov, N.A., Naumova, E.A. & Eskin, D.G. (1999) *Mater. Sci. Eng. A*, **A271**, pp 134–142.

Belov, N.A., Aksenov, A.A. & Eskin, D.G. (2002a) *Iron in Aluminum Alloys: Impurity and Alloying Element*, Taylor and Francis, London, New York.

Belov, N.A., Koltsov, A.V. & Eskin, D.G. (2002b) *Mater. Sci. Forum*, **396–402**, pt. 2, pp 929–934.

Belov, N.A., Alabin, A.N. & Eskin, D.G. (2004) *Scr. Mater.*, **50**(1), pp 89–94.

Boom, E.A. (1963) *Dokl. Akad. Nauk SSSR*, **151**, pp 96–97.

Bouchaud, E., Khan, T., Kubic, L. & Octor, H. (1990) in *Advanced Aluminium and Magnesium Alloys*, Eds. Khan, T. & Effenberg, G., ASM, Amsterdam, pp 251–258.

Budberg, P.B. (1993) in *Ternary Alloys*, Eds. Petzow, G. & Effenberg, G., **Vol. 8**, VCH, Weinheim, pp 7–21.

Budberg, P. & Price, A. (1992) in *Ternary Alloys*, Eds. Petzow, G. & Effenberg, G., **Vol. 5**, VCH, Weinheim, pp 309–323.

Cacciamani, G., De Negri, S., Saccone, A. & Ferro, R. (2003) *Intermetallics*, **11**, pp 1135–1151.

Chakrabarti, D.J. & Laughlin, D.E. (2004) *Progr. Mater. Sci.*, **49**(3–4), pp 389–410.

Chakrabarti, D.J. & Murray, J.L. (1996) *Mater. Sci. Forum*, **217–222**, pt. 1, pp 177–182.

Champier, G. & Samuel, F.G. (1986) in *Aluminium–Lithium Alloys III*, pp 131–136.

Charai, A., Walther, T., Alfonso, C., Zahra, A.-M. & Zahra, C.Y. (2000) *Acta Mater.*, **48**, pp 2751–2764.

Chen, R., Huang, Z., Chen, C.Q., Shen, J.Y. & Zhang, Y.G. (2000) *Mater. Sci. Eng. A*, **A280**, pp 146–150.

Chen, Z., Zheng, Z. & Wang, Z. (2004) *J. Mater. Sci.*, **39**, pp 2253–2255.

Chen, S.-L., Zuo, Y., Liang, H. & Chang, Y.A. (1997) *Metall. Mater. Trans.*, **28A**, pp 435–446.

Cooper, M. (1967) *Acta Crystall.*, **23**, pp 1106–1107.

Cornish, L.A. & Witcomb, M.J. (1999) *J. Alloys Comp.*, **291**, pp 145–166.

Davydov, V.G. (1984) Doctor Sci. thesis, VILS, Moscow.

Davydov, V.G., Zakharov, V.V., Sakharov, E.D. & Novikov, I.I. (1973) *Diagrammy izotermicheskogo raspada v aluminievykh splavakh* (Diagrams of Isothermal Transformations in Aluminum Alloys), Metallurgiya, Moscow.

Davignon, G., Serneels, A., Verlinden, B. & Delaey, L., L. (1996) *Metall. Trans. A*, **27A**, pp 3357–3361.

Davis, J.R., Ed. (1993) *Aluminum and Aluminum Alloys*, ASM Specialty Handbook, ASM International, Materials Park.

Davydov, V.G., Ber, L.B. & Ananiev, V.N. (1996) *Mater. Sci. Forum*, **317–222**, pp 859–864.

Davydov, V.G., Ber, L.B., Kaputkin, E.Ya., Komov, V.I., Ukolova, O.G. & Lukina, E.A. (2000a) *Mater. Sci. Eng. A*, **A280**, pp 72–82.

Davydov, V.G., Rostova, T.D., Zakharov, V.V., Filatov, Yu.A. & Yelagin, V.I. (2000b) *Mater. Sci. Eng. A*, **A280**, pp 30–36.

Demarkar, S. (1986) *Proc. Intern. Conf. Rapidly Solidified Powder Aluminum Alloys*, ASTM STP 890, Eds. Fine, M.E. & Starke, E.A. (Jr.), ASTM, Philadelphia, pp 154–165.

De Negri, S., Saccone, A., Cacciamani, G. & Ferro, R. (2003) *Intermetallics*, **11**, pp 1125–1134.
Denholm, W.T., Esdaile, J.D., Siviour, N.G. & Wilson, B.W. (1984) *Metall. Trans. A*, **15A**, pp 1311–1317.
Ding, J.J., Rogl, P. & Schmidt, H. (2001) *J. Alloys Comp.*, **317–318**, pp 379–384.
Dobatkin, V.I., Elagin, V.I. & Fedorov, V.M. (1995) *Bystrozakristallizovannye alyuminievye splavy* (Rapidly Solidified Aluminum Alloys), VILS, Moscow.
Doboleg, S.A. (1963) *Karbid Kremniya* (Silicon Carbide), Gos. Izd. Tekhn. Lit, Kiev.
Dons, A.L. (1984) *Z. Metallkde.*, **75**, pp 171–174.
Dons, A.L. (1985) *Z. Metallkde.*, **76**, pp 609–612.
Dons, A.L. (2002) "Alstruc" – a model for solidification and homogenization of industrial aluminum alloys. PhD thesis, NTNU, Trondheim.
Dorward, R.C. (1988) *Metall. Trans. A*, **19A**, pp 1631–1634.
Drits, M.E., Ed. (1997) *Svoystva Elementov* (Properties of Elements), Metallurgiya, Moscow.
Drits, M.E., Kadaner, E.S., Padezhnova, E.M., Rokhlin, L.L., Sviderskaya, Z.A. & Turkina, N.I. (1977) in *Phase Diagrams of Aluminum- and Magnesium-Based Systems*, Ed. Abrikosov, N.Kh., Nauka, Moscow.
Du, Y., Schuster, J., Wetser, F., Krendelsberger, N., Huang, B., Jin, Z., Gong, W., Yuan, Z. & Xu, H. (2004) *Metall. Mater. Trans. A*, **35A**, pp 1613–1628.
Edwards, G.A., Stiller, K., Dunlop, G.L. & Couper, M.J. (1998) *Acta Mater.*, **48**, pp 3893–3904.
Elliott, R.P. (1965) *Constitution of Binary Alloys, First Supplement*, McGraw-Hill, New York.
Eskin, D.G. (1992) *Z. Metallkde.*, **83**, pp 762–765.
Eskin, D.G. (1995) *Z. Metallkde.*, **86**, pp 60–63.
Eskin, D.G. (2003) *J. Mater. Sci.*, **38**, pp 279–290.
Eskin, D.G., Zolotorevskii, V.S., Istomin-Kastrovskii, V.V. & Aksenov, A.A. (1989) *Izv. Akad. Nauk Met.*, no. 2, pp 114–118.
Eskin, D.G., Massardier, V. & Merle, P. (1999) *J. Mater. Sci.*, **34**, pp 811–820.
Ferro, R., Saccone, A. & Delfino, S. (1991a) in *Ternary Alloys*, Eds. Petzow, G. & Effenberg, G., **Vol. 4**, VCH, Weinheim, pp 307–310.
Ferro, R., Zanicchi, G. & Marazza, R. (1991b) in *Ternary Alloys*, Eds. Petzow, G. & Effenberg, G., **Vol. 4**, VCH, Weinheim, pp 86–90.
Ferro, R., Zanicchi, G. & Marazza, R. (1992) in *Ternary Alloys*, Eds. Petzow, G. & Effenberg, G., **Vol. 5**, VCH, Weinheim, pp 677–681.
Ferro, R., Zanicchi, G. & Marazza, R. (1993a) in *Ternary Alloys*, Eds. Petzow, G. & Effenberg, G., **Vol. 6**, VCH, Weinheim, pp 318–327.
Ferro, R., Zanicchi, G. & Marazza, R. (1993b) in *Ternary Alloys*, Eds. Petzow, G. & Effenberg, G., **Vol. 7**, VCH, Weinheim, pp 412–415.
Ferro, R., Zanicchi, G. & Marazza, R. (1993c) in *Ternary Alloys*, Eds. Petzow, G. & Effenberg, G., **Vol. 8**, VCH, Weinheim, pp 58–62.
Fjellstedt, J. & Jarfors, A.E.W. (2001) *Z. Metallkde.*, **92**, pp 563–571.
Fjellstedt, J., Jarfors, A.E.W. & Svendsen, L. (1999) *J. Alloys Comp.*, **283**, pp 192–197.
Flower, H.M. & Gregson, P.J. (1987) *Mater. Sci. Technol.*, **3**, pp 81–90.
Frage, N., Frumin, N.N., Levin, L., Polak, M. & Dariel, M.P. (1998) *Metall. Mater. Trans. A*, **29A**, pp 1341–1345.
Fridlyander, I.N., Rokhlin, L.L., Dobatkina, T.V. & Nikitina, N.I. (1993) *Metalloved. Term. Obrab. Met.*, no. 10, pp 16–19.

Fridlyander, I.N., Rokhlin, L.L., Dobatkina, T.V., Bochvar, N.R., Lysova, E.V. & Korolkova, I.G. (2001) in *Metallovedenie I tekhnologiya legkikh spalvov* (Physical Metallurgy and Technology of Light Alloys), VILS, Moscow, pp 39–51.

Gao, M.C., Hackenberg, R.E. & Shiflet, G.J. (2003) *J. Alloys Comp.*, **353**, pp 114–123.

Ghosh, G. (1991) in *Ternary Alloys*, Eds. Petzow, G. & Effenberg, G., **Vol. 4**, VCH, Weinheim, pp 311–319.

Ghosh, G. (1992a) in *Ternary Alloys*, Eds. Petzow, G. & Effenberg, G., **Vol. 5**, VCH, Weinheim, pp 394–438.

Ghosh, G. (1992b) in *Ternary Alloys*, Eds. Petzow, G. & Effenberg, G., **Vol. 5**, VCH, Weinheim, pp 492–497.

Ghosh, G. (1993a) in *Ternary Alloys*, Eds. Petzow, G. & Effenberg, G., **Vol. 7**, VCH, Weinheim, pp 176–186.

Ghosh, G. (1993b) in *Ternary Alloys*, Eds. Petzow, G. & Effenberg, G., **Vol. 8**, VCH, Weinheim, pp 71–78.

Ghosh, G. (1993c) Ternary Alloys, Eds. Petzow, G. & Effenberg, G. **Vol. 6**, VCH, Weinheim, pp 356–375.

Gödecke, T., Sun, W., Luck, R. & Lu, K. (2001a) *Z. Metallkde.*, **92**, pp 723–730.

Gödecke, T., Sun, W., Luck, R. & Lu, K. (2001b) *Z. Metallkde.*, **92**, pp 717–722.

Gopienko, V.G. & Smagorinskii, M.E. (1993) *Spechennye Materiali iz Aluminievikh Poroshkov* (Sintered Materials from Aluminum Powders), Metallurgiya, Moscow.

Graf, R. (1957) *C.R. Academic Sciences*, **244**(8), pp 337–340.

Grieb, B. (1991a) in *Ternary Alloys*, Eds. Petzow, G. & Effenberg, G. **Vol. 4**, VCH, Weinheim, pp 48–50.

Grieb, B. (1991b) in *Ternary Alloys*, Eds. Petzow, G. & Effenberg, G. **Vol. 4**, VCH, Weinheim, pp 59–63.

Grieb, B. (1991c) in *Ternary Alloys*, Eds. Petzow, G. & Effenberg, G. **Vol. 4**, VCH, Weinheim, pp 365–384.

Grieb, B. (1992) in *Ternary Alloys*, Eds. Petzow, G. & Effenberg, G. **Vol. 5**, VCH, Weinheim, pp 236–238.

Gröbner, J., Kevorkov, D. & Schmid-Fetzer, R. (2001) *Z. Metallkde.*, **92**, pp 22–27.

Gröbner, J., Kevorkov, D. & Schmid-Fetzer, R. (2002) *Intermetallics*, **10**, pp 415–422.

Grushko, B., Mi, S. & Highfield, J.G. (2002) *J. Alloys Comp.*, **334**, pp 187–191.

Grushko, B. & Velikanova, T.Ya. (2004) *J. Alloys Comp.*, **367**, pp 58–63.

Gupta, A.K., Lloyd, D.J. & Court, S.A. (2001) *Mater. Sci. Eng. A*, **A301**, pp 140–146.

Guzei, L. (1991) in *Ternary Alloys*, Eds. Petzow, G. & Effenberg, G. **Vol. 4**, VCH, Weinheim, pp 455–462.

Hackenberg, R.E., Gao, M.C., Kaufman, L. & Shiflet, G.J. (2002) *Acta Mater.*, **50**, pp 2245–2258.

Harmelin, M. & Legendre, B. (1991a) in *Ternary Alloys*, Eds. Petzow, G. & Effenberg, G. **Vol. 4**, VCH, Weinheim, pp 538–545.

Harmelin, M. & Legendre, B. (1991b) in *Ternary Alloys*, Eds. Petzow, G. & Effenberg, G. **Vol. 4**, VCH, Weinheim, pp 585–587.

J.E. Hatch, Ed. (1984) *Aluminum: Properties and Physical Metallurgy*, ASM, Metals park.

He, G., Shek, C.H., Lai, J.K.L., Bian, Z., Hui, X.D. & Chen, G.L. (2002) *J. Mater. Res.*, **17**, pp 814–820.

Hohls, J., Cornish, L.A., Ellis, P. & Witcomb, M.J. (2000) *J. Alloys Comp.*, **308**, pp 205–215.

Hollingsworth, E.H., Frank, G.R., Jr. & Willet, R.E. (1962) *Trans. Metall. Soc. AIME*, **224**, pp 188–189.

Hsu, C., O'Reilly, K.A.Q., Cantor, B. & Hamerton, R. (2001) *Mater. Sci. Eng. A*, **304–305**, pp 119–124.

Huang, G.C. & Ardell, A.J. (1986) in *Aluminium–Lithium Alloys III, Proc. 3rd Intern. Al–Li Conf.*, Eds. Baker, C., Gregson, P.J., Harris, S.J. & Peel, C.J., Oxford, July 1985, The Institute of Metals, London, pp 455–470.

Hutchinson, C.R. & Ringer, S.P. (2000) *Metall. Mater. Trans. A*, **31A**, pp 2721–2733.

Inoue, A. (1998) *Progr. Mater. Sci.*, **43**, pp 365–520.

Jena, A.K., Gupta, A.K. & Chaturvedi, M.C. (1993) *Metall. Trans. A*, **24A**, pp 2181–2187.

Katgerman, L. & Eskin, D. (2003) Hardening, Annealing, and Aging, in *Handbook of Aluminum. Physical Metallurgy and Processes,* Eds. Totten, G.E. & MacKenzie, D.S. **Vol. 1**, Marcel Dekker, New York, pp 259–303.

Kaufman, L. & Bernstein, H. (1970) *Computer Calculation of Phase Diagrams*, Academic Press, New York and London.

Kharakterova, M.L., Eskin, D.G. & Toropova, L.S. (1994) *Acta Metall. Mater.*, **42**, pp 2285–2290.

Kolobnev, I.F. (1973) *Zharoprochnost Liteinykh Alyuminievykh Splavov* (Heat Resistance of Foundry Aluminum Alloys), Metallurgiya, Moscow.

Kosuge, H. & Mizukami, I. (1972) *J. Jpn. Inst. Light Met.*, **22**, pp 437–444.

Kotelnikov, Z.B., Bashlov, S.N. & Galiakbarov, Z.G. (1968) *Osobo Tugoplavkie Elementy i Soedineniya* (High-Melting Elements and Compounds), Metallurgiya, Moscow.

Kubaschewski, O. (1991) in *Ternary Alloys*, Eds. Petzow, G. & Effenberg, G. **Vol. 4**, VCH, Weinheim, pp 195–197.

Kubaschewski, O. (1993) in *Ternary Alloys*, Eds. Petzow, G. & Effenberg, G. **Vol. 7**, VCH, Weinheim, pp 199–218.

Kubicek, L., Tagijev, E., Zolotorevskij, V., Belov, N., Holicek, S. & Procio, M. (1993) *Fyzikalni Metalurgie* (Bratislava), **31**(4), pp 329–336.

Kuznetsov, G.M., Krivosheeva, G.B., Puzakov, K.S. & Aksenov, A.A. (1992) in *Metallovedenie i obraborka tsvetnykh splavov* (Physical Metallurgy and Processing of Nonferrous Alloys), Nauka, Moscow, pp 24–32.

Langsrud, Y. (1990) in *Proc. Worshop Effect of Iron and Silicon in Aluminum and Its Alloys*, Balatonfured, Hungary, May, pp 95–116.

Lawson-Jack, S.G., Flower, H.M. & West, D.R. (1993) *Mater. Sci. Technol.*, **9**, pp 562–571.

Lee, Y.S. (1998) *J. Mater. Sci.*, **17**, pp 1161–1164.

Lee, K.H., Lee, Y.J. & Hiraga, K. (1999) *J. Mater. Res.*, **14**, pp 384–389.

Legendre, B. & Harmelin, M. (1991) in *Ternary Alloys*, Eds. Petzow, G. & Effenberg, G. **Vol. 4**, VCH, Weinheim, pp 475–489.

Lide, D.R., Ed. (1992) *CRC Handbook of Chemistry and Physics*, 74th Edition, CRC Press, Boca Raton.

Lin, J.C., Zolotorevsky, V.S., Glazoff, M.V., Murtha, M.J. & Belov, N.A. (2004) Patent US 6,783,730 B2 Aug 31, *Al–Ni–Mn casting alloys for automotive and aerospace structural components.*

Liu, Y.L., Kang, S.B. & Kim, H.W. (1999) *Mater. Lett.*, **41**, pp 267–272.

Lukas, H.L. (1990) in *Ternary Alloys*, Eds. Petzow, G. & Effenberg, G. **Vol. 3**, VCH, Weinheim, pp 540–548.

Lysova, E.V. (1991) in *Ternary Alloys*, Eds. Petzow, G. & Effenberg, G. **Vol. 4**, VCH, Weinheim, pp 504–506.

Marioara, C.D., Andersen, S.J., Jansen, J. & Zandbergen, H.W. (2001) *Acta Mater.*, **49**, pp 321–328.

Martin, J.W., Doherty, R.D. & Cantor, B. (1997) *Stability of Microstructure in Metallic Systems*, Cambridge University Press, Cambridge.

Maruyama, N., Uemori, R., Hashimoto, N., Saga, M. & Kikuchi, M. (1997) *Scr. Mater.*, **36**, pp 89–93.

Matsuda, K., Sakaguchi, Y., Miyata, Y., Uetani, Y., Sato, T., Kamio, A. & Ikeno, S. (2000) *J. Mater. Sci.*, **35**, pp 179–189.

Matsuda, K., Uetani, Y., Sato, T. & Ikeno, S. (2001) *Metall. Mater. Trans.* A, **32A**, pp 1293–1299.

Miura, Y., Horikawa, K., Yamada, K. & Nakayama, M. (1994) in *Proc. 4th Intern. Conf. on Aluminum Alloys: Their Physical and Mechanical Properties*, Eds. Sanders, T.H., Jr. & Starke, E.A., Jr., **Vol. 2**, GIT, Atlanta, pp 161–168.

Mondolfo, L.F. (1976) *Aluminum Alloys: Structure and Properties*, Butterworths, London.

Mondolfo, L.F., Gjostein, N.A. & Levinson, D.W. (1959) *Trans. AIME, J. Metals,* **206**, pp 557–568.

Montoya, K.A., Heubaum, F.H., Kumar, K.S. & Pickens, J.R. (1991) *Scr. Metall. Mater.,* **25**, pp 1489–1494.

Morrissey, K.J., Czanderna, K.K., Merrill, R.P. & Carter, C.B. (1985) *Ultramicroscopy,* **18**, pp 379–385.

Mukhopadhyay, A.K., Rama Rao, V.V., Ghosal, P. & Ramachandra Rao, N. (2000) *Z. Metalldkde.*, **91**, pp 483–488.

Murayama, M., Hono, K., Miao, W.F. & Laughlin, D.E. (2001) *Metall. Mater. Trans. A*, **32A**, pp 239–246.

Norman, A.F., Prangnell, P.B., McEwen, R.S. (1998) *Acta Mater.*, **46**, pp 5715–5732.

Novikov, I.I. (1982) *Teoriya termicheskoi obrabotki metallov* (Theory of Heat Treatment of Metals), Metallurgiya, Moscow.

Oden, L.L & McCune, R.A. (1987) *Metall. Trans. A*, **18A**, pp 2005–2014.

Odinaev, K.O., Ganiev, I.N. & Ikromov, A.Z. (1996) *Russ. Metall.,* no. 3, pp 122–125.

Petrov, D.A. (1991) *Chetvernye Sisstemy* (Quaternary Systems), Metallurgiya, Moscow.

Phillips, H.W.L. (1959) *Annotated Equilibrium Diagrams of Some Aluminium Alloy Systems*, Monograph and Report Series no. 25, The Institute of Metals, London.

Phillips, H.W.L. (1961) *Equilibrium Diagrams of Aluminium Alloy Systems*, The Aluminium Development Association, Information Bull, London, 25 December, pp 105–108.

Phragmen, G. (1950) *J. Inst. Met.,* **77**, pp 489–552.

Pisch, A., Gröbner, J. & Scmid-Fetzer, R. (2000) *Mater. Sci. Eng. A*, **A289**, pp 123–129.

Polkin, I.S., Zolotorevskii, V.S., Romanova, V.S., Aksenov, A.A., Ponomorenko, A.M. & Churmukov, E.A. (1993) *Tekhnologiya Legkikh Splavov*, no. 12, pp 59–64.

Prigunova, A.G., Belov, N.A., Taran Yu.N., Zolotorevskii, V.S., Napalkov, V.I. & Petrov, S.S. (1996) *Silumins. Atlas of Structures and Fractograms. Handbook*, Eds. Y.N. Taran & Zolotorevskii, V.S., MIS&A, Moscow.

Prince, A. & Effenberg, G. (1991) in *Ternary Alloys*, Eds. Petzow, G. & Effenberg, G. **Vol. 4**, VCH, Weinheim, pp 547–566.

Ramanan, V.R.V., Skinner, D.J. & Zedalis, M.S. (1991) *Mater. Sci. Eng. A*, **A134**, pp 912–916.

Ran, Q. (1991a) in *Ternary Alloys*, Eds. Petzow, G. & Effenberg, G. **Vol. 4**, VCH, Weinheim pp 168–174.

Ran, Q. (1991b) in *Ternary Alloys*, Eds. Petzow, G. & Effenberg, G. **Vol. 4**, VCH, Weinheim, pp 527–537.

Ran, Q. (1992) in *Ternary Alloys*, Eds. Petzow, G. & Effenberg, G. **Vol. 5**, VCH, Weinheim, pp 250–264.

Ran, Q. & Stadelmaier, H.H. (1992) in *Ternary Alloys*, Eds. Petzow, G. & Effenberg, G. **Vol. 5**, VCH, Weinheim, pp 51–66.

Ran, Q. (1993) in *Ternary Alloys*, Eds. Petzow, G. & Effenberg, G. **Vol. 6**, VCH, Weinheim pp 437–443.

Red Book (1955–1994) *Constitutional Data and Phase Diagrams of Metallic Systems*, Moscow, VINITI (1955–1990), Stuttgart, MSI (from 1990).

Riola, R.J. & Ludwiczak, E.A. (1986) in *Aluminium–Lithium Alloys III, Proc. 3rd Intern. Al–Li Conf.*, Eds. Baker, C., Gregson, P.J., Harris, S.J. & Peel, C.J., Oxford, July 1985, The Institute of Metals, London, pp 471–482.

Robinson, K. & Black, P. J. (1953) *Philos. Mag.*, **44**, pp 1392–1397.

Rogl, P. (1991a) in *Ternary Alloys*, Eds. Petzow, G. & Effenberg, G. **Vol. 4**, VCH, Weinheim, pp 52–56.

Rogl, P. (1991b) in *Ternary Alloys*, Eds. Petzow, G. & Effenberg, G. **Vol. 4**, VCH, Weinheim, pp 401–415.

Rokhlin, L.L., Dobatkina, T.V., Muratova, E.V. & Korol'kova, I.G. (1994a) *Izv. RAN, Metally*, **1**, pp 113–118.

Rokhlin, L.L., Nikitina, N.I. & Oreshkina, A.A. (1994b) *Izv. RAN, Metally*, **1**, pp 168–171.

Rokhlin, L.L., Bochvar, N.R. & Lysova, E.V. (1997) *Russ. Metall.*, no. 5, pp 144–150.

Rokhlin, L.L., Dobatkina, T.V., Eskin, D.G., Nikitina, N.I. & Kharakterova, M.L. (1998) *Physico-Chemical Investigations of New Aluminum and Magnesium Alloys*, Transactions of Baikov Institute of Metallurgy and Materials Science on the Occasion of the 60th Anniversary, Ed. Lyakishev, N.P., Eliz, Moscow, pp 235–242.

Rokhlin, L.L., Dobatkina, T.V., Bochvar, N.F. & Lysova, E.V. (2004) *J. Alloys Comp.*, **367**, pp 10–16.

Rosell-Laclau, E., Durnad-Charre, M. & Audier, M. (1996) *J. Alloys Comp.*, **233**, pp 246–263.

Røyset, J., Hovland, H. & Ryum, N. (2002) *Mater. Sci. Forum*, **396–402**, pt. 4, pp 619–624.

Samuel, F.H., Champier, G., Todeschini, P. & Torres, J.H. (1992) *J. Mater. Sci.*, **27**, pp 4917–4926.

Saunders, N. (1989) *Z. Metallkde.*, **80**(12), pp 894–903.

Saunders, N. (1993) in *Ternary Alloys*, Eds. Petzow, G. & Effenberg, G. **Vol. 7**, VCH, Weinheim, pp 348–357.

Schmid, E.E. (1991) in *Ternary Alloys*, Eds. Petzow, G. & Effenberg, G. **Vol. 4**, VCH, Weinheim, pp 264–277.

Schmid, E.E. (1992) in *Ternary Alloys*, Eds. Petzow, G. & Effenberg, G. **Vol. 5**, VCH, Weinheim, pp 230–232.

Schuster, J.C. (1991) *J. Phase Equilibria*, **12**, pp 546–549.

Shepherd, E.S., Rankin, G.A. & Wright, W. (1909) *Amer. J. Sci.*, **28**, p 301.

Shunk, F.A. (1969) *Constitution of Binary Alloys, Second Supplement*, McGraw-Hill, New York.

Sigli, C. (2004) in *Proc. 9th Intern. Conf. Aluminium Alloys, ICAA9*, Eds. Nie, J.F., Morton, A.J. & Muddle, B.C. Brisbane, Inst. Mater. Eng., Australasia, August 2004, pp 1353–1358.

Simensen, C.J. & Vellasamy, R. (1977) *Z. Metallkde.*, **68**, pp 428–431.
Skjerpe, P. (1987) *Metall. Trans. A*, **18A**, pp 189–200.
Srivastava, A.K. & Ranganathan, S. (2001) *J. Mater. Res.*, **16**, pp 2103–2117.
Staley, J.T. (1985) *Can. Aeronautics Space J.* **31**, pp 14–29.
Stéfaniay, V., Lendvai, A. & Turmezey, T. (1996) *J. Mater. Sci.*, **31**, pp 3763–3768.
Stiltz, S. (1993) in *Ternary Alloys*, Eds. Petzow, G. & Effenberg, G. **Vol. 6**, VCH, Weinheim, pp 411–417.
Svendsen, L. & Jarfors, A. (1993) *Mater. Sci. Technol.*, **9**, pp 948–957.
Tagiev, E., Holicek, S. & Belov, N. (1996) *Mater. Sci. Forum*, **217–222,** pt. 1, pp 289–292.
Thiede, V.M.T., Ebel, T. & Jeitschko, W. (1998) *J. Mater. Chem.*, **8**, pp 125–130.
Thompson, S., Cockroft, S.L. & Wells, M.A. (2004) *Mater. Sci. Technol.*, **20**, pp 497–504.
Tiryakioglu, M. & Staley, J.T. (2003) in *Handbook of Aluminum*, Eds. Totten, G.E. & MacKenzie, D.S., **Vol. 1**, Marcel Dekker, New York, pp 156–157.
Toropova, L.S. (1987) *Bull. Tsvetn. Metall.*, no. 12, pp 17–20.
Toropova, L.S. & Kamardinkin, A.N. (1989) *Fiz. Met. Metalloved.* **67**, pp 147–151.
Toropov, N.A., Barzakovskiy, V.P., Lapin, V.V. & Kurtseva, N.N. (1969) *Diagrammi Sostoaniya Silikatnih Sistem* (Phase Diagrams of Silicate Systems), Nauka, Leningrad.
Toropova, L.S., Eskin, D.G., Kharakterova, M.L. & Dobatkina, T.V. (1998) *Advanced Aluminum Alloys Containing Scandium: Structure and Properties*, Gordon and Breach Science Publishers, Amsterdam.
Tretyachenko, L.A. (1992) in *Ternary Alloys*, Eds. Petzow, G. & Effenberg, G., **Vol. 5**, VCH, Weinheim, pp 113–121.
Tretyachenko, L., Sheftel, E., Ibe, G., Grieb, B. & Rogl, P. (1993) in *Ternary Alloys*, Eds. Petzow, G. & Effenberg, G., **Vol. 7**, VCH, Weinheim, pp 452–460.
Trømborg, E., Dons, A.L., Ørcsund, R., Jensen, E.K. & Arnberg, L. (1993) *Modeling of Casting, Welding and Advanced Solidification Processes VI*, Eds. Piwonka, T.S., Voller, V. & Katgerman, L., TMS, Warrendale, pp 103–111.
Tsubukino, H., Ushihara, U. and Yamamoto, A. (1996) *Mater. Sci. Forum*, **217–222**, pt. 2, pp 907–912.
Viala, J.C., Fortier, P. & Bouix, J. (1990) *J. Mater. Sci.*, **25**, pp 1842–1850.
Villars, P. & Calvert, L.D. (1985) *Pearsons Handbook of Crystallographic Data for Intermettalic Phases*, ASM, Metals Park.
Wang, Q.G. (2001) *J. Mater. Sci.*, **36**, pp 739–750.
Wang, J.Q., Tseng, M.K., Hao, Y.Y., Hu, Z.Q. & Wang, Z.G. (1998) *Mater. Sci. Eng.* A, **A247**, pp 51–57.
Westengen, H. (1982) *Z. Metallkde.*, **73**, pp 360–368.
Williams, D.B. (1981) *Aluminum–Lithium Alloys, Proc. 1st Intern. Al–Li Conf.*, Eds. Sanders, T.H., Jr. & Starke, E.A., Jr., Georgia (USA), May 1980, TMS, Warrendale, pp 89–100.
Yan, X.-Y., Chang, Y.A., Xie, F.-Y., Chen, S.-L., Zhang, F. & Daniel, S. (2001) *J. Alloys and Comp.*, **320**, pp 151–160.
Young, R.M.K. & Clyne, T.W. (1981) *Scr. Metall.*, **15**, pp 1211–1216.
Zakharov, A.M. (1964) *Chetvernye Diagrammy Sostoyaniya* (Quaternary Phase Diagrams), Metallurgiya, Moscow.
Zakharov, A.M. (1980) *Promyshlennye Splavy Tsvetnykh Metallov. Fazovyi Sostav i Strukturniye Sostavlyayushchie* (Commercial Alloys of Nonferrous Metals: Phase Composition and Structure Constituents), Metallurgiya, Moscow.

Zakharov, A.M. (1985) *Mnogokomponentnye Metallicheskie Sistemy s Promezhutochnymi Fazami* (Multicomponent Metallic Systems with Intermediate Phases), Metallurgiya, Moscow.

Zakharov, A.M. (1990) *Diagrammy Sostoyaniya Dvoinykh i Troinykh Sistem* (Phase Diagrams of Binary and Ternary Systems), Metallurgiya, Moscow.

Zakharov, A.M., Fridlyander, I.N. & Edelman, N.M. (1961) *Zhurn. Neorg. Khim.,* **6**, pp 1165–1171.

Zakharov, A.M., Guldin, I.T., Arnold, A.A. & Matsenko, Yu.A. (1988) *Izv. Akad. Nauk SSSR, Metally,* no. 3, pp 178–181.

Zakharov, A.M., Guldin, I.T., Arnold, A.A. & Chernova, E.P. (1989a) *Izv. Akad. Nauk SSSR, Metally,* no. 2, pp 216–218.

Zakharov, A.M., Guldin, I.T., Arnold, A.A. & Matsenko, Yu.A. (1989b) *Izv. Akad. Nauk SSSR, Metally,* no. 4, pp 214–218.

Zakharov, A.M., Guldin, I.T., Arnold, A.A. & Matsenko, Yu.A. (1992) in *Metallovedenie i Obrabotka Tsvetnykh Splavov* (Physcial Metallurgy and Processing of Nonferrous Alloys), Nauka, Moscow, pp 6–17.

Zhang, L.M. & Gille, P. (2004) *J. Alloys Comp.,* **370**, pp 198–205.

Zhang, L.C., He, A.Q., Ye, H.Q., Huang, C. & Zhang, Y.C. (2002) *J. Mater. Sci.,* **24**, pp 5183–5189.

Zolotorevskii, V.S., Istomin-Kastrovskii, V.V., Bakirov., G.T. & Rokhlina, A.L. (1984) *Izv. Akad. Nauk SSSR, Met.,* no. 4, pp 208–209.

Zolotorevskii, V.S., Belov, N.A. & Yu Mansurov, N. (1986) *Izv. Vyssh. Uchebn. Zaved., Tsvetn. Metall.,* no. 4, pp 85–90.

Zolotorevskii, V.S., Istomin-Kastrovskii, V.V. & Eskin, D.G. (1987) *Izv. Akad. Nauk SSSR, Met.,* no. 6, pp 87–92.

Zolotorevskii, V.S., Aksenov, A.A., Belov, N.A., Mansurov, Yu.N. & Istomin-Kastrovskii, V.V. (1988) *Izv. Akad. Nauk SSSR. Metally,* no. **1**, pp 114–120.

Zolotorevskii, V.S., Belov, N.A. & Kurdyumova, T.A. (1989) *Izv. Vyssh. Uchebn. Zaved., Tsvetn. Metall.,* no. 1, pp 76–88.

Zupanič, F., Spaić, S. & Križman, A. (1998) *Mater. Sci. Technol.,* **14**, pp 1203–1212.

Index

Crystal Growth –

From Fundamentals to Technology

Edited by

Georg Müller

Crystal Growth Laboratory,
Department of Materials Science, University Erlangen-Nürnberg,
Fraunhofer Institute of Integrated Systems and Device Technology
Erlangen, Germany

Jean-Jacques Métois

Centre de Recherche de la Matière Condensée et des Nanosciences
Campus Luminy
Marseille, France

Peter Rudolph

Institute for Crystal Growth
Berlin, Germany

2004
ELSEVIER

Amsterdam – Boston – Heidelberg – London – New York – Oxford
Paris – San Diego – San Francisco – Singapore – Sydney – Tokyo

ELSEVIER B.V.
Radarweg 29
P.O. Box 211, 1000 AE
Amsterdam, The Netherlands

ELSEVIER Inc.
525 B Street
Suite 1900, San Diego
CA 92101-4495, USA

ELSEVIER Ltd
The Boulevard
Langford Lane, Kidlington,
Oxford OX5 1GB, UK

ELSEVIER Ltd
84 Theobalds Road
London WC1X 8RR
UK

First edition 2004
Reprinted 2005

British Library Cataloguing in Publication Data
A catalogue record is available from the British Library.

Library of Congress Cataloging in Publication Data
A catalog record is available from the Library of Congress.

ISBN: 0-444-51386-8

♾ The paper used in this publication meets the requirements of ANSI/NISO Z39.48-1992 (Permanence of Paper).
Printed in

Preface

Crystals have fascinated mankind for thousands of years – be it snowflakes, minerals or jewels. For at least 50 years crystals have fascinated engineers as the key materials of modern electronics, optoelectronics and other technical fields of application. The formation of crystals in nature, like snowflakes and minerals , as well as the preparation of crystals in laboratories and factories for technical applications is called "crystal growth".

Due to the fact that many of today's technological systems in the fields of information, communication, energy, transportation, medical and safety technologies depend critically on the availability of suitable crystals with tailored properties, their fabrication – crystal growth – has become an important technology.

From this point of view it is evident that teaching crystal growth with the aid of books and schools is of great importance for the further development and dissemination of knowledge and expertise in crystal growth. This idea is one of the reasons why, approximately 50 years ago, a group of scientists working world wide in the field of crystal growth decided to organize themselves within the IOCG and hold International Schools of Crystal Growth. These schools were to be held during the summer, in parallel with their triannual conferences (ICCG). The big success of the previous International Summer Schools of Crystal Growth (ISSCG1 - 11) since is started in the Netherlands in the year 1971, proves the wisdom of the founders of IOCG in this decision.

In 2004, from August 1 to 7, the 12th Int. Summer School of Crystal Growth ISSCG 12 will be held in Berlin (Germany). ISSCG 12 is organized in conjunction with the 14th Int. Conference on Crystal Growth (August 9 - 13 in Grenoble, France) in a close collaboration of the German and French associations of crystal growth, Deutsche Gesellschaft für Kristallwachstum und Kristallzüchtung (DGKK) and the Groupe Francaie de Croissance Cristalline (GFCC).

Concerning the selection of the topics and subjects to be treated in the lectures and seminars we had a rather difficult task due to the duration (one week) of the school. In this limited time frame it is necessary to cover the vast field of crystal growth, from fundamentals to technology to characterization. Another problem is the different interests of the participants and their previous knowledge and experience – ranging from beginners to experts. The latter issue we have addressed by offering introductory seminars on various topics for beginners, one day before the official lecture program starts.

During the school the program is composed of expert lectures each morning and tutorial seminars in the afternoon. The new idea of the tutorial seminars, which are held in parallel for 3 topics, is to offer the participants the possibility of selecting certain topics and to give the lecturers the opportunity to organize the style and the contents according to the level of knowledge of the group of participants.

Nevertheless, the selection of the topics to some degree reflects the background of the editors and we can only apologise for the omission of topics which participants of ISSCG 12 and readers of this book would have wished to have been included – unfortunately no school and no book can be exhaustive.

We would also like to emphasize that the lecturers were solely selected on the basis of their knowledge and expertise in their particular fields and on their ability to contribute within the imposed time frame, but not on their nationality, seniority or other considerations.

This book contains the edited versions of the expert lectures presented during ISSCG 12 in Berlin. In most cases the contents of the manuscripts considerably exceeded the amount of material which could be presented within an one hour lecture. This gives them more the character of a hand book article than just a lecture note. The outline with five chapters is organized according to the lecture program as it was presented during the school, starting with fundamentals (chpt. 1), modeling (chpt. 2), future topics (chpt. 3), crystal growth technology (chpt. 4) and finally crystal defects and characterization (chpt. 5).

As already mentioned, a series of introductory and tutorial seminars were held during the school which are not contained in this book:

- Introduction into Crystal Growth (R. Fornari)
- Crystal Growth Techniques (G. Müller)
- Growth Kinetics (J.-J. Métois)
- Crystal Defects (P. Rudolph)
- Modeling of Crystal Growth (J. Neugebauer and W. Miller)
- Course on Numerical Simulation of Crystal Growth Processes on PCs (T. Jung)
- Growth of Biological Crystals (S. Veesler)
- Phase Diagrams: Analysis and Modeling (M. Mühlberg and D. Klimm)
- MOCVD of Compound Semiconductors (M. Heuken)
- Fundamentals of Epitaxial Growth (A. Pimpinelli)
- Electron Microscopy (A. Albrecht)
- X-ray Methods (H. Leipner)
- Point Defect Analysis (K. Irmscher)

In summary, the aim of the school and this book is to provide the students and readers with advanced knowledge concerning the technologically important subject of inorganic and organic single crystal growth. We hope that this book will be highly valuable for the entire crystal growth community as well as physics, chemistry, materials science engineering, biology, crystallography and remain as an important source for crystal growers, beginners and specialists alike.

July, 2004

The Editors
Georg Müller
Jean-Jacques Métois
Peter Rudolph

Acknowledgement

The editors greatly acknowledge the contributions of all the lecturers of ISSCG 12, especially the authors of this book. In addition we thank the directors of the Institute of Crystal Growth (IKZ, Berlin)), Prof. Roberto Fornari and his predecessor Prof. Winfried Schröder. Both generously supported the preparation and organization of ISSCG 12 in many ways, from personal to financial support!

Also many thanks to our colleagues from the organization team, Sabine Bergmann, Dr. Thorsten Boek, Dr. Anke Lüdge, Dr. Uwe Rehse, Dr. Detlef Schulz, Dr. Jürgen Warneke from the IKZ and Dr. Jochen Friedrich from the Erlangen Crystal Growth Laboratory for their tireless efforts and activities before and during ISSCG 12.

We gratefully acknowledge the support of our esteemed friends and colleagues from the French Association of Crystal Growth (GFCC), Dr. Jean-Jacques Favier, Dr. Stéphane Veesler, Dr. Denis Camel and from the Grenoble team of ICCG 14, Prof. Jacques Villain, Prof. Thierry Duffar and Dr. Peter Wellmann.

Last but not least we acknowledge the permanent support and advice from the President of IOCG, Prof. Robert Sekerka.

Both the ISSCG 12 technical program and the extension of about 60 student participation grants could not be accomplished without major financial contributions from several organizations, public institutions and companies.

The organizers acknowledge the contributions of:

DGKK, ICCG 14/GFCC, European Space Agency., Int. Union of Crystallography, German Association for Crystallography, Crystal Growth Laboratory of FhG Erlangen, InWent GmbH, IKZ-Berlin, DFG.

Companies: Kistler, ECM Infra Fours Physitherm Grenoble, Schunk Kohlenstoff, Linn High Term, Aixtron, Heraeus, Netzsch–Gerätebau, Anteam Industrielektronik, FLP Stähli, Saint-Gobain Winter, Siltronic AG, Freiberger Compound Materials, GTT-Technologies, SiCrystal AG, Röntec, PV Silicon, Matech, Crystec, Sterenat Elektrowärme, Hüttinger Elektronik GmbH.

The preparation of this book was supported by Dr. Jakob Fainberg, Marc Hainke, Stefan Hussy, Dr. Paul Sonda and by the secretary of the Erlangen Crystal Growth Laboratory, Elisabeth Henneberger.

Finally, our thanks to all those from the ELSEVIER publication team who were associated with the production of this book, especially to Jeanette Bakker, John Lardee, Linda Versteeg and Donna Wilson-Wheerd.

The Editors

Georg Müller
Jean- Jacques Métois
Peter Rudolph

Contents

Chapter 2: Modeling

Chapter 3: Crystal Growth Technology

Crystal Growth – from Fundamentals to Technology
G. Müller, J.-J. Métois and P. Rudolph (Editors)

Thermodynamics of modern epitaxial growth processes

Prof. G.B. Stringfellow

College of Engineering, University of Utah, Salt Lake City, Utah 84112, U.S.A.

The purpose of this lecture is to review the basic thermodynamic concepts that lead to a better understanding of: 1) the epitaxial processes used for the growth of semiconductor materials and structures used for photonic and electronic devices and 2) the properties of the materials produced by these techniques.

All epitaxial growth processes are driven by the free energy difference between the nutrient and solid phases. Thus, thermodynamics determines the maximum possible growth rate. For conditions where the rates of interface processes are sufficiently rapid to make mass transport the rate limiting step, a condition that can be approximated as thermodynamic equilibrium exists between the solid and the nutrient right at the interface. Thus, thermodynamics can be used to describe the solid composition in terms of the independent growth parameters. This is true for the near-equilibrium liquid phase epitaxial growth techniques as well as for the vapor phase techniques operating with a very large supersaturation, such as organometallic vapor phase epitaxy (OMVPE) and molecular beam epitaxy (MBE). In addition to determining the overall solid composition of semiconductor alloys, thermodynamics may also determine the stoichiometry of the solid and, consequently, dopant incorporation.

Many important properties of semiconductor materials and alloys produced epitaxially are determined by thermodynamics. Since thermodynamics controls the stoichiometry of the solids produced epitaxially, it controls the densities of native defects as well as deep levels acting as non-radiative recombination centers. The solid composition in alloys is thermodynamically controlled; thus, the most important parameters such as lattice constant and energy band gap are determined by thermodynamic factors. In addition, many semiconductor alloys have miscibility gaps due to the large, positive enthalpy of mixing. Thus, some alloys are not thermodynamically stable. By understanding the epitaxial growth processes, such metastable alloys can often be grown epitaxially. However, the distribution of the constituent atoms is often not random for such alloys. The resulting clustering has deleterious effects on the properties. In addition, under some growth conditions, surface thermodynamic factors drive the system to produce ordered alloys, with a composition modulation with monolayer periodicity along a particular crystallographic direction. Ordering has profound effects on the electrical and optical properties. For instance, the band gap energy can be reduced significantly.

Recently, the role of surface thermodynamics has begun to be recognized. Often, equilibrium is maintained only between the nutrient phase and the surface layers of the epitaxial film. This results in the growth of layers that are metastable in the bulk form. This means that the surface composition and bonding, i.e., the surface reconstruction, can dramatically affect the growth process and the properties of the epitaxial layer. This has led to the use of surfactants to control OMVPE and MBE growth and the resulting epilayer properties. Surfactant effects have been demonstrated on the microstructure of alloys, dopant incorporation, and the growth of low dimensional structures.

1. INTRODUCTION

Today, many semiconductor devices and circuits require epitaxial processes. In Si, this is due to the extremely fine geometries and thin layers required. In the world of compound semiconductors, nearly all devices have always required epitaxy due to the use of alloys, the extremely high quality needed for minority carrier devices, and the fine geometries, especially now, when bandgap engineered structures require quantum wells, wires, and dots.

A number of epitaxial growth techniques has been developed for semiconductors over the last 50 years. The first processes used chlorides for transporting the material for both Si and III/V semiconductors. Liquid phase epitaxy was also extremely useful for the production of layers of high purity III/V semiconductor materials. However, in recent years these techniques have been largely displaced by more flexible techniques for the growth of a large range of materials and special structures, such as organometallic vapor phase epitaxy (OMVPE) [or equivalently metal organic VPE (MOVPE) or MO chemical vapor deposition (MOCVD)], molecular beam epitaxy (MBE), and chemical beam epitaxy (CBE). Each of these techniques will be described very briefly below.

OMVPE has come to be the leading technique for the production of III/V materials, especially for photonic devices. It is particularly successful for devices requiring large areas, such as solar cells, cold cathodes, and light emitting diodes (LEDs). For example, OMVPE is used for commercial scale production of a number of materials and devices including: AlGaInP alloys for visible LEDs and injection lasers[1], the III/V nitrides used for blue and green LEDs and blue injection lasers[2], AlGaInP for the highest efficiency tandem solar cells [3,4], and GaInAsP and AlGaAs structures for infrared lasers and detectors[5,6]. Today, commercially available reactors can be purchased for both laboratory-scale and large production-scale applications from several manufacturers. For in-depth reviews of the OMVPE technique see refs [7-9].

MBE has, for decades, been the leading technique for the production of fine-scale structures. It was the first technique to produce layers showing quantum confinement and has been at the forefront of the development of bandgap engineered structures. It is widely used in production operations, especially for electron switching devices and circuits such as high electron mobility transistors and integrated circuits. Reviews and books are available for in-depth reviews of the technique and applications[10-12].

CBE is essentially a hybrid of OMVPE and MBE. It uses an ultra-high vacuum chamber, as for MBE, but used organometallic or hydride precursors rather than the elemental sources typically used in MBE. This gives certain advantages[13], but this technique remains mainly a laboratory technique which is used less frequently than OMVPE and MBE in production operations.

Each of the epitaxial growth processes is exquisitely complex when viewed in detail at the atomic level. Due to these complexities, even after many thousands of man years of effort we are still nowhere near a complete understanding. For each process, the fundamentals are often broken down into the separate categories of thermodynamics, kinetics, and hydrodynamics and mass transport. In fact, all of these separate factors and processes are tightly coupled.

As a result of the complexity of epitaxy processes, early crystal growth studies were largely empirical, giving epitaxy the appearance of an art. This is partly because of the complex, multicomponent, multiphase systems that are normally of interest and partly because the processes are dynamic. The pursuit of a more perfect understanding of the epitaxial processes has often involved the development of increasingly sophisticated, often *in-situ*, characterization tools. However, with every step in increasing the sophistication of the tools used to study the processes, it becomes clearer that they are even more complex than we had thought. Thus, a complete understanding of any of these processes remains elusive.

In this review, we will concentrate nearly entirely on the thermodynamic aspects of epitaxy. In fact, the epitaxial processes are essentially controlled phase transitions, and, thus, thermodynamic aspects control many of the resulting characteristics of both the processes and the resulting epitaxial layers. Having said that, it must also be acknowledged that the very process of growing an epitaxial layer demonstrates that the system cannot be completely at equilibrium. It is the intentional deviation from equilibrium that gives the driving force for epitaxy. Nevertheless, the extremely slow growth rates and relatively high temperatures typically used for the production of the highly perfect, thin layers we are interested in can give a condition approximating equilibrium between the growing solid and the nutrient phase right at the interface. This guarantees a dominant role for thermodynamics in most cases. As will be seen, the insights gained by thermodynamic analysis are most useful. Thermodynamics defines the maximum growth rate, often determines the composition of the solid being grown including alloy composition, stoichiometry, and dopant incorporation. Thermodynamics also defines the equilibrium microstructure of the solid and so can be used to predict whether the solid alloy will be uniform (random), clustered, or ordered.

2. THERMODYNAMIC DRIVING FORCE FOR EPITAXY

As mentioned above, thermodynamics underlies the epitaxial growth process. At low growth rates and relatively high temperatures, the chemical reaction kinetics play less of a role than in very rapid crystal growth processes. In the limit of infinitely slow growth rates, thermodynamics defines the relationships between the concentrations of the constituents in the vapor and solid phases. Thermodynamics also determines the driving force for any crystal growth process, hence defining the maximum growth rate. To make this section concrete, we will focus on a particular process, namely the OMVPE growth of the III/V semiconductor GaAs.

The equilibrium state for a two phase, $\alpha + \beta$, system is defined in terms of the chemical potentials,

$$\mu_i^{\alpha} = \mu_i^{\beta} \tag{1}$$

where the subscript i indicates the ith component and the superscripts indicate the phase. The chemical potential is usually written in terms of the chemical potential in an arbitrary standard state, denoted by the superscript zero,

$$\mu = \mu^{o} + RT \ln p/p^{o} \tag{2}$$

For an ideal gas mixture,

$$\mu_i = \mu_i^{o} + RT \ln p_i/p_i^{o} \tag{3}$$

where p_i is the partial pressure, equal to the mole fraction x_i multiplied by P, the total pressure, and the standard state is usually pure component i.

For an ideal solid solution, the same expression holds with p_i/p_i^{o} replaced by x_i/x_i^{o}. However, the standard state is pure i, so $x_i^{o} = 1$. The form of eq. (3) is so useful that it is retained even for non-ideal solutions with x_i replaced by the activity, a_i, which may also be considered a product of x_i multiplied by a non-ideality factor, γ_i, the activity coefficient.

For the OMVPE growth of GaAs using trimethylgallium (TMGa) and arsine (AsH_3) the overall reaction is

$$(CH_3)_3Ga(v) + AsH_3(v) = GaAs(s) \tag{4}$$

Assuming the TMGa and AsH_3 to completely decompose in the gas phase to give Ga and As_4, the reaction can be simplified:

$$Ga(v) + 1/4\ As_4(v) = GaAs(s) \tag{5}$$

The equilibrium condition is

$$\mu^v{}_{Ga} + 1/4\ \mu^v{}_{As4} = \mu^s{}_{GaAs} \tag{6}$$

or

$$\mu^{ov}{}_{Ga} + 1/4\ \mu^{ov}{}_{As4} + RT \ln p^e{}_{Ga}(p^e{}_{As4})^{1/4} = \mu^{os}{}_{GaAs} + RT \ln a_{GaAs} \tag{7}$$

where the superscript "e" denotes the equilibrium value of partial pressure. Thus,

$$a_{GaAs} / p^e{}_{Ga}\ (p^e{}_{As4})^{1/4} = K_{GaAs} \tag{8}$$

where K is the equilibrium constant. This is the basic law of mass action.

When the system is not at equilibrium, the thermodynamic driving force to restore equilibrium is

$$\Delta\mu = \mu^v{}_{Ga} + 1/4\ \mu^v{}_{As4} - \mu^s{}_{GaAs} \tag{9}$$

or

$$\Delta\mu = RT \ln [p_{Ga}\ p_{As4}{}^{1/4} / p^e{}_{Ga}\ p^e{}_{As4}{}^{1/4}] \tag{10}$$

This is the driving force for epitaxy. A situation is intentionally created where higher than equilibrium reactant partial pressures drive the system to produce the GaAs solid desired. The maximum quantity of GaAs solid that can be produced is simply the amount (the supersaturation) that would establish equilibrium, and is, thus, fundamentally limited by thermodynamics and the total amount of gas transported through the OMVPE reactor.

For the OMVPE growth of GaAs using arsine and TMGa, the thermodynamic driving force at 1000 K is approximately 80 kcal/mol [14]. The driving forces for the epitaxial growth of GaAs are compared for several epitaxial processes in Fig. 1 [14]. This very high driving force to form solid GaAs is due to the instability of both arsine and TMGa at 1000 K. It will be noted in Fig. 1 that MBE also has a high thermodynamic driving force. In this case, this is due to the instability of elemental Ga and As in the vapor at typical growth temperatures. These high driving forces for formation of the solid have prompted many researchers to dub both OMVPE and MBE as "highly non-equilibrium" growth processes.

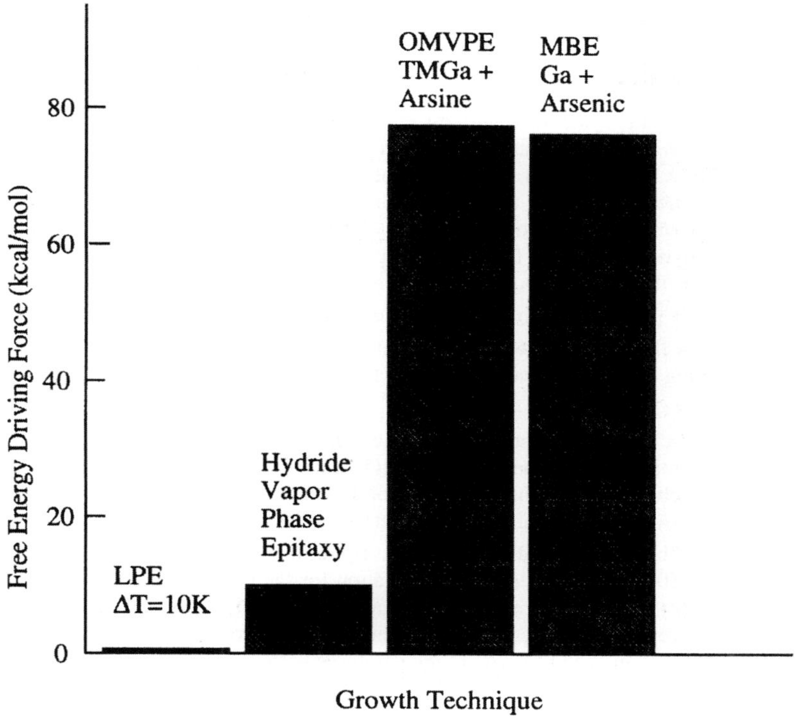

Figure 1. Estimated thermodynamic driving force for several epitaxial growth processes. All calculations are for the growth of GaAs at 1000K. (After Stringfellow [14], reprinted with permission from Elsevier Science.)

This raises the question: how does thermodynamics relate to epitaxial growth for OMVPE and MBE where the driving force is extremely high? The connection between thermodynamics and growth results has long been recognized for liquid phase epitaxy (LPE), where, as seen in Fig. 1, the thermodynamic driving force is relatively small. Here, it seems entirely natural to treat the growth process using thermodynamics and, in fact, phase diagrams are well-known to give vital information about alloy composition, solid stoichiometry, and dopant incorporation. However, even for MBE and OMVPE powerful thermodynamic factors still control much of the growth process. This is because, even for a system with a high supersaturation of the input vapor phase, near equilibrium conditions may prevail near the solid/vapor interface. This means that thermodynamics can provide important information about the growth process and the properties of the resultant materials. However, it may prove necessary to consider the thermodynamic properties of the surface in addition to the vapor and bulk solid phases. Thermodynamic factors largely determine the equilibrium structure of the surface, leading to surface phase diagrams, as discussed below, that give the surface reconstruction as a function of the extensive parameters, such as temperature and the group v partial pressure, as discussed below. Surface reconstruction has profound effects on both the epitaxial growth processes and the properties of the resulting layer.

Ordinarily, in the OMVPE system the growth rate is considerably less than that calculated from thermodynamics. Kinetics, both surface reaction rates (at low temperatures) and diffusion through the gas phase (at higher temperatures), are not rapid enough to allow equilibrium to be established throughout the system at all times. This situation is illustrated by Fig. 2a, where $\Delta\mu$ from eq. (9) is plotted versus reaction coordinate [15]. This allows the schematic representation of the overall, thermodynamic driving force for the growth reaction, represented as $\Delta\mu^*$. The superscript "* " denotes the chemical potential in the input gas phase, where for all reactants $p_i = p^*_i$. The growth rate is proportional to the flux of atoms diffusing through the boundary layer, which is identical to the flux of atoms crossing the interface into the solid. The diagram shows schematically the driving forces necessary to sustain this flux for the diffusion process ($\Delta\mu_D$) and the surface reactions ($\Delta\mu_S$).

Even in cases with a large supersaturation in the input vapor phase, i.e., $\Delta\mu^* >> 0$, near equilibrium conditions may exist at the growing solid surface. This simply requires that the interface kinetics be much more rapid than the diffusion kinetics, i.e., the two processes proceed at the same rate with $\Delta\mu_S << \Delta\mu_D$. This situation, termed diffusion limited growth, is shown schematically in Fig. 2b. Using ordinary growth conditions, with temperatures between approximately 550 and 800 °C, this is the normal situation for the OMVPE growth of GaAs, as deduced from the nearly temperature independent growth rate [16]. For surface kinetically limited processes, the growth rate increases exponentially with increasing temperature [17,18]. This occurs for the OMVPE growth of GaAs at temperatures below approximately 550 °C when TMGa is the Ga precursor, but this temperature depends on the group III precursor used.

In the diffusion limited case, illustrated schematically in Fig. 2b, the interfacial partial pressures, p^i_i, nearly satisfy the equilibrium relationship (assuming the dominant species to be the As tetramer),

$$a_{GaAs} / p^i_{Ga}(p^i_{As4})^{1/4} = K_{GaAs} \qquad (11)$$

where a_{GaAs}, the activity of GaAs in the solid, is unity for the growth of pure GaAs. K_{GaAs} is the equilibrium constant at the particular temperature chosen. Since the input vapor is highly supersaturated,

$$p^*_{Ga}(p^*_{As4})^{1/4} >> p^i_{Ga}(p^i_{As4})^{1/4} \qquad (12)$$

This is equivalent to stating that $\Delta\mu^* >> 0$. For the typical OMVPE case

$$p^*_{Ga} << 4p^*_{As4} \qquad (13)$$

i.e., the V/III ratio is >>1, as will be discussed below. This means that the Ga is nearly depleted at the interface,

$$p^i_{Ga} << p^*_{Ga} \qquad (14)$$

while the As_4 partial pressure is hardly diminished,

$$p^i_{As4} \approx p^*_{As4}, \tag{15}$$

since the same number of As and Ga atoms are removed from the vapor phase to produce GaAs. This situation makes the analysis of growth rate and solid composition particularly simple.

The growth rate is proportional to the flux of Ga and As atoms diffusing through the vapor to the growing interface. For simplicity, this can be analyzed in terms of diffusion through a boundary layer of thickness d. A more complete description is given in reference [19]. The two fluxes are equal, since stoichiometric GaAs is the only product. The flux may be expressed,

$$J = D_{Ga}(p^*_{Ga} - p^i_{Ga})/RTd \tag{16}$$

where D_{Ga} is the diffusion coefficient of Ga, in whatever form it may appear while diffusing through the boundary layer. In light of eq. (14), the Ga flux and the GaAs growth rate are proportional to p^*_{Ga}, as observed experimentally [18]. Equally clear is that the ratio of the concentrations of A and B for alloys with mixing on the group III sublattice, $A_{1-x}B_xC$, will be the same as the ratio p^*_A/p^*_B, assuming the diffusion coefficients for the A and B species are nearly the same [20]. This will be discussed in more detail below.

For MBE growth, the situation is quite similar. The growth rate is typically determined by the rate of arrival of Ga atoms at the solid-vapor interface [10]. As is incorporated from the vapor in the amount needed to produce stoichiometric GaAs. Again, at low temperatures, where the group III atoms cannot re-evaporate from the surface, the ratio of group III elements incorporated into a solid alloy with mixing on the group III sublattice is the same as the flux of the group III atoms to the surface. For both OMVPE and MBE, as the temperature is raised to the point that group III atoms can re-evaporate from the surface, thermodynamics begins to control the solid composition [20]. For mixing on the group V sublattice, thermodynamics typically controls the solid composition under typical growth conditions [20].

3. BINARY PHASE DIAGRAMS

The III/V compounds melt congruently, i.e., a liquid and solid having identical compositions are in equilibrium at the melting point, as illustrated in the temperature-composition (T-x) phase diagram for GaAs seen in Fig. 3(a). Thus, the III/V compounds can be grown directly from the melt, a process commonly used to produce large boules of GaAs, GaP, InP, InAs, InSb, and GaSb. This phase diagram shows that only a single III/V compound is stable, with melting point typically well above the melting points of the group III and group V elements. In this diagram the III/V solid appears as an infinitely thin vertical line. This is indicative of the high degree of stoichiometry discussed above. In fact there is a slight, temperature-dependent deviation from stoichiometry, indicated in the inset. The non-stoichiometry is so small that it is extremely difficult to measure, and, in fact, has not been reliably measured for most III/V compounds. However, the deviation from stoichiometry has profound effects on the electrical and optical properties of the material as well as bulk diffusion coefficients, and related to this, the operating lives of many devices.

During the LPE growth of GaAs, the solid is in equilibrium with a liquid phase specified by the liquidus curve shown in Fig. 3(a). At a temperature T_2 the Ga-rich liquid in equilibrium with the GaAs solid has a composition x_2 on the liquidus curve. A horizontal "tie-line" (not shown) joins the liquid and solid compositions in equilibrium. The stoichiometry is controlled by the composition of the nutrient phase. For example, for LPE growth from the Ga-rich liquid, with

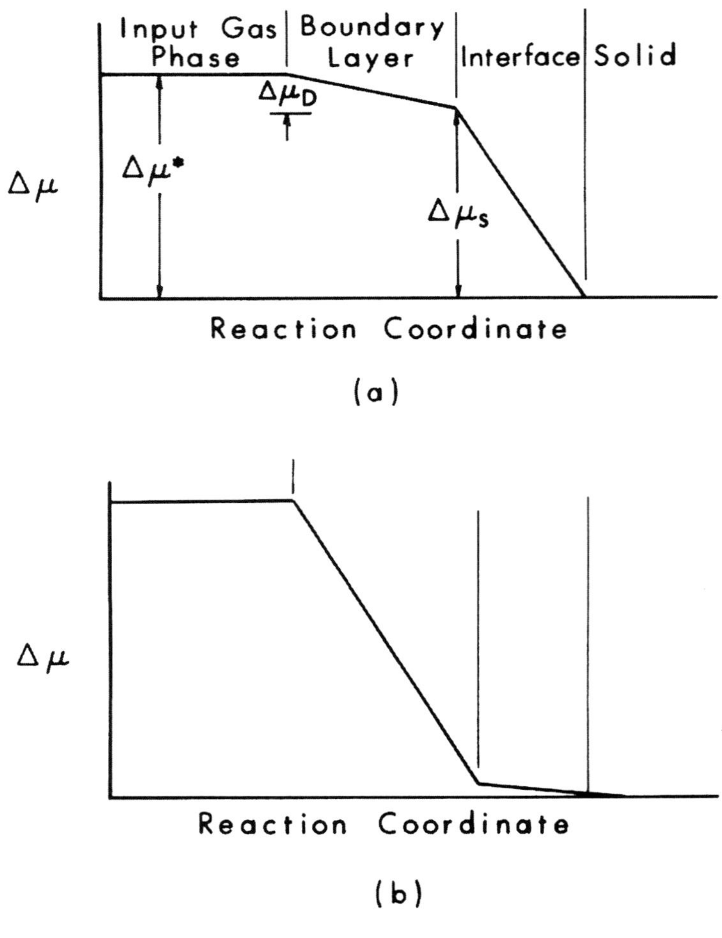

Figure 2. Schematic diagram of chemical potential versus reaction coordinate, showing the drop in chemical potential required for each step in the growth sequence to keep all rates equal: (a) the general case and (b) the case of rapid surface kinetics. (after Stringfellow [15], reprinted from the Journal of Crystal Growth with permission from Elsevier Science.)

composition x_2, the resulting GaAs is on the extreme Ga-rich side of the (very narrow) GaAs phase field. Because of the high As pressures required, GaAs would not typically be grown from the As-rich liquid, with composition x_2' in this example. The resulting GaAs would be on the extreme As-rich side of the GaAs phase field. The difference in the As/Ga ratio in the solid may be only a few parts per million from one side of the phase field to the other, but this small difference has dramatic effects on the semiconductor properties. Very recently the As-rich GaAs grown at low temperatures by MBE has become of some interest because of its semi-insulating properties. Because of the extremely low temperatures, <200 °C, an As-rich second phase is formed. Thus, growth occurs in the two-phase "GaAs + As" region of Fig. 3(a). It should be noted that the OMVPE growth of GaAs typically produces material near the As-rich side of the solid stability region due to the very high As/Ga ratios at the interface, as discussed below.

In Fig. 3(a) only the condensed phases are shown. However, the system also necessarily contains both Ga and As vapor species. The As (both dimer and tetramer forms) and Ga partial pressures in equilibrium with the condensed phases as a function of composition, at a temperature of approximately 980°C, are shown in Fig 3(b). The stoichiometry of the solid produced at a fixed temperature during LPE growth is not variable. The phase rule specifies that the system is invariant when three phases are in equilibrium in a two component system when the temperature is fixed [21]. However, during vapor phase growth, since the number of phases is reduced by one, the phase rule specifies that one degree of freedom exists even when the temperature is fixed. The stoichiometry can be controlled by either defining the Ga or the As pressure at the solid/vapor interface. Since there is frequently a large excess of the group V

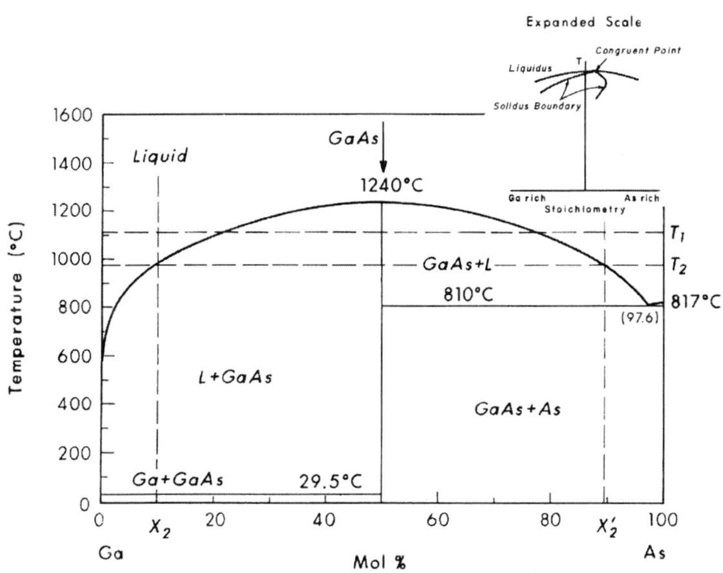

(a)

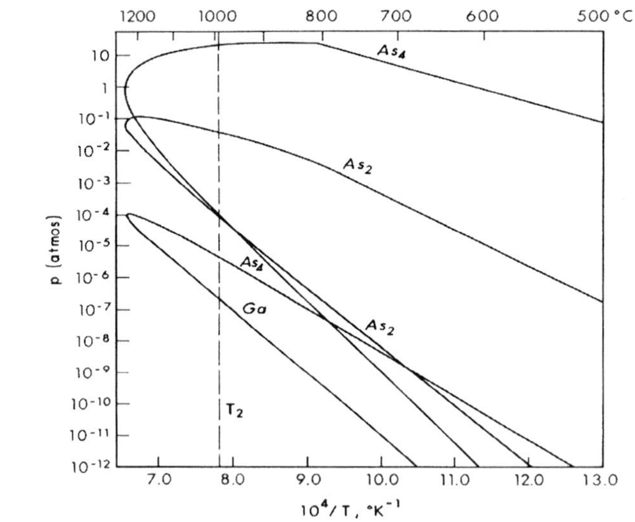

(b)

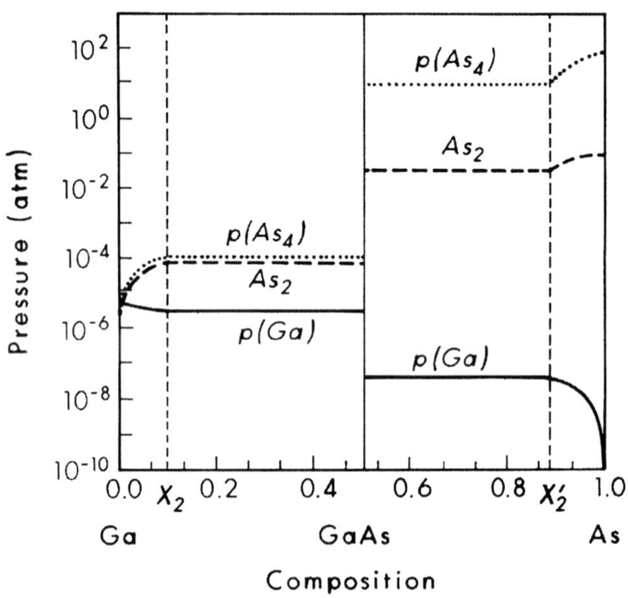

(c)

Figure 3. GaAs phase diagrams: (a) temperature-composition, (b) pressure-temperature, and (c) pressure-composition. The latter is drawn at a temperature T_2 of approximately 980 °C, where the Ga- and As-rich liquidus compositions are x_2 and x_2'. (after Stringfellow [7]).

element during VPE, the stoichiometry is typically controlled by the group V partial pressure input to the system. For the OMVPE growth of GaAs, for example, the As is depleted very little at the interface and the Ga is nearly all used up, as discussed above. In this case the Ga partial pressure at the interface, P^{i}_{Ga}, is determined by the mass action expression, Eqn (11).

A word of caution is in order here. The literature is filled with papers where the V/III ratio is used improperly to analyze the thermodynamic state of the system, due to a confusion between the input V/III ratio and that at the interface. For example, the former has no obvious connection to the stoichiometry of the solid. A common mistake is to change the "V/III ratio" during growth by changing the input group III flow rate. For conditions where thermodynamic equilibrium is approximately satisfied at the interface, changing the group III flow rate has no effect on the V/III ratio at the interface, the quantity determining the solid stoichiometry. Even more common is a series of experiments where the temperature is varied and the input V/III ratio is held constant. Of course, since the equilibrium constant in Eqn. (11) is strongly dependent on temperature, the V/III ratio at the interface will certainly not be constant.

A slightly different form of the phase diagram can be used to describe the OMVPE growth process, where the independent variables are typically the temperature and the input group III and group V partial pressures. For the growth of GaAs using trimethylgallium (TMGa) and AsH_3 at normal growth temperatures, the precursors can be assumed to be decomposed to elemental Ga and As_4, although incomplete decomposition and the production of As dimers can also be considered, where appropriate. Again, equilibrium is assumed to prevail at the interface, as described above. This results in the phase diagram shown in Fig. 4 [22]. It explains many of the most elementary constraints, first discovered empirically, on the growth of GaAs having a good surface morphology. For input V/III ratios of less than 1 (III/V >1 in Fig. 4) poor surface morphologies are invariably produced. A close examination of the surface for such conditions reveals shiny, hemispherical liquid metal droplets. These frequently reside at the tops of long whiskers protruding from the surface. This is because, as the phase diagram indicates, growth occurs in a two (condensed) phase region. The input vapor is highly supersaturated, i.e., the product of the input As and Ga partial pressures is much larger than a_{GaAs}/K_{GaAs} in Eqn (11). As and Ga are taken from the vapor in a 1:1 ratio to form GaAs. Thus, if the input V/III ratio is not unity, one of the elements is strongly depleted at the growing interface in order to establish equilibrium. When the V/III ratio is less than unity, the group V element is depleted. Because of the low volatility of the group III element, it accumulates on the surface as a second condensed phase. Much better results are obtained for V/III ratios of much greater than unity. It is impossible to form a condensed As phase except for input As partial pressures in excess of the vapor pressure of elemental arsenic, approximately 28 atmospheres at a temperature of 800°C. Thus, the rectangular region labeled GaAs(s) defines the combination of Ga and As input partial pressures yielding single phase GaAs at a particular temperature. Of course, the solid stoichiometry is completely controlled by the input As partial pressure, as described above. For temperatures more likely to be used for OMVPE growth the only change in this diagram would be a reduction in the vapor pressure of pure, elemental As.

As the pressure in the system is reduced into the regime of MBE growth, the free energy of the vapor phase is reduced, which results in major changes in the phase diagram, as discussed in detail by Tsao [12]. For example, the GaAs temperature-composition diagram at a total pressure of 10^{-5} torr is shown in Fig. 5. Under these conditions, the GaAs will sublime before it reaches its melting point. The decreased stability of the liquid phase results in a phase diagram with a fairly extensive region labeled c + v (shaded in Fig. 5) where GaAs is in equilibrium with the As-rich vapor phase. Note that if the stoichiometry of the system is slighty Ga-rich, a liquid phase, essentially pure Ga, is formed. Thus, similar to OMVPE, MBE growth of GaAs is carried out with an excess of As in the vapor phase. The stoichiometry of the solid is then controlled by the As partial pressure at the interface. The pressure-composition phase diagram, Fig. 6, gives additional information defining the range of total pressure leading to successful MBE growth of GaAs. The two phase diagrams define the MBE "window" [12], i.e., the combinations of temperature, total pressure, and V/III ratio leading to growth of a single condensed GaAs phase.

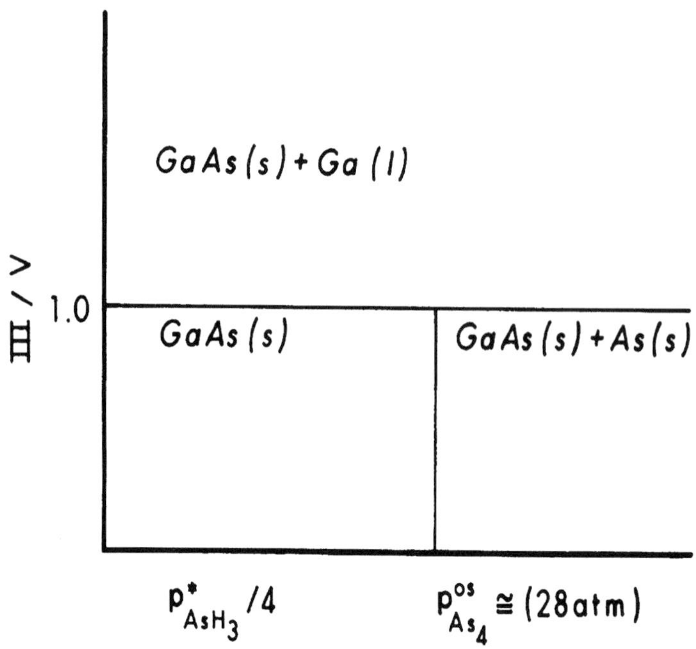

Figure 4. GaAs phase diagram appropriate for OMVPE growth. The regions containing a single solid GaAs phase and the two two-phase regions are shown. (after Stringfellow [22].)

4. SURFACE PHASE DIAGRAMS

The actual surface of a III/V semiconductor is not that expected from a simple termination of the crystalline lattice. This would leave a large concentration of high energy dangling bonds. Thermodynamic factors determine the equilibrium structure of the surface, as for the bulk solid.

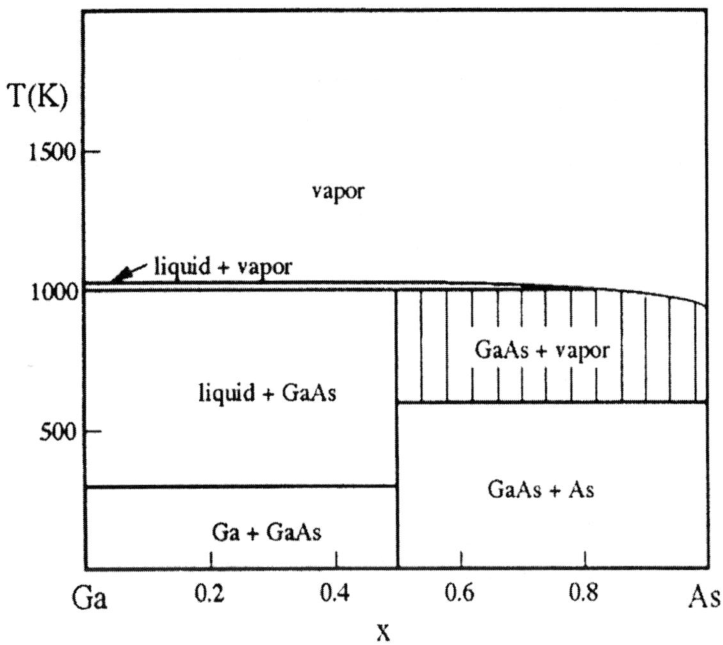

Figure 5. GaAs phase diagram at a total pressure of 10^{-5} torr, applicable to MBE growth. (after Tsao [12].)

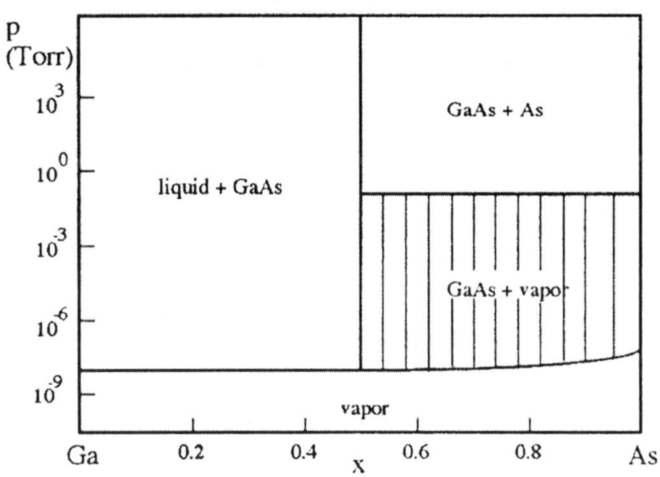

Figure 6. Pressure-composition phase diagram for GaAs under MBE growth conditions at 850 K (after Tsao [12].)

Considering specifically the (001) surface, that most commonly used for the epitaxial growth of III/V semiconductors, termination of the lattice would produce a configuration with two dangling bonds per surface atom, an extremely high energy configuration. Both theoretical calculations [23] and experimental results [24] clearly show that the equilibrium configuration for the As-rich (001) GaAs surface simply involves the joining of pairs of adjacent dangling bonds to form As dimer rows on the surface, similar to the Si dimer rows observed on the reconstructed (001) Si surface [25]. The (2x4) structure has every fourth As dimer missing. The surface atoms also buckle to produce a non-planar arrangement [23]. The reconstructed surface is stable, which gives rise to high surface mobilities for adsorbed atoms, with diffusion lengths reported to be as large as a micron [26]. This must be the key to obtaining the nearly atomically abrupt interfaces reported for OMVPE and MBE growth of quantum well structures widely reported in the literature. Adsorbed atoms that could make two bonds to surface atoms would obviously not be at all mobile. This would lead to statistically rough, three dimensional growth, precluding the possibility of producing quantum wells and other nano-structures.

The surface structures produced during MBE growth of (001) GaAs have been determined by both reflection high energy electron diffraction (RHEED) and scanning tunneling microscopy (STM) [24]. The predominant structure for As-rich conditions is the (2x4) structure, discussed above. For Ga-rich conditions, the (4x2) structure, consisting largely of Ga dimers, is formed. Several other structures are observed, depending primarily on the As/Ga ratio at the surface [27]. Thus, a surface phase diagram can be constructed, as shown in Fig. 7 [28]. Structures similar to those discussed above are also observed for other III/V semiconductors. The major feature appears to be the formation of either group III or group V dimers at the surface to reduce the density of dangling bonds.

The surface structures produced during OMVPE growth are revealed by x-ray scattering [29] and optical probes [28]. The tools most useful for typical laboratories are the optical techniques of reflection-difference spectroscopy (RDS) [28] and surface photo absorption(SPA) [30]. Kamiya et al. [28] used RDS to determine the surface phase diagram, i.e., the reconstruction versus substrate temperature and input AsH_3 flow rate for OMVPE growth. These results are added to the MBE results in Fig. 7. They found that for As-rich conditions, typically used for OMVPE growth, the GaAs surface consists of As dimer rows, producing either the (2x4) structure discussed above or the c(4x4) structure consisting of two layers of As at the surface. Apparently, at a given As flux and growth temperature, the surface structures obtained by MBE and OMVPE are similar, as expected for a thermodynamically controlled phenomenon.

5. SOLUTION THERMODYNAMICS

The relation between the composition of the vapor at the solid/vapor interface and the solid can be approximated very accurately by assuming a condition of thermodynamic equilibrium [31], as described above. On an even finer scale, the same logic can be applied to the surface processes. If the kinetic barrier at the step edge doesn't limit the growth rate, the process is limited by surface diffusion and thermodynamics can be used to describe the surface processes in OMVPE as well as MBE growth [31, 32, 33, 34].

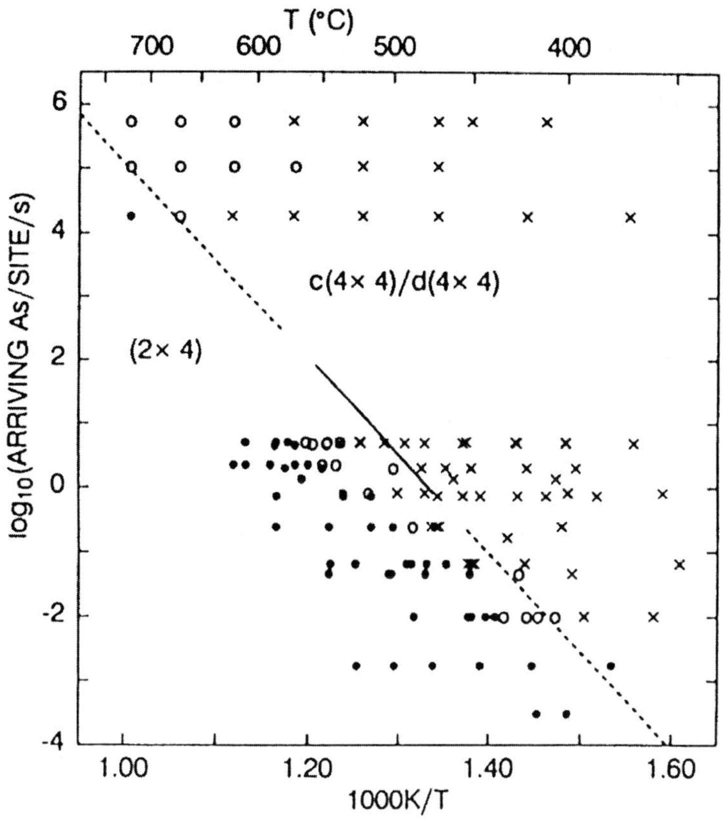

Figure 7. Phase diagram showing the surface reconstruction of (001) GaAs as a function of substrate temperature and the arriving flux of As atoms at the surface. Dots, crosses, and circles represent (2x4), c(4x4), and marginal structures, respectively. (after Kamiya et al [28].)

For example, considering the OMVPE growth of alloys, the equilibrium approximation allows the composition of the solid phase to be accurately calculated as a function of temperature and the composition of the vapor. An important consideration in such calculations is the deviation from ideality of the solid phase. For the III/V alloys, such as $Ga_xIn_{1-x}P$ [35], as well as for the II/VI and IV alloys, for example Ge_xSi_{1-x}, the semi-empirical delta-lattice parameter (DLP) model gives a good representation of the enthalpy of mixing [35]. The valence force field (VFF) model [36] as well as first principles calculation s [37], give accurate estimates of the enthalpy of mixing without adjustable parameters.

The microscopic strain energy associated with stretching and bending the bonds is the most important factor in the thermodynamics of mixing of semiconductor alloys. Thus, the enthalpy of mixing is always positive, and increases with the square of the difference in lattice parameters of the two constituent compounds (or elements for group IV alloys) [35]. This dictates that at equilibrium a random alloy in a certain composition range will decompose into a mixture of two phases when the difference in lattice constants of the component compounds is large, i.e., the phase diagram contains a miscibility gap.

In general, the simpler DLP model can accurately be used for alloy phase diagram calculations. The entropy of mixing for a pseudobinary solution of the type $A_{1-x}B_xC$ is typically simply taken to be the ideal configurational entropy of mixing,

$$\Delta S^M = -R[x \ln x + (1-x)\ln(1-x)] \tag{17}$$

Using these concepts, applied to the solid-vapor equilibria of concern for vapor phase growth, including OMVPE [38] and MBE [39], we can calculate the composition of a multicomponent solid alloy from the temperature and the concentrations of the various components in the vapor phase. Deviations from ideality for the vapor phase are commonly neglected.

The equilibrium conditions for the ternary(or pseudobinary) system may be obtained in exactly the same way as described above for binary systems (Eqn (6)), by equating the chemical potentials of the 2 components in the 2 phases:

$$\mu^V{}_A + \mu^V{}_C = \mu^S{}_{AC} \tag{21a}$$

$$\mu^V{}_B + \mu^V{}_C = \mu^S{}_{BC} \tag{21b}$$

This leads to two mass action expressions, similar to Eqq. (11). As discussed above, equilibrium is assumed to be established at the interface.

As an example of the use of such calculations to understand epitaxial processes, consider the OMVPE growth of $GaAs_{1-x}Sb_x$. The 2 mass action expressions, one for GaAs and one for GaSb, are solved simultaneously with 2 conservation equations, one for solid stoichiometry and one for solid composition [41]. Complete pyrolysis of the source molecules is assumed. The activity coefficients of GaAs and GaSb in the solid are calculated as described above using the DLP model.

The calculation can be performed with no adjustable parameters, yielding solid composition versus vapor composition and substrate temperature during growth. The calculated results are compared with experimental data in Fig. 8 [41]. Several important aspects of VPE are illustrated in this rather complex figure. First, consider the open data points, obtained for an input V/III ratio (the ratio of the input group V to group III molar flow rates) of 2.0. Notice that the calculated curve for V/III = 2.0 fits the data well. The Sb distribution coefficient, defined as kSb = xsSb/xvSb, where xvSb = p*TMSb/(p*TMSb+p*AsH3), is seen to be less than unity. GaAs is more stable than GaSb, thus As is more likely to bond to the Ga on the surface and be incorporated into the solid. The excess Sb evaporates from the surface. An additional important point is that the calculation for a V/III ratio of less than unity yields an antimony distribution coefficient of unity. As discussed above for the case of alloys with mixing on the group III sublattice, when V/III>1, essentially all of the group III elements reaching the interface are incorporated. The case of GaAsSb with mixing on the group V sublattice with V/III ≤ 1 is completely analogous. The establishment of equilibrium at the interface while the input vapor is highly supersaturated requires that the group V elements be virtually exhausted at the interface. A final point relative to Fig. 8 is the presence of a two solid phase region or miscibility gap. Because of the large difference in lattice constant between GaAs and GaSb a miscibility gap exists, as discussed above. However, when the V/III ratio is less than unity, the As and Sb atoms arriving in a random pattern at the surface do not have time to redistribute themselves into GaAs and GaSb rich areas before being covered over by the next layer. Thus, we are able to grow metastable $GaAs_{1-x}Sb_x$ alloys throughout the entire range of solid composition as shown by the solid data points in Fig. 8. Other, even less stable alloys, such as GaPSb and InPSb can also be grown in this way [42].

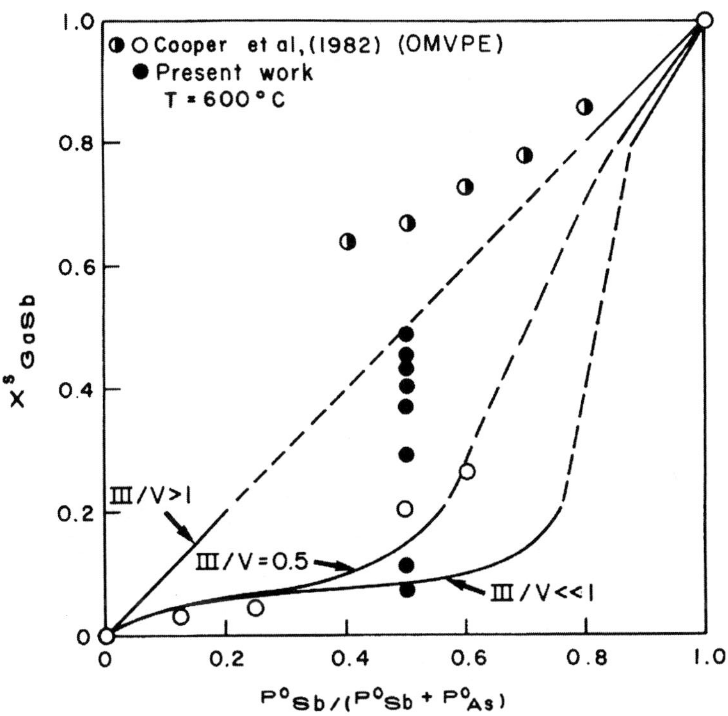

Figure 8. Solid versus vapor composition for the alloy GaAsSb. The data are from the work of Cooper et al [40] and Stringfellow and Cherng [41]. The curves were calculated for various V/III ratios. The broken sections of each curve represent the calculated regions of solid immiscibility. (after Stringfellow and Cherng [41]).

The magnitude of the enthalpy of mixing has significant consequences for OMVPE growth. As mentioned above, it has a direct impact on the composition of the solid alloy grown from a given vapor and, when large, gives rise to a miscibility gap. A dramatic example of this phenomenon involves alloys where N, an extremely small group V element, is used to replace a much larger element such as As or P in an effort to produce 1.3 micron alloys lattice matched to GaAs substrates for photonic devices [43]. For these interesting and increasingly important "dilute N" alloys, the amount of N that can be added, at equilibrium, is limited to values of much less than 1% [36]. Even the important alloy GaInN, used for short wavelength LEDs and lasers, is predicted to have a significant miscibility gap, although the solubility of In in GaN is predicted to be 6% at 850°C [36].

5.1. Surface Thermodynamics

The importance of bulk thermodynamics in determining the solid composition and stoichiometry for semiconductors grown epitaxially, as described above, has been known for decades. In recent years, it has become apparent that the thermodynamics of the surface often plays the dominant role in many aspects of the OMVPE growth process. After all, the growth occurs entirely at the surface. As discussed above, the vapor and surface are nearly in equilibrium for OMVPE and MBE growth of semiconductor materials under most conditions. If the entire system were completely at equilibrium, it would not be important to consider the surface, since the bulk solid would also be in equilibrium with the vapor at the interface. However, the diffusion coefficients in these highly stoichiometric solids, where the vacancy concentrations are extremely low, are so small that the bulk solid is often not able to attain equilibrium with the surface. In this case, the atomic structure and stoichiometry of the solid may be determined by the thermodynamics of the surface phase.

For growth on the common (100) orientation two of the four sp^3 bonds in the bulk are unsatisfied for surface atoms. This would be expected to give very low adatom mobilities, resulting in rough surfaces, as discussed above. This appears to be contrary to experimental observations of large diffusion lengths [44] and the ability to grow quantum well structures with abrupt and smooth interfaces. These are the first indications that each surface atom does not have two dangling bonds. As discussed above, elementary thermodynamic considerations suggest that the surface atoms will rearrange themselves to reduce the surface energy by reducing the number of dangling bonds as well as changing the bonding geometries. First principles calculations have given detailed information about the way in which this might occur [23]. RHEED experiments on semiconductor surfaces have given corresponding experimental information about the surface configuration versus temperature and the group V flux arriving at the surface [45]. The results are found to correspond closely to those obtained using RDS [46]. The surface reconstructions seen for GaAs under conditions used for growth by atmospheric pressure OMVPE are surprisingly similar to those for growth by MBE, as discussed above and seen in Fig. 7.

Typical OMVPE growth conditions, using V/III ratios well in excess of unity, will yield a group V rich surface, producing a (2x4) or (4x4) reconstruction for (001) surfaces, as seen in Fig. 7. Very low group V partial pressures result in the group III rich (4x2) surface. This condition is not normally observed during OMVPE growth where V/III ratios at the interface are essentially always >>1.

A very clear indication of the importance of surface thermodynamics for III/V alloys(as well as Si-Ge and II/VI alloys) is the effect of the surface on the microscopic arrangement of the atoms in the solid, e.g., the production of ordered structures spontaneously during OMVPE growth. Thermodynamic calculations indicate that the (2x4) reconstruction will favor formation of the CuPt structure, with ordering on {111} planes, which is *not* stable in the bulk. Empirically, nearly all III/V alloys are observed to form the CuPt structure during OMPVE growth using typical growth conditions [47]. SPA studies of the effects of temperature, group V flow rate, and substrate misorientation give a clear correlation between the density of $[1,\bar{1},0]$ P dimers and formation of the CuPt ordered structure during OMVPE growth of GaInP[48].

The thermodynamics at the surface are also expected to have a direct influence on other materials properties, such as dopant incorporation and the concentrations of native defects. The surface reconstruction will also almost certainly affect the surface mobilities of adatoms and the density and nature of steps and kinks on the surface, all of which will affect surface morphologies and defect densities. However, understanding of these phenomena is limited at this time.

5.2. Effect of Surface on Growth Processes

The physical nature of the surface, as described above, has significant consequences for other epitaxial growth phenomena. For example, the surface structure affects adsorption/desorption phenomena. Naturally, the surface reconstruction affects the binding of adatoms at the surface and, hence, the adsorption energy. Thus, it will affect both adsorption/desorption rates as well as heterogeneous reaction rates.

A major consequence of reconstuction on the growth process is the flat interfaces achieved for growth on (001) surfaces. As mentioned above, the adatom mobilities would be extremely small for growth on an unreconstructed (001) surface. Reconstruction dramatically lowers the energy of the surface and thus weakens the bonding to adatoms. This accounts for one extremely significant feature of the OMVPE (and MBE and CBE) growth process, namely that extremely abrupt interfaces can be formed separating layers of different compositions and/or doping levels. Abrupt interfaces are possible only for low energy surfaces, where the adatoms are not strongly bound. The resulting layer-by-layer and step-flow growth processes facilitate the growth of the low dimensionality structures, quantum wells, superlattices, quantum wires, and quantum dots, that form the basis of many of the modern high performance photonic and electronic devices.

Perhaps the most well-defined effect of steps on the growth process relates to ordering, i.e., the spontaneous formation of monolayer superlattice structures during OMVPE growth. This effect has been extensively studied for GaInP alloys grown lattice matched to GaAs substrates. The CuPt structure is formed for most growth conditions. This ordered structure is also formed for most other III/V alloys grown by OMVPE. The driving force for formation of the CuPt structure is related to the thermodynamics of the surface phase. It is driven by the [110] rows of P-P dimers on the (2x4) reconstructed surface. The degree of order is directly related to the density of the $[1,\bar{1},0]$ P dimers on the surface.

First principles calculations have been used to estimate the relative thermodynamic stabilities of the various ordered structures, including the CuPt structure, in III/V alloys [49]. For the bulk alloys, i.e., ignoring surface effects, the CuPt structure is not found to be stable relative to the disordered alloy. This was initially surprising, since CuPt is nearly the only ordered structure observed experimentally. However, this dilemma is resolved by considering the stabilities of the various ordered structures at the reconstructed surface [50].

For the most commonly observed (2xn), typically (2x4), reconstruction on group V-terminated (001) surfaces, VFF calculations [50] indicate that the B variants of the CuPt structure with ordering on $(\bar{1},1,1)$ and $(1,1,\bar{1})$ planes) are the most stable in the layers just beneath the (001) surface. The [110] rows of [1, ¯1,0,] oriented group V dimers lead to alternating [110] rows of compressive and tensile strain in the 3rd buried layer. For alloys with mixing on the group III sublattice, such as GaInP, this produces the [110] rows of alternating large and small atoms that comprise the CuPt-B variants. These calculations also predict that the surface structure of alloys with mixing on the group V sublattice, such as GaAsP, will also produce the CuPt-B variants, in agreement with experimental observations [51].

Clearly, the ordering phenomenon is very sensitive to the surface structure. For this reason, ordering is a perfect vehicle for the study of surface processes during epitaxial growth, in general. The occurrence and mechanism of ordering are fascinating materials science problems that reveal much about the important general features of the surface processes occurring during vapor phase epitaxial growth. This paper will concentrate on a review of recent advances in our understanding of the effects of surface processes on ordering.

The flow rate of the group V precursor and growth temperature were found to have a significant effect on the ordering process. The concentration of [-110] P dimers decreases both as the group V partial pressure decreases and as the growth temperature increases. The degree of

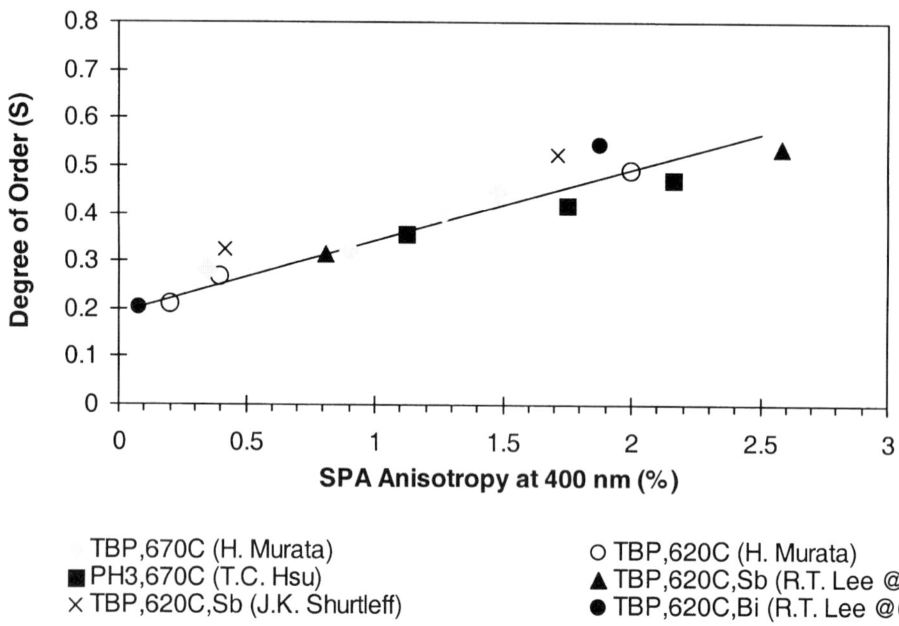

Figure 9. Degree of CuPt order versus SPA difference signal at 400 nm due to P dimers. All the data is for GaInP grown by OMVPE on singular (001) GaAs substrates. (after Lee et a l[53].)

order has been closely correlated with the surface reconstruction [52]. Clearly, the loss of CuPt ordering at low V/III ratios correlates closely with the loss of the (2x4)-like reconstruction. Data from the studies of changing temperature and TBP partial pressure are combined with data for surfactant effects, to be described below, for the plot of the degree of order versus the SPA signal in Fig. 9 [53]. A one-to-one relationship between the degree of order and the concentration of $[1,\bar{1},0,]$ P dimers on the surface is observed for changes in both temperature (620-720°C) and TBP partial pressure (<200 Pa).

6. EFFECTS OF SURFACTANTS

The discussion in the last section indicates that the surface structure can be controlled during growth by the use of temperature and the partial pressure of the P precursor. More recent experiments indicate that this can also be accomplished via the use of surfactants during the epitaxial growth process.

The original use of surfactants during the epitaxial growth (mainly MBE) of semiconductors was for control of the structure and morphology of highly strained layers in both elemental [54] and III/V semiconductors [55]. The addition of dopants during MBE growth was shown to affect surface reconstruction[56]. It also appears to affect the the diffusion of adatoms on the surface[57].The use of isoelectronic surfactants was practically unknown until the recent work described below. This also represented early work on the use of surfactants during OMVPE growth.

7. ANTIMONY

Perhaps the most interesting surfactant, shown to strongly affect the ordering in GaInP, is the isoelectronic dopant Sb. Since it has the same number of valence electrons as P, no significant change in the Fermi level position at the growth temperature is expected. Because it is much larger than P, the solubility is small [58]. Thus, it is likely to accumulate at the surface during growth. The amount of TESb added to the system was sufficient to produce only 10^{18} cm^{-3} ($x = 10^{-4}$) of Sb in the solid [59]. However, the addition of this tiny amount of Sb produces a marked increase in the low temperature PL peak energy [53]. The results indicate that the layers are highly ordered without Sb and essentially disordered when grown with a small amount of Sb present, as indicated in Fig. 10.

The surface photo absorption (SPA) anisotropy intensity at 400 nm decreases markedly versus TESb flow rate during growth. The results indicate a dramatic change in the surface reconstruction induced by the addition of Sb to the system. As discussed above, the intensity of the peak near 400 nm in the SPA anisotropy spectra directly correlates to the concentration of $[1,\bar{1},0,]$ P dimers on the surface.

Comparison of the spectra for the undoped $Ga_{0.52}In_{0.48}P$ layers and layers with Sb clearly indicates that the P dimer concentration on the surface of the samples grown with Sb present has been significantly reduced [53]. This was the first experimental evidence that an isoelectronic dopant, such as Sb, can act as a surfactant to change the surface reconstruction of a III/V semiconductor layer during OMVPE growth.

Since the $[1,\bar{1},0,]$ P dimers produce the surface thermodynamic driving force for formation of the CuPt structure during growth, it is not surprising that the replacement of $[1,\bar{1},0,]$ P dimers by Sb dimers eliminates ordering. The Sb dimer spacing is larger than for P dimers, thus the driving force for ordering is expected to be dramatically reduced. These results confirm the

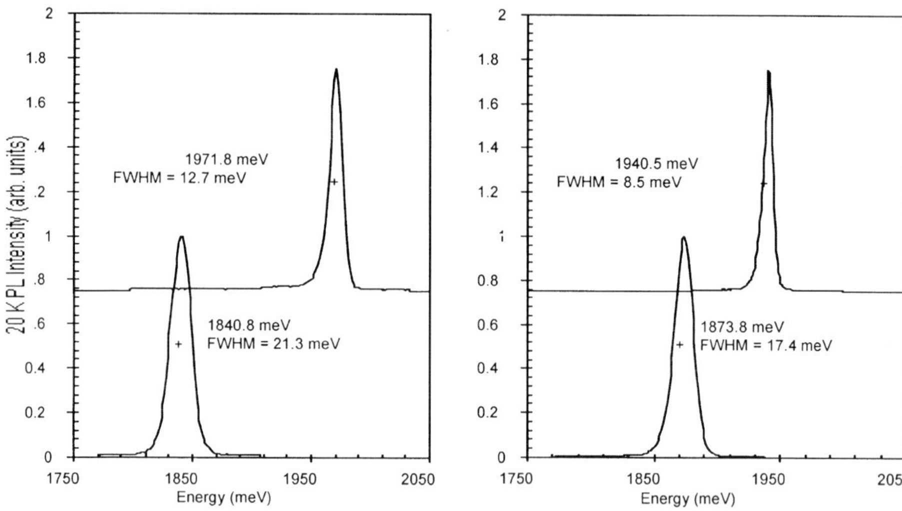

Figure 10. The effect of TESb flow rate on the degree of order and SPA difference signal at 400 nm. (after Lee et al [53].)

observation that the degree of order in $Ga_{0.52}In_{0.48}P$ layers is directly related to the P dimer concentration. In fact, the degree of order versus 400 nm SPA anisotropy signal for the layers grown using Sb fall very near the data discussed above for the effects of temperature and flow rate of the P precursor, as indicated in Figure 9.

The ability to control the ordering and, thus, the bandgap energy of GaInP layers by the addition of Sb during growth suggests the possibility of producing heterostructures and QWs. To grow a D/O heterostructure, the surface of the undoped, ordered $Ga_{0.52}In_{0.48}P$ layer grown first was exposed to tertiary butyl phosphine (TBP) and TESb for five minutes, without growth, to allow Sb to accumulate. The disordered layer, grown with Sb present, followed. Fig. 11 shows the PL spectrum for an order/disorder heterostructures grown by this procedure. A remarkable difference in the bandgap energybetween the two layers of 135 meV was observed. This is the first demonstration of a useful method for the production of atomically engineered structures for advanced electronic and photonic devices. The ability to independently modulate the bandgap and the Fermi level position is the key to the potential usefulness of this technique. The 135 meV bandgap discontinuity is more than 5 kT at room temperature, which should be sufficient for many devices.

These results suggest a new and powerful concept: determination of the surface reconstruction during growth, using a surfactant, as a method of controlling the properties of the resultant epitaxial layer. This has been clearly demonstrated for the control of the bandgap energy in GaInP. However, the implications may be much broader. Surface reconstruction will certainly affect morphology, and hence the ability to grow low dimensional structures. In fact, recent studies have shown that both surface diffusion and attachment kinetics at steps are significantly

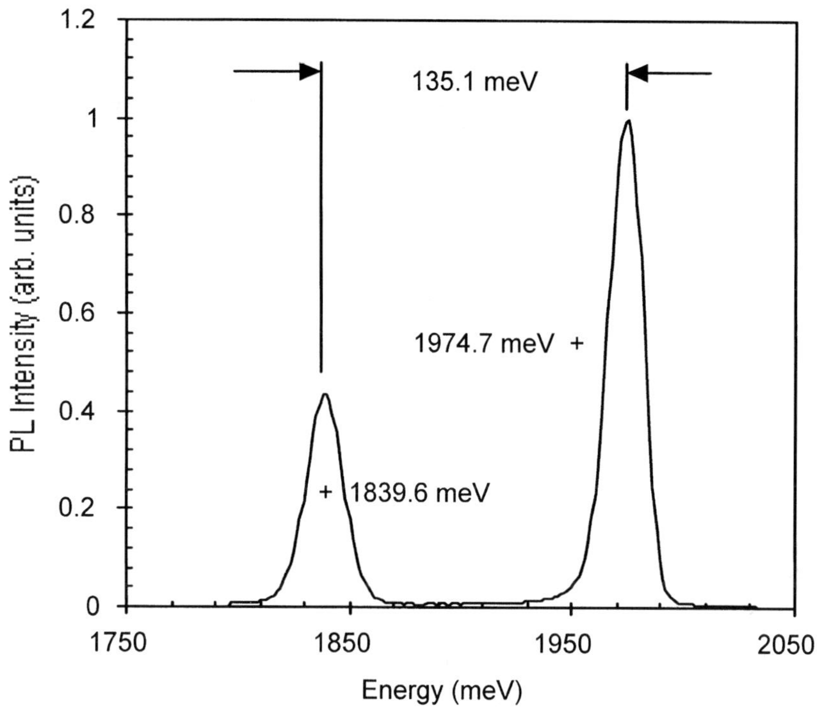

Figure 11. 20-K PL spectrum for a GaInP order/disorder heterostructure grown by OMVPE, with TESb present during growth of the top layer. (after Shurtleff et al [60].)

changed by surfactants such as Sb, Bi, and Te [61]. It would not be surprising to find that the surface reconstruction during growth also controls other important processes such as spinodal decomposition, a process known to occur at the surface during growth. It has also been demonstrated that the surface reconstruction during growth affects the dopant incorporation in GaAs [59]. Thus, surfactants, such as Sb, can be used to control, not only the bandgap energy in alloys, but also the conductivities, mobilities, and minority carrier lifetimes of layers grown by vapor phase processes.

SUMMARY

Thermodynamics plays a central role in determining many aspects of the growth process for many epitaxial techniques such as growth rate and solid composition. It also is the determining factor in many aspects of the extrinsic materials properties associated with the incorporation and distribution of the constituent and dopant elements in the resultant solid epitaxial layer. Only recently have the thermodynamic effects of the surface come to be realized. These factors have been shown to control dopant incorporation and ordering. The use of surfactants to control the structure of the surface during growth is only beginning to be exploited. However, this technique has already proven useful in the control of ordering and dopant incorporation in GaInP grown by OMVPE.

ACKNOWLEDGEMENTS

It is a pleasure to acknowledge the long-term support of the Department of Energy.

REFERENCES

1. C.H. Chen, S.A. Stockman, M.J. Peanasky, and C.P. Kuo, In High Brightness LEDs, Vol. 48 (eds, G. B. Stringfellow and M. G. Craford) Academic Press, Boston,1997.
2. S. Nakamura and G. Fasol. The Blue Laser Diode, Springer-Verlag, Berlin,1997.
3. T. Takamota, E. Ikeda, H. Kurita and M. Ohmori, Appl. Phys. Lett., 70 (1997) 381.
4. K.A. Bertness, S.R. Kurtz, D.J. Friedman, A.E. Kibbler, C. Kramer and J.M. Olson, Appl. Phys. Lett., 65 (1994) 989.
5. N. Grote, In Materials for Optoelectronics(ed, Quillec, M.) Kluwer Academic Publishers, Boston, 1996.
6. M. Razeghi, The MOCVD Challenge, Institute of Physics, Bristol, 1995.
7. G.B. Stringfellow, Organometallic Vapor Phase Epitaxy: Theory and Practice, 2nd Edition, Academic Press, Boston, 1999.
8. D.W. Kisker and T.F. Kuech, Handbook of Crystal Growth, Vol. 3, ed, D.T.J. Hurle, Elsevier, Amsterdam, 1994, pp. 93.
9. G.B. Stringfellow, Materials Science and Engineering (B) 87 (2001) 97.
10. M.A. Herman and H. Sitter, Molecular Beam Epitaxy: Fundamentals and Current Status, Springer-Verlag, Berlin, 1989.
11. C.T. Foxon, Handbook of Crystal Growth, Vol. 3 (ed, D.T.J. Hurle) Elsevier, Amsterdam, 1994, pp. 155.
12. J.Y. Tsao, Materials Fundamentals of Molecular Beam Epitaxy Academic Press, Boston, 1993.
13. H. Hirayama and H. Asahi, Handbook of Crystal Growth, Vol. 3 (ed, D T.J. Hurle) Elsevier, Amsterdam 1994, pp. 185; W.T. Tsang, J. Crystal Growth 120 (1992) 1.
14. G.B. Stringfellow, J. Crystal Growth 115 (1991) 1.
15. G.B. Stringfellow, J. Crystal Growth 68 (1984) 111.
16. GB. Stringfellow, (1999) Organometallic Vapor Phase Epitaxy: Theory and Practice, 2nd Edition, Academic Press, Boston, Section 7.3.

17. D.W. Shaw, Treatise on Solid State Chemistry, Vol 5, ed. N.B. Hannay, Plenum, New York, 1975, p. 283.
18. G.B. Stringfellow, Organometallic Vapor Phase Epitaxy: Theory and Practice, 2nd Edition, Academic Press, Boston, 1999, Chapter 1.
19. K.F. Jensen, Handbook of Crystal Growth, Vol. 3 (ed, D.T.J. Hurle) Elsevier, Amsterdam, 1994, pp. 541.
20. G.B. Stringfellow, (1999) Organometallic Vapor Phase Epitaxy: Theory and Practice, 2nd Edition, Academic Press, Boston, Section 2.4.
21. ibid, Section 2.2.
22. G.B. Stringfellow, in Advanced Crystal Growth, ed. P.M. Dryburgh, B. Cockayne, and K.G. Barraclough, Cambridge: Prentice Hall, 1987, p. 289.
23. D.J. Chadi, J. Vac. Sci. Technol. A, 5 (1987) 834.
24. B.A. Joyce, Advanced Crystal Growth, ed. P.M. Dryburgh, B. Cockayne, and K.G. Barraclough, Cambridge: Prentice Hall, 1987, p. 337-385; M.D. Pashley, J.W. Haberern, W. Friday, J.M. Woodall, and P.D. Kirchner, Phys. Rev. Lett. 60 (1988) 2176; D.K. Biegelsen, R.D. Brignans, J.E. Northrup, and L.E. Swartz, Phys. Rev. B 41 (1990) 5701.
25. M.G. Legally, Y.W. Mo, R. Kariotis, B.S. Swartzentruber, and M.B. Webb, Kinetics of Ordering and Growth at Surfaces, ed. M.G. Lagally, Plenum Press, New York, 1990, p. 145.
26. T. Isu, M. Hata, Y. Morishita, Y. Nomura and Y. Katayama, J. Crystal Growth 115 (1991) 423.
27. C. Deparis and J. Massies, J. Crystal Growth 108 (1991) 157.
28 I. Kamiya, H. Tanaka, D.E. Aspnes, L.T. Florez, L.T.E. Colar, J.P. Harbison and R. Bhat, Appl. Phys. Lett. 60 (1992) 1238.
29. D.W. Kisker, G.B. Stephenson, P.H. Fuoss, F.J. Lamelas, S. Brennan and P. Imperatori, J. Crystal Growth 124 (1992) 1.
30. N. Kobayashi, Y. Kobayashi and K. Uwai, J. Crystal Growth 174 (1997) 544.
31. G.B. Stringfellow, J. Crystal Growth, 68 (1984) 111.
32. T. Nishinaga, T. and T. Suzuki, J. Crystal Growth, 115 (1991) 398.
33. J. Tersoff, M.D. Johnson and B.G. Orr, Phys. Rev. Lett., 78 (1997) 282.
34. J.Y. Tsao, Materials Fundamentals of Molecular Beam Epitaxy, Academic Press, Boston, 1993, Chapter 7.
35. G.B. Stringfellow, J. Crystal Growth, 27 (1974) 21.
36. I.H. Ho and G.B. Stringfellow, Appl. Phys. Lett., 69 (1996) 2701; I.H. Ho and G.B. Stringfellow, Proceedings of MRS, 449 (1997) 871; I.H. Ho and G.B. Stringfellow, J.Crystal Growth, 178 (1997) 1.
37. S.H. Wei and A. Zunger, Phys. Rev. B, 39 (1989) 3279.
38. G.B. Stringfellow, J. Crystal Growth, 62 (1983) 225.
39. A. Koukitu, H. Nakai, T. Suzuki and H. Seki, J. Crystal Growth 84 (1987) 425.
40. C.B. Cooper, R.R. Saxena and M.J. Ludowise, J. Electron. Mater. 11 (1982) 1001.
41. G.B. Stringfellow and M.J. Cherng, J. Crystal Growth 64 (1983) 413.
42. M.J. Jou, and G.B. Stringfellow, J. Crystal Growth 98 (1989) 679.
43. D.J. Friedman, J.F. Geisz, S.R. Kurtz and J.M. Olson, J. Crystal Growth, 195 (1998) 409; S.R. Kurtz, A.A. Allerman, E.D. Jones, J.M. Gee, J.J. Banas and B.E. Hammons, Appl. Phys. Lett., 74 (1999) 729; X. Yang, J.B. Heroux, M.J. Jurkovic and W.I. Wang, Appl. Phys. Lett., 76 (2000) 795; S. Sato and S. Satoh, J. Crystal Growth, 192 (1998) 381.
44. M. Hata, T. Isu, A. Watanabe, Y. Kajikawa and Y. Katayama, J. Crystal Growth, 114 (1991) 203.

45. J.M.V. Hove, P.I. Cohen and C.S. Lent, J. Vacuum Sci. and Technol. A, **1** (1983) 546.
46. D.E. Aspnes, R. Bhat, C. Caneau, E. Colas, L.T. Florez, S. Gregory, J.P. Harbison, I. Kamiya, V.G. Keramidas, M.A. Koza, M.A.A. Pudensi, W.E. Quinn, S.A. Schwarz, M.C. Tamargo and H.J. Tanaka, J. Crystal Growth, 120 (1992) 71.
47. G.B. Stringfellow, Mat. Res. Soc. Symp. Proc., 312 (1993) 35.
48. H. Murata, I. H. Ho, L.C. Su, Y. Hosokawa and G.B. Stringfellow, J. Appl. Phys., 79 (1996) 6895.
49. A. Zunger and S. Mahajan, Handbook on Semiconductors(Ed, T.S. Moss.) Elsevier Science B.V., Amsterdam, 1994, pp. 1399.
50. S.B. Zhang, S. Froyen and A. Zunger, Appl. Phys. Lett. 67 (1995) 3141.
51. G.S. Chen, D.H. Jaw and G.B. Stringfellow, J. Appl. Phys. 69, 4263 (1991).
52. H. Murata, S.H. Lee, I.H. Ho and G.B. Stringfellow, J. Vac. Sci. Technol. B 14 (1996) 3013.
53. R.T. Lee, C.M. Fetzer, J.K. Shurtleff and G.B. Stringfellow, J. Appl. Phys. 87 (2000) 3730.
54. E. Tournie and K.H. Ploog, Thin Solid Films 231 (1993) 43.
55. B.R.A. Neves, M.S. Andrade, W.N. Rodrigues, G.A.M. Safar, M.V.B. Moreira and A.G. de Oliveira, Appl. Phys. Lett. 72 (1998) 1712.
56. H. Oigawa, M. Wassermeier, J. Behrend, L. Daweritz and K.H. Ploog, Surf. Sci. 376 (1997) 185.
57. B. Voightlander, A. Zinner, T. Weber and H.P. Bonzel, Phys. Rev. B 51 (1995) 7583.
58. G.B. Stringfellow, J. Crystal Growth 27 (1974) 21; M.J. Jou, D.H. Jaw, Z.M. Fang, and G. B. Stringfellow, J. Crystal Growth 190 (1990) 208.
59. J.K. Shurtleff, S.W. Jun, and G.B. Stringfellow, Appl. Phys. Lett. 78 (2001) 3038.
60. J.K. Shurtleff, R.T. Lee, C.M. Fetzer and G.B. Stringfellow, Appl. Phys. Lett. 75 (1999) 1914.
61. G.B. Stringfellow, J.K. Shurtleff, R.T. Lee, C.M. Fetzer and S.W. Jun, J. Crystal Growth 221 (2000) 1.

Crystal Growth – from Fundamentals to Technology
G. Müller, J.-J. Métois and P. Rudolph (Editors)

Actual Concepts of Interface Kinetics

Kenneth A. Jackson

Department of Materials Science and Engineering
University of Arizona, 4715 East Fort Lowell Rd., Tucson, AZ 85712, USA

1. GENERAL CONSIDERATIONS

The crystal growth process is usually reversible: the net growth rate is the difference between the arrival rate and departure rate of atoms at the crystal surface At equilibrium these two rates are equal. When both phases are present, the crystal will grow when the interface is below the equilibrium temperature, and it will melt or dissolve when then temperature is above the equilibrium temperature.

1.1. Atoms at an Interface

In general, crystals grow by the addition of atoms or molecules from the mother phase, one at a time. The path of the atoms or molecules between the two phases is reversible, which is known as "microscopic reversibility". The rate at which an atom or molecule joins or leaves the crystal depends on the local environment of the atom or molecule in the growth phase as well as on how many of the nearest neighbor lattice sites are occupied with atoms of the crystal. This determines how tightly the solid atom or molecule is bound to the crystal, and how readily an atom the growth phase can join or leave the crystal.

In growth from the vapor phase or from a solution, it is fairly obvious whether an atom or molecule is part of the crystal or not. Surface atoms or molecules can arrive from the vapor phase or from the solution, move around on the surface by diffusion, and then perhaps leave the surface again. Indeed, if it were not possible to distinguish the atoms of the crystal from those of the neighboring phase, it is difficult to imagine how there could be a nucleation barrier to the formation of new layers, as there usually is in vapor phase or solution growth.

In melt growth the situation is not so obvious. For metals, there is only a few percent difference in density between the melt and the crystal, so it is possible that there could be a gradual transition from the crystal to the melt. But molecular dynamics simulations indicate that atoms which belong to the crystal can be distinguished from those which are part of the melt. The distinction can be made in two ways.

One is using the time trajectory of the atoms. Atoms which are part of the crystal vibrate about a lattice site. Atoms which are part of the liquid migrate randomly. An individual atom at the interface will wander randomly for a while, then perhaps join the crystal where it vibrates about a fixed position, and then perhaps leave the crystal again to wander randomly. There is a clear distinction between these two modes. When an atom joins the crystal, it typically stays there for many vibrational periods, and when it is in the liquid, it typically wanders for many vibrational periods. The vibrational mode can be clearly distinguished from the wandering mode.

The other method of distinguishing between solid and liquid atoms is using the radial distribution function of the atom, which describes the locations of neighboring atoms. The radial distribution function of solid atoms has a distinct minimum which is not present for liquid atoms. That the state of individual atoms at a crystal/melt interface can be distinguished is perhaps a surprising result, since a metal crystal and its melt have such similar properties. It appears that the atoms of the crystal at the interface create a potential well because of their regular positions. Atoms in the liquid can fall into this well, stay there for a while and then hop out again when they have enough energy. The latent heat of the transformation is associated with the potential energy of the atom falling into or hopping out of this energy well.

There has been some discussion in the literature about the importance of the observation in molecular dynamics simulations that there is density variation in a liquid next to a plane wall. The period of the density variations is the atomic diameter, and they extend several atomic diameters into the liquid away from the wall. However, there is no evident dependence of the radial distribution function of the atoms on their distance from the wall. These density variations are apparently due to minor variations in the atom positions near the flat wall. Similar density variations are found in the liquid next to crystal surfaces, and it has been suggested that these density waves influence the crystallization of the liquid atoms. But it is difficult to understand how these minor variations in position of the atoms could influence the energy or momentum distributions of the atoms in such a way as influence the rate at which the atoms join the crystal.

Although atoms are either solid or liquid, there can be a gradual transition from the liquid to the solid in one sense. For the simple metals and some other materials, the interface is rough at the atomic level, as discussed below. There are a few atomic layers at the interface in which there are both solid and liquid atoms. The atoms are either solid or liquid, and these layers contain both.

So in general, we can assume that there are atoms which belong to each phase, and that the transition between the two phases occurs atom by atom or molecule by molecule. This statement is not meant to imply that there is no interaction between adjacent atoms during the crystallization process.

Some crystal interfaces are smooth at the atomic level, and some are rough at the atomic level. The theory of the surface roughening transition is based on the assumption that the atoms of the crystal can be distinguished from the atoms in the phase it is in contact with. It concludes that some interfaces are smooth at the atomic level, and that some are rough at the atomic level, depending in large measure on the difference between the properties of the two phases. If the density of the atoms or molecules is very different in the two phases, as it is for vapor phase growth or solution growth, then the interface is likely to be smooth. If the difference is small, as it is for a metal in contact with its melt, then the interface is likely to be rough.

1.2. General Equation for the Growth Rate of Crystals

The rate of crystallization can be expressed as the product of four terms: a length, a frequency, a term which depends on the structure of the interface, and the free energy difference between the two phases.

$$v = \boldsymbol{a}\, v^{+} \boldsymbol{f}\, u_k \qquad (1)$$

The first term, $\boldsymbol{a}$, is a distance which is related to the atomic or molecular diameter of the growth unit. The second term, v^{+}, is the rate at which atoms join the crystal at active growth sites on the surface. The third term, ***f***, depends on the roughness of the interface, and is the fraction of interface sites which are active growth sites. The difference between the rate at which atoms or molecules join and leave the crystal depends on the local free energy or chemical potential difference between the two phases, which is the driving force for the transformation. The normalized difference between the rate at which atoms or molecules join and leave the crystal is contained in the fourth term u_k:

$$u_k = 1 - \exp\left(\frac{\Delta G}{kT}\right) \tag{2}$$

Here ΔG is the free energy difference between the two phases, k is Boltzmann's constant, and T is the temperature. ΔG can be expressed as an undercooling, as a vapor pressure difference, or as a supersaturation.

The atoms or molecules at the interface join and leave the crystal at rates which depend only on their local environment and on the local departure from equilibrium. Their motion depends on their individual kinetic energies and the local potentials to which they are subject. There is no action at a distance.

There is an ambiguity in defining the area occupied by a growth site on the interface for different orientations of the crystal. We can get around this ambiguity by defining the distance $\boldsymbol{a}$ to be the cube root of the atomic volume, $\boldsymbol{a}^3 = \Omega$, and the area of a growth site on the interface to be $\boldsymbol{a}^2$. Then if an atom is added to each growth site, the interface will advance a distance $\boldsymbol{a}$. This is numerically equivalent to adding a volume Ω to the crystal with each atom, and then defining ***f*** to be the density of growth sites on the interface.

Not all of the surface sites on a crystal are active growth sites. The factor ***f*** is the probability of finding an active growth site in an area $\boldsymbol{a}^2$ of the interface. Atoms leave and join the crystal surface at many sites, but net growth occurs only at active growth sites. The density of active growth sites can vary widely, and is strongly temperature dependent if the growth depends on the nucleation of new layers. This will be discussed in more detail below. ***f*** is typically about ¼ for a rough surface, but can be very small on a smooth surface

The rate at which atoms or molecules join the crystal at active growth sites, v^+, depends on the phase into which the crystal is growing, and on the mobility of growth units in that phase.

The classical theory for vapor growth was developed by Knudsen [1] and for melt growth by Wilson [2] and Frenkel [3]. In these models, the density of growth sites, ***f***, was not discussed. It was assumed implicitly that atoms or molecules could join a crystal at any surface site. But these models contain the essence of the physical description of the rate at which atoms or molecules can join the crystal at active growth sites, v^+.

Below is a brief review of two models for growth on smooth surfaces: surface nucleation, and growth aided by screw dislocations. These are the two classical models for ***f***, the fraction of interface sites which are active growth sites. These models apply only to growth on smooth surfaces, and their applicability depends on how smooth the surface is.

Reliable information about the distribution of active growth sites on a surface can be obtained from Monte Carlo computer simulations, and detailed information about growth rates for specific materials and orientations requires molecular dynamics modeling.

1.3. Entropy Change on Crystallization

A crystal invariably has a lower enthalpy than its melt. The difference is latent heat. The atoms of a crystal must have sufficient energy to leave the crystal and join the higher enthalpy growth phase.

The rate at which atoms leave the crystal to return to the growth phase at active growth sites must contain this energy difference as a Boltzmann factor:

$$\nu^- = \nu_0^- \exp\left(-\frac{L}{kT}\right) \tag{3}$$

At the melting point, T_M, the rate at which atoms join the crystal must contain a similar factor:

$$\nu^+ = \nu_0^- \exp\left(-\frac{L}{kT_M}\right) \tag{4}$$

This exponential factor is implicit in ν^+ in Eqn. (1), so that the difference between ν^+ and ν^- is normalized in u_K. $L/T_M = \Delta S$ is the entropy difference between the two phases. This entropy term compensates for the increase in energy when atoms leave the crystal. Its magnitude depends on the difference in order between the crystal and the growth phase.

An atom or molecule in the crystal must increase its energy to leave the crystal, but it is relatively easy for it to find a place where it can join the growth phase. An atom or molecule in the growth phase decreases its energy on joining the crystal, but it is geometrically more difficult for it to join the crystal, because it can only do so at a lattice site. This represents and increase in its state of order. These two effects, one energetic, and the other geometrical, compensate to make the net rate of transition equal to zero at equilibrium.

For melt growth of metals, $\Delta S/k$ is typically about one, so this factor is about 1/3. For small organic molecules, $\Delta S/k$ is typically about 6. The difference in order between the crystal and the melt should account for one unit of this, as for a metal. The other five units come from the rotational disorder of the melt. The molecule must have the correct orientation in order to join the crystal. This suggests that less than 1% of the molecules are in the right orientation to join the crystal.

1.4. Early Models for Melt Growth

The first analysis for the growth rate of a crystal from a liquid was published by Wilson [2] in 1900. He related the rate at which atoms join the crystal using the diffusion coefficient of the liquid. About 30 years later, Frenkel [3] expressed the growth rate in terms of the viscosity of the liquid. The two treatments are equivalent since the diffusion coefficient of a liquid, D, is related to its viscosity, η, by the Stokes-Einstein relationship:

$$D = \frac{kT}{3\pi\eta a} \tag{5}$$

In this equation, a is the diameter of the diffusing atom or molecule.

It is found experimentally that the diffusion coefficient and the reciprocal of the viscosity have the same temperature dependence. Experiments on glass-forming systems, where it is relatively easy to measure the growth rate as a function of temperature, indicate that the same activation energy also applies for crystal growth.

These early expressions for melt growth assumed that all surface sites are active growth sites. The modern version of these equations includes the factor ***f***, as well as the entropy factor of Eqn. (4).

In addition, the diffusion jump distance in the liquid is not the atomic diameter as was assumed in this early model. An atom in the liquid diffuses by a series of small motions of length Λ, which is the average distance which an atom moves during a diffusive motion.

In some cases, a rearrangement of the configuration and positions of other atoms in the liquid is required in order for the an atom or molecule to join the crystal. In these cases, it is the average time required for an atom to move a diffusion jump distance, Λ, which determines the rate at which an atom or molecule can join the crystal. An atom which moves a distance Λ to join a crystal, advances the interface by an atomic diameter. Taking this into account, and adding the factor ***f*** as well as entropy term from Eqn. (4) and expressing the growth rate in terms of the diffusion coefficient in the liquid, results in the modern version of the Wilson-Frenkel expression for the growth rate:

$$v = \frac{6\,a\,D}{\Lambda^2} f \exp\left(-\frac{\Delta S}{k}\right)\left[1 - \exp\frac{\Delta G}{kT}\right] \tag{6}$$

This is the appropriate expression for the crystallization rate for many materials, where the mobility of the atoms limits the growth rate. For glass forming materials, the growth rate can be measured rather readily. In these materials, at large undercooling, the term in the square bracket changes slowly with temperature, so the temperature dependence of the growth rate is dominated by the temperature dependence of the molecular mobility. This equation also agrees with the results of molecular dynamics simulations of the crystallization of silicon, as discussed below.

1.5. Growth Rate from the Melt

Metal crystals grow very rapidly at small very small undercoolings, and so growth usually occurs with the interface close to the equilibrium temperature. In this case, the term in the square brackets can be approximated to give a linear relationship between the growth rate and the interface undercooling, which is often written in the simple form:

$$v = \mu\,\Delta T \tag{7}$$

here $\Delta T = T_M - T$, and μ is called the kinetic coefficient.

The growth rate of metals is usually limited by diffusive processes. For pure metals, the growth rate is limited by heat flow, that is, by how fast the latent heat can diffuse away from the interface. For alloys, compositional diffusion is also important. For a metal growing from the melt, μ is typically about 0.5 m/s.K.

It is important to remember that growth processes take place only at the interface. The atoms at the interface are subject to the conditions at the interface: the local temperature, the local composition, the local curvature of the interface. The thermal fields and compositional fields in the sample as a whole influence the local conditions at the interface, but the kinetic processes at the interface on the atomic scale depend only on the local conditions there. The atoms at the interface respond only to their immediate environment.

1.6. Nucleation of Layers

When the surface of a crystal is essentially atomically flat, with only a few adatoms in the next atomic layer, there is a kinetic barrier to the formation of new layers surface. On the surface of a defect free crystal, new layers must form by a nucleation process. Applying standard nucleation theory to the formation of new layers gives a rate of nucleation which can be written as:

$$I = I_0 \exp\left(-\frac{T^*}{\Delta T}\right) \tag{8}$$

Here T^* depends on the edge free energy of the surface step, and ΔT is the undercooling. On a smooth surface, the density of active growth sites, f in Eqn. (1), depends on the density of growth sites provided by the nucleation process.

1.7. Growth on Screw Dislocations

Charles Frank (see [4]) suggested that defects in the crystal structure could help to form new layers, so that a nucleation process was not necessary for growth. He pointed out that a screw dislocation which ended at a surface could provide a continuous step on a surface for growth. He suggested that the surface step would wind up into a spiral, with the spacing between the steps on the spiral given by the critical radius, which is proportional to the undercooling. This means that the factor f is proportional to ΔT, and since u_K is also proportional to ΔT at small undercoolings, Frank predicted that the growth rate of a crystal should be proportional to ΔT^2.

Dislocations are not necessary for crystal growth: large dislocation-free crystals of silicon are grown every day. However, growth on defects can be important when there is a barrier to the formation of new layers. When the rate at which new layers nucleate is very slow, then growth can occur at the growth sites provided by defects. A growth rate which is proportional to ΔT^2, as Frank predicted, is often observed at small undercoolings on surfaces where there is a nucleation barrier to growth. At larger undercoolings, the nucleation rate increases so that the nucleation of new layers takes over the growth process, and the growth rate increases dramatically.

2. MOLECULAR DYNAMICS SIMULATIONS OF CRYSTAL GROWTH

Molecular dynamics simulations have provided detailed information about the process of crystal growth at the atomic level.

2.1. Crystallization from the Melt

The diffusion coefficients in molecular dynamics (MD) simulations of the liquid phase have been determined using both the Lennard-Jones (LJ) potential, which is valid for argon and works reasonably well for many metals, and for the Stillinger-Weber (SW) potential, which is valid for silicon. The diffusion processes in the liquid for both LJ and the SW simulations are very similar. The activation energies for diffusion scale with the melting points. Liquid silicon is metallic, and the atoms in all these liquids behave essentially as spheres. Although both the LJ and SW liquids are very similar, the crystals which form are quite different.

For the LJ potential, the force which any two atoms exert on each other depends only on the distance between them, and not on the location of any other atoms. This is called a pair potential. With this potential, the atoms crystallize with a face centered cubic structure, which is a close-packed structure.

The Stillinger-Weber [5] potential is a three-body potential developed especially for silicon. Each atom in crystalline silicon has four nearest neighbors, and the angle between the bonds to the nearest neighbors is important. And so the potential depends on not only the distance from each neighboring atom, but also on the angle between each pair of neighboring atoms. The potential provides a good approximation for the structure and properties of liquid silicon, and, below the melting point, a crystal forms with the diamond cubic structure, as the potential was designed to do.

The crystallization behavior of the two is quite different.

MD simulations using the SW potential [6] are shown in Fig. 1. The solid line is the Wilson-Frenkel expression, Eqn. (6), fitted to the data using the activation energy derived from the diffusion data for liquid silicon. This correctly describes the computer simulation data. This curve is similar to the experimental data for the crystallization of SiO_2 and GeO_2.

MD simulations using the LJ potential [7] are shown in Fig. 2. The open circles are the simulations results. The solid line is Eqn. (6), using the activation energy for liquid diffusion from simulation. It is obvious that the data are not described by this equation. The growth is not thermally activated.

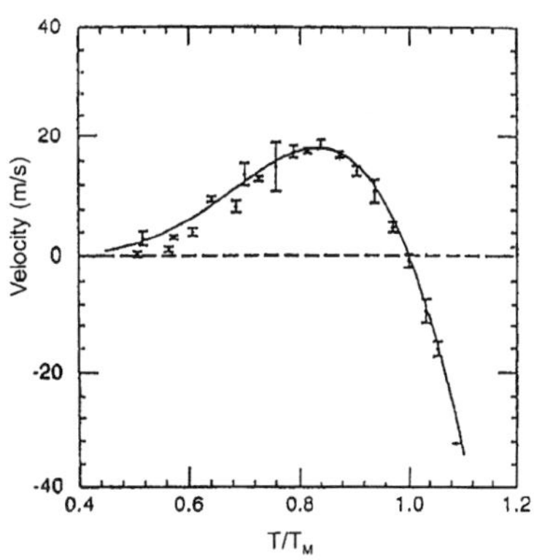

Figure 1. Molecular dynamics growth rates for the (100) face of silicon, using the Stillinger-Weber potential. The solid line is Eqn. (6).

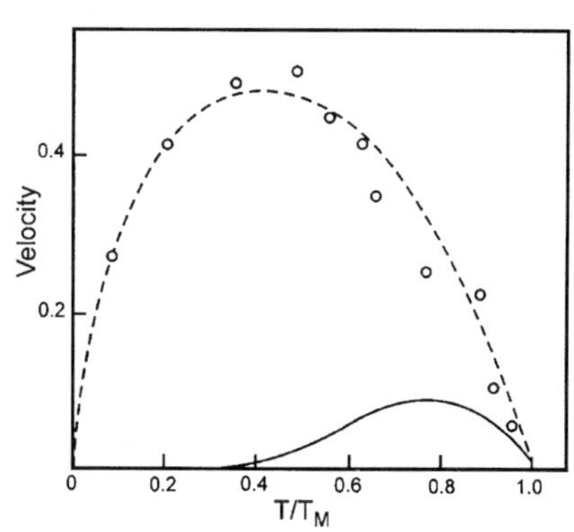

Figure 2. Molecular dynamics growth rates for the (100) face of an fcc crystal, using the Lennard-Jones potential. The solid line is Eqn. (6), the dashed line is Eqn. (10).

This result suggests that an atom can move a small distance to join the crystal without interference from other atoms. In this case, it is the mean free time, which is the time required for the atom to traverse its mean free path, λ, which determines the growth rate.

An atom can traverse its mean free path to join the crystal at the average thermal velocity, $(3kT/m)^{1/2}$. v^+ in Eqn. (1) is then the reciprocal of the mean free time between collisions in the liquid,

$$v^{+} = 1/\tau = \bar{v}/\lambda = \sqrt{3kT/m}/\lambda \tag{9}$$

The growth rate can be written:

$$v = \frac{a}{\lambda}\sqrt{\frac{3kT}{m}}\exp\left(-\frac{\Delta S}{k}\right)\left[1-\exp\left(-\frac{L\Delta T}{kT_M T}\right)\right]f \tag{10}$$

The dashed line through the data was calculated using Eqn. (10), with $\lambda = 0.1a$, and $f = 0.25$. This implies that the atoms run unobstructed from their positions in the liquid at the interface into the neighboring solid site at the interface, with the average thermal velocity of the atoms in the liquid. This growth rate does not depend on the rate at which atoms can move around in the liquid by diffusion. Unlike the diffusion jump distance, Λ, the mean free path, λ, is not strongly temperature dependent. This is the appropriate expression for the solidification rate of metal crystals, and also for the crystallization of the inert gases.

There are potential wells formed at the interface by the regular positions of the solid atoms at the interface. In the simulations, a liquid atom falls into one of these wells, stays there for a while, vibrating about an average position, and then hops to and moves randomly in the liquid for a while until it falls back into the same, or a different, well.

For silicon, and for molecular materials in general, there is an activation energy associated with rearrangement of the liquid structure. This activation energy determines the temperature dependence of the diffusion coefficient in the liquid, the viscosity of the liquid, and the crystallization rate. This is unlike the results from simulations with the LJ potential, where the atoms are essentially spheres. In this case, the activation energy for liquid diffusion and viscosity are the same, but the crystallization process is not thermally activated.

These results suggest that materials can be divided into two groups based on their crystallization behavior:

(i) One group contains materials for which no rearrangement of the liquid structure is required for any individual atom to join the crystal. The crystallization is not thermally activated. These materials can crystallize at very low temperatures, and they do not form glasses.

(ii) The other group contains materials where there must be some structural rearrangement of the liquid around an atom or molecule before it can join the crystal. The structural rearrangement process is thermally activated with the same activation energy as liquid diffusion and viscosity. These materials will not crystallize at very low temperatures, so that they can, in principle, be made into glasses.

In the discussion above, the number of active surface sites, f, has been assumed to be constant. This is usually the case for metals and the inert gases. However, for many materials, surface nucleation is required for growth, in which case f is strongly temperature dependent.

3. THE KOSSEL-STRANSKI MODEL

Stranski assumed that the strength of binding of an atom to the surface depends on its number of nearest neighbors. He applied this to a simple cubic crystal, which is also known as a Kossel crystal [8], and so this is now known as the Kossel-Stranski model. This is, of course, a simple approximation to how the binding depends on the number of neighbors, but this model does contain the essence of an explanation of why there is a strong anisotropy in the growth rate of some crystals. This model is still the basis for modern statistical mechanical models of crystal growth.

3.1. Bonding at an Interface

Stranski [9] suggested that the reason crystals grow with a crystalline shape is because there is a barrier to the formation of new layers on the closest packed faces. He suggested that the barrier exists because the rate at which individual atoms leave the surface of a crystal depends on how tightly they are bound to the surface. In this model, the binding of an atom to the surface of a crystal depends on how many nearest neighbor bonds it has. An isolated atom sitting on a flat surface has only one bond to the crystal, and so will leave relatively rapidly. An atom at a kink in a step has three bonds to the crystal, and so will leave more slowly. An atom which is part of the flat face of the crystal has five bonds to the crystal, and will leave it much more slowly. The surface of a simple cubic crystal where the surface atoms have various numbers of nearest neighbors in the crystal is illustrated in Fig. 3.

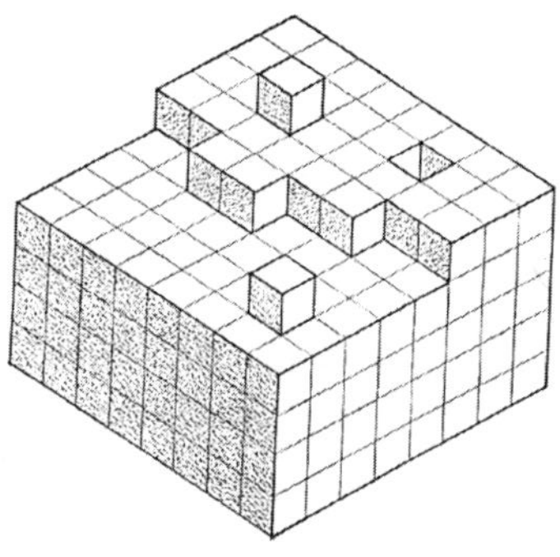

Figure 3. Configurations on the surface of a simple cubic crystal.

The rate at which atoms join the crystal is assumed to be independent of the local configuration, but the rate at which an atom leaves the crystal depends on how many of its nearest neighbor sites are occupied by atoms of the crystal. The normalized net rate at which atoms join and leave the crystal at a site with n nearest neighbors is:

$$u_k^n = 1 - \exp\left(\frac{L}{kT_M} - \frac{n\phi}{kT}\right) \tag{11}$$

Here ϕ is the bond energy, given by $\phi = 2L/Z$, where Z is the number of nearest neighbor sites in the crystal. n is the actual number of nearest neighbors of the atom at the surface of the crystal. Because each bond is shared between two atoms in the bulk of the crystal, the binding energy per atom, L is equal to $Z\phi/2$.

An important concept introduced by Stranski is the repeatable step site, which is also called a "kink site" because it is the site at a kink in a surface step. It is a site, as can be seen in Fig. 3, where an atom has half of its nearest neighbors, $n = Z/2$. An atom in the kink site breaks these Z/2 bonds when it leaves the surface, so the energy for it to leave is $Z\phi/2 = L$, which is the latent heat per atom. When an atom joins or leaves a kink site, the repeatable step site moves along the step, so the free energy associated with the surface configuration does not change. Except for edge effects, the entire crystal can be built up by the motion of these kink sites.

At equilibrium, the rate of arrival of atoms at a kink site is the same as the rate of departure from the kink site, as can be seen from the Eqn. (11) by substituting $n = Z/2$, so that $n\phi = Z\phi/2 = L$. The normalized rate of departure of an atom from a kink site as given by Eqn. (11) is identical to normalized average rate of departure of atoms from the crystal as in Eqn. (2).

In the general expression for the growth rate, Eqn. (1), ***f*** is the fraction of surface sites which are active growth sites. All active growth sites are kink sites. But on very rough surfaces, not all of the kink sites are active growth sites. For growth on a rough crystal surface, the factor *f* is more or less a constant, approximately ¼, a value which depends on the configuration of a rough surface. A surface cannot be constructed which has only repeatable step sites, and computer modeling suggests that the value of ¼ for ***f*** is about the best that can be done. The surface can be made rougher, but the effective number of growth sites does not increase much.

If we knew the site distribution function for the surface, that is, the probability of an atom on the surface having **n** nearest neighbors, we could replace the terms ***f*** and u_K in Eqn. (1) with a sum of the probability for each type of site, times the net rate of addition of atoms at that type as in Eqn.(11). The site distribution function can be estimated at equilibrium, but it is best determined by computer simulation.

3.2. Surface Roughness

It turns out that the mathematics of the Kossel-Stranki model is identical to the mathematics of the model which was originally developed by Ising [10] for the ferromagnetic Curie point, see for example [11]. The same mathematical model also applies to ordering in an alloy. In the Ising model, there is one interaction energy if the neighboring atom has the same spin, and a different interaction energy if it does not. For crystal growth, an atom has one interaction energy with an occupied neighboring site, and a different interaction energy with an empty neighboring site. The same mathematics applies to each of these cases, and so they are said to belong to the same Ising Universality Class. The Ising model is simple in principle, but becomes very complex in detail. It has been extensively studied in statistical mechanics.

Monte Carlo computer simulations can be readily performed on the Ising system [12]. And then the cluster distributions and the other behavior obtained from the computer simulations can scaled and used to describe the cluster distributions and the behavior of real systems. In applying the Ising model to crystal growth, it is assumed that there is a lattice in both phases, in order to enable the analysis, but that does not affect the universality of the result. The distribution of cluster sizes is the same for each of the systems in the Ising universality class, independent of whether or not there is a lattice. But the cluster distributions can be determined most readily by doing a calculation or a computer simulation using a lattice.

At first it seems strange that the cluster distributions should be so universal. But consider standard nucleation theory. The presence or absence of a lattice is not mentioned. The resulting equations for the cluster distributions are applied to the nucleation of a crystal in a liquid, a precipitate in a crystal, or a liquid in a vapor, etc. Similarly, the cluster distributions derived by analysis on a lattice can be applied universally to all the members of the Ising universality class.

It is worth emphasizing that the critical temperature of a member of a universality class cannot be predicted from an analysis of the cooperative behavior, because in the past it has been assumed that the critical point for the surface roughening temperature in the Ising model can be used to predict the surface roughening transition for real crystals. The Ising model describes the behavior of the system in the vicinity of the critical point, which is similar for all members of the same universality class. But, for example, the Curie temperature of a ferromagnetic material cannot be predicted from the Ising model. Nor can the roughening transition of the surface of any particular crystal be predicted from the Kossel-Stranski model.

The Ising model contains nucleation behavior as one limit. For example, in a magnetized ferromagnet at a low temperature, all the spins are aligned. In a demagnetized sample, all the spins are locally aligned, but there are magnetic domains of the opposite alignment, so there is no net magnetization of the sample. If a magnetic field is applied to a magnetized sample in a direction which requires the spins to flip, then there will be a nucleation barrier to the formation of the new magnetic domains. At a low temperature, there will be a critical size for new domains, which depends on the strength of the applied magnetic field. The formation of domains of reverse magnetization can be described by the standard equations for nucleation. But at temperatures approaching the Curie temperature, many flipped spins occur spontaneously, and small sub-critical regions of flipped spins will increase in size and number. These will interact and merge. Close to the curie temperature, standard nucleation theory does not apply to the formation of a domain of reverse magnetization, because the density of flipped spins is much too high. But the Ising model does incorporate this behavior, even though it is so difficult to describe analytically.

3.3. Monte Carlo Simulations of Crystallization

In these simulations, which are based on the Kossel-Stranski model, atoms arrive randomly at sites on the surface at some specified rate, and atoms leave the surface at a rate which is determined by a probability factor which depends on the number of adjacent crystalline atoms. If the departure probability is bigger than a random number, then the atom leaves, and if not, it does not. The arrival probability, P^+, and departure probability, P_n^-, can be written as:

$$P^+ = v^+$$

$$P_n^- = v^+ \exp\left(\frac{L}{kT_M} - \frac{n\phi}{kT}\right) \tag{12}$$

These are similar to Eqn. (11), but in the computer the arrival rate at a repeatable step, v^+, must be set small enough so that the departure rate for a site with n = 1, P_1^-, is less that one. The simulation is conducted by atoms individually joining and leaving the crystal at each site on the interface with these probabilities. n is the actual number of nearest neighbors of the surface atom. At a repeatable step site where n = Z/2, the arrival rate is equal to the departure rate at equilibrium. In the computer, atoms join and leave the crystal according to these simple rules and so the computer simulation provides a statistical analysis of this model. The statistics are never exact, since the simulated systems are finite in size and simulation time. But the atoms form clusters and nucleate new layers, and so on, just as we suppose happens in the real world. When the bond energy is comparable to kT then there are many adsorbed atoms on the surface, there is no nucleation barrier to the formation of new layers. When the bond energy is large compared to kT, a single atom on the surface has a high probability of leaving the surface, and so there are few on the surface. In this regime, the formation of nuclei on the surface is difficult. Many atoms have to get together on the surface to form a cluster of critical size and the growth of new layers involves the lateral spreading of these clusters. In this regime, growth occurs by nucleation and the motion of steps on the surface.

3.4. Equilibrium Surface Structure

The density of growth sites on the surface of a crystal depends above all on the roughness of the crystal surface. This is an intrinsic property of the surface at equilibrium. It depends on the nature of the two phases separated by the interface and on the atomic structure of the interface.

A simple two-dimensional model for the equilibrium structure of a crystal surface or interface [13] identifies an important parameter, which is known as the Jackson α-factor, on which the surface roughness depends:

$$\alpha = \frac{L}{kT_M}\frac{\eta_1}{Z} \tag{13}$$

Here η_1 is the number of nearest neighbor sites in the substrate layer and in the surface layer respectively, and Z is the total number of nearest neighbors. The two-dimensional mean field model predicts a singularity in the surface roughness at a critical value of α, $\alpha_C = 2$. This value for the surface roughening transition is in very good agreement with the observed behavior of melt-growing crystals. That is, for surfaces with a small α-factor, the surface is rough and the crystals can grow readily without surface nucleation. Surfaces with a large α-factor are smooth, and there is a nucleation barrier to the growth of each layer.

The α-factor has two components, one of which depends on the change in entropy of the transformation, $\Delta S = L/T_M$, and the other, η_1/Z, on the geometry of the crystal face. The factor η_1/Z is largest for the closest-packed face and smaller for the other faces of the same crystal.

For materials with a small entropy change on melting, all the crystal faces will be rough. These materials will exhibit relatively isotropic growth, and will grow rapidly at small interface undercoolings. For materials with a large entropy change on melting, the closest packed faces will be smooth and so there is a nucleation barrier to growth on those faces.

3.5. Monte Carlo Computer Simulation Results

The above discussion outlines the basic story of surface roughening, but there are many more details that have been worked out primarily using Monte Carlo computer modeling, but also mathematical analyses of approximations to the Ising model. A full analytical treatment of a multi-layer interface has been done for one special case, but there is no general solution. Multilevel cluster expansion models have been used to demonstrate that there is an analytical singularity in the thermodynamic properties of the surface of an Ising crystal at the surface roughening transition, but it is a very weak singularity. The surface roughening transition in the three-dimensional Ising Model is at $\alpha_C = 3.2$, rather that at $\alpha_C = 2$, as in the two-dimensional Bragg-Williams model. The surface tension and most of the other properties of the surface are continuous through this transition[14]. Fig. 4 is a plot of the surface roughness [15], which is defined as the number of unsatisfied bonds on the surface.

The surface roughness increases as the alpha factor decreases, but it is difficult to locate the surface roughening transition from this plot. The surface roughness is similar for each of the models: the three dimensional Bragg-Williams (0th order cluster expansion), the 1st order cluster expansion, and for Monte Carlo modeling (three dimensional Ising model), even though the analytical singularity which defines the surface roughening transition is different for each of these models. $8\exp(-\alpha)$ provides a reasonable approximation for the surface roughness. As the surface becomes rougher, the roughness is spread over more atomic layers. But in all the models, nothing much happens to the surface roughness at the roughening transition. The surface roughness increases continuously through the roughening transition.

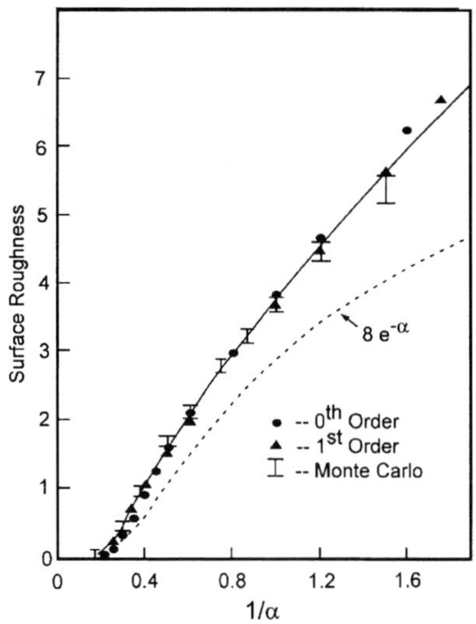

Figure 4. Surface roughness versus $1/\alpha$.

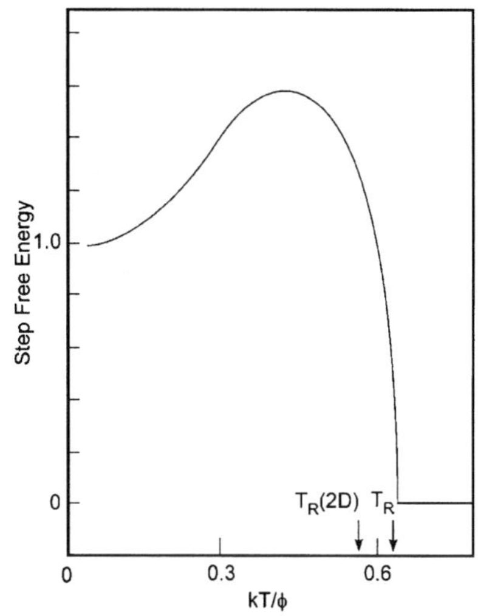

Figure 5. Step free energy versus kT/ϕ.

This is not true for the free energy of a step on the surface. The free energy of a step decreases as the surface roughening transition is approached, and it goes to zero at the surface roughening transition [16] as shown in Fig. 5.

Far above the surface roughening transition, the surface is smooth, although steps on the surface may be jagged. As the surface roughening transition is approached, the steps become increasingly jagged. There is a major contribution to the free energy of the steps from the entropy associated with this jaggedness. When the surface is rough, the steps are lost in the roughness, as illustrated in Fig. 6.

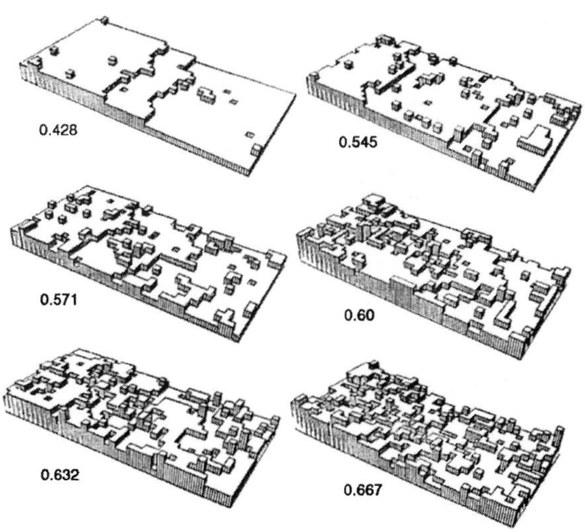

Figure 6. Configurations of surfaces from Monte Carlo simulations. The configuration on the lower right is a rough surface, below the surface roughening transition. The other configurations are for surfaces progressively farther above the roughening transition [16]. The parameter is kT/ϕ.

It has been shown that above the surface roughening transition, the interface is like a fluid/fluid interface [17]; the interface ignores the crystal lattice. On the rough side of the surface roughening transition, there is no difficulty in forming new layers. There is no barrier to continuous growth, and so the growth rate is linear with undercooling. On the smooth side of the surface roughening transition, steps on the surface have a finite free energy, and so the growth rate depends on the rate of nucleation of new layers. This is illustrated in Fig. 7, where growth rates from Monte Carlo computer simulation studies are shown. The curve labeled 1.08 is just on the rough side of the transition, and the growth rate is linear with undercooling. The other curves are for smooth surfaces. They come into the origin with zero slope [18], and farther from the roughening transition the curves are flatter and more "nucleation-like".

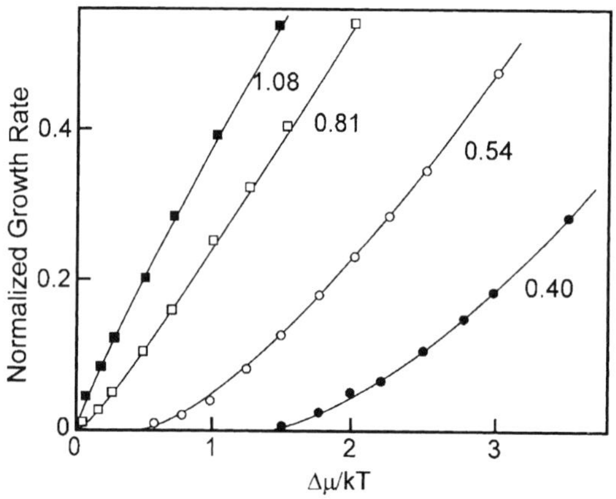

Figure 7. Normalized growth rate versus undercooling above and below the roughening transition from Monte Carlo computer simulations. The parameter is α_C/α.

As suggested by Fig. 7, it has been shown, using linear response theory, that the growth rate on rough interfaces is linear with undercooling, whereas the growth rate on smooth surfaces approaches equilibrium with zero slope. And so there is a discontinuity in the kinetic properties of the interface at the roughening transition.

The free energy of a step depends on the proximity of the surface roughening transition, and so the difficulty in nucleating new layers increases with the distance from the roughening transition. The growth rate for crystals with large entropies of transformation is highly anisotropic.

3.6. Simulations of Silicon Growth

As an example of growth anisotropy, Monte Carlo simulations have been carried out for silicon, with the roughening transition in the simulations scaled to the roughening transition for silicon [19]. The entropy change for melt growth of silicon, $\Delta S/k = 3.6$, so the α-factor for (111) is 2.7, and for (100) is 1.8, and it is lower for all the other faces. The crystallization rate for the (100) face, as determined by both experiment and MD simulations, is approximately $v = 0.12\ \Delta T$ m/sec. In Monte Carlo simulations, growth rate on the (111) face is limited by the rate of nucleation of new layers, whereas the growth rate on the (100) face is much more rapid, and linear with undercooling, as illustrated in Fig. 8.